OXFORD
UNIVERSITY PRESS

Complete Geography for Cambridge IGCSE® & O Level

Second Edition

David Kelly
Muriel Fretwell

Oxford excellence for Cambridge IGCSE® & O Level

OXFORD
UNIVERSITY PRESS

OXFORD
UNIVERSITY PRESS

Great Clarendon Street, Oxford, OX2 6DP, United Kingdom

Oxford University Press is a department of the University of Oxford. It furthers the University's objective of excellence in research, scholarship, and education by publishing worldwide. Oxford is a registered trade mark of Oxford University Press in the UK and in certain other countries

First edition published 2012
This edition published 2018

British Library Cataloguing in Publication Data

Data available

ISBN 978-0-19-842495-6

10

The manufacturing process conforms to the environmental regulations of the country of origin.

Printed in the UK by Bell & Bain Ltd.

Acknowledgements

The publishers would like to thank the following for permissions to use their photographs:

Cover: imageBROKER/Alamy Stock Photo; Dennis Hallinan/Alamy Stock Photo; 4kodiak/iStockphoto; Bartosz Hadyniak/iStockphoto; R. Fassbind/Shutterstock; nikkytok/Shutterstock.

p1: OUP/Photodisc; **p11**: Mike Goldwater/Alamy Stock Photo; **p13:** Tim Graham/Alamy Stock Photo; **p15:** Sean Sprague; **p18(b):** John Moore/Getty Images; **p18(t):** Bloomberg/Getty Images; **p34:** ssguy/Shutterstock; **p36:** Susan Montgomery/Shutterstock; **p38:** Joe Doylem/Alamy Stock Photo; **p43:** Friedrich Stark/Alamy Stock Photo; **p50(t):** Marco Cristofori/Getty Images; **p50(b):** Paulo Fridman/Getty Images; **p56(tl):** OUP/Corel; **p56(tr):** OUP/Digital Vision; **p58(t):** OUP/Image Source; **p62(l):** OUP/James Hardy/PhotoAlto; **p62(r):** OUP/Robert Stainforth; **p68:** OUP/Photodisc; **p69(tl):** Dinodia Photos/Alamy Stock Photo; **p69(b):** Daniel Berehulak/Getty Images; **p69(tr):** Fredrik Renander/Alamy Stock Photo; **p70(t):** Adrian Fisk; **p70(b):** Punit Paranjpe/Reuters; **p71:** Fly Fernandez/Getty Images; **p72:** SEBASTIAN D'SOUZA/Getty Images; **p73(l):** SEBASTIAN D'SOUZA/Getty Images; **p73(r):** Bloomberg/Getty Images; **p74:** OUP/Photodisc; **p75(t, b):** OUP/Photodisc; **p76(l, r):** OUP/Photodisc; **p77(l):** OUP/Digital Vision; **p77(r):** OUP/Fuse; **p78(l):** OUP/Keith Levit; **p78(l, tr, br):** OUP/Photodisc; **p79:** OUP/Photodisc; **p80:** OUP/FOTOG; **p81(t, b):** OUP/Photodisc; **p82(m):** Peter Adams Photography Ltd/Alamy Stock Photo; **p85:** Bloomberg/Getty Images; **p86:** OUP/Photodisc; **p89:** Tim Clayton - Corbis/Getty Images; **p93:** OUP/Photodisc; **p94(bl):** OUP/Photodisc; **p95(b):** OUP/Photodisc; **p104:** Bloomberg/Getty Images; **p107:** OUP/Photodisc; **p109(t):** imageBROKER/Alamy Stock Photo; **p113(b):** PETER BOWATER/SCIENCE PHOTO LIBRARY; **p122:** Asianet-Pakistan/Alamy Stock Photo; **p124(t):** Andrew Findlay Alamy Stock Photo; **p124(b):** Rafael Garea-Balado/Alamy Stock Photo; **p125(t):** Global Warming Images/Alamy Stock Photo; **p125(b):** © chrisstockphotography/Alamy Stock Photo; **p142(br):** Dan Burton Photo/Alamy Stock Photo; **p157:** OUP/Corbis; **p163(l):** OUP/Photodisc; **p163(tr):** sciencephotos/Alamy Stock Photo; **p163(br):** Hornbil Images/Alamy Stock Photo; **p164:** OUP/John Cartwright; **p165(l):** Hugh Threlfall/Alamy Stock Photo; **p165(r):** David J. Green - technology/Alamy Stock Photo; **p167:** jan suttle/Alamy Stock Photo; **p168(t):** GIPhotoStock X/Alamy Stock Photo; **p168(b):** Peter Jordan_EU/Alamy Stock Photo; **p174(tr):** Getty Images/Photodisc; **p178:** NASA; **p197(t):** OUP/Photodisc; **p202(l):** Martin Harvey/Getty Images; **p202(tr):** Frans Lanting Studio/Alamy Stock Photo; **p202(br):** nelzajamal/Shutterstock; **p203:** Photofusion Picture Library/Alamy Stock Photo; **p204(t):** KSTFoto/Alamy Stock Photo; **p210(t):** imageBROKER/Alamy Stock Photo; **p210(bl):** Bildagentur-online/Begsteiger/Alamy Stock Photo; **p210(br):** blickwinkel/Alamy Stock Photo; **p212:** adrian arbib/Alamy Stock Photo; **p213(t):** Aurora Photos/Alamy Stock Photo; **p213(b):** NASA; **p214(t):** adrian arbib/Alamy Stock Photo; **p214(b):** Jim West/Alamy Stock Photo; **p216:** Shutterstock; **p218(t, b):** Shutterstock; **p219:** Shutterstock; **p220(tl, tr, bl, br):** Shutterstock; **p223(l):** Shutterstock; **p223(r):** Shutterstock; **p224:** Volkswagen Group; **p229:** Ant Clausen/Shutterstock; **p232(t, l, mr):** OUP/Photodisc; **p232(br):** Horizon International Images Limited/Alamy Stock Photo; **p233:** OUP/Clearviewstock; **p235:** OUP/Photodisc; **p239**: Shutterstock; **p240:** Worldwide Picture Library/Alamy Stock Photo; **p241(t, b):** OUP/Photodisc; **p242:** Tyler Olson/Alamy Stock Photo; **p244:** Neil Cooper/Alamy Stock Photo; **p255:** OUP/Photodisc; **p258:** OUP/Digital Vision; **p259(t):** Aerial Archives/Alamy Stock Photo; **p259(b):** Dave Porter/Alamy Stock Photo; **p260:** Visions of America, LLC/Alamy Stock Photo; **p261(t):** Dorling Kindersley ltd/Alamy Stock Photo; **p261(b):** Iain Masterton/Alamy Stock Photo; **p262:** OUP/Photodisc; **p268:** Dave Pratt/PA Archive/PA Images; **p272:** Hemis/Alamy Stock Photo; **p273:** David Cumming/Getty Images; **p274(t):** Rolf Richardson/Alamy Stock Photo; **p277:** khd/Shutterstock; **p280(t):** OUP/Photodisc; **p280(m):** OUP/Digital Vision; **p280(b):** OUP/Tips Italia RF; **p281(l):** OUP/White; **p281(r):** OUP/Photodisc; **p282:** OUP/Stockbyte; **p284:** Paul Glendell/Alamy Stock Photo; **p285:** Earl & Nazima Kowall/Getty Images; **p288:** OUP/Nick Dolding; **p289:** OUP/Photodisc; **p290:** Ashley Cooper/Getty Images; **p292(t):** WIS Bernard/Getty Images; **p292(b):** Geof Kirby/Alamy Stock Photo; **p293:** CHRISTIAN RIZZI/Getty Images; **p295:** Mohsen Shandiz/Getty Images; **p296:** Prisma by Dukas Presseagentur GmbH/Alamy Stock Photo; **p297:** Régis BOSSU/Getty Images; **p298:** LOOK Die Bildagentur der Fotografen GmbH/Alamy Stock Photo; **p303(t):** Friedrich Stark/Alamy Stock Photo; **p303(b):** Friedrich Stark/Alamy Stock Photo; **p304:** Corbis RF/Alamy Stock Photo; **p330:** Dorothy Burrows/LGPL/Alamy Stock Photo.

40, 41, 45, 46(l, r), 47, 55(t, l, r), 56(bl), 57(t, b), 58(b), 82(t, b), 83, 97(t), 100, 102, 109(b), 110, 111(t, m, b), 112, 113(t, m), 114, 115, 120(t), 147, 156(t, m), 230(l, r), 230(r), 246, 310(b), 312(bl), 313(1, 3), 314(r), 321(t), 323(tl, tr, ml, mr, bl ,br), reproduced with permission from David Kelly.

56(br), 84, 94(t, br), 95(t), 96, 97(bl, br), 126, 129(t, b), 131(l, r), 132, 133(tl, tr, br), 134(t, m), 135, 136(tl, tr, br), 137(tl, tr, bl), 138, 140(t, b), 141, 142(bl, tr), 143, 144(t, m, b), 145, 146(l, r), 148(l, r), 149(t, b), 150(t, b), 151(l, r), 154, 156(t, m, b), 169, 170(tl, tr), 170(ml, mr, bl, br), 171(tl, tr, bl, br), 172(ml, mr, bl, br), 174 (all except tr), 180, 185, 186, 191, 194, 196, 197(b), 198, 199(t, b), 200, 204(b), 205, 206, 207(tl, tr, bl, br), 208, 209(tl, tr, bl, br), 211, 215(t, b), 247, 252, 266(tl, tr, bl, br), 267(tl, m, tr, bl, br), 269(l, r), 270(tl, tr, m, bl, br), 271, 274(b), 275, 276(t, m b), 287, 332, 347, reproduced with permission from Muriel Fretwell.

Ordnance Survey maps reproduced with permission of Ordnance Survey.
Illustrations are by: Barking Dog Art and Thomson Inc.

The publishers are grateful to the following for permission to reproduce copyright material:

Enerdata for map of 'Global Total Energy Consumption' in *Global Energy Statistical Yearbook 2017*, https://yearbook.enerdata.net/total-energy/world-consumption-statistics.html

Global Wind Energy Council (GWEC) for graph of world wind power generating capacity at www.gwec.net

World Tourism Organization (UNWTO) for extracts from table 'The top ten countries for tourist arrivals in 2016' in *UN Tourism Highlights 2017*, copyright © UNWTO, 92844/56/17.

World Trade Organization for graph 'The growth in world trade between 2005 and 2015' in *International Trade Statistics 2015*, copyright © World Trade Organization 2015.

This Student Book refers to the Cambridge IGCSE® Geography (0460), Cambridge IGCSE® (9-1) Geography (0976) and Cambridge O Level Geography (2217) Syllabuses published by Cambridge Assessment International Education.

Contents

Answers, mark schemes and additional exam tips can be found on your free support website. Access the support website here: www.oxfordsecondary.com/9780198424956

What's on the support website?

Welcome to your Geography course. The material on the support website has been specially written to support your learning. On this page you can see what you will find on the website. Everything in the book and website has been designed to help you prepare for your examination and achieve your best.

Additional questions for every chapter in the book

Pull together all the aspects of the work that you have considered earlier, and test your knowledge on this material.

A glossary

A comprehensive revision tool that unpacks the vocabulary of the subject and carefully explains tricky terms.

Interactive multiple-choice tests

Test your knowledge on every chapter of the book with interactive multiple-choice tests that encourage reflection and revision.

Revision checklists

Check that you have covered all the essentials by printing out these handy revision tools and ticking off those topics that you are confident about.

Notes on using mathematical skills in geography

Additional guidance is provided on ways to correctly apply mathematical skills in different geographical situations.

Answers to all questions

Every question in the book and website, both from activities and exam-style questions, has a model answer included to enhance learning.

Exam practice

An extensive range of exam-style questions for the most effective revision and practice.

Introduction

The purpose of this book is to prepare you for Cambridge IGCSE® (0460 and 0976) and Cambridge O Level (2217) Geography. It is intended to provide you with a good preparation for studying geography at a higher level.

- **Chapters 1-11** provide the knowledge and understanding needed for IGCSE® and O Level Paper 1 (Geographical Themes) and Paper 2 (Geographical Skills).
- **Chapter 12** provides preparation in the skills and analysis needed for Paper 2 (Geographical Skills).
- **Chapter 13** provides preparation in the investigative skills needed for IGCSE® Component 3 (Coursework) or Paper 4 (Alternative to Coursework), and O Level Paper 3 (Geographical Investigations).

This book provides an active approach to the subject, with questions for you to answer and tasks for you to do both within and at the end of each chapter. Specimen answers to all of the questions can be found on the accompanying website.

The final part of each question on the Cambridge IGCSE® and O Level Paper 1 asks you to describe an example that you have studied in detail - a case study. Each chapter contains case studies to help you answer these questions. You may, of course, choose to do different case studies that are more relevant to the area where you live.

As well as answers to all of the questions in the book, the accompanying website also contains:

- a full glossary of key words related to each chapter
- additional questions based on the material in the book (some of them interactive)
- additional Cambridge IGCSE® and O Level exam-style questions, which you can print off.

A note about terms

This book uses the terms 'more economically developed countries' (MEDCs) and 'less economically developed countries' (LEDCs). These terms are often referred to in examination questions. This classification is used throughout the book but it must be remembered that there is no generally accepted, up-to-date definition of these terms and not all countries are easy to classify.

Two other terms used throughout the book - which often cause confusion - are *physical* and *human* when referred to in geography. Physical geography is the natural features of relief, drainage and vegetation, while human geography covers non-natural features such as settlement, agriculture, industry and transport.

CAIE syllabus matching grid

Matching grid for Cambridge IGCSE® syllabuses 0460 and 0976, and Cambridge O Level syllabus 2217, for examination from 2020.

Syllabus section	Student Book
Theme 1: Population and settlement	
1.1 Population dynamics	Chapter 1
• Describe and give reasons for the rapid increase in the world's population	Pages 2-5
• Show an understanding of over-population and under-population	Pages 7-9
• Understand the main causes of a change in population size	Pages 6-7, 10-12
• Give reasons for contrasting rates of natural population change	Pages 10-12
• Describe and evaluate population policies	Pages 12-17
Case studies	
• A country which is over-populated	Page 9
• A country which is under-populated	Page 9

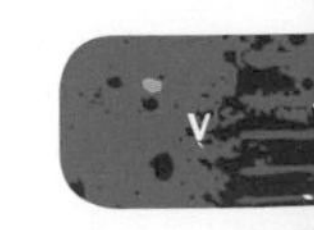

Syllabus section	Student Book
• A country with a high rate of natural population growth	Pages 11–12
• A country with a low rate of population growth (or population decline)	Pages 16, 17
1.2 Migration	Chapter 1
• Explain and give reasons for population migration	Pages 17–18
• Demonstrate an understanding of the impacts of migration	Pages 19–23
Case study	
• An international migration	Pages 19–23
1.3 Population structure	Chapter 1
• Identify and give reasons for and implications of different types of population structure	Pages 24–26
Case study	
• A country with a high dependent population	Pages 26–29
1.4 Population density and distribution	Chapter 1
• Describe the factors influencing the density and distribution of population	Pages 29–33
Case studies	
• A densely populated country or area (at any scale from local to regional)	Page 31
• A sparsely populated country or area (at any scale from local to regional)	Pages 31–33
1.5 Settlements (rural and urban) and service provision	Chapter 2
• Explain the patterns of settlement	Page 39
• Describe and explain the factors which may influence the sites, growth and functions of settlements	Pages 40–42
• Give reasons for the hierarchy of settlements and services	Pages 35–38
Case study	
• Settlement and service provision in an area	Pages 43–47
1.6 Urban settlements	Chapter 2
• Describe and give reasons for the characteristics of, and changes in, land use in urban areas	Pages 55–60
• Explain the problems of urban areas, their causes and possible solutions	Pages 61–66
Case study	
• An urban area or urban areas	Pages 64–66
1.7 Urbanisation	Chapter 2
• Identify and suggest reasons for rapid urban growth	Pages 48–52
• Describe the impacts of urban growth on both rural and urban areas, along with possible solutions to reduce the negative impacts	Pages 67–74
Case study	
• A rapidly growing urban area in a developing country and migration to it	Pages 69–73
Theme 2: The natural environment	
2.1 Earthquakes and volcanoes	Chapter 3
• Describe the main types and features of volcanoes and earthquakes	Pages 88–95
• Describe and explain the distribution of earthquakes and volcanoes	Pages 86–87

Syllabus section	Student Book
• Describe the causes of earthquakes and volcanic eruptions and their effects on people and the environment	Pages 88–92, 101–102
• Demonstrate an understanding that volcanoes present hazards and offer opportunities for people	Pages 95–97
• Explain what can be done to reduce the impacts of earthquakes and volcanoes	Pages 96, 102
Case studies	
• An earthquake	Pages 103–105
• A volcano	Pages 99–100
2.2 Rivers	Chapter 4
• Explain the main hydrological characteristics and processes which operate in rivers and drainage basins	Pages 108–116
• Demonstrate an understanding of the work of a river in eroding, transporting and depositing	Pages 108–116
• Describe and explain the formation of the landforms associated with these processes	Pages 108–116
• Demonstrate an understanding that rivers present hazards and offer opportunities for people	Pages 117–120
• Explain what can be done to manage the impacts of river flooding	Page 119
Case study	
• The opportunities presented by a river or rivers, the associated hazards and their management	Pages 121–125
2.3 Coasts	Chapter 5
• Demonstrate an understanding of the work of the sea and wind in eroding, transporting and depositing	Pages 127–132
• Describe and explain the formation of the landforms associated with these processes	Pages 132–137, 140–144
• Describe coral reefs and mangrove swamps and the conditions required for their development	Pages 143–144, 150–151
• Demonstrate an understanding that coasts present hazards and offer opportunities for people	Pages 138, 152
• Explain what can be done to manage the impacts of coastal erosion	Page 139
Case study	
• The opportunities presented by an area or areas of coastline, the associated hazards and their management	Pages 151, 153–156
2.4 Weather	Chapter 6
• Describe how weather data are collected	Pages 159–165, 167–175
• Make calculations using information from weather instruments	Pages 160–162
• Use and interpret graphs and other diagrams showing weather and climate data	Pages 166–167
2.5 Climate and natural vegetation	Chapter 7
• Describe and explain the characteristics of two climates: equatorial	Pages 187–191
• Describe and explain the characteristics of two climates: hot desert	Pages 191–197
• Describe and explain the characteristics of tropical rainforest and hot desert ecosystems	Pages 198–201, 208–210
• Describe the causes and effects of deforestation of tropical rainforest	Pages 202–205
Case studies	
• An area of tropical rainforest	Pages 201–208
• An area of hot desert	Page 211

Syllabus section	Student Book
Theme 3: Economic development	
3.1 Development	Chapter 8
• Use a variety of indicators to assess the level of development of a country	Page 217
• Identify and explain inequalities between and within countries	Pages 218–219
• Classify production into different sectors and give illustrations of each	Page 220
• Describe and explain how the proportions employed in each sector vary according to the level of development	Pages 221–222
• Describe and explain the process of globalisation, and consider its impacts	Pages 222–223
Case study	
• A transnational corporation and its global links	Pages 225–228
3.2 Food production	Chapter 9
• Describe and explain the main features of an agricultural system: inputs, processes and outputs	Pages 230–248
• Recognise the causes and effects of food shortages and describe possible solutions to this problem	248–251
Case studies	
• A farm or agricultural system	Pages 233–235, 238–242, 244–248
• A country or region suffering from food shortages	Page 251
3.3 Industry	Chapter 10
• Demonstrate an understanding of an industrial system: inputs, processes and outputs (products and waste)	Pages 253–254, 256–257
• Describe and explain the factors influencing the distribution and location of factories and industrial zones	Pages 254–255, 257–261
Case study	
• An industrial zone or factory	Pages 258–261
3.4 Tourism	Chapter 10
• Describe and explain the growth of tourism in relation to the main attractions of the physical and human landscape	Pages 266–269
• Evaluate the benefits and disadvantages of tourism to receiving areas	Pages 269–274
• Demonstrate an understanding that careful management of tourism is required in order for it to be sustainable	Pages 269–276
Case study	
• An area where tourism is important	Pages 269–276
3.5 Energy	Chapter 11
• Describe the importance of non-renewable fossil fuels, renewable energy supplies, nuclear power and fuelwood; globally and in different countries at different levels of development	Pages 278–285
• Evaluate the benefits and disadvantages of nuclear power and renewable energy sources	Pages 285–298
Case study	
• Energy supply in a country or area	Pages 296–298

Syllabus section	Student Book
3.6 Water	Chapter 11
• Describe methods of water supply and the proportions of water used for agriculture, domestic and industrial purposes in countries at different levels of economic development	Pages 299–300
• Explain why there are water shortages in some areas and demonstrate that careful management is required to ensure future supplies	Pages 300–301
Case study	
• Water supply in a country or area	Pages 302–303
3.7 Environmental risks of economic development	Chapters 2, 6, 7, 9, 10, 11
• Describe how economic activities may pose threats to the natural environment and people, locally and globally	Pages 67–73, 179–185, 201–205, 206–208, 261–263
• Demonstrate the need for sustainable development and management	Pages 183–184, 205, 208, 212–215, 236–237, 263–265
• Understand the importance of resource conservation	Pages 183–184, 205, 208, 236–237
Case study	
• An area where economic development is taking place and causing the environment to be at risk	Pages 69–73, 182–185, 201–205, 206–208, 212–215, 238–242, 262–263, 266, 272–274, 296–298
Mathematical skills in geography	Throughout the book and on the support website
Geographical skills (needed for Cambridge IGCSE® and O Level Paper 2)	Chapter 12
Coursework and geographical investigations skills (needed for Cambridge IGCSE® Component 3/Paper 4, and O Level Paper 3)	Chapter 13

1 Population

This chapter covers the following Cambridge IGCSE® and O Level topics:

- **1.1 Population dynamics**
- **1.2 Migration**
- **1.3 Population structure**
- **1.4 Population density and distribution**

- Did you know that, around the world, more than 100 babies are born every minute?
- Are you worried that there are now more than 7.5 billion people on the planet, and that this total is predicted to rise to 9 billion by 2050?
- Will global food supplies be enough to feed us all?
- Why is contraception not universally available?
- How do governments attempt to control population size?
- Did you know that the world's population is getting older?
- How will we look after all of our old people in the future?

In this chapter you will learn about:

- the rapid increase in the world's population, and the factors influencing it
- the relationship between population growth and resources
- contrasting patterns of population growth in different parts of the world
- the benefits and problems caused by different patterns of population growth
- different types of population structure, and how they relate to the demographic transition model
- differing population densities and distributions, and the factors which influence them
- migrations of people
- graphs, diagrams and maps used to display information about population.

Population dynamics

Human population growth

How fast is the world's population growing?

As you can see in Fig. 1.1, human population growth was very slow until around 1500. Three hundred years later, the population had doubled to 1 billion. Then, the rate of increase quickened until the 1960s, when the planet supported more than 3 billion people. By 2011 this figure had more than doubled and now exceeds 7 billion. A glance at Fig. 1.1 suggests that the bigger the population, the faster it has grown - but is that still the case?

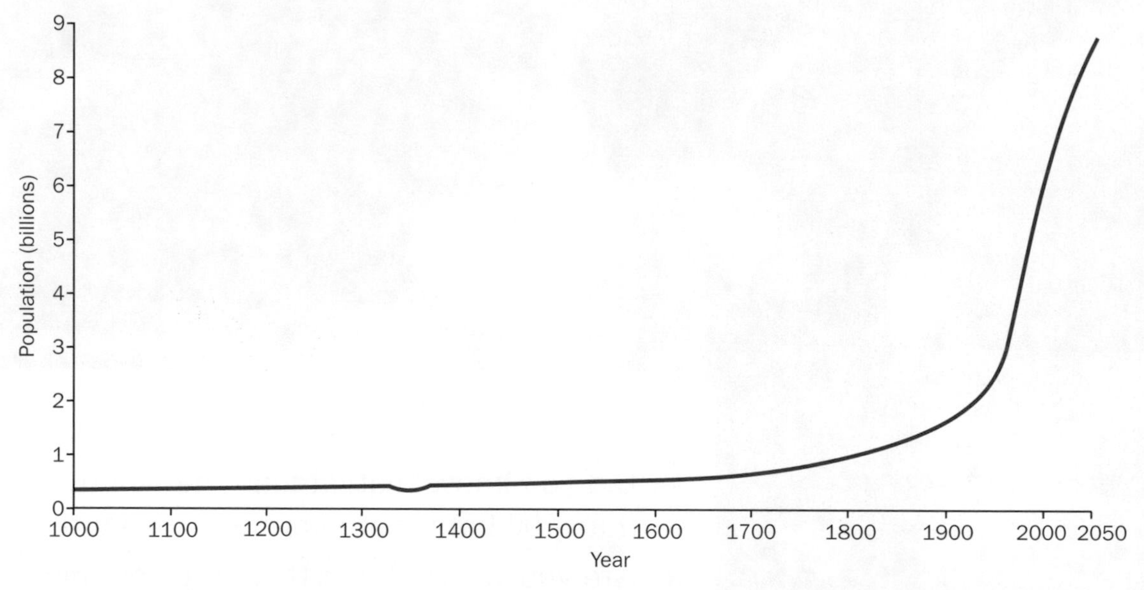

Fig. 1.1 World population growth since the year 1000

The changes shown in Table 1.1 show that the rate of population growth is slowing down. In fact, the highest growth rate in world population was 2.2% in the 1960s - when there was an increase in the global population of more than 200 people a minute. By 2016, the growth rate had fallen to 1.13%. It is difficult to estimate future population numbers, but they are likely to continue rising for many years - with serious implications for the world's resources. The United Nations expects world population to reach 11 billion around the year 2100.

2009		2016
6.79 billion	**Total population**	7.23 billion
	Extra people every...	
77 760 000	**... year**	83 190 000
216 000	**... day**	227 917
1.23%	**... Annual growth rate**	1.13%

Table 1.1 A comparison of world population growth in 2009 and 2016

LEARNING TIP Be clear about the difference between growth in total numbers and growth rates. Always look carefully at the units used on graph axes. Remember that a fall in the growth rate does not result in a fall in the total population until the growth rate becomes negative.

Discussion point

How can the young people of today influence how many people will be living in the world in 2050?

Why has the population growth rate changed?

The **natural increase** or **decrease** of population depends on the difference between the **birth** and **death rates**. The birth rate is the number of babies born each year per 1000 people, and the death rate is the number of people who die each year per 1000 people.

The rate of natural increase, or decrease, is the birth rate minus the death rate. For 2016, the calculation is 18.5 per 1000 - 7.8 per 1000 = 10.7 per thousand increase, which can be expressed as a percentage growth rate of 1.07%.

Population change depends on how birth and death rates are influenced by social, economic, and political factors.

1 **a** Look again at Fig. 1.1. Describe what the relationship between the birth and death rates would have been like in (i) 1100, (ii) 1800, and (iii) 1960.

b **i** Copy and complete Table 1.2 by calculating how many years it is estimated it will take to add each extra billion people.

ii If this projection is correct, what trend does it show?

c Look at Table 1.3.

i Calculate the natural population growth rates of Niger and Russia in 2016.

ii What do your answers tell you about the populations in those countries?

iii Compare the birth and death rates of Niger and Russia.

Year	Estimated human population	Number of years to add one billion people
2024	8 billion	-
2038	9 billion	
2056	10 billion	
2100	11 billion	

Table 1.2 Estimated time to add each extra billion people

Country	Birth rate per 1000	Death rate per 1000
Niger	44.8	12.1
Russia	11.3	13.6

Table 1.3 Birth and death rates for Niger and Russia in 2016

Niger has the highest birth rate in the world, and Russia has one of the highest death rates. Niger has a positive population growth rate but Russia's is negative because its death rate exceeds its birth rate. Although now there are differences between countries, until the 1960s almost every country had an increasing population.

Population growth rates, 1700-1939

Population growth rates started to rise significantly in the late 1700s, as a result of the Agricultural and Industrial Revolutions in North America and Europe. New agricultural machinery led to greater crop yields with fewer workers. Many rural people moved to urban areas to work in the rapidly expanding industries there.

As countries became more prosperous, they were able to support larger populations. Families with more than ten children were common in the late 1800s. Then death rates began to fall rapidly - on account of tremendous advances in medical care and the supply of clean water and sanitation - particularly in Europe and North America. Because the birth rate was still high, there was a 'population explosion'.

Population growth rates since 1939

The slowing down of population growth was partly due to women in **MEDCs** (More Economically Developed Countries) joining the permanent workforce after the Second World War (1939-45), during which they had temporarily taken over the work of absent servicemen. More women also extended their education and delayed having children - aided by the availability of the contraceptive pill. Family sizes began to fall.

Even though **life expectancy** was increasing as a result of improved medical care, the total populations of countries such as Italy and Sweden began to fall. People realised that they could have a better standard of living if they had a smaller number of children to support. As the desire for material possessions grew, the number of children being born decreased.

In some countries, many women are delaying having children until they are in their forties. MEDCs are now experiencing a rise in death rates because the elderly are an increasing proportion of their populations. This has been countered by a rise in birth rates in many European countries as a result of a recent inflow of young immigrants from LEDCs.

Many **LEDCs** (Less Economically Developed Countries) have been experiencing the same pattern, but their changes started later and they are not so far along the path to low birth rates and low death rates as MEDCs. From the 1970s, the birth rate dropped in many LEDCs, where the falling death rate meant that more children survived, so there was no need to have such large families. Also, agricultural machinery replaced the need to have many children to work on farms. Some countries have introduced population policies to control birth rates.

In other countries, attitudes to women are slowly changing, so that a woman has the right to work and decide whether or not to put her career before having children. In many countries, however, the status of women is still inferior and they do not have the right to decide how many children they will have, and when. Populations are still growing rapidly in many countries, particularly in Africa and South East Asia - regions that are expected to hold more than 60% of the total world population by 2050.

The birth rate depends mainly on the **fertility rate** (the average number of children born per woman) and the age structure of the population. If the fertility rate is less than 2.1 (the official replacement level), the population will fall (discounting migration patterns).

2 Look at Table 1.4.

a In which continents were the countries with **(i)** the five highest birth rates and **(ii)** the five lowest birth rates in 2016?

b How have the birth rates changed in general since 2000, and which country is the main anomaly?

c In which two countries did the birth rate decrease the most?

Rank	Country	Highest birth rate per 1000
1	Niger	44.8 (51.5)
2	Mali	44.4 (49.2)
3	Uganda	43.4 (48.0)
4	Zambia	41.8 (41.9)
5	Burundi	41.7 (40.5)

Rank	Country	Lowest birth rate per 1000
1	Japan	7.8 (9.7)
2	Slovenia	8.3 (9.4)
3	Singapore	8.4 (12.8)
4	South Korea	8.4 (15.1)
5	Greece	8.5 (9.8)

Table 1.4 The countries with the highest and lowest birth rates per 1000 in 2016. The numbers in brackets show their birth rates in 2000.

LEARNING TIP Question 2b asks for your analysis of the general situation. If the same question were asked on an examination paper without the inclusion of 'general', you should answer by generalising and mention any anomalies to the normal situation. Do not state the change in each country, although you should mention the ones with the greatest and smallest changes.

Discussion point

1 Is it right that in some societies women are not allowed an education?

2 Why do attitudes to women vary in different cultures, and how might the spread of the Internet and ready access to other media lead to a breakdown of traditional practices?

Reasons for high birth rates

Cultural and social reasons

- In many cultures the greater the number of children a man has, the greater is his prestige and standing in society.
- A desire for a son to carry on the family name is important in many cultures, and parents will keep having more children until a son is born.
- In countries without good care services for the elderly and without adequate pension provision, people have children to ensure that they are looked after in their old age.
- In polygamous societies, a man might have children with more than one wife. One of the most extreme examples of this was the former King Sobhuza of Swaziland, who had 70 wives and 210 children!
- In many societies girls marry and start giving birth at a young age, so that they produce many children in their lifetimes.

Religious reasons

Some religions oppose any form of contraception and encourage families to have children. LEDC countries with high Catholic, Hindu and Muslim populations often have particularly high birth rates. However, this is not always the case, as natural birth control is permitted.

Demographic reasons

Countries with a high proportion of females of child-bearing age will tend to have higher birth rates.

Economic reasons

- One of the reasons why parents have children in LEDCs is to provide labour for the family's farm, or extra workers to boost the family's income.
- In poorer economies, the chances of a good education are limited to a privileged few. Without the ability to read, many people have no knowledge about contraception, especially if they live in rural areas. Other people are too poor to buy contraceptives, while some countries are too poor to develop family-planning clinics and subsidise contraception.
- Another reason why people in the poorest economies have many children is to ensure that some survive into adulthood, because the medical provision is so inadequate. Also, where poverty prevails, people are undernourished and too weak to fight infections. There is high child and **infant mortality** (children who die in their first year of life).

3 List as many reasons as you can why the birth rate might fall as a country develops its economy.

Reasons for falling death rates

4 Look at Table 1.5.

a Describe the general change in death rates between 2000 and 2016 and state an anomaly.

b The countries with the highest death rates in 2009 were all in Africa, south of the Sahara. Describe their location in 2016.

c What do the countries with the five lowest death rates have in common?

d Suggest reasons why countries in Europe and North America are not in the top five for the lowest death rates.

Rank	Country	Highest death rate per 1000
1	Lesotho	14.9 (20.4)
2	Bulgaria	14.5 (14.6)
3	Lithuania	14.5 (12.9)
4	Ukraine	14.4 (16.5)
5	Latvia	14.4 (14.9)

Rank	Country	Lowest death rate per 1000
1	Qatar	1.5 (4.2)
2	United Arab Emirates	2.0 (3.7)
3	Kuwait	2.2 (2.5)
4	Bahrain	2.7 (3.9)
5	Saudi Arabia	3.2 (6.0)

Table 1.5 The countries with the highest and lowest death rates per thousand in 2016. The figures in brackets show their death rates in 2000. Note: the effect of HIV/AIDS treatment is shown by the change in death rate in Lesotho since 2000.

With the exception of countries ravaged by HIV/AIDS and wars, death rates have generally been falling steadily. Reasons for this include:

- the development of new medical knowledge and medicines, better-trained doctors and greater access to clinics - even in rural areas. Smallpox has now been eradicated and polio almost eradicated by vaccines. There are also better treatments for typhoid, cholera and HIV/AIDS. There have been major attempts to reduce the incidence of malaria through drugs, and by providing people in affected areas with sleeping nets.
- programmes in many LEDCs to increase access to clean water and proper sanitation. Aid agencies from MEDCs, such as WaterAid, assist these projects.
- the spread of knowledge about what constitutes a better diet and a healthy lifestyle.
- a general improvement in access to food supplies.

Reasons for high death rates

Reasons for high death rates are complex and can change rapidly. War and natural disasters have an effect for a limited time, whereas changes in standards of living, health, and nutrition affect death rates more slowly and are likely to be more permanent.

As time goes on, most countries should continue to experience falling mortality as people's health improves. However, they will also experience a rise in death rates, due to a greater proportion of their populations living into old age. Also, not all countries have populations that take care of their health:

- The death rate in Russia has actually increased since 1990, partly because of problems with alcoholism and smoking that have led to high rates of cancer.
- Obesity, resulting from a fast food diet, is also likely to result in higher incidences of heart disease in the USA.
- The high death rates in Latvia and Lithuania are thought to result from the mental stresses of coping with changes resulting from their independence after the break-up of the Soviet Union. The high numbers of male deaths are linked to smoking and alcohol.

Discussion point

People in other parts of the world are increasingly eating a western diet, high in fat, salt, and refined sugar. How can they stay healthy?

The influence of migration on population growth rates

Changes in overall world population and *natural* population growth in countries have only two influences - birth and death rates - but the population growth rates of countries, or areas within them, are often influenced by another factor - **migration**. Migration is the movement of people from one place to another. The source area loses the people who leave, the **emigrants**, and the host area to which they move has people added to it, the **immigrants** or **in-migrants**.

Net migration for an area is calculated by:

number of immigrants - number of emigrants

If more people come in than go out, the net migration is a positive figure. If more leave than come in, it is a negative figure expressed either per 1000 or as a percentage change.

The population growth (or fall) of an area is calculated by:

natural change + net migration

The formula, showing all components, is:

(birth rate - death rate) + (number of immigrants - number of emigrants)

5 Look at Table 1.6.

a Calculate, showing your working, the population growth rate per 1000 and as a percentage for (i) Italy and (ii) Mexico.

b Describe the contrasts in natural increase and net migration between Italy (an MEDC) and Mexico (an LEDC).

c Explain the contrasts you described in (b).

Country	Birth rate per 1000	Death rate per 1000	Net migration rate per 1000	Population growth rate
Italy	8.7	10.3	3.9	?
Mexico	18.5	5.3	-1.7	?

Table 1.6 The population growth rates for Italy and Mexico in 2016

CASE STUDY

Botswana – The impact of HIV/AIDS on the population

Southern Africa has much lower population growth rates than the rest of the continent. This is the result of the spread of the HIV virus and the AIDS illness that it causes. AIDS became a problem in southern Africa in the late 1980s. It led to many deaths because those infected had their immune systems destroyed, leaving them unable to resist secondary illnesses like tuberculosis and pneumonia.

- AIDs also reduced birth rates as it caused the deaths of many women of childbearing age.
- Botswana's first case of HIV/AIDS was identified in 1985.
- The infection spread rapidly. By 1998, one in four Botswanans between the ages of 15 and 49 was HIV-positive. This was still the case in 2009.
- In 2000 44% of pregnant women in Francistown, the second largest settlement, had HIV/AIDS, compared with only 7% nine years earlier.
- Deaths from AIDS have fallen since their peak in 2001 and the natural population growth rate in 2016 was 1.25%. Botswana still has 22% of adults with AIDS, third highest after Swaziland and Lesotho.

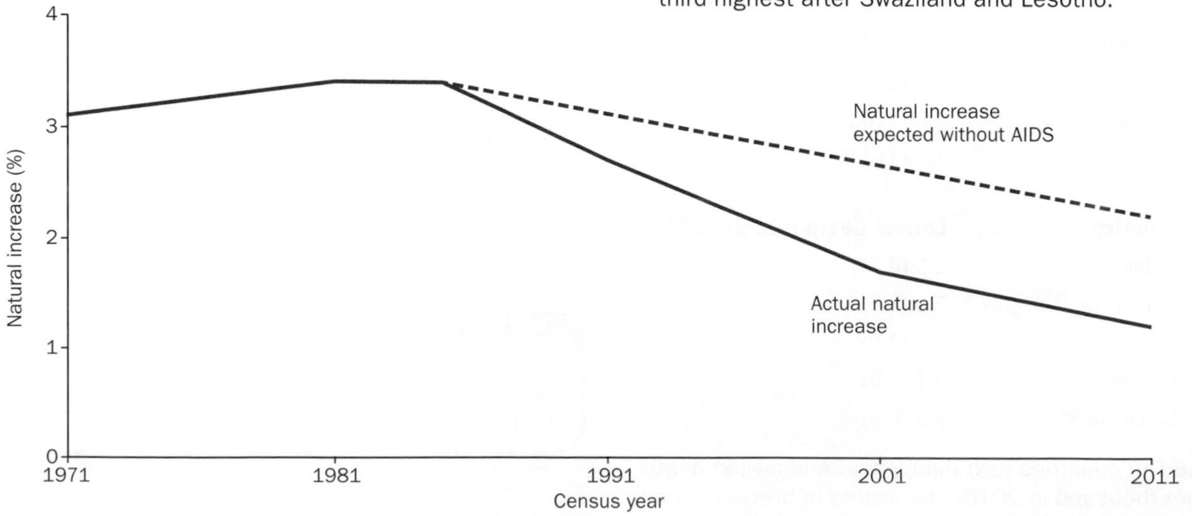

Fig. 1.2 Natural population increase in Botswana, 1971-2011

Death rate without AIDS (per 1000)	4.8
Death rate with AIDS (per 1000)	28.6
Infant mortality rate without AIDS (per 1000 live births)	20.0
Infant mortality rate with AIDS (per 1000 live births)	64.8
Life expectancy at birth without AIDS (years)	72.4
Life expectancy at birth with AIDS (years)	33.9

Table 1.7 The effect of HIV/AIDS on death rate, infant mortality, and life expectancy in Botswana in 2002

Population characteristics	1971	1981	1991	2001	2011
Birth rate (per 1000)	45.3	47.7	39.3	28.9	22.3
Death rate (per 1000)	13.7	13.9	11.5	12.4	10.6
Infant mortality rate (per 1000 live births)	97	71	48	56	11.1
Life expectancy at birth (years)	55	56	65	55	58

Table 1.8 Changes to population characteristics in Botswana, 1971–2011

What is being done to deal with Botswana's problem?

- From 2002, antiretroviral drugs were given to pregnant women with HIV in Botswana (the virus can be passed from mother to baby in the womb). But a shortage of medical staff reduced the number who could be treated.
- Despite the high costs involved for the government, the numbers of people being treated rose steadily until 170 000 had received the drug by 2010.
- Botswana also carries out routine testing for the HIV virus, and has an awareness campaign to promote the effectiveness of the antiretroviral treatment.
- Use of contraception is encouraged and is on the increase. Roadside slogans have been erected to encourage people to practise safe sex.

Why is HIV/AIDS such a problem in Botswana?

- Only about 18% of Botswanans are married. Many have more than one sexual partner and do not use protection. Some men are polygamous.
- There is considerable ignorance about HIV/AIDS and its prevention in rural areas. especially where belief in traditional healers is strong.
- Botswana lacked sufficient medical staff to cope with the numbers needing treatment. Many doctors and nurses had to be recruited from abroad.
- Poverty caused some vulnerable young women to become prostitutes, which increased the spread of the virus.
- The incidence of HIV/AIDS is particularly high in mining towns, such as Selibi-Phikwe, where 52% of the population was HIV-positive in 2003.

HIV/AIDS has had serious consequences for Botswana's economy and society:

- The country was fortunate that it had a strong economy (largely due to diamond mining) when HIV/AIDS first emerged. Nevertheless, funding the antiretroviral drug programme has been very costly. Also, bringing in doctors from other countries raised wages, which had to be paid.
- People who are ill cannot work or contribute to the economy. There has been a particular shortage of skilled labour in the workforce.
- HIV/AIDS has also caused very severe reductions in the wealth of individual families, as adults become too ill to work but medical and funeral expenses still have to be paid.
- There are now nearly 130 000 AIDS orphans in Botswana. Many have to care for siblings and can no longer find time to go to school. There is now welfare support for them until they are 18.

6 **a** Which population characteristic has reduced over time because of AIDS and for other reasons? Explain your answer.

b Describe the other impacts of HIV/AIDS on the population of Botswana since 1991.

c Note any ways in which the statistics suggest that Botswana might be winning the battle to reduce deaths from HIV/AIDS.

Discussion point

Why is it important to both individuals and governments that all people should be made aware of how to avoid catching HIV/AIDS? Consider the best ways of raising awareness in your own country.

Over-population and under-population

These conditions depend on the relationship between the population of an area and its resources. In an ideal situation, the entire population of a country should have a good standard of living by using the country's resources to build a strong economy. Over-population is where **there are too many people to be supported to a good standard of living by the resources** of the country. **If there are too few people to use all the resources of a country to maximum efficiency**, the country is said to be under-populated. These descriptions can be applied to areas as well as to countries.

RESEARCH Find out about the population of your own country, or another country you are interested in. The main source of your information might be its government's website. The CIA World Factbook on the CIA website contains relevant information about all countries. Try to find out the following information about your chosen country. Because this is quite a large research task, you could start the work now and add to it as you work your way through this chapter.

- The total population in 1900, 1950, 2000 and last year. (Show the information in graph form, so that you can see the trend or trends clearly and write a note about it.)
- The birth rate, fertility rate and death rate. Use this information to calculate the natural increase (or decrease) of the population. How does it compare with the world average of 10.7 per 1000?
- Use the birth and death rates to find out where your country fits into the demographic transition model (pages 24–5). What does this suggest will happen to the population of your country in the future?
- The population structure – how many males and females are there in each five-year age group? Plot a population pyramid (see pages 21 and 25) using five-year age ranges. Draw horizontal lines on it at 15 and 65 years. Do the working population (aged 15–64) have to support a lot of young people or elderly people, or both?
- How many people moved in from other countries last year (the number of immigrants)? Which two countries were the main sources of these people? Why did they leave their countries to live in your chosen country?
- How many people left the country (emigrants) to live in another country? To which two main countries did they move? Why did they leave and why did they choose their destination countries, do you think?
- Does the country have a population policy to try to control population growth, or to increase it? If not, should it have one and what do you think it should be?

Discussion point

What would be the problems for you if you lived in an overpopulated region of your country? What could you do to overcome them?

LEARNING TIP Do not think that all sparsely populated areas are under-populated, or that all densely populated areas are over-populated. It is the balance between population and resources that must be considered.

7 Look at the following two case studies.

a Copy and complete Table 1.9 to highlight the differences between the two countries.

b Which statistics indicate that Bangladesh has few resources but Australia has many?

	Under-populated Australia	Over-populated Bangladesh
GDP per person (US$)		
Value of exports (US$)		
Population increase per year (%)		
Labour force		
Main sector of employment		
Net migration rate (per 1000)		
Years of education		
Literacy rate (%)		
Infant mortality rate (%)		

Table 1.9 Comparing an under-populated and an over-populated country

CASE STUDY

Bangladesh, an over-populated country

Almost the whole of Bangladesh is made up of the Ganges delta and the wide floodplains of the Ganges and Brahmaputra rivers. This means that the country is frequently flooded – both by river floods and by coastal floods, which occur as a result of storm surges caused by cyclones approaching from the Bay of Bengal. Floods have caused the deaths of more than a million people there in the last 200 years.

The causes of its overpopulation are:

- → Rapid population growth on the fertile plains in the past because the mainly Muslim population used little contraception so had a very high birth rate
- → Few natural resources and reliance on farming. Of the 73.8 million people in the labour force, 45% work in agriculture – mainly as subsistence farmers. Minerals like iron ore are the basis of the manufacturing industry, but such raw materials are so expensive that industries based on imports would not be profitable.

Bangladesh has the eighth largest population in the world – 165 million – but only ranks 94th in the world in terms of land area, so it has a high population density of more than 1200 people per square kilometre. The net migration rate is negative at –3.1 per 1000.

Bangladesh has far more people than its resources can support. Its **gross domestic product (GDP) ppp** is only US$3900, which is far too low to provide a good standard of living. Its exports of garments, tea, seafood, jute and leather are worth only US$33 billion a year. An estimated 40% of the population are under-employed. Many people exist on low wages for a few hours' work a week and 31.5% live below the poverty line.

- **GDP (gross domestic product)** is the value, in dollars, of the goods and services that a country produces in a year.
- GDP is divided by the country's population to give **GDP per capita**.
- **ppp** means **purchasing power parity**. GDP is adjusted because a dollar buys more in some countries than in others.

There are also not enough schools and hospitals. Only 61% of the population are literate, and education is provided for only eight years of a person's life. Most people have no qualifications. Access to health care is also poor. For example, the infant mortality rate is 3.3%.

The outlook for the future is not good:

- → The agricultural land on the floodplains of the Ganges and Brahmaputra rivers is already over-cultivated.
- → There has been widespread deforestation for firewood on the foothills of the Himalayas – increasing the flood risk.
- → The capital, Dhaka, is heavily congested with traffic and has overcrowded housing – often lacking basic amenities.
- → The cost of repairing damage to infrastructure after flooding is a frequent drain on the economy.

On the positive side, the population growth rate had fallen to 1.6% by 2016.

CASE STUDY

Australia, an under-populated country

With a population of just under 23 million and a labour force of only 12.6 million, Australia, the sixth largest country in the world, is under-populated. It is very rich in resources, with large reserves of iron ore, coal, gold, copper, natural gas and uranium – and abundant potential for solar and wind power development. Australia therefore needs to attract migrant workers to exploit fully its many resources. The positive net migration rate of over 5.6 per 1000 is one of the highest in the world.

The quantities of many of Australia's resources are greater than the country's needs, so any surpluses can be exported – particularly coal from Newcastle, iron ore from Iron Knob and gold from Kalgoorlie, meat, wheat, machinery and transport equipment. Australia's exports were worth US$184 billion in 2016. Its GDP per person was US$48 800. The service sector employs 75% of Australians. The unemployment rate is low (5.8%).

Australia's low birth and death rates give a natural population increase of only 1.05% a year. Its population density is only 2.9 people per square kilometre and, although a large proportion of the country is desert or semi-desert, there is ample suitable land for an increase in settlements in the outback.

Education standards in Australia are high. 99% of Australians are literate, and education is available for 20 years of a person's life. Health care is also good: e.g. the infant mortality rate is only 0.46%.

With all of these different factors combined, Australia could support a larger population.

Under-populated countries have insufficient workers to exploit their resources efficiently, support their retired populations and provide enough services. As their total domestic markets are small, it is difficult to attract foreign investors to promote industrial growth.

They also have a shortage of workers to produce food and goods, so many items have to be imported - which increases their cost and adversely affects the countries' balance of payments. As a result, under-populated countries often encourage immigration.

Rapid natural population growth

Worldwide, the human population is still growing - but is it the same everywhere? The short answer is no, as Fig. 1.3 shows. Generally, higher levels of population growth are happening in developing or poorer countries, and lower levels of growth, population balance - or even decline - are happening in developed or richer countries.

Look at Table 1.10 and concentrate on the columns for population growth rate and GDP per capita (GDP shows how wealthy a country is). You should notice that there's a link between them.

Problems of rapid population growth

For many poorer countries, rapid population growth is slowing down their development. They're struggling to earn enough money from farming and basic industry to provide for more and more people. The ever-growing population puts too much pressure on their resources.

Some countries:

→ find it difficult to feed everyone - but the population keeps on growing. The result: millions of people go hungry.

→ can't afford to provide enough schools and teachers. The result: millions of people don't get the education and skills that would help to raise them out of poverty and help their countries to develop.

→ can't afford to provide good basic health care, with enough doctors and hospitals. The result: millions of people suffer and die from illnesses and diseases that could have been cured or prevented.

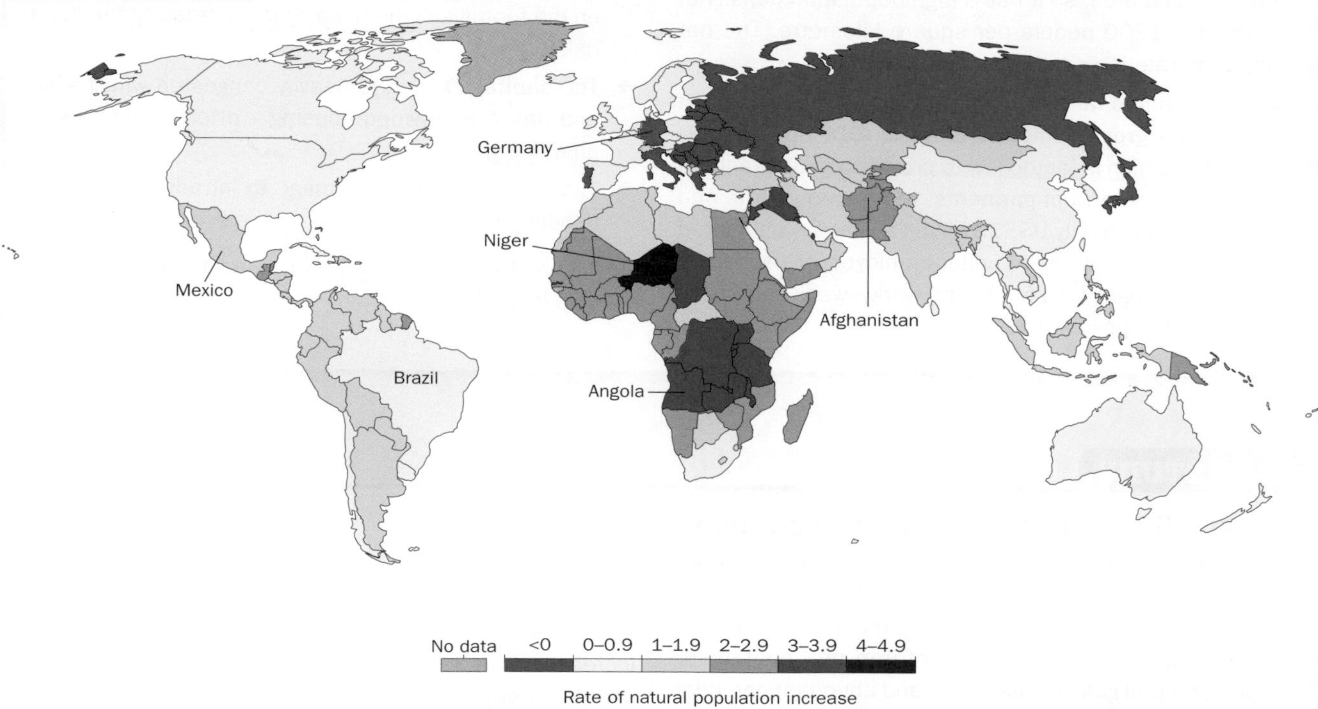

Fig. 1.3 The natural increase in population around the world in 2015. The twelve fastest growing populations were in Africa. (Source: Max Roser and Esteban Ortiz-Ospina (2017) 'World Population Growth' published online at OurWorldInData.org)

Country	Infant mortality per 1000	Natural population growth rate (%)	Fertility rate per 1000	Life expectancy (years)	GDP per capita (US$ ppp)
Niger	82.8	4.0	6.62	55.5	$1100
Angola	76.5	3.2	5.31	56.0	$6800
Afghanistan	112.8	2.5	5.22	51.3	$2000
Mexico	11.9	1.3	2.25	75.9	$18 900
Brazil	18.0	0.9	1.76	73.8	$14 800
Germany	3.4	0.2	1.44	80.7	$48 200

Table 1.10 Countries with different rates of population change, 2015

Sustainable development is defined as: 'meeting the needs of the present without compromising the ability of future generations to meet their own needs'. But what is the link between population growth and sustainable development? For a population to be sustainable, the rate at which it grows must not threaten the survival of future generations. You can probably see that a population that is growing too rapidly, or one that is falling, won't be sustainable.

Niger is a dangerous place in which to be born. More than 82 babies out of every 1000 born will die before they reach their first birthday. As a comparison, in Japan only 2 babies out of every 1000 die before their first birthday.

8 **a** Look again at Fig. 1.3.

- **i** Describe where population is growing fastest (i.e. countries where the natural increase is over 2%).
- **ii** Describe where population is declining (i.e. countries where the natural increase is below 0%).

b **i** Create a diagram to show how rapid population growth can affect a country's development.

- **ii** Highlight the social impacts in one colour and the economic impacts in another colour.

9 **a** Describe and explain the relationship between natural population growth rates and (i) fertility rates (the average number of births to a woman) and (ii) life expectancy.

b Use information in Table 1.10 to show how wealth (GDP per capita) has an effect on:

- **i** infant mortality rates
- **ii** population growth rates
- **iii** life expectancy.

CASE STUDY

Niger, a country with rapid population growth

Reasons for Niger's rapid natural population explosion are mainly social.

- Niger had the highest birth rate in the world in 2016, with a reducing death rate but high infant mortality rate.
- Niger has the highest total fertility rate in the world – almost 7 children per woman in 2016 – with a desire for large families (see pages 4–5 for the reasons for this).
- There is little use of contraception, especially in rural areas where there are few family planning clinics.
- More than 80% of the population are Muslim. Procreation is encouraged in Islam. Many men do not use contraception, believing that the more children a man has, the greater his prestige in society; some men practise polygamy, having children with several wives.
- Many are farmers and run small businesses, so need children for labour.
- Niger has a traditional and mainly rural society in which men dominate (they want 12 children on average) and women lack education.
- Early marriage is common – 60% of marriages involve children under 16.
- Niger has a young population with a large proportion of childbearing age.

Fig. 1.4 Malnourished children with their mothers

The population of Niger is expected to double in less than 25 years. Already the problems typical of rapid population growth are acute:

- 45% live below the poverty line and the economy is the second poorest in the world.
- Farms become much smaller because they are split between many children when the father dies, so food production does not keep pace with population growth and 38% of children under five are underweight. Frequent droughts also reduce food production.
- Agricultural land is over-used, causing soil exhaustion and erosion.

- ➔ There are too few teachers, so children only attend school for five years and the literacy rate is only 19%.
- ➔ Health care is inadequate, with only two doctors per 100 000 people.
- ➔ Nearly 90% lack access to proper sanitation and about 50% of the rural population lack safe drinking water.
- ➔ Deforestation is a major problem.
- ➔ There are not enough jobs for everyone of working age.

The government acknowledges that these economic problems result from the population explosion but have no policy to deal with it.

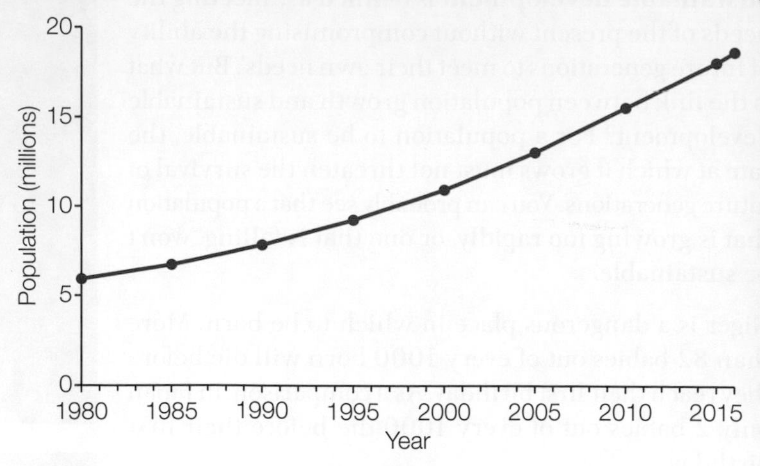

Fig. 1.5 Population growth in Niger, 1980 to 2016

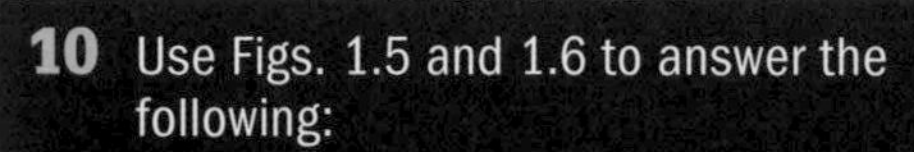

10 Use Figs. 1.5 and 1.6 to answer the following:

- **a** In which year was Niger's total population 9 million?
- **b** How many years did it take for Niger's population to double to 18 million in 2015?
- **c** If the growth trend between 1980 and 1990 had continued, what would the approximate population have been in 2016?
- **d** Using figures, explain why the total population grew more quickly after 1990 than before it.

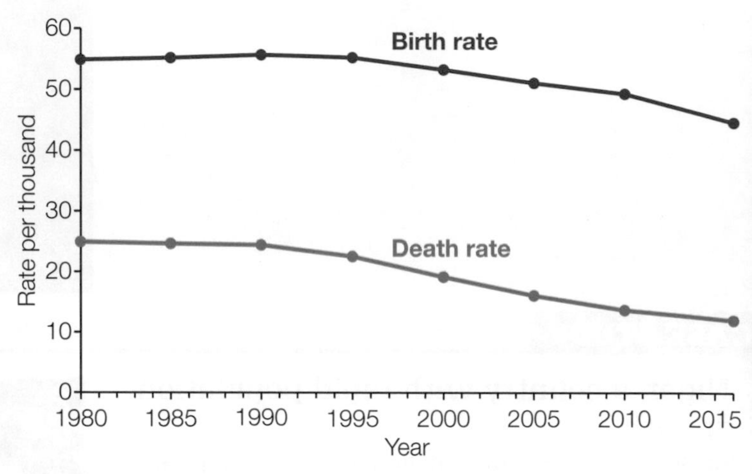

Fig. 1.6 Birth and death rates for Niger, 1980 to 2016

Anti-natalist policies to cope with high rates of population growth

CASE STUDY

China's one-child population policy

During the 1950s and 1960s, China's population grew rapidly – the birth rate reached 48 per 1000, which was seen as unsustainable. China didn't have enough food, water, and energy to provide for such a rapidly growing population. Therefore, in 1979, the Chinese government introduced rules to limit population growth – its one-child policy. Couples who had only one child received financial rewards and welfare benefits.

Positive impacts of the one-child policy

- ➔ The policy reduced the fertility rate from 3 births per woman in 1980 to only 1.5 in 2011. This is well below the 2.1 rate needed to keep a population stable and has successfully reduced the rate of population growth.
- ➔ By reducing the number of children born, China has also reduced the problems of overpopulation in its most crowded regions. There is less pressure on social services, waste disposal, and housing, and less danger of epidemics spreading.

Has China's policy led to sustainable development? The policy has prevented around 300 million babies being born, so China's population now – and going into the future – is lower than it would have been. However, by controlling one problem, has China just succeeded in creating other problems?

Negative impacts of China's one-child policy

China's chosen method of population control has had a range of social and economic impacts.

Social impacts

- A typical Chinese child today will have two parents and four grandparents to look after when they reach old age (a married couple might have up to four parents and eight grandparents to look after). So more old people's homes will be needed whereas in 2015 most elderly Chinese lived with their families.

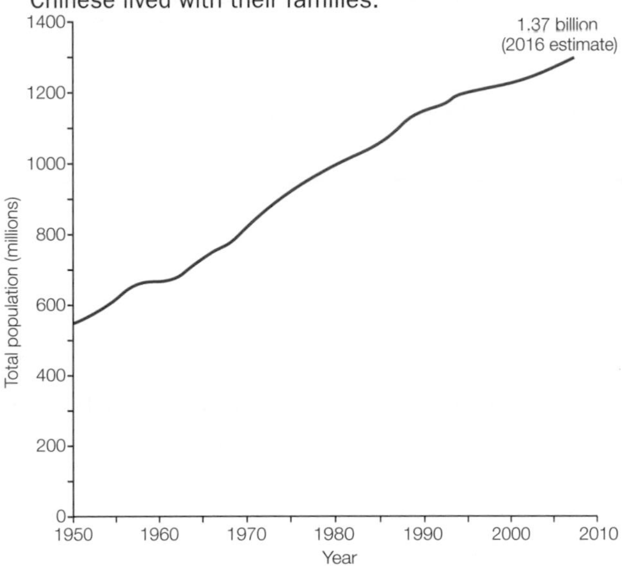

Fig. 1.7 China's population growth since 1950

- By 2020, it is estimated that men in China will outnumber women by 30 million, which might lead to social tension and unrest as more and more men find themselves unable to get married.

Economic impacts

- China's population is ageing rapidly. People will all need supporting financially in their old age, which includes an increasing need for expensive health care.

Fig. 1.8 One of China's problems – there are now fewer young people to support an ageing population

- The percentage of people aged over 65, compared to people of working age, is going to increase rapidly – from 10% in 2009 to 40% by 2050. From 2025, China is expected to have more elderly people than children (see Fig. 1.9).
- Many experts feel that China's growing economy won't have enough workers in the future to keep it expanding, while also supporting the growing number of non-workers in the population.

CASE STUDY

One-child policy replaced

Since January 2016 a two-child policy for married couples has been in force. It is expected to result in the Chinese population declining from about 2030 from a peak of 1.4 billion.

11 a Why did China implement a one-child policy?

b To what extent has China's one-child policy achieved its aim?

c What problems caused it to be changed to a two-child policy?

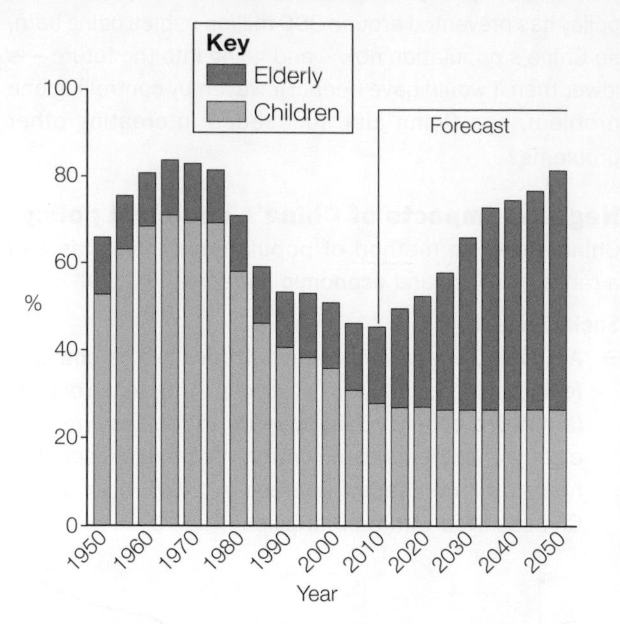

Fig. 1.9 The changing proportions of elderly and children in China's population

Discussion point

Could China's one-child policy be considered as fair and necessary?

An evaluation of policies for reducing population growth and size

Free family planning

Success in poor countries of Africa and Asia has been limited because:

- some people in rural areas are unaware of the free contraception
- governments have many other problems to address and may - especially in Muslim and Roman Catholic countries - be reluctant to spend money on family planning
- introducing family planning is very difficult in societies where males make all the decisions. (In most MEDCs family planning has been very successful and emancipating for educated women.)

Improved health and education

- This has been very successful in Kerala. However, countries with limited spending power may prioritise other needs, such as combatting HIV/AIDS and malaria. There is a shortage of trained doctors and nurses as some move to countries where they can earn more.
- Education is vital to improving health care, as people who cannot read remain unaware of advice in print. In some parts of the world education for females remains a hope rather than a reality. Many parents cannot afford to pay for their children's schooling.

Other anti-natalist policies

- Limiting family size by law, as in China, which is successful but leads to problems.
- Incentives, including cash and priority for school places, for people with small families.
- Withdrawing tax allowances from people with large families.
- Advertising the need to limit family size through the media and posters.

CASE STUDY

Population control in the Indian state of Kerala

The south Indian state of Kerala has the country's lowest birth rate. Its population growth rate was less than half the Indian average in 2008.

What made Kerala different from the rest of India was its focus on health care and education. Its female literacy rate is 85%, compared with 57% in India as a whole.

This success story is the result of two things:

- Political decisions to invest in education and women's health; almost all villages have access to a school and a modern health clinic within 2.5 km.
- Economics; Kerala relies less on farming and more on service industries than other Indian states, especially tourism.

How Kerala compares with India as a whole

- Women's health and education are the best in India. Food programmes focus on mothers and children, using ration cards and free school lunches.
- Attitudes towards women are positive. There are more girls than boys in higher education, and women hold some of the top jobs. An educated woman knows how to keep her children healthy and how to limit family size to have a higher standard of living.
- Women in Kerala marry later, and have their first child later, than other Indian women. They have fewer than two children on average.
- Over 95% of babies are born in hospital. Investing in health facilities reduces infant mortality and, therefore, the need for more children.

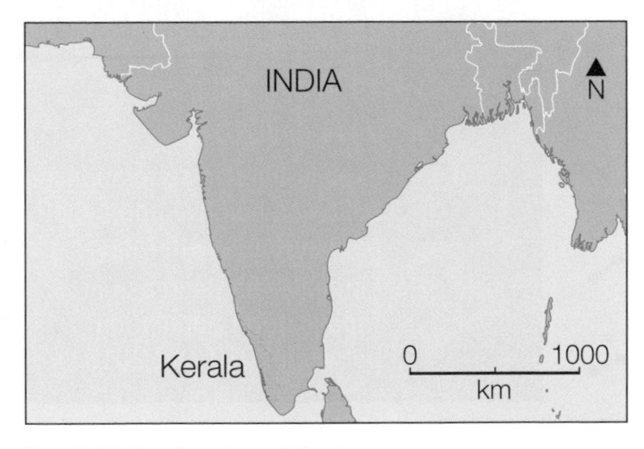

Fig. 1.10 The location of Kerala

Fig. 1.11 Most children in Kerala complete 10 years at school

Factor	Kerala	India
Infant mortality (%)	14	70
Female literacy rate (%)	85	57
Use of contraceptives by married women (%)	64	48
Fertility rate (average number of children per woman	1.8	3.2

Table 1.11 Kerala compared with India as a whole, in factors affecting population growth rates

India followed Kerala's example and has now reduced its population growth rate considerably.

Has Kerala's approach led to sustainable development? Kerala has managed to control its population growth by investing in health care and education – while still allowing people the freedom to choose their own family size. However, it looks as if Kerala's population could stop growing altogether within 30 years. This could create new problems associated with an ageing population.

Reasons for low rates of population growth

Population growth becomes low when birth rates are only a little higher than death rates. A negative growth rate occurs when death rates exceed birth rates, so the population declines.

CASE STUDY

Japan, a country with a declining population

Japan's population has been declining at a fairly steady rate since 1977. The graphs show information about Japan's population since 1990.

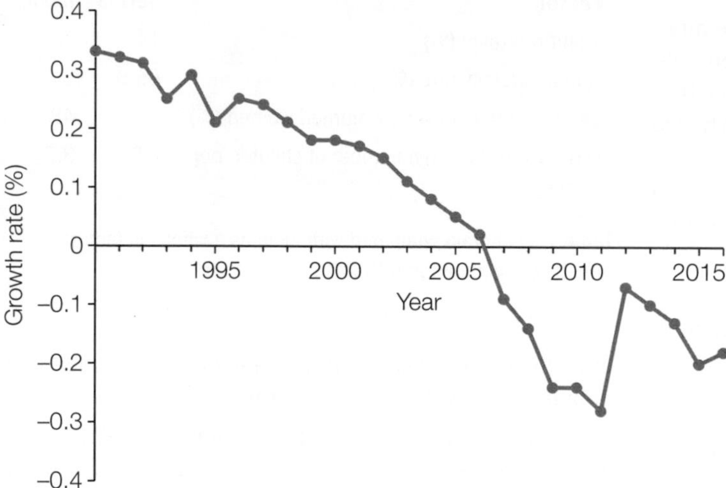

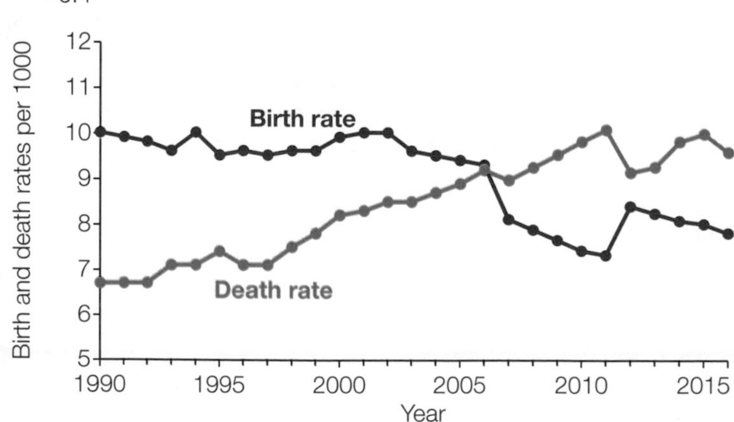

Fig. 1.12 Japan's total population and factors influencing it, 1990–2016

Japan's slow population growth between 1996 and 2005 was caused by the steadily increasing death rate as the birth rate remained fairly static. The decline in the total population since 2005 is caused by the death rate being constantly higher than the birth rate. Both are low and fluctuating. The cost of living in Japan is high. People have to consider whether or not they can afford to bring up a child and many choose not to. In addition, the proportion of people of childbearing age in the total population is decreasing.

Meanwhile, the death rate has been increasing because the population has an increasing proportion in the older age groups in which deaths occur most frequently.

12 Use Fig. 1.12 to answer the following:

a Identify the year in which Japan had zero population growth. Explain why there was zero population growth in the year you have identified.

b Describe and explain the changes in total population in Japan between

i 1990 and 1996

ii 1996 and 2005

iii 2005 and 2016.

c Describe the relationship between Japan's birth rate and death rate between

i 1990 and 2011

ii 2011 and 2016.

d Explain how these relationships influence positive and negative population growth rates.

e Estimate the decrease in Japan's total population between 2010 and 2016.

Policies to increase the population

A population policy started in 1994 to encourage more births had little effect.

In 2009 a programme began to enable mothers to remain in the workforce to earn enough income to maintain a good standard of living, while also bringing up children. Men are encouraged to spend more time at home and share the childcare.

CASE STUDY

Sweden's policies for coping with slow growth

Sweden is trying to increase the birth rate, so that it will have a larger working population in the future. The Swedish government has offered incentives, such as:

- paying a generous benefit for each child born
- giving fathers 13 months' paid leave after the birth of a child, at 80% of their salaries (and 120 paid days off work a year to care for sick children)
- providing all-day childcare or all-day schools, which is a great benefit for working parents.

However, by 2011, Sweden's birth rate had hardly altered since 2000. The situation changed as a result of a large influx of immigrants from Syria and the Middle East after 2012. They are the main reason for the recent population growth, as the immigrants have a considerably higher birth rate than Swedish-born inhabitants.

An evaluation of pro-natalist population policies

Pro-natalist policies are used if a country's population growth is too slow or not replacing itself, leading to fewer workers paying taxes, reducing economic growth and increasing poverty.

The different policies of Japan and Sweden to encourage more births have been less successful than the policies used in France. France gives incentives to encourage couples to have three children:

- the more children a person has, the less tax they pay
- women paid to stay off work for their third baby's first year
- paid leave for one parent for 36 months after having a child
- subsidised childcare for children under the age of three
- free schooling for children aged over three.

France's policies have been successful, as its fertility rate is now one of the highest in Europe. The success is partly because immigrants in France from former colonies in Africa have higher birth rates.

Countries such as Iran and Mexico have had to alternate anti-natalist and pro-natalist policies, to rectify the problems caused by their successes.

Migration

Migrations are either **internal** (within the country) or **international** (from one country to another). They can be further classified as **voluntary** (where the individual decides to move) or **involuntary** (where there is little choice but to move).

Push and pull factors

The reasons why people migrate are often described as push and pull factors.

Push factors that can force people to leave their own country include:

- not enough jobs
- low wages
- poor educational opportunities
- poor health care
- war with another country
- civil war and lawlessness
- drought and famine.

Pull factors that can attract people to a new country include:

- hope of finding a job
- higher wages than at home
- better health care
- chance of a better education
- a better standard of living
- family and friends have moved there already
- lower levels of crime and safety from conflict.

Reasons for internal migrations are similar, with the addition of movements away from natural disasters, which are often temporary.

LEARNING TIP Push and pull factors are often direct opposites. Do not use both in examination answers. For example, 'People move to towns for better-paid jobs, because they get low pay in the countryside' is stating the same factor twice. You might be asked to write about only push or only pull factors, so be clear about what the difference is between them.

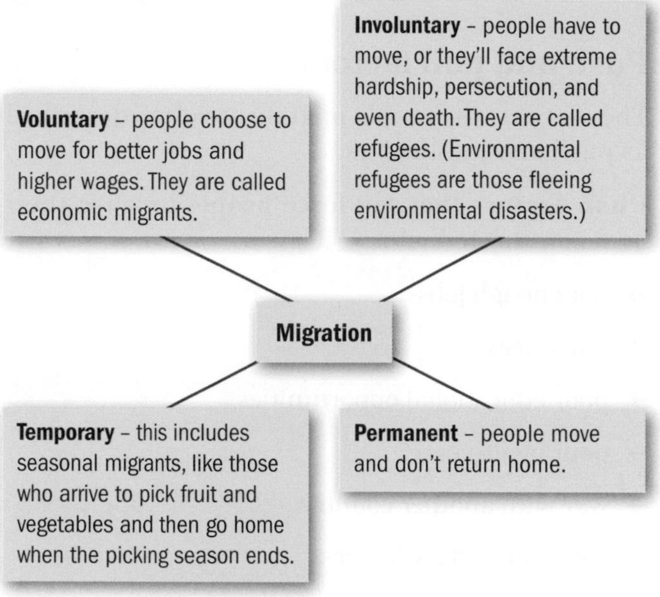

Voluntary – people choose to move for better jobs and higher wages. They are called economic migrants.

Involuntary – people have to move, or they'll face extreme hardship, persecution, and even death. They are called refugees. (Environmental refugees are those fleeing environmental disasters.)

Migration

Temporary – this includes seasonal migrants, like those who arrive to pick fruit and vegetables and then go home when the picking season ends.

Permanent – people move and don't return home.

Fig. 1.13 Classifications of migration

Fig. 1.14 Conflict, or the threat of conflict (as here in Afghanistan), often forces people to flee their homes and they may never return

Fig. 1.15 Seasonal work, such as cutting celery, attracts migrants for a short period of time

Voluntary international migration

Economic international migrations have increased with globalisation. In 2015 the UAE had 8 million foreign migrant workers in its population of 9 million but the top country for immigration was the USA – with nearly 43 million of its citizens born in another country. The top country for emigration was Mexico.

The mass migration of Mexicans to the USA started in about 1950, when they went to work in California and Texas as farm labourers at harvest time. They were also employed as maids or factory workers, or in any poorly paid or dirty job which Americans didn't want to do themselves. Then a recession hit and many Americans became unemployed. Resentment soon built up against Mexicans who held jobs that Americans now wanted. Laws were passed to make migration into the USA from Mexico more difficult. Many Mexicans were deported, but a large unknown number still cross the border illegally. In 2016 the net movement was back to Mexico.

13 **a** List the pull factors which an MEDC (like the USA) might have to attract migrants from LEDC countries (like Mexico).

b What do the major migration routes from Mexico to USA and Bangladesh to India have in common? (Use an atlas to find the answer.)

Discussion point

Should people be free to live in whichever country they please? Why would such a worldwide policy cause problems for some countries but not others?

CASE STUDY

Migration from Syria into Germany and other European Union countries, 2014–16

Migration to the European Union (EU) has been increasing since 2000 as a result of globalisation, with more than 200 000 applications for asylum each year. This reached crisis proportions in 2014 when more than half a million applied for asylum. First-time asylum applications increased greatly to a peak of more than 1 300 000 in 2015, and was almost as large in 2016 (see Fig. 1.16). In addition, many migrants entered the EU illegally – either without permission or without the required documents. Eurostat statistics state that the number of immigrants from non-member countries in 2015 was 2.7 million but actual totals are difficult to determine.

Very large numbers of refugees escaping civil war in Syria tried to reach Germany after the German Chancellor announced in September 2015 that all Syrian refugees would be welcome. Germany had a negative population growth rate and a shortage of workers. This led to an immediate influx that grew rapidly to numbers that could not be managed easily. Refugees from other countries and economic migrants from Africa and the Balkans were also able to enter as it was difficult to process selectively such large numbers. Chancellor Merkel urged other EU states to accept their fair share of the migrant burden.

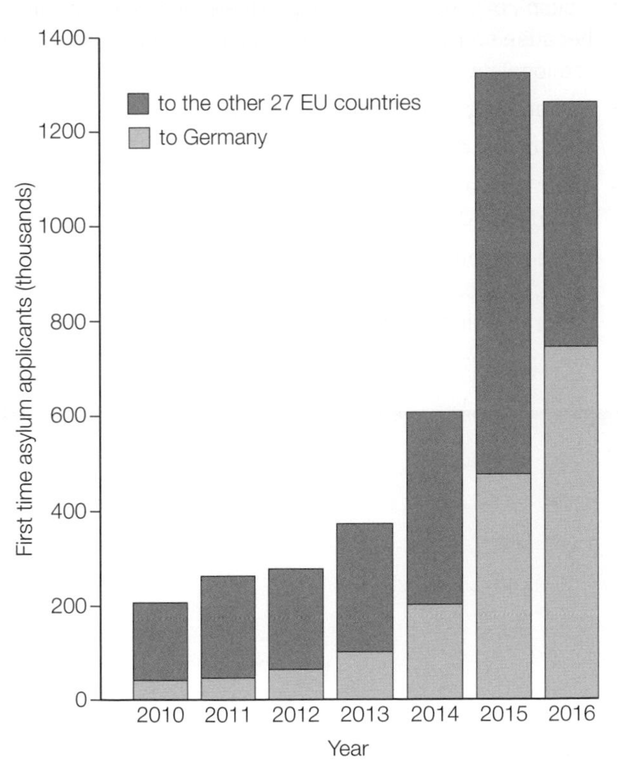

Fig. 1.16 First-time asylum applicants to Germany and other European Union countries, 2010–16 (source: Eurostat)

Reasons for mass migration

Push factors

- Refugees were seeking safety from war and persecution in Syria, Afghanistan, Iraq, Eritrea, and northeast Nigeria.

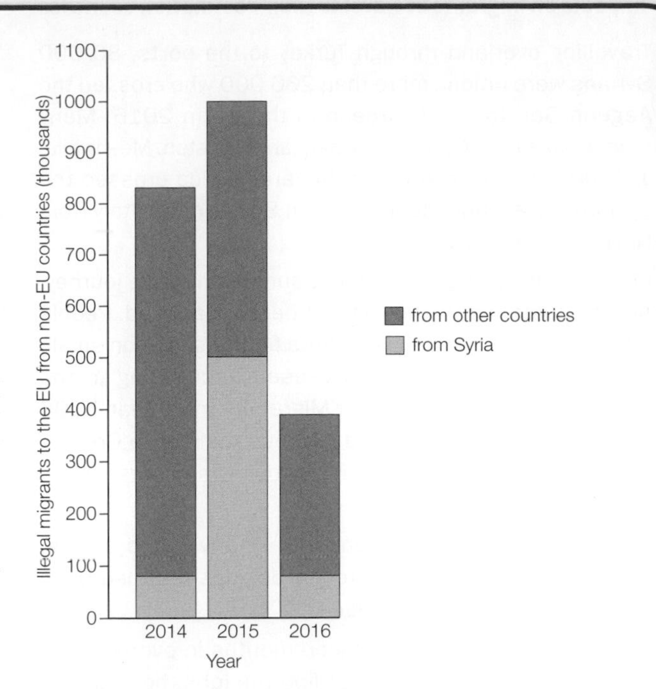

Fig. 1.17 Illegal migrants to the EU from non-EU countries, 2014–16

- A lack of jobs and extreme poverty motivated economic migrants to leave Nigeria and other African countries and Kosovo and Albania in the Balkans.
- Refugees from Somalia and Sudan were fleeing from famine caused by desertification.

Pull factors

- Europe is politically stable.
- There is more chance of a job, as there is a shortage of workers in European countries with declining and ageing populations.
- Wages and living standards are higher in Europe.
- European societies are generally peaceful, law-abiding and tolerant of different faiths.
- People want to join family members already in EU countries.

By 2017 more than one million Syrian refugees were living in the EU (600 000 of them in Germany) but 4.5 million were in camps in Syria's neighbouring countries – Turkey, Lebanon and Jordan – from where they can return home more easily when the war ends.

Impacts of migration

Impacts on the country of origin

- As most economic migrants and many political refugees are young able-bodied men, their home communities lose workers and providers for families left behind. Women are over-burdened and the economy suffers.
- As the most educated usually migrate, education, health, and other services decline because of shortages of skilled teachers, doctors, and nurses.

Impacts on migrants of travel to their destination countries

Travelling overland through Turkey to the ports, 80 000 Syrians were among more than 266 000 who crossed the Aegean Sea to reach Greece in the EU in 2016. Many others came from Afghanistan, Iraq, and Pakistan. Meanwhile, 136 000 migrants from sub-Saharan Africa crossed the Sahara Desert and Mediterranean Sea to reach Italy from North African ports.

Many of these migrants did not survive the long journey. At the coasts they were often robbed and abused. People smugglers charged large amounts for passages on small, unsuitable and overloaded vessels. According to the International Organisation for Migration, more than 5000 drowned in 2016 attempting to cross the sea to Greece, Italy, and Spain. Some, taken by people-traffickers, ended up in slavery.

Those determined to go on to Germany faced a long, arduous walk and some border crossings that had been shut to prevent their passage.

At their destination, many faced months in overcrowded reception centres and did not find the jobs, homes, and welcome they had hoped for as a third of them were low skilled. Nearly half were at risk of unemployment and poverty. Migrants in general tend to be exploited in low-paid jobs, have language difficulties and face discrimination.

Genuine refugees were regarded with suspicion or hostility when it became apparent that some terrorists had arrived with them, disguised as migrants.

By 2017 there were so many migrants in the EU that:

- several countries severely restricted the numbers they allowed to pass through. Macedonia blocked the route north towards Germany from Greece. Many were unable to move on.
- more citizens voted for anti-immigration parties in 2017 elections in Germany and other EU countries
- the EU paid Turkey and Libya to hold migrants and repatriate those who could be sent back home safely.

Negative impacts on the receiving countries

- Immediate shelter, food and other necessities had to be provided.
- There was extreme pressure on housing, health services, and schools. Many school classes have children speaking numerous languages.
- Maternity services were severely stretched as the migrants are mainly of childbearing age (their median age is 27). As their culture is to have more children than the average EU family, the impact on birth rates and total population size will be felt for generations.
- Most migrants were Muslims and their destination countries were Christian, causing culture clashes. By 2017 there were nearly six million Muslims in the German population. As migrants tend to live together and keep to their own ways instead of integrating into the local community, racial tension can result.
- The willingness of immigrants to work for lower wages depressed incomes and the increased competition for jobs resulted in unemployment.
- A few migrants did not respect the culture and laws of their adopted country. There was a particular problem if they came from areas where females were not given an equal standing in society.

Positive impacts on the receiving countries

- A supply of needed labour helped the economy, especially because some migrants were willing to undertake jobs the locals would rather not do.
- Migrants work and then pay taxes.
- They created a larger market for local businesses.
- Their arrival resulted in the provision of services, including a diversity of food in ethnic restaurants.
- Migrants made available a wealth of cultural experiences in art, music, and literature, as well as creating a better understanding of different cultures.

Discussion point

Should the EU have restricted immigration and given more financial assistance to countries near Syria to allow them to look after the migrants?

CASE STUDY

Internal and international migrations in Botswana

Botswana has experienced every type of migration since 1966, except **transmigration**.

Temporary international economic emigration

When Botswana gained its independence from the UK in 1966, it was one of the world's poorest countries. The standard of living was very low. Botswana's economy was weak and based on subsistence farming, which suffered in drought years. Many men migrated to neighbouring South Africa to work in the gold mines between 1970 and 1980. This benefited Botswana because the men's earnings helped the economy to grow.

However, the migration of so many young Botswanan men to South Africa caused problems for the families they left behind. Children were deprived of their fathers, except for short visits once or twice a year. The women were left to do the work on the subsistence farm as well as look after the children – so agricultural productivity remained low.

Botswana's economy eventually became one of the fastest growing in the world, especially after diamond production started in 1971 (followed by copper and nickel mining at Selebi-Phikwe). This economic growth was helped because, during the 1980s, South Africa reduced the number of migrant workers that it allowed in from neighbouring countries. As a result, Botswana suddenly had a greater number of skilled men aged 15 to 50 available for its own workforce.

Internal permanent rural-to-urban migration

In 1964, only 21 000 people, or 1% of the population, lived in towns. This figure rapidly increased as the economy grew – reaching 61% by 2017. The change has been caused by **rural-to-urban migration** – the movement of people from rural areas to live in towns. The term used for the increasing proportion of a country's population who live in towns is urbanisation. Botswana's capital, Gaborone, has experienced enormous population growth because it has the most employment opportunities.

Push factors from rural areas

- Droughts, pests and poor farming skills have resulted in low productivity in agriculture, the main occupation in rural areas.
- Young people want to get away from traditional lifestyles.

Almost a third of Botswana's population now live within 50 km of Gaborone.

Work is available in the capital city's administrative offices and in the headquarters of numerous international companies that have set up in Gaborone. In addition, employment opportunities occur in financial institutions, the University of Botswana, Debswana (diamond sorting), and shopping centres.

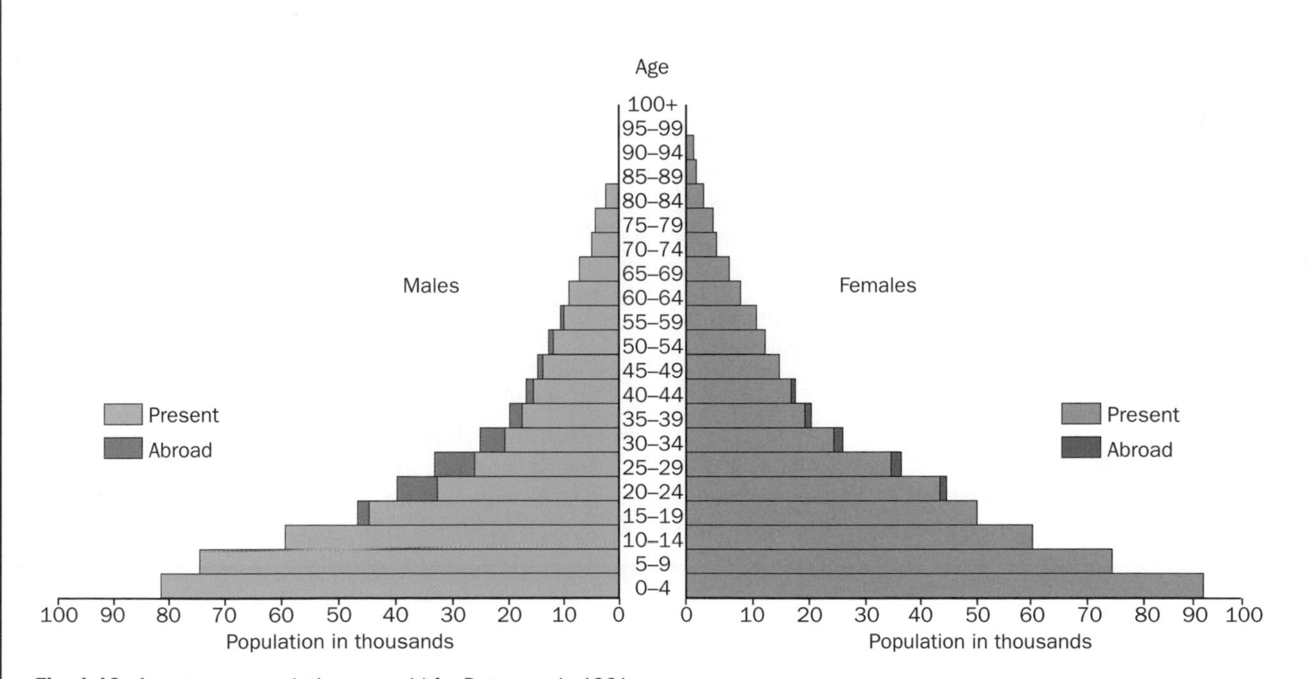

Fig. 1.18 An age–sex population pyramid for Botswana in 1981

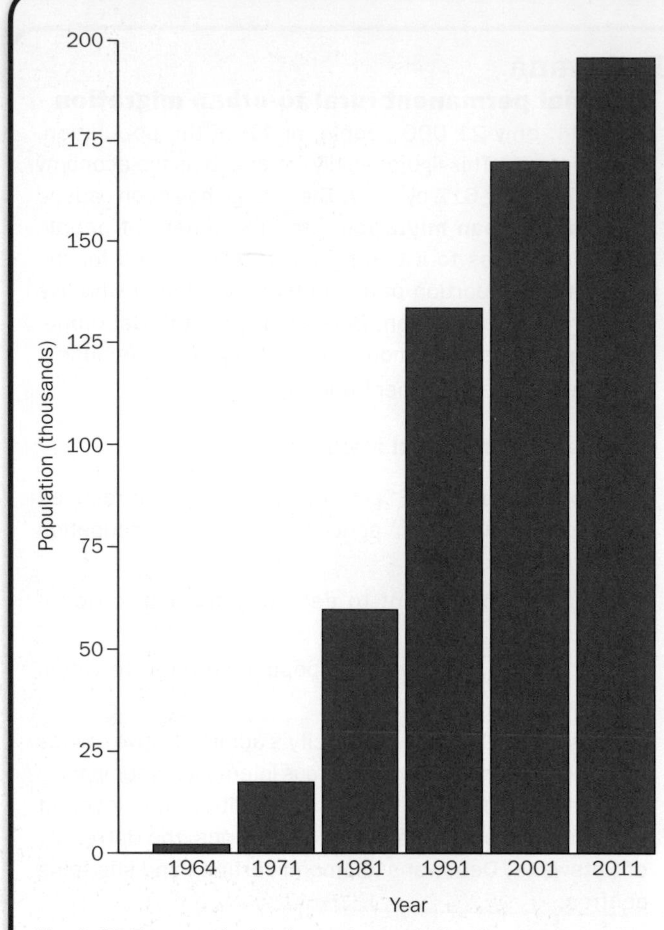

Fig. 1.19 The growth of Gaborone's population; it was expected to reach 264 000 in 2017

It is mainly the 15 to 50 age groups that move to the city – drawn by job prospects, housing with better services, nearby schools and clinics and the desire to have access to more entertainment (the bright-lights effect). More than 70% of the migrants have been educated to secondary level or higher. They have the skills needed to work in the industrialised urban areas.

About 20% of Gaborone's population are young dependents, with only 2% elderly dependents. The proportions of both young and elderly dependents are higher in rural than in urban areas.

As in other developing countries, not all incoming migrants get jobs. Many live in improvised shelters made of scrap materials on the edge of the urban area. A shanty town called Old Naledi grew up on the edge of Gaborone – occupied by both unemployed and low-income workers. It lacked sanitation and electricity until the city council supplied tarred roads, a sewerage network and street lighting.

Other internal migrations

Urban-to-urban migration is also common in Botswana now, but the main direction of movement for people aged over 55 is back to rural areas. This is **urban-to-rural migration**.

Some people move daily from their rural homes to work in cities like Gaborone, and then return home at night. This temporary type of voluntary movement is known as **commuting**. It is much more important in large cities in MEDCs where many workers prefer to live out of the congested cities in the more peaceful rural areas surrounding them.

Recent international immigration

Voluntary

Since 1980, Botswana has experienced net immigration. Many of the migrants are skilled professionals from other African countries, Europe, Asia, the Americas, and the Caribbean. They have been attracted by Botswana's fast-growing economy and shortage of skilled workers.

In the 1970s and 1980s Botswana was under-populated, so the country had an open immigration policy to make sufficient skilled labour available. Incentives, such as free education for children, were offered to attract immigrants.

Involuntary

- In the 1970s and 1980s, large numbers of **refugees** entered Botswana. They came from Namibia, South Africa, and Zimbabwe, whose governments had restricted the freedom of their people. Large numbers of refugees also flooded in to escape the long civil war in Angola.
- By 1996, almost all of those refugees had returned home. Then there was another influx of refugees from Namibia and from Angola (where the civil war started up again).
- After 2000, Botswana was swamped with refugees fleeing political oppression and catastrophic economic decline in Zimbabwe. Some estimates put the number as high as 800 000 by 2004.

The Botswanan government responded by tightening border controls and issuing harsher punishments for illegal immigration, but net migration was still high at 4.2 per 1000 in 2011.

The government withdrew the incentives that had previously been given to skilled workers, but **economic migrants** still increased in numbers from Zimbabwe. The Botswanan people became increasingly anti-immigration, and disputes about the higher wages given to non-Botswanan workers occurred. Immigrants in such high numbers can cause problems, especially if they cannot find work or if they compete with the local people for work.

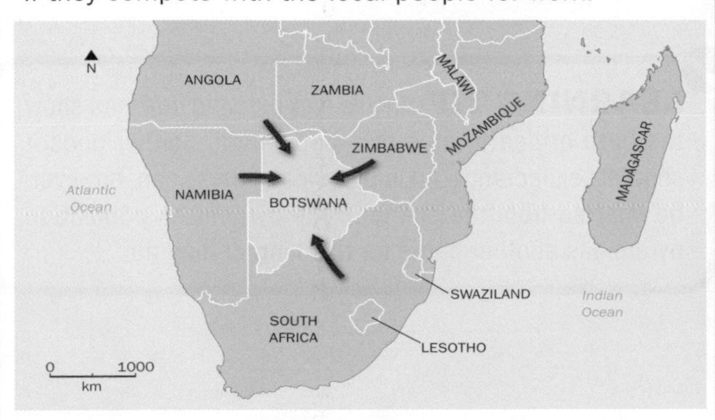

Fig. 1.20 International migrations into Botswana since the 1970s

14 a Suggest as many reasons as you can to explain involuntary migrations. Divide them into two groups: those likely to be temporary and those likely to be permanent.

b What difficulties are international migrants likely to experience (i) before they move and (ii) after they have moved?

c What are the benefits of immigration for the host country?

Internal migrations

Type of internal migration	Push factors	Pull factors
Rural-to-urban Mainly by young males, with or without families. Occurs in LEDCs to any large town or city, e.g. in Botswana to Gaborone.	Jobs mainly in farming. Mechanisation causes job losses. Poor harvests from impoverished soils. Drought and famine. Poverty. Lack of services. Often need to fetch water. Mainly earth roads. Poor accessibility.	Variety of jobs, e.g. in shops, offices, factories. Less likely to be affected by natural disasters. Jobs are better paid. Schools, hospitals, electricity, and water supplies available. Many paved roads and railways. Entertainment.
Urban-to-rural In MEDCs the retired elderly often move from the city to the surrounding country-side, e.g. from London to Surrey	Polluted, noisy cities with high traffic volumes. Expensive rents and rates.	Pleasant countryside. Cleaner air. Healthier. Less expensive houses.
Transmigration An involuntary economic movement of 2.5 million people by the government of Indonesia, 1979–2015	Population pressure in overpopulated Java pushed the government to spread the country's population more evenly by relocating them in its other islands.	None for those who were moved. (It caused violent conflicts with the local people and many migrants have moved back to Java.)

Table 1.12 Types of internal migration

15 How could Indonesia have taken a different approach to the problem of Java's over-population?

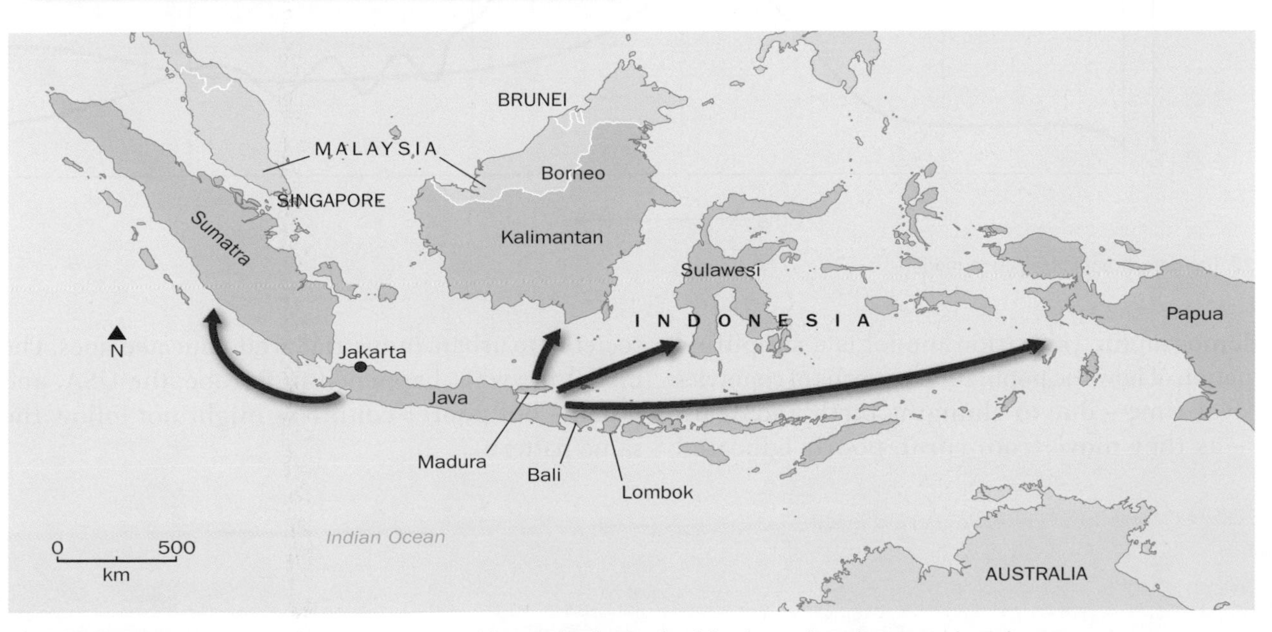

Fig. 1.21 Transmigration in Indonesia: the communities on the destination islands now have better roads and more hospitals and schools but more rainforest has been destroyed

Population structure

Age–sex pyramids

Age-sex pyramids are diagrams designed to show **population structure** (the composition of a country's population). This means the proportions or numbers of males and females in three broad age bands - usually subdivided into five-year age ranges.

When describing the population structure, we can divide it into these three broad bands:

→ the **young dependents** below 15 years of age
→ the **economically active** or working population aged from 15 to 64
→ **elderly dependents** aged 65 and above.

Taxes paid to the government by the economically active are needed to support the needs of the two dependent groups. The population pyramid is a population history for the time since the oldest age group on it was born.

Constructing an age–sex pyramid

- Draw a vertical scale showing age ranges in five-year bands, with the youngest group at the bottom. If there are insufficient numbers in the highest age ranges to show on the diagram, combine two or more bands. For example, you could have an 'over 75' band. The vertical scale is often placed in the centre of the diagram.
- Draw a horizontal scale to cover the biggest age group size on each side.
- Plot bars for males in each age range on the left and for females on the right.

LEARNING TIP An age–sex pyramid does not show the birth or death rates of a population, neither does it show life expectancy. An impression of these can, however, be inferred from the shape of the pyramid. A population pyramid is another name for this kind of diagram.

The demographic transition model

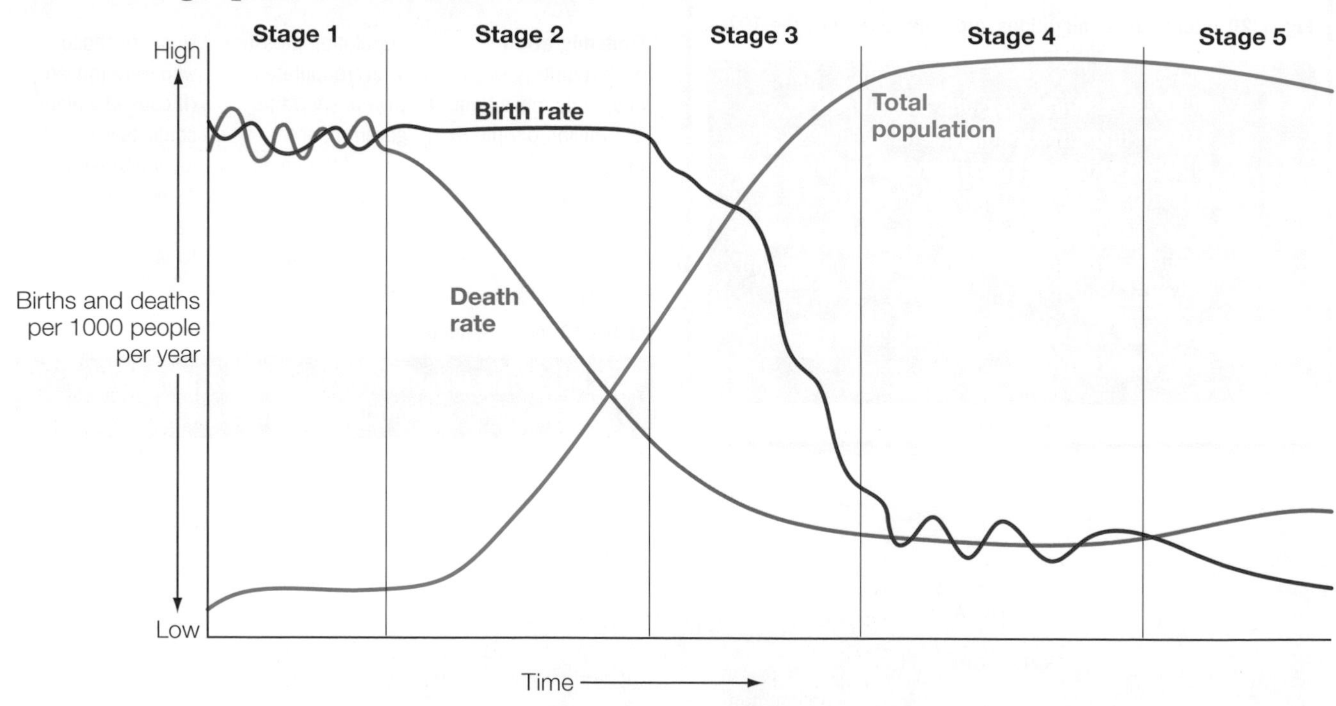

Fig. 1.22 The demographic transition model

The demographic transition model is a simplified explanation of how the population patterns of countries alter over time - due to changing birth and death rates - as they move from rural, poorly educated societies to urban, industrial, well-educated ones. The model fits what happened in Europe, the USA, and Japan - but poorer countries might not follow the same pattern.

Age–sex pyramids and the demographic transition model

Countries at different stages of the demographic transition model have different-shaped age-sex pyramids. If you can recognise the different basic shapes, and understand what they're showing, then you can tell which stage of the model a country is at (see Fig. 1.23).

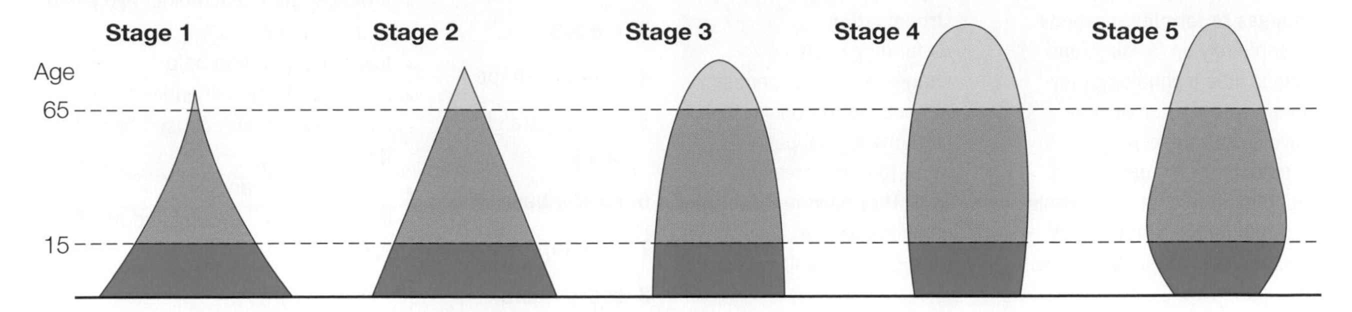

Fig. 1.23 The changing shapes of typical age–sex pyramids at different stages of the demographic transition model

	Stage 1	**Stage 2**	**Stage 3**	**Stage 4**	**Stage 5**
Death rate	High because of disease, famine, lack of clean water, lack of medical care.	Starting to fall because of improved medicine, cleaner water, more and better food, improved sanitation.	Still falling, for the same reasons as Stage 2.	Remains low.	Goes up slightly because more of the population is elderly.
Birth rate	High, due to a lack of birth control; women also marry very young; children are needed to work in the fields to support the family's income. They are also needed to care for parents in their old age.	Still high, for the same reasons as Stage 1.	Starting to fall, because fewer people are farmers who need children to work; birth control is now available; numbers of infant deaths are falling; women are staying in education longer and marrying later.	Low, because of birth control – people are now having the number of children they want.	Remains low, and can fall below the death rate; changes in lifestyle mean people have fewer children later.
This means that ...	... natural increase is low: population doesn't increase much.	... natural increase is high: population increases quickly.	... there's still some natural increase, but it's lower than it was: overall population increase is slowing down.	... there is little or no natural increase: population doesn't increase much.	... if more people die than are born, the total population will probably fall (depending on migration patterns).
Places at this stage today	Perhaps just a few remote tribes in tropical rainforests, isolated from the rest of the world.	Poor countries with low levels of economic development, such as Niger and Mali.	Countries where economic development is improving, like India and Brazil.	Richer countries which are more economically developed, such as the USA and France.	A few richer countries, like Japan, Italy and Germany.

Table 1.13 What happens at each stage of the demographic transition model?

16 Look at Fig. 1.23 and Table 1.13.

a At which stage of the demographic transition model are most LEDCs?

b What is happening to the birth rate, death rate and natural increase at that stage?

c At which stage of the demographic transition model are most MEDCs?

d What is happening to the birth rate, death rate and natural increase at that stage?

The demographic transition model shows how population patterns can change over time. Below is a summary of some of the factors that help to explain this change.

Changes to farming methods
If people rely on farming, and there is little technology, they often have large families to provide extra workers.
As technology increases, and countries develop, fewer people are employed in farming and the need for large families declines.

Urbanisation
As farming methods change, and fewer people are needed to work on the land, many rural people move to urban areas to work. They need fewer children there, so they have smaller families.

Factors affecting population change

Education and women
As society and the economy develop, women tend to stay in education longer. This means that they get married and start having children later, and usually have fewer children as a result. Educated women also know more about birth control, and so can limit their families more effectively.

Fig. 1.24 Factors affecting natural population change

Discussion point

Why might some LEDCs not pass through all of the stages of the demographic transition model?

Pyramid shape	What it indicates and the reasons for it
1 wide base **2** narrow top **3** low top **4** pyramid shape **5** almost straight sides **6** narrow base **7** wide top **8** shorter bars between longer ones above 15 years **9** high top	***What it indicates:*** • few elderly • fewer people in each older age group • many children • fewer children than before • more elderly than in earlier times • low life expectancy caused by high death rate • long life expectancy • low death rate in young and middle aged ***Reasons:*** • deaths in war • low birth rate • high birth rate • low death rate • high death rate • increased death rate with each older age group

Table 1.14 Reasons for the shapes of age–sex pyramids

17 Look at Table 1.14.

a The shapes of age–sex pyramids differ. Match the descriptions on the left with the statements on the right, in order to show what each pyramid shape indicates and what might be causing it.

b Look at the age–sex pyramids of Japan's population in Fig. 1.25. Describe and explain the structure of Japan's population in 1950.

CASE STUDY

Japan, a country with a high elderly population

Japan has a population that is ageing – and getting smaller. Japan has the oldest population in the world – over-65s make up nearly 28% (with under-15s just 12.8%). The average age is almost 47 (the highest of any country).

Japan's population structure is changing because:

- people are living longer. The average life expectancy in Japan is 81 for men and 88 for women. This is due to a healthy diet (low in fat and salt) and a good quality of life. Japan is one of the richest countries in the world and has good health care and welfare systems. There are 230 doctors for every 100 000 people.
- the birth rate in Japan has been declining since 1975. This is partly due to the rise in the average age at which women have their first child. This rose from 25.6 years in 1970 to 30 in 2012. Throughout this period, the number of couples getting married has fallen, and the age at which they get married has risen.

By July 2016, Japan had a population made up of 13% young (aged under 15), 60% economically active (aged 15–64), and 27% elderly (aged 65 and over).

The **dependency ratio** is calculated using the number of people in the three groups. Japan's dependency ratio in 2011 was:

$$\frac{16\,237\,000\ \text{(young)} + 65\,245\,000\ \text{(elderly)}}{74\,969\,000\ \text{(economically active)}} \times 100 = 68.7$$

(Percentages could be substituted for numbers in the calculation.)

This means that, in 2016, every 100 workers in Japan had to support the needs of almost 69 people. The taxes that workers pay to the government need to cover what should be provided by the State for the care of the young and the elderly, with more costs on the elderly (38 per 100 compared with 30 young per 100 of working age).

As the age–sex pyramids for Japan in Fig. 1.25 show, there are now fewer children growing up to become workers, so the dependency ratio will get worse. Meanwhile, more people will become pensioners – a lot more because the post-war baby bulge has reached 65. Japan ranks third in life expectancy in the world at 82.25 years. Japan – in common with European countries like France, Germany, and Italy – has low birth and population growth rates and high and increasing life expectancies.

Japan had a negative growth rate in 2016 of –0.19%, so not only is the country's population getting older, but it is also shrinking. This will make the problem worse. By 2055, Japan's population is expected to shrink from 126 million in 2016 to only 90 million – and by then the elderly are expected to make up 41% of the total population. Their children will have to save instead of spending to help support them. Reduced spending is already causing retailers in Japan to drop their prices. The Japanese people will be unable to maintain their high standard of living as the dependency ratio worsens.

Having a smaller proportion of young people in the population has already caused a number of problems for Japan:

- Some underused – and therefore uneconomical – schools and colleges have had to be closed, so students have to travel further.
- A shortage of recruits for the armed forces has weakened Japan's ability to defend itself.
- A shortage of labour, especially of innovative workers, has caused Japan's high-tech electronics industries to stagnate. As a result, companies like Sony have had to increase their rates of pay to attract foreign workers. These high wages might deter foreign investment in Japan in the future.

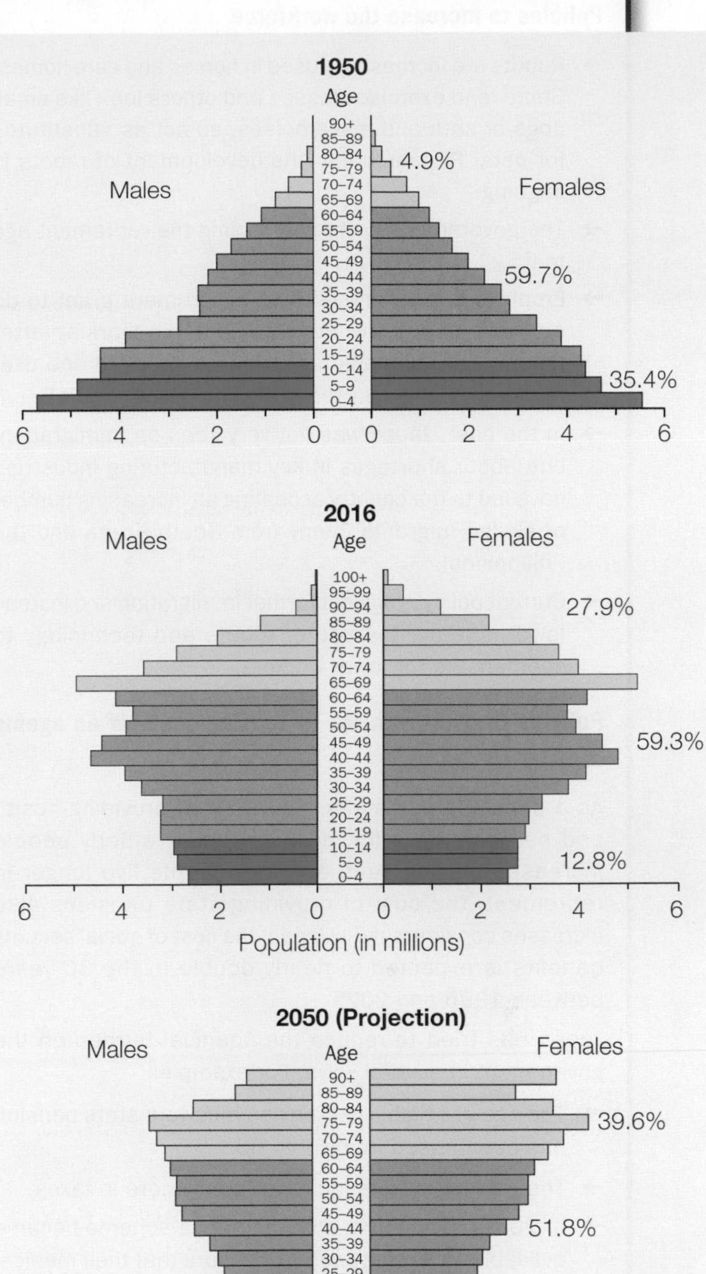

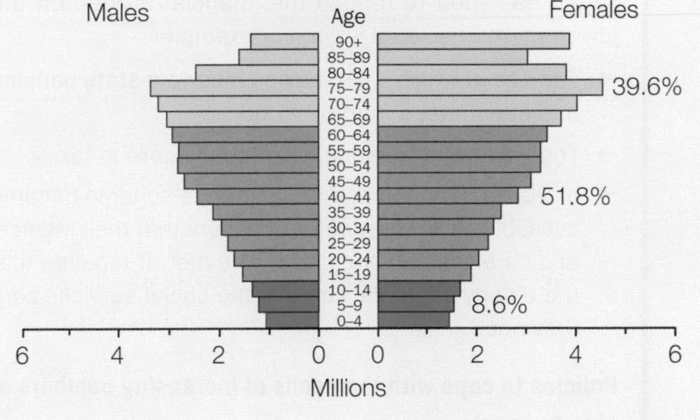

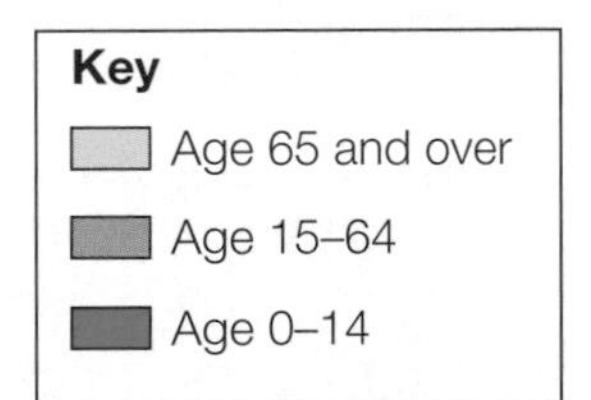

Fig. 1.25 Japan's population structure in 1950, 2016 and 2050 (projection)

Coping with an ageing dependent population

Policies to increase the workforce

- Robots are increasingly used in homes and care homes. Some lead exercise classes and others look like small dogs or cats and make noises, so act as substitutes for pets. Research into the development of robots is ongoing.
- The government is steadily raising the retirement age to 65.
- Employers, encouraged by a government grant to do so, have allowed workers to continue working after reaching pensionable age. People aged 65 and over made up over a quarter of the workforce in 2015.
- In the past, Japan was not very keen on immigration, but labour shortages in key manufacturing industries have led to the country accepting an increasing number of skilled migrants (many from South Korea and the Philippines).
- Current policy is against further immigration and instead is to increase the use of robots and technology to compensate for fewer workers.

Policies to cope with the increasing costs of an ageing society

As a population gets older, the cost of providing health and personal care for more and more elderly people increases quickly. And, as more people live longer in retirement, the cost of providing state pensions also increases considerably. In Japan, the cost of social security benefits is expected to nearly double in the 30 years between 1995 and 2025.

Japan has tried to reduce the financial burden on the government in several ways. For example:

- The age at which a person can receive a state pension has been raised from 60 to 65.
- The working population have to pay more in taxes.
- In 2000, a long-term care insurance scheme became available – allowing people to ensure that their medical and care costs in later life will be met. It requires that the elderly share the costs of the social services care they need.

Policies to cope with the needs of increasing numbers of elderly people

- An ageing population is more prone to degenerative diseases like cancer, dementia, arthritis, and heart disease. So increasing access to specialist health care is required.
- Many elderly people also need homes without stairs, or with adaptations for wheelchairs.
- As they age, most people need more care and may have to move to live in a care home. Japan has had to build more care homes and provide more health care. Unfortunately, this is not happening fast enough to meet demand.

The cost of providing for all of the needs of its ageing population is proving to be a real challenge for the Japanese government, which is now receiving less tax income because of the declining size of the workforce.

The government's ability to improve the spending on care for the elderly was reduced by the need to repair damage caused by the 2011 earthquake and tsunami. Care for the elderly is a pressing problem because the workforce is being reduced as thousands of Japanese workers have given up their jobs to look after elderly relatives.

Discussion point

What problems would Japan's population structure cause for the government in 1950 and in 2050?

18 Look at Fig. 1.25.

- **a** Japan was involved in the Second World War from 1941 to 1945. How does that show up in the population pyramid for 1950?
- **b** People who served in Japan's armed forces between 1941 and 1945 would have been over 85 in 2016. How did the Second World War influence the 2016 pyramid?
- **c** There is often a growth in the birth rate after a war. How does this show up in Japan's 2016 pyramid?

19 At which stage, or between which two stages, of the demographic transition model was Japan in (a) 1950 and (b) 2016? Explain your answers.

20
- **a** How does an ageing population affect an economy?
- **b** How has Japan's elderly population been affected by government measures to reduce its costs?
- **c** Explain why an increase in the dependency ratio can lower the standard of living.

CASE STUDY

Niger, a country with a high young dependent population

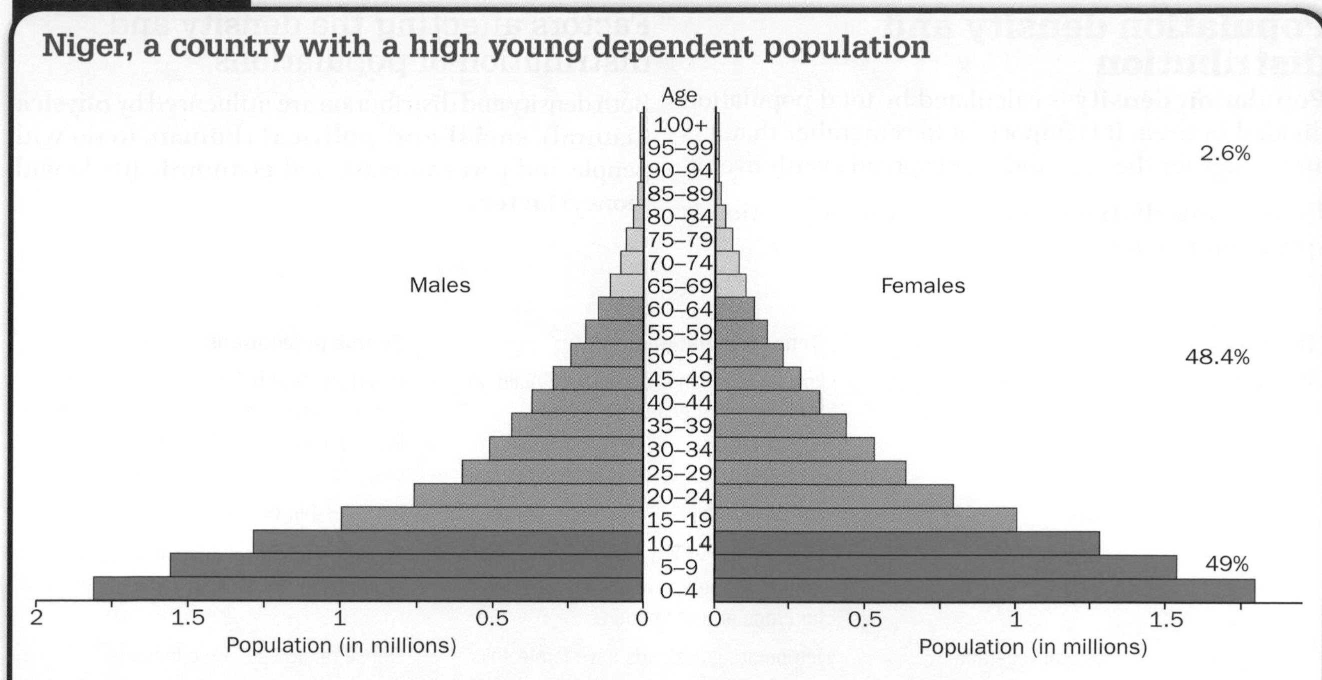

Fig. 1.26 Niger's population structure in 2016. The median age in Japan was almost 47; in Niger it is only 15

21 Use Fig. 1.26 to answer the following:

- **a** What is the evidence from the shape of the age-sex pyramid that Niger has a high dependent population?
- **b** What stage of the demographic transition model is Niger in? State the evidence for your answer.
- **c** Calculate the total population in the 0-4 age group in Niger and compare it with the 50-54 age group.
- **d** Niger's total dependency ratio is 106 (100: 106.6). How many people of working age are there to support nearly 1.7 people of non-working age?
- **e** How does your answer explain why child labour is frequently used in countries such as Niger?
- **f** Calculate the dependency ratio for the young people in Niger and explain what this means in theory for every person of working age in Niger.

22

- **a** Write about the impacts of a high number of young dependents for the country. Refer to present and future birth rates, food supply, health services, schools, future unemployment.
- **b** Name the type of population policy the government needs to adopt to prevent the problems increasing.

Variations in dependent populations and standards of living

There are various indicators of standards of living, including GDP per person **ppp (purchasing power parity)**, literacy levels, energy consumption, calorie intake, and so on.

- → The above indicators are highest in MEDCs. However, MEDCs face the costs of dealing with increasing elderly dependency ratios.
- → On the other hand, LEDCs have low numbers of elderly dependents but many young dependents instead.

An increase in either group will result in lower living standards for the economically active in the country.

Discussion point

How do you expect your own country's dependency ratio to change by 2050?

Population density and distribution

Population density is calculated by total population divided by area. It is important to remember that it is an average for the area and is not spread evenly over it.

Population distribution is how the population is spread over an area.

Factors affecting the density and distribution of populations

Both density and distribution are influenced by **physical** (natural), **social and political** (human, to do with people and governments), and **economic** (to do with money) **factors**.

Factor		Dense populations	Sparse populations
Physical	Relief	Low and flat lands are easy to build on and to use for economic activity.	It is difficult to build houses, transport routes and to farm on steep slopes. High lands are cold and the thin air lacks oxygen.
	Soils	Fertile and thick.	Infertile, thin or stony.
	Climate	Warm or hot with a rainy season or all-year rainfall so there is a long growing season for crops and vegetation.	Too wet – transport difficult in muddy ground. Too dry, lacking water supplies.
	Natural vegetation	Temperate grasslands have fertile soils.	Dense rainforests are difficult to penetrate.
Economic	Farming	Areas with high crop yields from intensive farming, which requires much labour.	Areas unsuitable for farming or with large extensive farms, both pastoral and arable.
	Transport	Good transport is essential for economic activity. People cluster at junctions and ports.	In areas lacking transport, settlement is restricted to being located near any route already present.
	Industry	Industrial and mining areas with many jobs have dense populations.	Areas lacking industry have few jobs to support populations.

Table 1.15 Physical and economic factors influencing population density

Social

Some tribes and family groups live close together. It is their culture to do so. Their homes are often grouped around that of the head of the tribe or family.

Political

A number of governments have built a new capital city and placed it in an undeveloped part of their country to stimulate growth. An example is Brasilia in central Brazil, which replaced Rio de Janeiro as the capital city in 1960. In the UK, when old areas were being redeveloped in London and the city was getting too large, people were moved out to new towns around, thus increasing the density of places such as Harlow and Crawley.

CASE STUDY

Population distribution and density in Bangladesh

Most of Bangladesh is lowland and slopes gently to the south; 80% of the area is the floodplain and delta of the Ganges and Brahmaputra rivers. The northern part of this fertile plain is slightly higher and less often flooded than the south. It is mainly rural. People live in villages on any land slightly raised above their rice fields.

- The population density of Bangladesh is about 1200 per km^2, which is the highest in the world (excluding very small countries). No region in the country can be described as having a low population density.
- The capital city, Dhaka, is the world's most densely populated city by far, with in excess of 50 000 people per km^2. Its squatter areas are reputed to have densities of over 1 million per km^2.
- In the area around Dhaka, densities are extremely high.
- Densities are very high in a broad band across the centre of the country, from the north-west to Chittagong in the south-east. Some of this rich arable area is slightly higher than the south of the country.
- Lower (but still high) densities occur in the very low delta south of Dhaka, in the higher north and north-west border zone and the low hills of the extreme north-east.
- The steep, high Chittagong Hills in the extreme south-east have the lowest densities of 80–100 km^2.
- Areas near the borders with India and Myanmar tend to have lower densities than regions further inland.

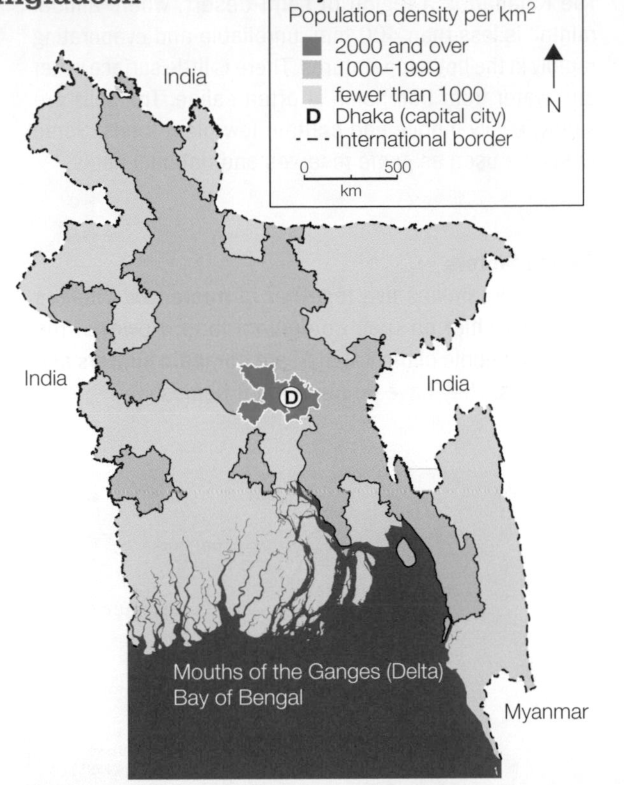

Fig. 1.27 Population densities in Bangladesh (2011)

23 Using the information given, describe, without mentioning density, the distribution of population in Bangladesh.

CASE STUDY

Population density and distribution in Botswana

Population distribution

Most Botswanans live in the east of the country, with very few living in the north, centre or west. Away from the east, the population is concentrated in small, widely scattered areas.

Population density

Most of Botswana has a very low population density of less than one person per km^2 – especially in the Kalahari Desert, which makes up a large part of the country.

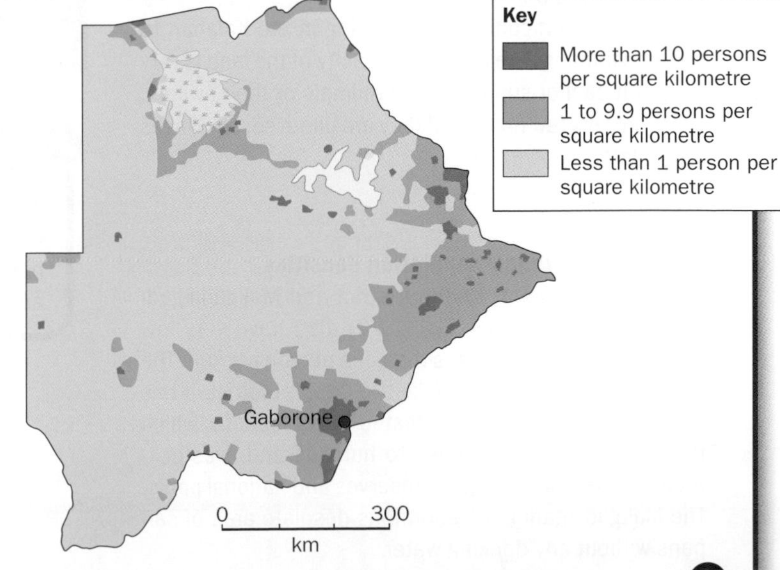

Fig. 1.28 Population density and distribution in Botswana

Physical factors
The Kalahari is a region of semi-desert, where annual rainfall is less than 400 mm, unreliable and evaporating rapidly in the hot temperatures. There is little surface water and water from boreholes is often saline. The soils are sandy, easily mobile and contain few plant foods. Some areas are used as game reserves and national parks.

Human factors
Sedentary families live together in **nucleated villages** with their chief on their communal land. However, the Basarwa people of the Kalahari are **nomadic hunters and gatherers**. They have no permanent homes.

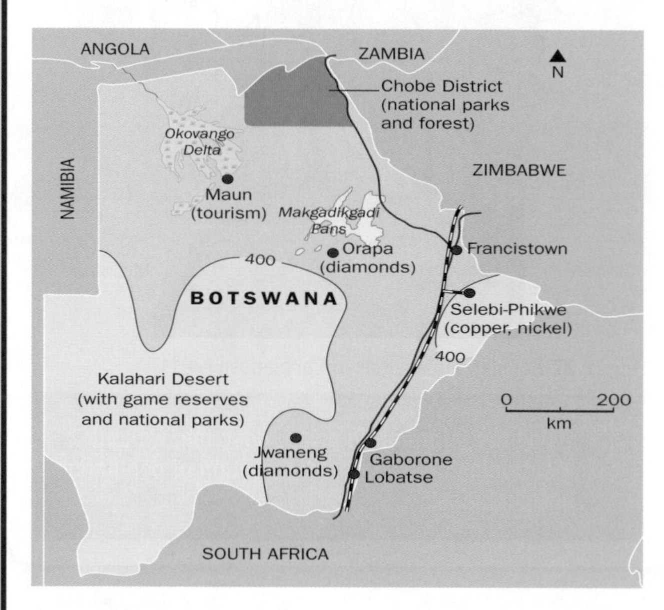

Fig. 1.29 Factors influencing population density and distribution in Botswana

Economic factors
The physical difficulties limit farming in the Kalahari to pastoral farming. The carrying capacity of the land is very low, so it cannot support many animals or their owners. There are very few roads and they are unpaved, so villages are isolated.

Other areas with low population densities
The Okavango Delta, Chobe District and Makgadikgadi Pans in the north (see Fig. 1.29) also have very low population densities of less than one person per km^2. The Delta is wet and swampy. Like the Chobe District, it has many wild animals and is infested with tsetse fly, which transmits sleeping sickness to humans and livestock. Areas are set aside for game reserves and national parks. The Makgadikgadi is an enormous desolate area of salt pans without any drinking water.

More densely populated areas

Although nowhere in Botswana can be considered densely populated, more people live where farming and mining are possible.

- → There are a number of villages on the southern and western edges of the Okavango Delta, where fertile soils are used for arable farming.
- → Maun is a tourist centre with a small airport for visitors who go to see the wildlife of the Delta.
- → Eastern Botswana receives over 400 millimetres of rain a year. It also has seasonal rivers, which flow in the wet summers, so crops can be grown and cattle raised. There is also access to water from boreholes.
- → Areas with a population density of over 20 people per km^2 are found only in the east, where the railway line, tarred road and nearness to the South African border are important factors for stimulating trade and industry (see Fig. 1.28). Electricity is also produced in the east and most of the towns there are connected to the transmission line.
- → The areas around the capital, Gaborone, and around the second largest settlement, Francistown, are the most densely populated. Employment is available in shops, offices and industry, such as the abattoir at Lobatse.
- → Diamond-mining towns, such as Orapa, are islands of denser population within low-density areas. On density shading maps, population is averaged over areas, so towns raise the population of the whole area but on dot distribution maps (where one dot represents a certain number of people and each dot is positioned in the area where the people live) they would stand out as a number of dots surrounded by a blank area.

24 List reasons for the higher population density in eastern Botswana under two headings: Physical and Economic.

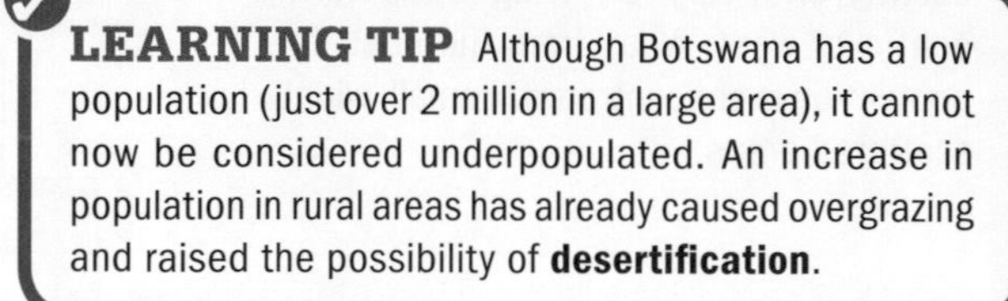

LEARNING TIP Although Botswana has a low population (just over 2 million in a large area), it cannot now be considered underpopulated. An increase in population in rural areas has already caused overgrazing and raised the possibility of **desertification**.

CASE STUDY

Population density and distribution in Canada

25 Look at Fig. 1.30. Use atlas maps and any other available means of research to explain the main reasons for the density and distribution of population in Canada. Include the terms north Canada, the Rocky Mountains, coniferous forest, the prairies of central Canada (a wheat-growing area), the St Lawrence Valley, and Vancouver.

Human factors influencing population density and distribution

One reason why the St Lawrence Valley has the highest population density in Canada is because this was the part that the first French and English colonists moved into. People of French origin tended to congregate in the province of Quebec, and those from England in Ontario and British Columbia. Since the 1990s, the majority of immigrants have originated from Asia, especially from China, the Philippines and India.

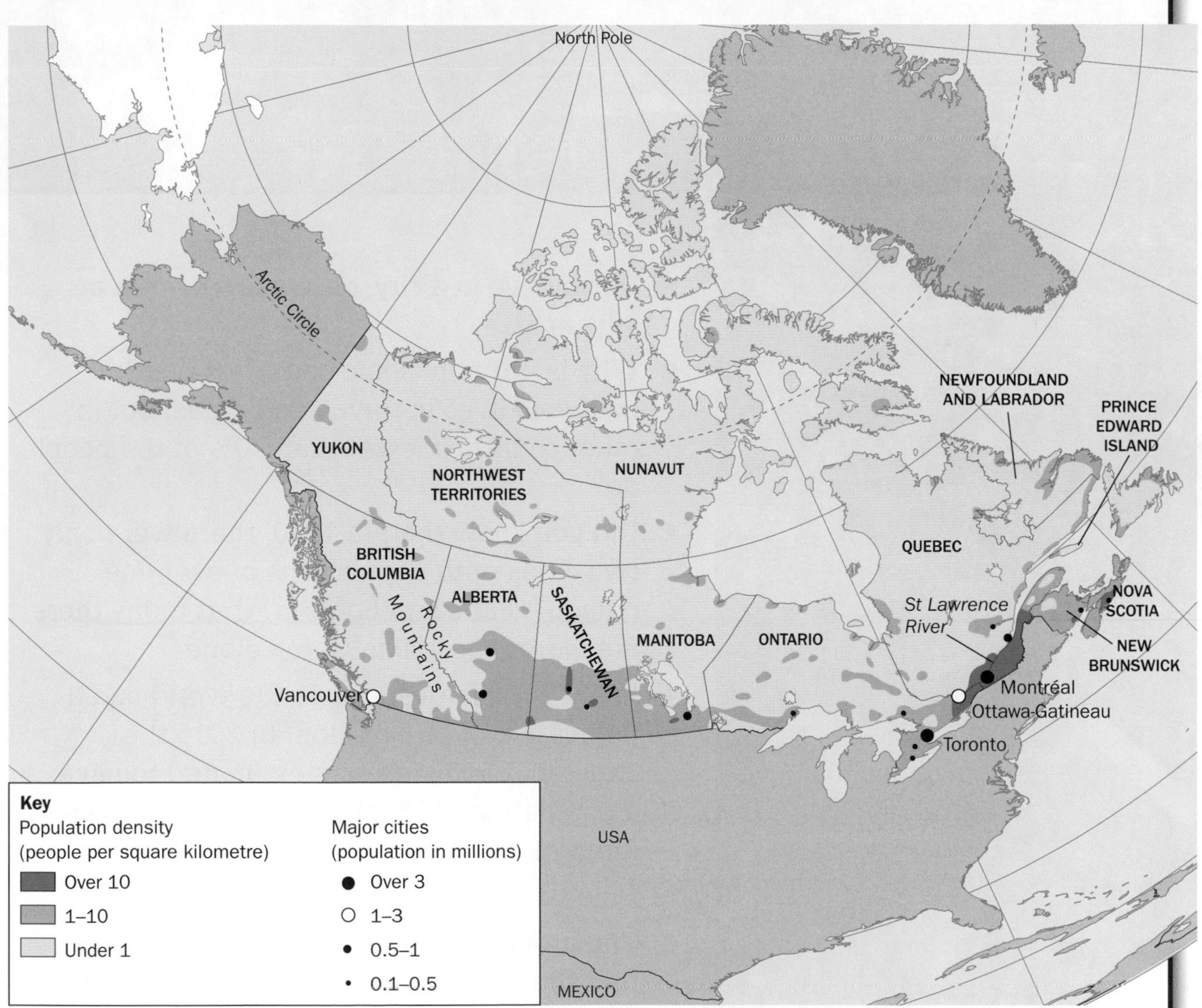

Fig. 1.30 Population distribution and density in Canada

LEARNING TIP Human factors include economic influences and also social ones.

Discussion point

What are the advantages and disadvantages of living in densely populated and sparsely populated areas?

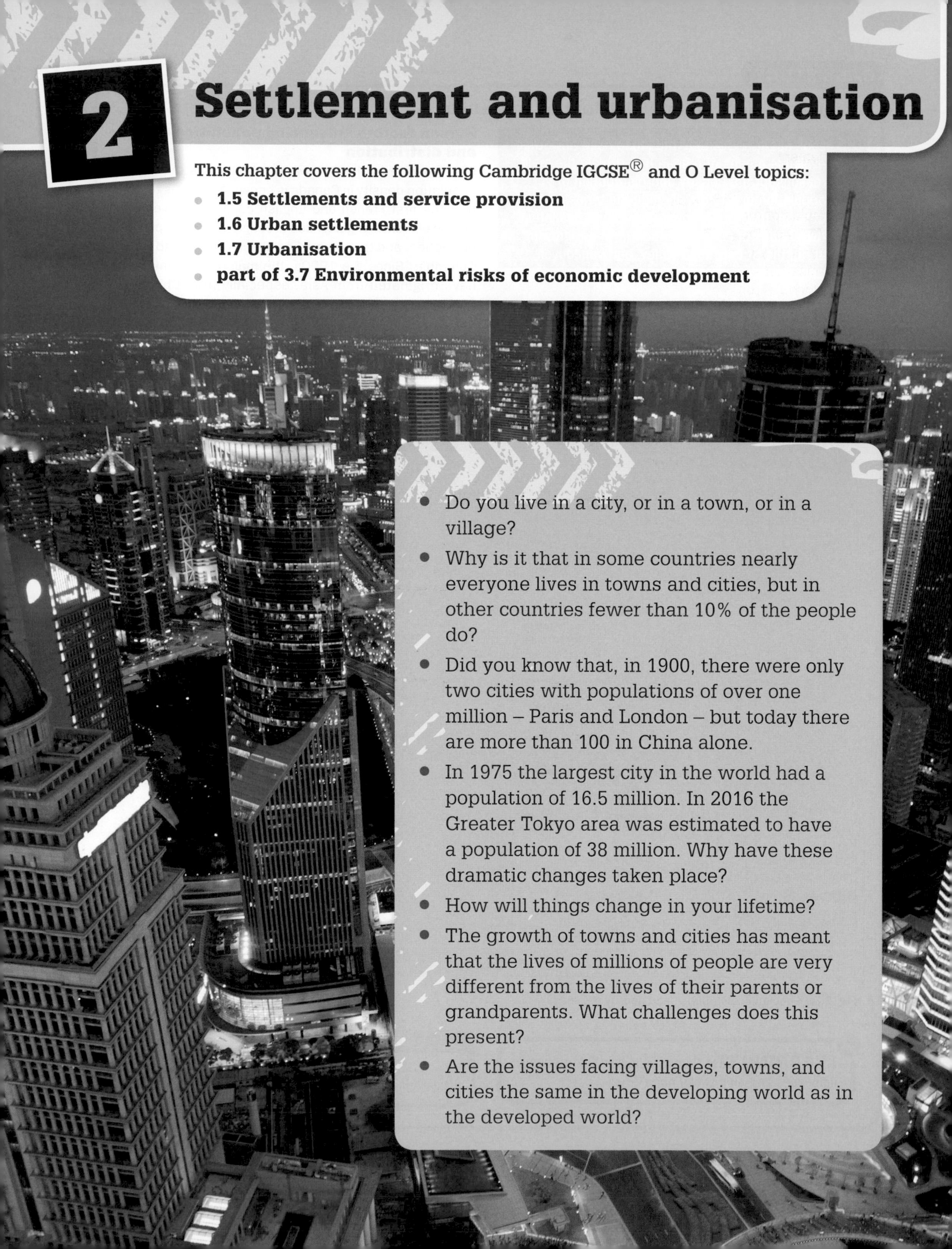

2 Settlement and urbanisation

This chapter covers the following Cambridge IGCSE® and O Level topics:

- **1.5 Settlements and service provision**
- **1.6 Urban settlements**
- **1.7 Urbanisation**
- **part of 3.7 Environmental risks of economic development**

- Do you live in a city, or in a town, or in a village?
- Why is it that in some countries nearly everyone lives in towns and cities, but in other countries fewer than 10% of the people do?
- Did you know that, in 1900, there were only two cities with populations of over one million – Paris and London – but today there are more than 100 in China alone.
- In 1975 the largest city in the world had a population of 16.5 million. In 2016 the Greater Tokyo area was estimated to have a population of 38 million. Why have these dramatic changes taken place?
- How will things change in your lifetime?
- The growth of towns and cities has meant that the lives of millions of people are very different from the lives of their parents or grandparents. What challenges does this present?
- Are the issues facing villages, towns, and cities the same in the developing world as in the developed world?

In this chapter you will learn about:

- the different types of settlement and settlement hierarchy
- the services provided by different types of settlement
- the location of rural settlements and rural settlement patterns
- examples of rural settlements in MEDCs and LEDCs, their location and the problems and changes they face
- urbanisation – how towns and cities have developed across the world
- the locations of towns and cities
- different types of land use in towns and cities
- the problems faced by towns and cities.

Hierarchy of settlements

A **settlement** is a place where people live. Settlements vary greatly in size - from a single dwelling to cities housing millions of people. A hierarchy puts items in order. A hierarchy of settlements would normally do this in order of population size. An example of a hierarchy of settlements is shown in Fig. 2.1.

Settlements higher up the hierarchy, e.g. large cities, are called **high-order settlements.** These are fewer in number and they are spaced further apart than settlements lower down the hierarchy, e.g. villages, which are called **low-order settlements**. There are more of these and they are more closely spaced. This is shown in Fig. 2.2.

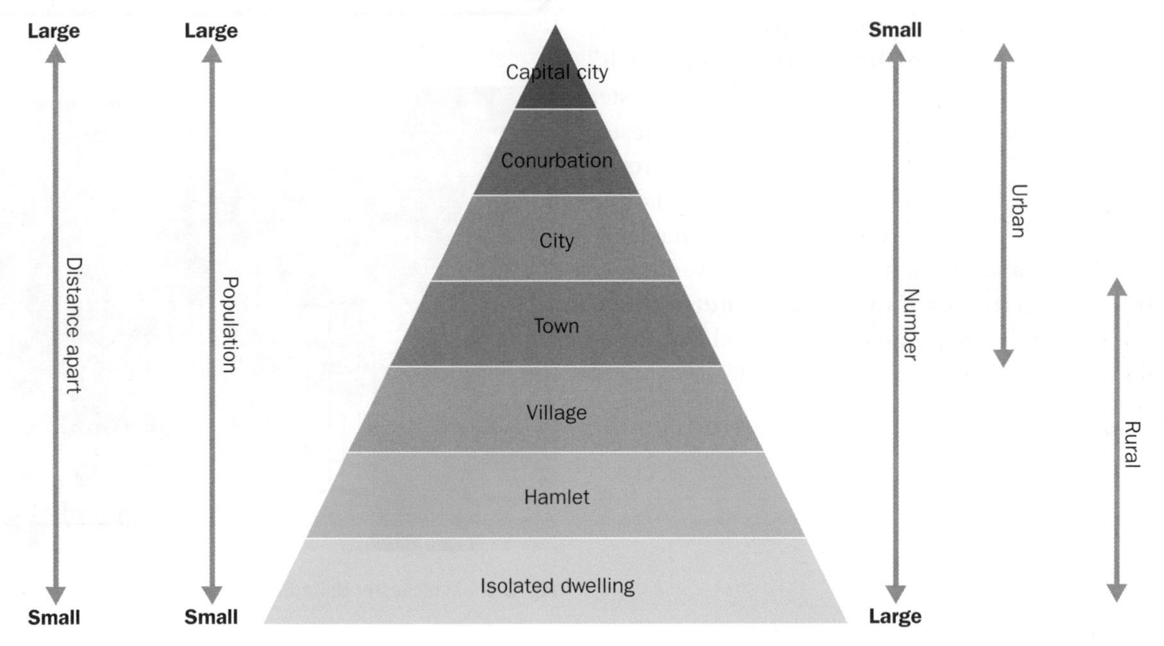

Fig. 2.1 A hierarchy of settlements

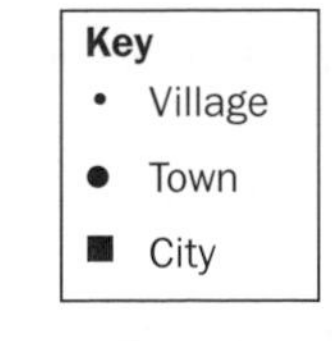

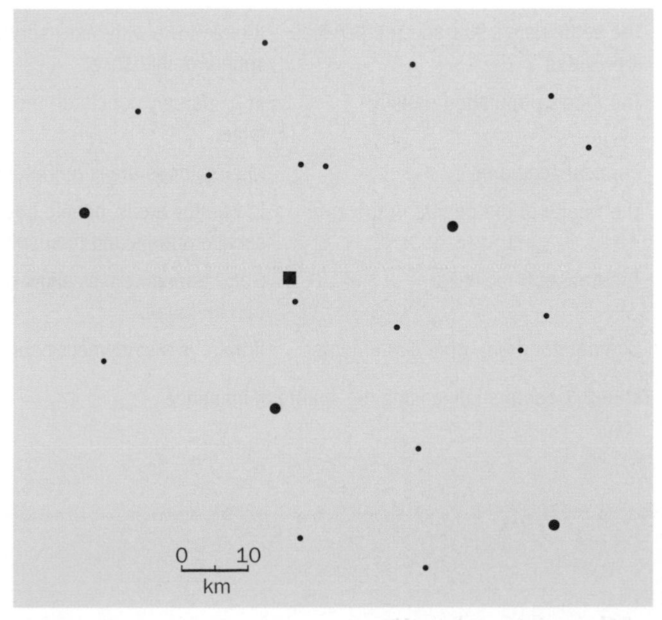

Fig. 2.2 The distribution of high-order and low-order settlements

What is a city, a town, or a village?

There are no internationally agreed definitions for any of these terms. For example, villages are normally thought of as fairly small settlements, but in some countries quite large settlements are still classed as 'villages'. For example, the village of Molepolole in south-east Botswana has a population of 70 000.

'Urban' and 'rural'

In broad terms, **urban** refers to towns and cities and **rural** refers to the countryside. However, this definition is also complicated, because rural countryside can contain large villages and small towns.

Hierarchy of services

Smaller settlements usually provide a limited range of services and goods for sale. These are likely to be services that people use almost every day, such as a primary school, or a small shop selling bread, milk and fresh vegetables. People tend to travel short distances to these. Large towns and cities provide a wider range of goods and services. People travel long distances for some of these services but less frequently, e.g. to buy furniture.

Services (sometimes called **functions**) are anything that is provided in a settlement for the population. They include goods that can be bought in shops and other retail outlets, e.g. food, petrol, and clothing. They also include businesses like hairdressers, which are sometimes called retail services, as well as public services like schools, hospitals, government, police, water, and electricity.

Sphere of influence

This is the area served by a settlement. The sphere of influence of a small village will be very small, but the sphere of influence of a city might be very large - and the sphere of influence of a capital city is the whole country.

Manchester United is a football club in north-west England. What do you think its sphere of influence is? Many years ago, the people who went to watch Manchester United's football matches lived in the city of Manchester and its surrounding area. But today - because of improved transport, a wealthier population and good marketing - people from all over the UK go to their matches. However, the club's sphere of influence is even larger than that. It also makes money from international activities. Its matches are watched on global TV and its football shirts are sold in every continent.

The size of the sphere of influence depends on the factors shown in this table.

Discussion point

Think about the people who live near you. How far do they travel to get different services? How frequently do they use these services? You should find that some services are usually found fairly locally and used frequently. However, others are used very infrequently and people will travel much further to use them.

Fig. 2.3 Manchester United, an international brand

The settlement's size and the services it provides	A large town with many services, and high-order services (see the next page), will have a large sphere of influence.
The area's population density	In a sparsely populated area, services will be widely spaced and spheres of influence will be very large.
Physical geography	Mountainous areas or marshland may be sparsely populated and spheres of influence will be large.
The wealth of the people in the area	In wealthy areas, people have more money to buy goods and services. Therefore there will be more service outlets and their spheres of influence will be smaller.
The transport facilities	Good transport links allow people to travel further to reach services, so the spheres of influence can be larger.
Competition from other settlements	If there is a settlement nearby that provides similar services, the sphere of influence will be smaller.

Table 2.1 Factors influencing the sphere of influence

For each town, different services will have different spheres of influence. This is shown in Fig. 2.4.

1 Look at Fig. 2.4.

a Why is the sphere of influence of the local shop so small?

b Why are the spheres of influence smaller and 'squashed' in the east?

c Why are the spheres of influence bigger in the north-west?

Spheres of influence can also overlap, as shown in Fig. 2.5.

2 Most people living in the area shown on Fig. 2.5 go to the nearest supermarket to buy goods. However, people living in area X go to either supermarket. Suggest why this might be so.

Threshold population

This is the minimum number of people needed to provide a large enough demand for a service. Some companies, such as supermarkets, require a minimum population size before setting up a store. Services with a small threshold population, such as a local shop or a primary school, are called **low-order services**. There are usually large numbers of these. Services with a large threshold population, such as a furniture store or a university, are called **high-order services**. There are usually small numbers of these.

High-order services will usually be found only in the larger settlements. These settlements will also have large numbers of low-order services.

Discussion point

In the area where you live, does everyone choose to go to the same place for a particular service? Are there spheres of influence that overlap? For example, do different people prefer different shops? Why is this? Do people support the same football team? You will probably find that the pattern of spheres of influence is quite complicated and is affected by a variety of factors.

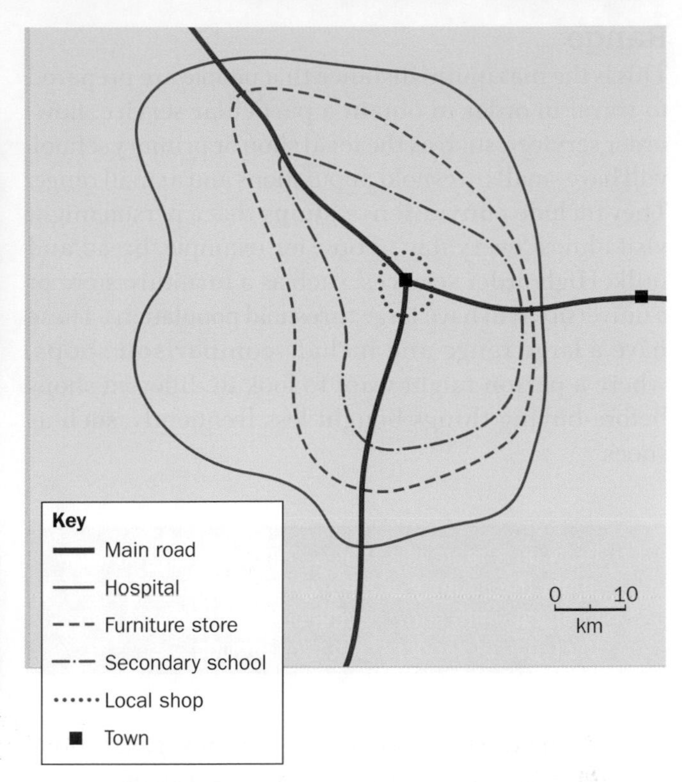

Fig. 2.4 The spheres of influence of different services in a town

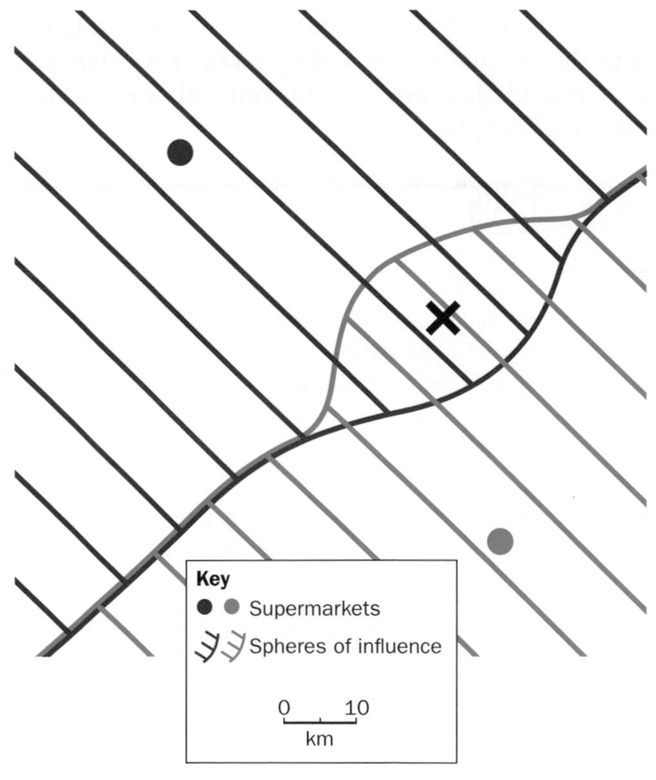

Fig. 2.5 The overlapping spheres of influence of two supermarkets

Range

This is the maximum distance that people are prepared to travel in order to obtain a particular service. Low-order services, such as the local shop or primary school, will have small threshold populations and a small range. They include **convenience shops** that a person might visit almost every day to buy, for example, bread and milk. High-order services, such as a furniture store or a university, will have large threshold populations. These have a large range and include **comparison shops**, where a person might want to look in different shops before buying things bought less frequently, such as shoes.

Type of settlement	Name(s) of settlements	Services provided
City		
Town		
Village		

Table 2.2 The settlement hierarchy in the area where I live

3 Draw up a table like Table 2.2. From what you know about the country or area where you live, add information to show examples of settlements and the types of services that they provide.

All over the world, petrol filling stations (gas stations in the USA) usually provide other services too. As well as carrying out car repairs, they may often act as convenience stores, providing basic foodstuffs for travellers and the local community. They do not fit into the normal pattern of spheres of influence as they provide low-order goods but sell to more than just local people.

Fig. 2.6 Petrol stations don't fit into the normal pattern of spheres of influence

RESEARCH Hierarchies of settlements and services are very easy to investigate and provide good coursework topics. You can find out about size and spacing of settlements using maps and population census information if available. You can find out about services and their spheres of influence by recording the number and types of service you can see in an area. You can produce questionnaires to find out about the range and sphere of influence of different services.

With the help of your teacher, find out where everyone in your class lives. Produce a map to show the sphere of influence of your geography class.

Discussion point

If you live in a town or in a village, where do the people who live there go for different services? How has this changed over time and what are the causes of this?

If you live in a city, how big is the sphere of influence of the city? Has this changed over time and, if so, why?

Rural settlements

Settlement patterns

The settlement pattern is the shape that a settlement forms on the map - and how clustered or scattered it is. Three common rural settlement patterns are **nucleated, dispersed** and **linear**. As well as being described below, they are also illustrated in the survey map questions in Chapter 12. These settlement patterns develop because of the physical geography of the area, the local culture or traditions, and the needs of the farmers in the rural area.

Type of settlement pattern	Reasons for this pattern
Nucleated Dwellings are clustered together as villages, with fewer isolated dwellings. The shape of the villages is compact and more square or circular. 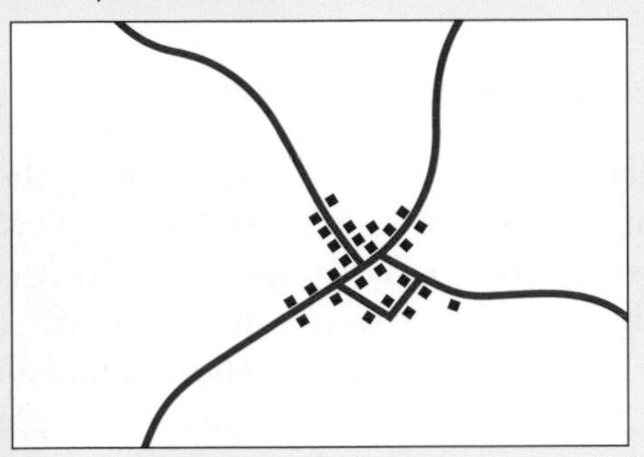	• People can enjoy the social benefits of living close to their neighbours. • They have easy access to services like shops and schools. • Being close to others may be helpful for defence in times of attack. • This pattern often develops in areas with rich agriculture, where farmers can live in the village but still be near their fields. • The culture of the people might favour this pattern. *Reasons for the location of the individual villages are given in the next section on site and situation.*
Dispersed There are scattered isolated dwellings and small hamlets - with few villages. 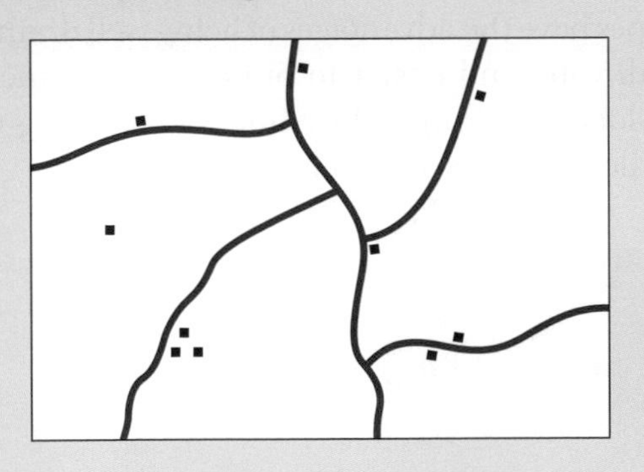	• Sometimes this pattern develops where the agricultural land is poor and people need to farm large areas, e.g. as grazing. It would be very hard for farmers to live in a village and still be within easy travelling distance of their land. • Cultural reasons can also play a part, where it is not the tradition to live grouped together in villages. • This settlement pattern is also found in some relatively modern settlements, such as on the Canadian Prairies, where commercial farms can be huge.
Linear Settlements are in long thin rows, often along roads or tracks. 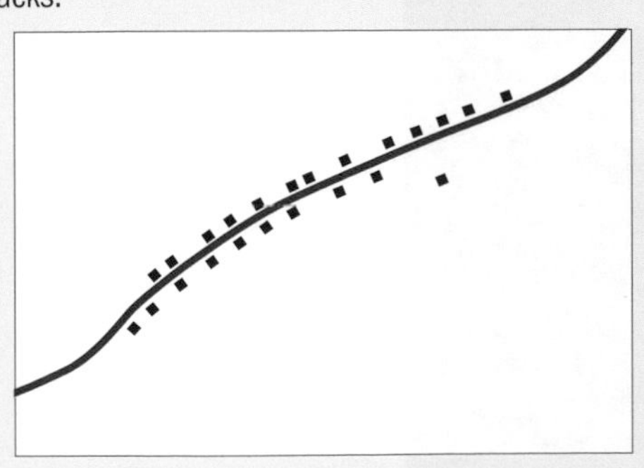	• This pattern allows each dwelling to have access to a road or track for transport, and also to an area of farming land at right angles to the road. • Physical geography can also play a part. The settlements might be along a river or a line of springs, for water supply, or along a valley floor to avoid the steep valley sides. • Settlements might also be in a line just above the floodplain of a river. • In other cases, people don't waste good farming land, e.g. irrigated land, for buildings but place their houses in a line next to the fertile land.

Table 2.3 Nucleated, dispersed and linear rural settlements

Site and situation

The terms 'site' and 'situation' mean slightly different things and should not be confused.

A **site** is the land that a settlement is actually built on. The site's features are very important when establishing the original settlement. They include:

- altitude
- gradient of the slope
- water supply
- crossing point of a river
- natural resources, such as minerals.

The **situation** means the position of the settlement in relation to the surrounding area. The features of the situation often allow a small settlement to grow into a larger town or city. The features include:

- transport routes
- the agricultural productivity of the area
- its position in relation to other settlements.

Factors influencing the sites and development of rural settlements

Agricultural land use

Most villages came into being as agricultural settlements. The surrounding land provided a food supply for the village. For this reason, the available agricultural land was a key feature of their origin and development. The relief, soils, drainage, and accessibility of the site had to allow agricultural land use to take place. In many cases, this was wholly or partly subsistence agriculture. However, it is also true of purely modern commercial farming systems, such as on the land reclaimed from the sea in the Netherlands.

Villages have also developed as a result of mining mineral deposits, or tourism. In MEDCs, villages have become desirable places to live. People often live in villages but work in nearby towns - **commuting** to work each day. These settlements are referred to as **dormitory villages**.

Relief

This includes altitude, gradient, and aspect.

Altitude alone is rarely a factor in influencing the site of settlements. However, in mountainous regions, the highest areas might be so cold that agriculture is extremely difficult - they could be covered in snow for much of the year. Therefore, these areas might be left uninhabited. By contrast, the lowest areas are sometimes sparsely populated because of poor drainage, diseases (see Drainage and flooding on page 42), and dense vegetation.

Gradient is an important factor in the site of settlements. Gentle slopes have the advantages of being well drained, easy to cultivate, and easier to build on. Settlements are often located along valley floors - avoiding the steeper valley sides (as shown in Fig. 2.7).

Fig. 2.7 Settlement along a valley floor in Franschhoek, in the Western Cape of South Africa

However, in the past, steep slopes often provided sites that were easier to defend against enemies. This has meant that, in areas where conflicts were frequent, hilltop villages often developed. Also, where flatter land is in short supply, people are forced to live on steep slopes, as, for example, in rural settlements on the island of Madeira in the Atlantic Ocean (Fig. 2.8). Steep slopes make transport more difficult and they can also be prone to landslides, which can destroy roads and buildings.

Fig. 2.8 Rural settlements on very steep slopes on the island of Madeira

Aspect is the direction in which a slope faces. It is an important factor in mountainous areas, especially in valleys that run east-west. The sun always rises in the east and sets in the west, but in the northern hemisphere it moves around the southern sky and in the southern hemisphere it moves around the northern sky. This is critical in areas further away from the equator (the high latitudes), particularly in the winter when the sun is so low in the sky.

Some slopes are warm and sunny and some are cold and shaded. Agriculture and settlements tend to be concentrated on the sunny slopes.

LEARNING TIP Remember that the south side of a valley is north facing, and the north side of a valley is south facing.

Soils

Areas with fertile soils are often more densely settled, because the greater productivity of the land can support a larger population. Examples of this include the alluvial soils found in river valleys and the soils that develop on certain volcanic rocks, such as basalt:

- The dense rural population of the Nile Valley, in Egypt, contrasts with the sparsely populated surrounding desert.
- Areas with rich volcanic soils, such as the island of Madeira in the North Atlantic Ocean and the island of Java in Indonesia, have relatively dense populations. The opposite is true of the areas of the savanna plains of Africa, which have infertile latosols (red or yellow soils rich in iron and aluminium) and sparse populations.

In the above examples, the density of rural settlements is greater but the soil rarely influences the actual sites of the settlements within the area. In some rural areas, such as desert oases, rich soils are too important to build houses on, so the settlements are built on less fertile land next to the fertile fields.

Water supply

Many settlements were first established next to a river, spring or well that could supply water for drinking, washing, and, in some cases, crop irrigation. Transporting water is hard, time-consuming work, so the settlements needed to have their own supply. Sites - in otherwise dry areas - with reliable water supplies from rivers, springs, and wells, are called **wet point** sites.

Drainage and flooding

Very low-lying areas, such as floodplains (see page 115), might have very fertile soils, but if they flood frequently they are not easy places to settle. They might also be waterlogged permanently and difficult to farm.

In tropical regions, these areas might also be prone to waterborne diseases (such as bilharzia and river blindness) or diseases carried by insects (such as malaria or sleeping sickness). Nevertheless, in areas with large populations and a shortage of land, such as Bangladesh, floodplains are sometimes densely settled.

Areas slightly higher than floodplains, with gentle well-drained slopes, are good for farming and provide good sites for rural settlements. Higher points in otherwise poorly drained areas are known as **dry point** sites.

Accessibility

Even the remotest settlements benefit from contact with other settlements to sell produce or to buy goods and services. This is one reason for the development of linear settlement patterns along roads or tracks.

Where roads meet (route centres), or at the bridging points of rivers, larger villages and regional service centres may develop. These are described on page 52 in the section on urban settlements.

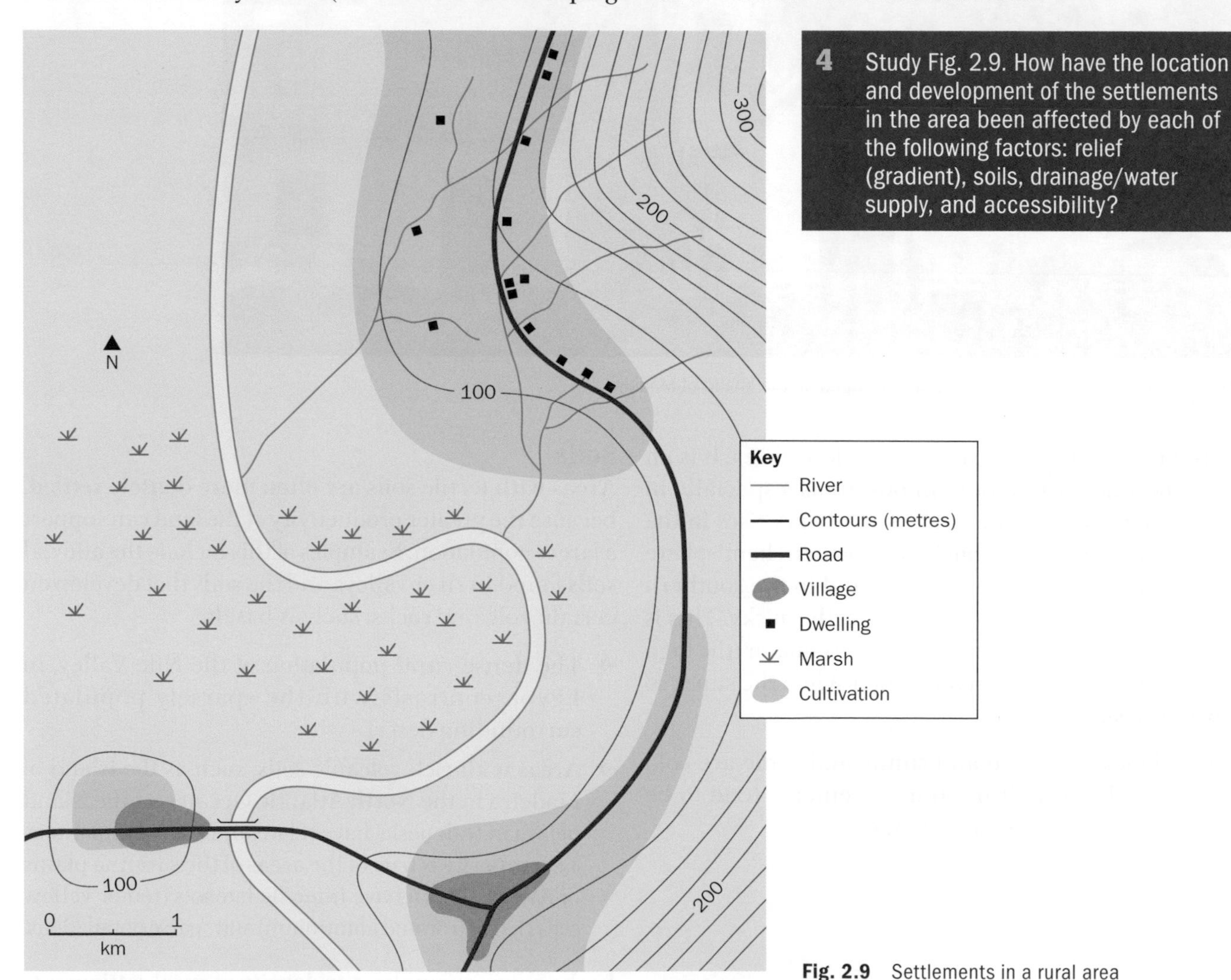

4 Study Fig. 2.9. How have the location and development of the settlements in the area been affected by each of the following factors: relief (gradient), soils, drainage/water supply, and accessibility?

Fig. 2.9 Settlements in a rural area

Discussion point

You have learned about the factors that affect the site and situation of settlements. Which of these factors are important today and which are of only historical importance?

CASE STUDY

Rural settlement in the Tsoelike Valley, Lesotho

Site and situation

Most rural settlements in Lesotho, an LEDC in southern Africa, have a nucleated or dispersed pattern.

The nucleated settlements originally developed partly because of the need for defence in the past. Today, it is easier and cheaper to provide services such as health clinics, schools, piped water and shops in nucleated settlements. Rural craft industries, such as weaving and basket making, are also more easily organised in nucleated settlements.

In the more mountainous areas, crops are grown and cattle keeping is also important. Settlements here are more often dispersed – consisting of a few huts and a kraal (enclosure) for cattle. The shortage of flat land is a problem for these settlements. Some can be reached only by tracks, rather than by proper roads.

Fig. 2.10 Traditional stone and thatched houses in Lesotho

- Most of the settlements in the Tsoelike Valley are below an altitude of 2250 metres, where it is warm enough to cultivate crops.
- The highest land can be used only for grazing. The settlements on the high ground are cattle posts, which are inhabited only in summer.
- Most settlements are on the north-facing slope – which receives more sunlight than the south-facing slope – making cultivation easier.
- The valley floor, which is boggy, poorly drained and liable to flooding, is avoided.
- Settlements are usually at the margins of the cultivated areas and along roads and tracks. Flatter areas, like benches on the valley sides, are common sites for settlement.
- Water supply is generally not a problem, because there are various small streams and springs in the area.

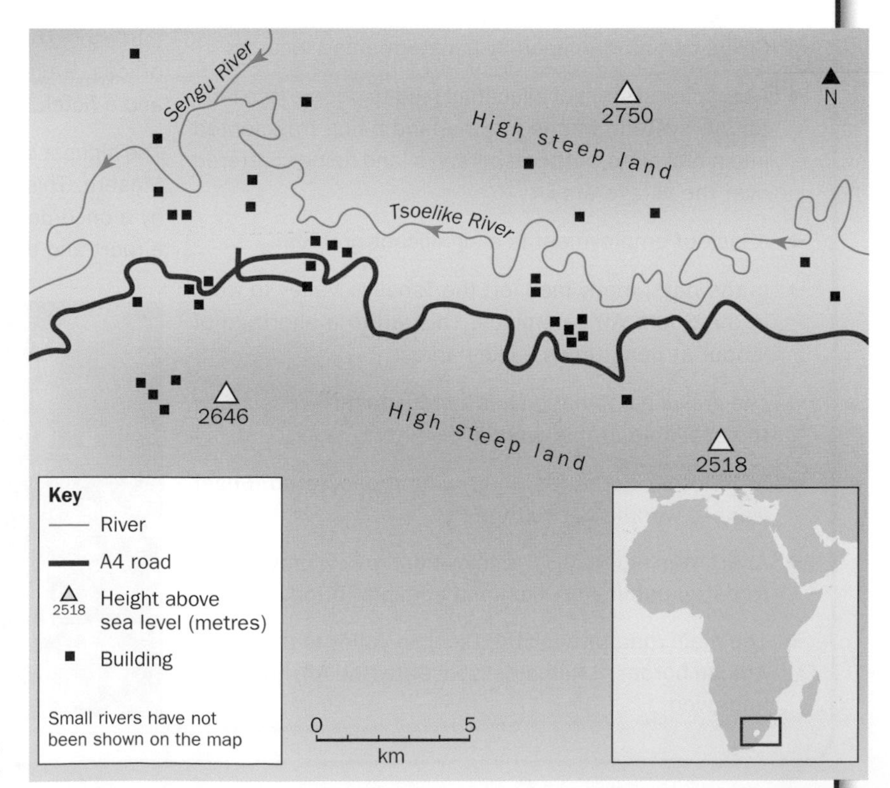

Fig. 2.11 The location of settlements in the Tsoelike Valley

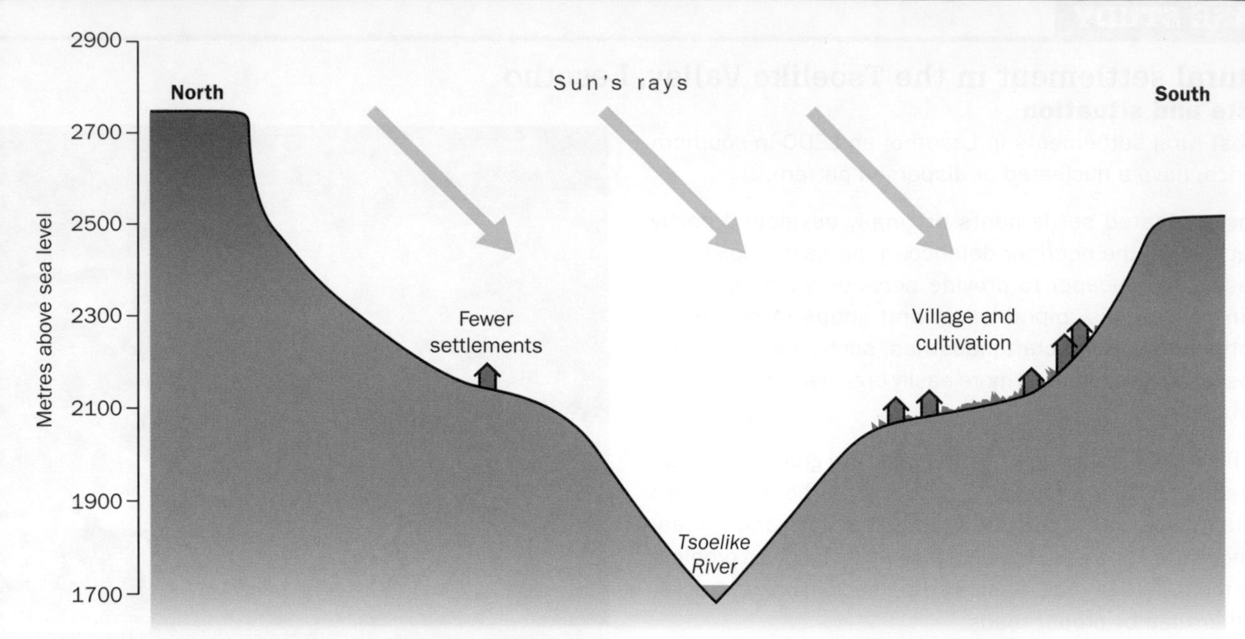

Fig. 2.12 A cross-section through the Tsoelike Valley in Lesotho

Problems and changes

- The steep, mountainous relief means that only a small percentage of the land is suitable for crop farming. It is more suitable for animal grazing.
- There is a short growing season, with a risk of late frosts. Communal grazing also discourages the growing of winter crops.
- Soil erosion by wind and running water is a constant threat.
- Crops can be damaged by hailstorms and locusts.
- A complex system of allocating land to people (the land tenure system) has led to the land being fragmented into small plots, with one person's land being scattered over the village area.
- A lack of employment is a significant problem.
- In the past, many men left the Tsoelike Valley to work in the South African mines. This led to a shortage of labour at peak times – like harvest.
- The growing population has increased the pressure on the available arable land.
- As in other countries, many young people no longer want to work in agriculture.
- Apart from agriculture, employment is now provided by forestry, public services, and administration.
- The main road through the Tsoelike Valley to the South African border at Ramatseliso's Gate (the A4) has been improved.
- The nearest town, Qacha's Nek, has a Farmer Training Centre, which aims to introduce better agricultural methods – especially methods of soil conservation.

Services and settlement hierarchy

The dispersed settlement in the Tsoelike Valley is the lowest in the settlement hierarchy. The only services are local general stores. The middle order in the hierarchy is the nearest town, Qacha's Nek, which is about 15 km away. The population of the town is about 8000. It provides shops, primary and secondary schools, the government hospital, the Lesotho Bank, the Farmer Training Centre, local government offices, a church, Forestry Division Offices and Nursery, and a hotel.

The highest order in the hierarchy is the capital of Lesotho, Maseru. This is over 250 km away and takes nine hours by a once-daily bus service. Maseru has an international airport and the national government.

5 **a** Look at Fig. 2.11. Which side of the valley has the most buildings? How far from the river is the nearest building?

b Is the settlement pattern in the Tsoelike Valley nucleated, linear or dispersed?

c Study Fig. 2.11 and Fig. 2.12. Explain how each of the following factors has affected the site and development of rural settlements in upland areas of Lesotho, like the Tsoelike Valley: agricultural land, relief (gradient and aspect), drainage/water supply, and accessibility.

CASE STUDY

Rural settlement in the Rio Poqueira Valley, Granada Province, Spain

Site and situation

This rural example from an MEDC shows a different settlement pattern to that of the Tsoelike Valley in Lesotho. Whereas the settlements in the highlands of Lesotho were mostly dispersed, in this area they are mainly nucleated.

LEARNING TIP Fig. 2.13 shows a contour map of a steeply sloping valley. Contour maps are explained in detail in Chapter 12. The contours on Fig. 2.13 are close together, which means that the slopes are very steep and there is no flat land.

Fig. 2.13 The location of villages in the Rio Poqueira Valley in southern Spain

There are three nucleated villages: Pampaneira, Bubión and Capileira, although isolated dwellings do occur as well. The villages were nucleated for the purposes of defence, ease of providing services, and cultural reasons. As in Lesotho, the valley floor has not been settled. The areas favoured for settlement are slightly flatter areas on the otherwise steep valley sides. The villages are surrounded by terraced plots, which are irrigated by a system of canals fed by melting snow from the mountains.

Fig. 2.14 The villages of Bubión (nearest to the camera) and Capileira in the Rio Poqueira Valley

Problems and changes

Common problems in villages in remoter areas of MEDCs

1. Because of low wages and a lack of jobs, young people leave their villages to work in urban areas.
2. The average age of the remaining population increases.
3. Services, such as schools, post offices, public transport, and shops, begin to close, because there are fewer people to use them.
4. Rich people from outside the area buy up properties as holiday and weekend homes, which drives up house prices. Local people, especially young adults, cannot afford these prices and more are forced to leave the area.
5. This encourages even more people to leave, and a downward spiral of rural depopulation occurs.

For years, agriculture in the Rio Poqueira Valley has been unable to employ everybody from its settlements. Many of the isolated dwellings have been abandoned and have fallen into disrepair. But, despite this, property prices are quite high.

Recently, tourism has been developed because of the sunny climate, nearby places of historical interest and outdoor activities like mountain biking and hill walking. This has led to some alternative employment in hotels and restaurants, and also in the provision of transport and tour guides.

Some properties have been bought (sometimes by foreigners) to use as holiday homes. Often these properties are empty for much of the year. As described in the panel on the left, this can have a major impact on a rural area's population structure and services, as many of the young people leave and the ageing population that remains puts more of a strain on medical and social services.

A significant number of the valley's residents now travel outside the area to work. The villages have become dormitory settlements for people who work in larger settlements further away, such as Órgiva and Lanjarón.

Fig. 2.15 Isolated houses in the Rio Poqueira Valley. Notice the disused cultivation terraces in the right-hand photo, showing the decline in agriculture.

Services and settlement hierarchy

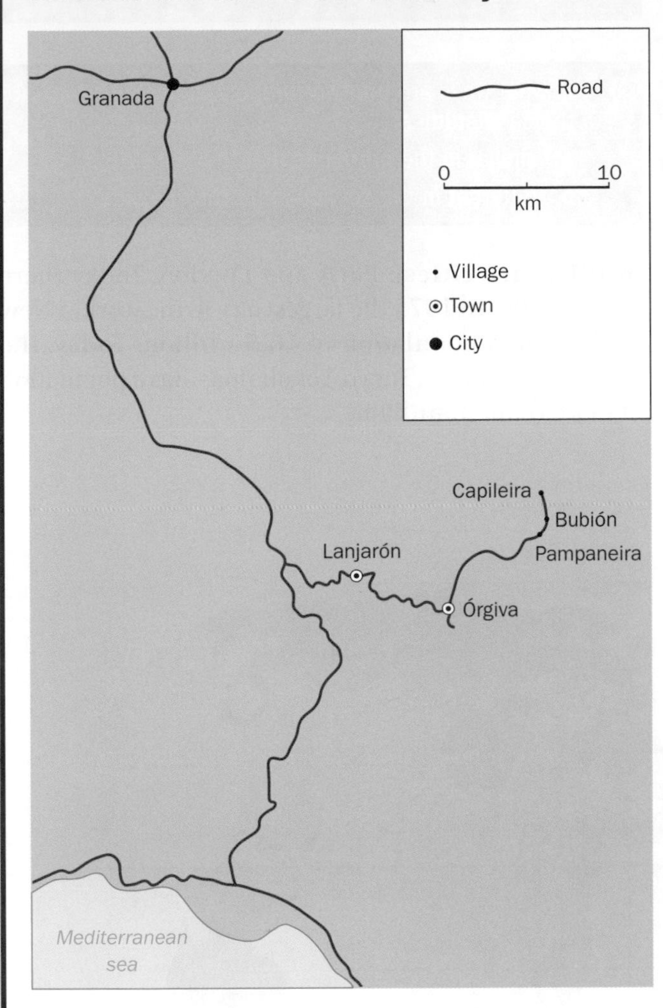

Fig. 2.16 The Rio Poqueira villages in the settlement hierarchy

The villages currently provide a full range of services, including shops, schools, churches, banks, and hotels. Only the higher-order services, such as department stores, are not present.

The roads are good, although the steep slopes and winding roads (see Fig. 2.13) make journeys slow. There is a regular bus service.

6 Look at Fig. 2.13 and Fig. 2.14.

- **a** In which compass direction was the camera pointing when Fig 2.14 was taken?
- **b** Suggest why the valley floor has not been settled.
- **c** Why are the sites of the two villages good for defence?

	Type of settlement	Examples, with population	Services
Lowest order	Isolated houses		None
Low order	Villages	Pampaniera (318) Bubión (294) Capileira (520)	Small shops, schools, churches, banks, hotels, bus service
Middle order	Towns	Órgiva (5543) Lanjaron (3587)	Weekly market, bars, restaurants, campsites, fiestas, secondary schools, bus services to the coast and Málaga, Granada, and to Pampaneira, Bubión, and Capileira.
High order	Provincial capital	Granada (236 982)	University, cathedral, important tourist sites, e.g. Alhambra Palace, cathedral, airport, national league (La Liga) football team.

Table 2.4 The settlement hierarchy in Granada Province, Spain

Fig. 2.17 The central part of Granada, the highest-order settlement in the hierarchy

Discussion point

In these pages you have learned about some of the problems affecting rural areas. Are the problems affecting rural areas in LEDCs and MEDCs the same or are they different? You should find that some problems apply to both types of area but that others apply to only one type of area.

7 Why is the main settlement in the Tsoelike Valley (Fig. 2.11) on the south side of the valley?

8 In what ways are the two rural areas described in the case studies:

a similar in the problems that they face?

b different in the services that they provide?

Urban settlements

Urbanisation

Urbanisation is the increase in the percentage of the population living in towns and cities. In 1900 there were only two cities with populations of over one million (**millionaire cities**): Paris and London. Today there are over 400. In 1975 the largest city in the world - New York - had a population of 16.5 million. Today, the largest urban area - Tokyo-Yokohama - has a population of more than 38 million.

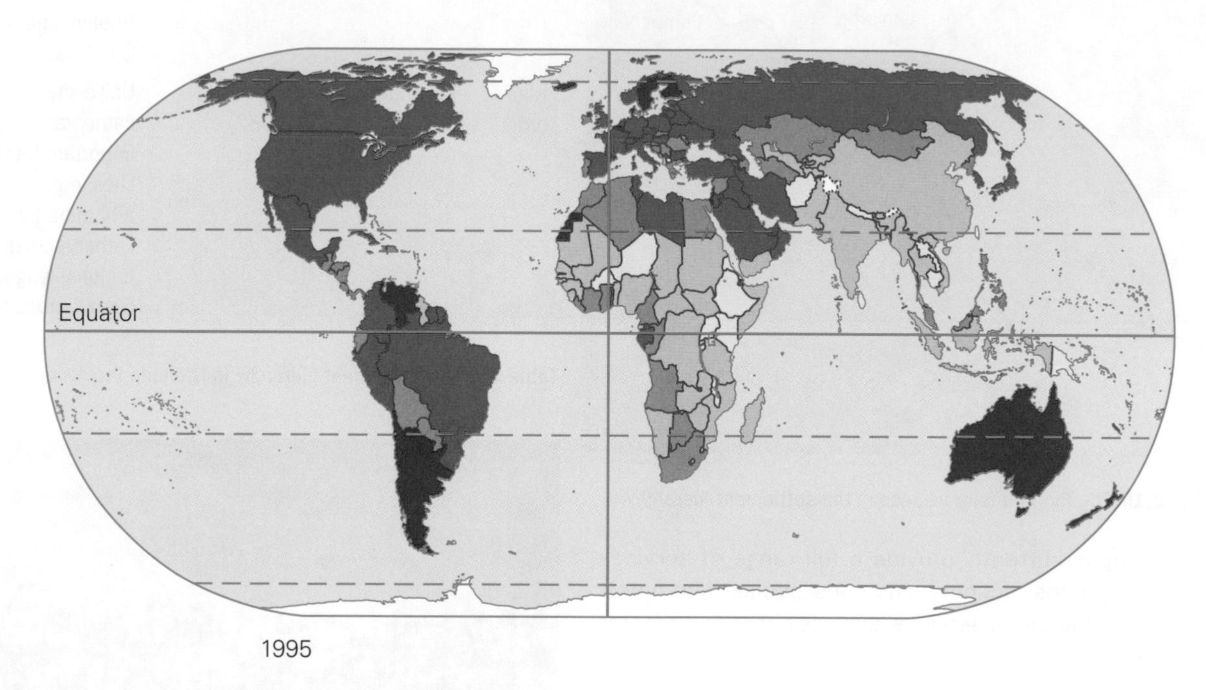

Fig 2.18 (part 1 of 2) The percentage of the population living in towns and cities between 1995 and 2025 (projected). *(Continue on next page)*

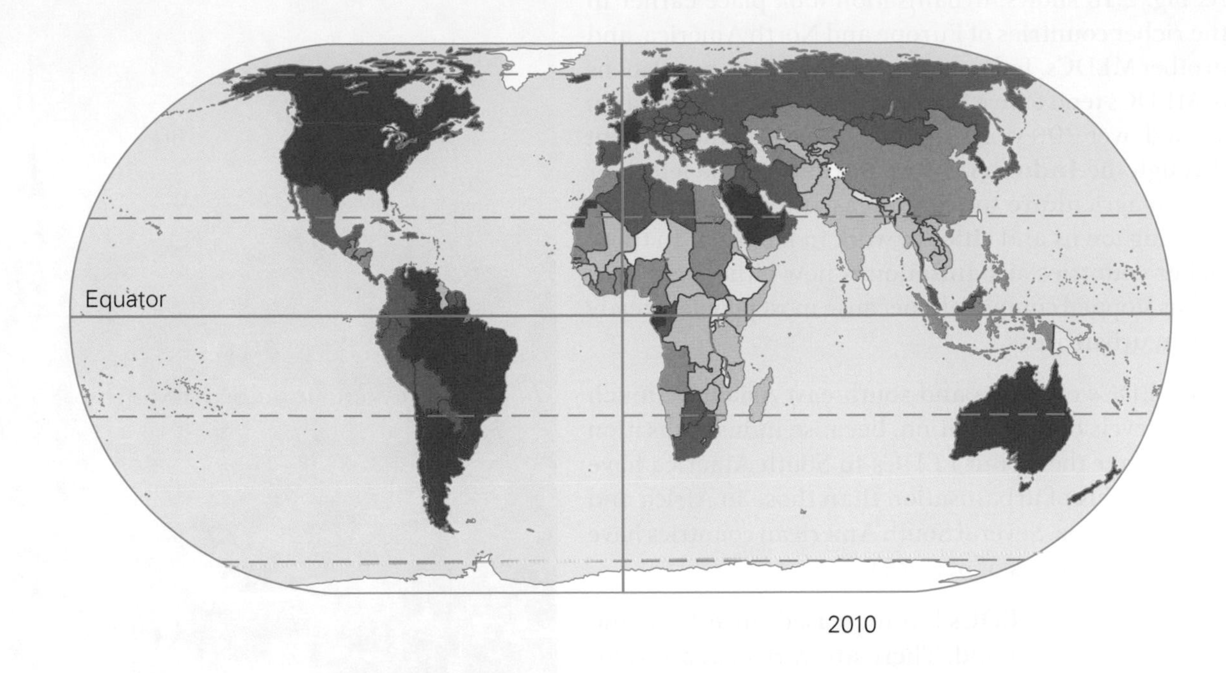

2010

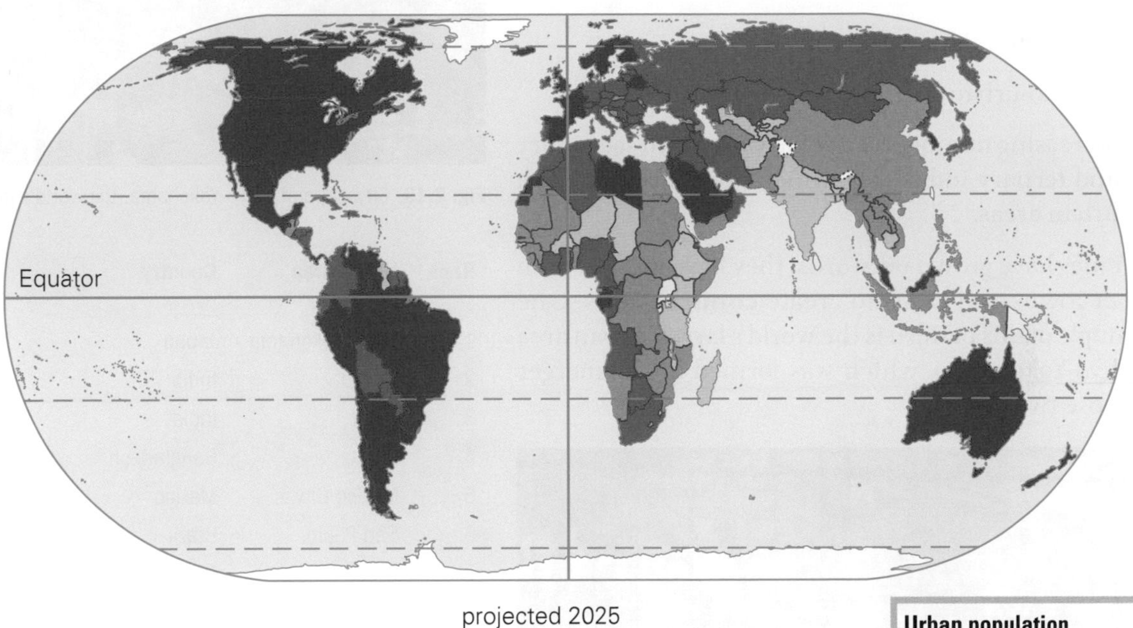

projected 2025

Fig. 2.18 (part 2 of 2) The percentage of the population living in towns and cities between 1995 and 2025 (projected)

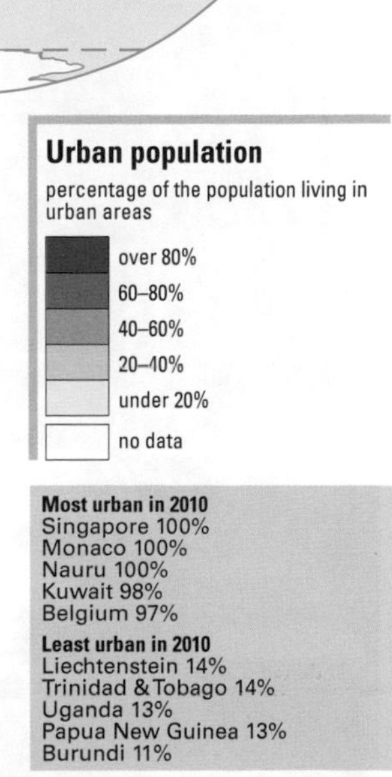

As Fig. 2.18 shows, urbanisation took place earlier in the richer countries of Europe and North America, and in other MEDCs. Today, more than 90% of the populations of MEDCs tend to live in towns and cities. Urbanisation started over 200 years ago when these countries went through the **Industrial Revolution**. People left their jobs in agriculture in rural areas and migrated to the growing towns and cities to work in factories. In these richer countries, urbanisation is now either very slow or has stopped completely, because most people already live in urban areas.

The LEDCs of Africa and south-east Asia have much lower levels of urbanisation, because industrialisation began later there. But LEDCs in South America have higher levels of urbanisation than those in Africa and south-east Asia. Several South American countries have urban populations of over 75%.

Urbanisation in LEDCs has increased since 1950 and is now extremely rapid. There are various reasons for this, including:

- overall population growth
- rural-to-urban migration (see page 21)
- increasing numbers of people working in secondary and tertiary industries, which are concentrated in urban areas.

As cities have grown outwards, they have merged with other towns and cities to create **conurbations**. One example of this pattern is the world's largest urban area (Tokyo-Yokohama), which was formed by the merger of those two cities.

Fig. 2.19 Cityscape of Tokyo-Yokohama, the world's largest urban area

Rank	Urban area	Country	Population (millions)
1	Tokyo-Yokohama	Japan	38
2	Mumbai	India	26
3	Delhi	India	26
4	Dhaka	Bangladesh	22
5	Mexico City	Mexico	22
6	São Paulo	Brazil	22
7	Lagos	Nigeria	22
8	Jakarta	Indonesia	21
9	New York	United States	20
10	Karachi	Pakistan	19

Table 2.5 An estimate of the ten largest urban areas in the world in 2020 according to population size

Fig. 2.20 The centre of São Paulo, Brazil

LEARNING TIP Population statistics for cities are sometimes confusing. This is often because some figures are for the area within the city's boundaries, while other figures include the whole conurbation. The limits of an urban area are not always easy to define.

The accuracy of censuses, including difficulties like illegal immigration, also lead to data problems. The figures in Table 2.5 show one estimate of the world's largest urban areas in 2020.

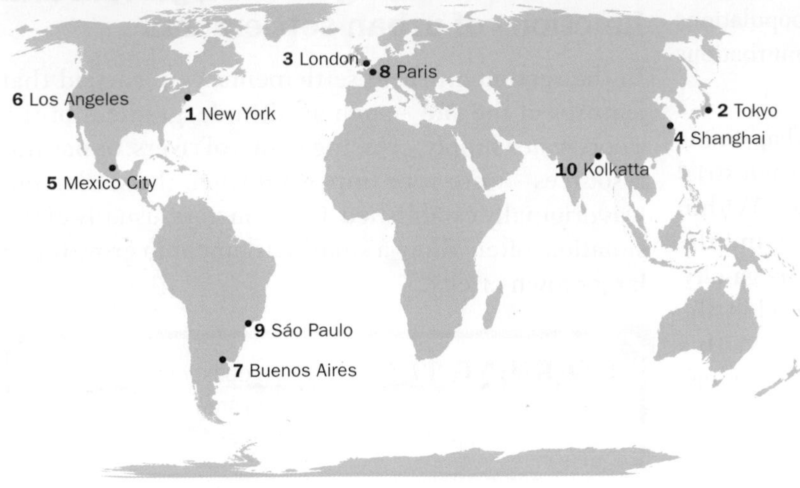

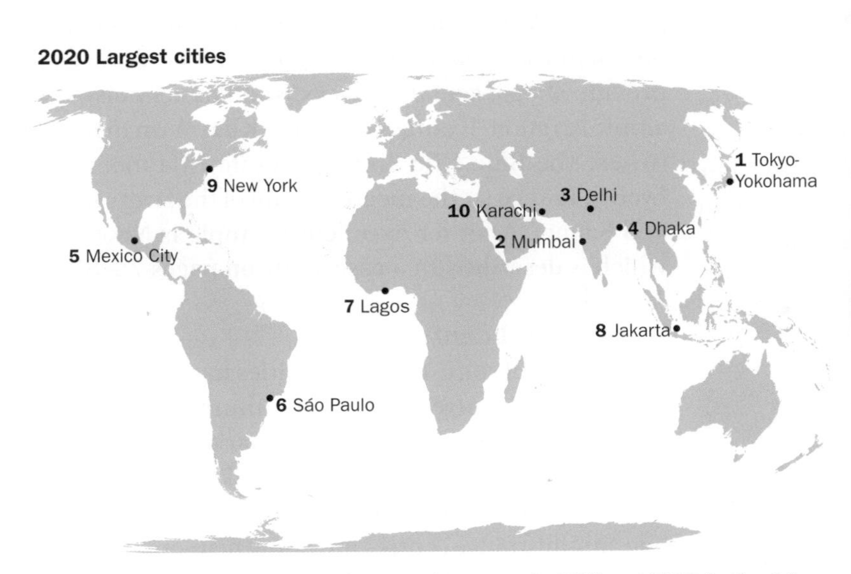

Fig. 2.21 The locations of the ten largest urban areas in 1975 and 2020 (estimate)

9 Look at Fig. 2.21 and the material on pages 49–50.

- **a** Of the world's ten largest cities in 1975, how many were in LEDCs and how many were in MEDCs?
- **b** How will this change by 2020?
- **c** How has the population of New York changed during this period, and how has its ranking changed?
- **d** Which continents are the most urbanised and which is the least urbanised?
- **e** In 2020 will there still be countries with less than 20% of the population living in urban areas?

Some of the fastest-growing cities today are in south-east Asia. Dhaka (Bangladesh), Karachi (Pakistan), Delhi (India), and Bangkok (Thailand) all doubled their populations between 1985 and 2000. Many of the fastest-growing cities are in the tropics.

Counter-urbanisation

In many MEDCs, e.g. the UK, urbanisation lasted until about 1970. Since then there has been some movement of people from urban areas back to rural areas. This has been called counter-urbanisation. Many of those who have moved are relatively wealthy. They have either retired, or still work in urban areas but now commute from their new homes in more rural areas.

This movement has often been due to the problems of urban life and the desire for the peace and quiet of rural areas. High housing costs in cities have allowed people to sell city properties and buy large houses at lower prices in the countryside. The effects of this on rural areas have already been discussed on page 46.

Discussion point

What do you think are:

- **a** the advantages and disadvantages of living in a city?
- **b** the advantages and disadvantages of living in a village?
- **c** Would an older person think differently about this compared with a teenager?

Mega-cities

This term has been used to describe cities with populations of over ten million, including extremely large conurbations like Tokyo-Yokohama.

The largest rural-to-urban migration in history is happening in China today. It began in the late 1980s, when industrial development occurred in China's coastal cities. Within the past decade, China has merged cities around the Pearl River Delta (see Fig. 2.22) to create a mega-city. The Pearl River Delta (PRD) has been declared by the World Bank as the world's largest mega-city. With a combined population of around 110 million (and growing), the PRD's population is about a third of that of the US.

The transport and telecommunications networks serving the mega-city are constantly evolving and improving. Today, the PRD is served by a number of civilian airports, some of which are international. All parts of the mega-city are also connected by metro systems, high-speed rail and intercity-rail services.

The Chinese government has proposed a new initiative which aims to transform the PRD into the Guangdong-Hong Kong-Macau "Greater Bay Area" by further integrating Hong Kong, Macau, and nine cities in Guangdong to become an important city cluster.

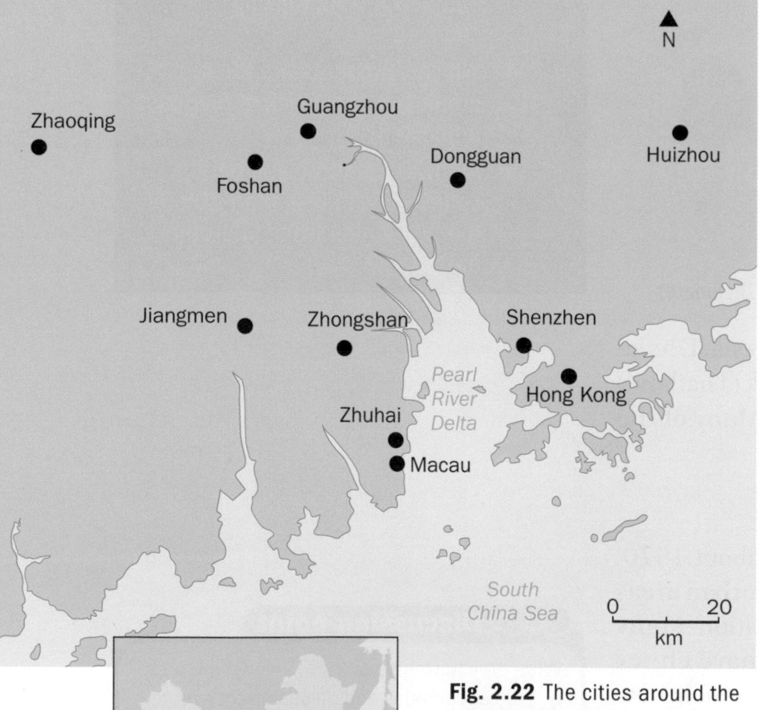

Fig. 2.22 The cities around the Pearl River Delta in southern China, soon to be part of the "Greater Bay Area"

Factors influencing the size, growth and functions of urban settlements

In the section on rural settlements, you learned that features of the site – such as altitude, gradient of the slope, water supply, crossing points of rivers, or natural resources – were very important when the settlement was originally established. Other factors, usually of the situation, often allow a small settlement to grow into a larger town or city.

LEARNING TIP Survey map questions on Paper 2 frequently ask about site, situation and growth of settlement. There are examples of these in Chapter 12.

Nodal points (route centres)

These are where natural routeways, such as river valleys, meet. These points often develop into important transport junctions. As a result, they become the most accessible points in the area, and are seen as the best places to provide shops, social services (such as schools), and administration. Examples include Khartoum in Sudan (where the Blue Nile and the White Nile meet) and Lyon in France (at the meeting point of the rivers Rhône and Saône). Another excellent example is New York, which is described in a case study on pages 75–80.

Agricultural centres

The growth of some towns and cities has been because they lie at the centres of rich agricultural regions. They have become the collection and marketing points for the produce of the area. Transport links have been built to focus on these points, and they have developed further into administrative centres providing high-order services for their surrounding areas.

The Prairies in Canada are important for commercial cereal farming (see Chapter 9) and provide good examples of cities that have grown up as agricultural centres. Each of the three Prairie provinces has its state capital centrally placed at the focus of main transport routes: Edmonton (in the province of Alberta), Regina (in the province of Saskatchewan), and Winnipeg (in the province of Manitoba).

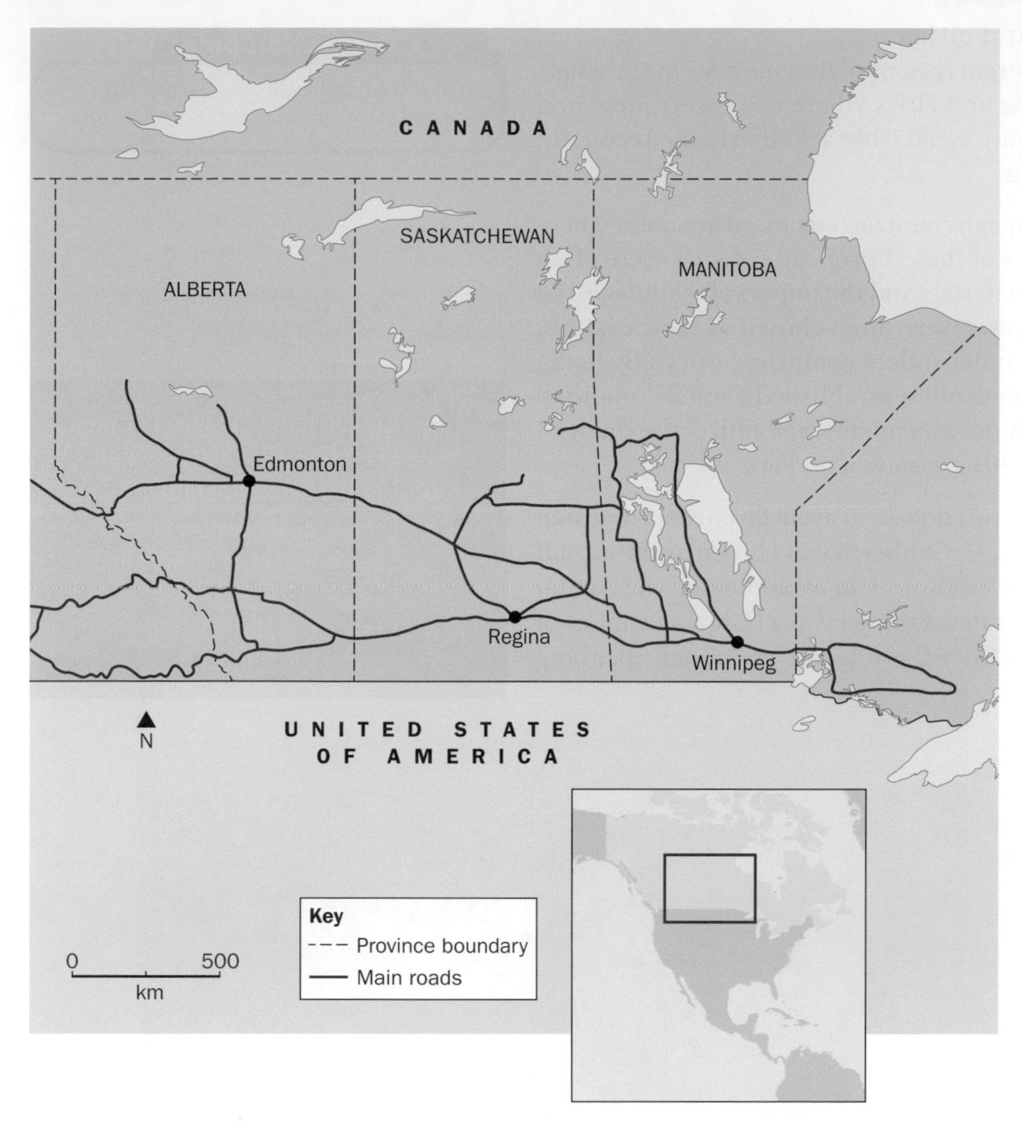

Fig. 2.23 The Canadian Prairie provinces: agricultural and route centres

Ports

A number of factors help to promote the growth and development of seaports:

- The presence of deep water close to the shore, where ships can anchor to unload their cargoes
- Shelter from strong winds and rough seas, provided by bays and river estuaries
- A gap into ports in the Tropics in a coral reef through which ships can pass
- The presence of a large area linked to the port from which goods are exported and to which goods are imported. This is known as the **hinterland** (meaning the land behind). Big ports are often **entrepôts** (places where goods are imported and then re-exported without paying taxes).
- Good transport links with the hinterland. The biggest port in Europe, Rotterdam in the Netherlands, is linked to much of Europe by road, rail, and the navigable River Rhine.
- A location at a strategic position on world shipping routes. This is illustrated by the Cape Town case study on pages 81-4.

Administrative towns and cities

These are the capitals of different regions of the country - or the whole country - and are the towns and cities where the government and civil service (administration) are based. They are often located centrally, or for other strategic reasons.

In the 1800s, powerful European countries colonised areas elsewhere in the world. The economies of these European colonies were often based on the export of raw materials and the import of manufactured goods. For this reason, seaports were often chosen as their capitals. When the colonies became independent countries, especially in the 1960s, their new governments often established capitals inland to unify the country, develop the local economy, and reflect the changed politics. Three examples of this are shown on Fig 2.24.

Many capitals have been neutral choices, to avoid favouring one group within the country. For example, Canberra was chosen as the capital of Australia - within its own territory - to avoid having the capital within one of the existing states. Ottawa was chosen as capital of Canada to avoid favouring either the English- or French-speaking communities.

Discussion point

What is the best location for the capital of a country? What factors should be important?

10 **a** Explain the difference between the terms 'site' and 'situation'.
b List the causes of urbanisation.
c Explain how urbanisation has been different in MEDCs and LEDCs.
d Draw a bar graph to represent the data in Table 2.5.

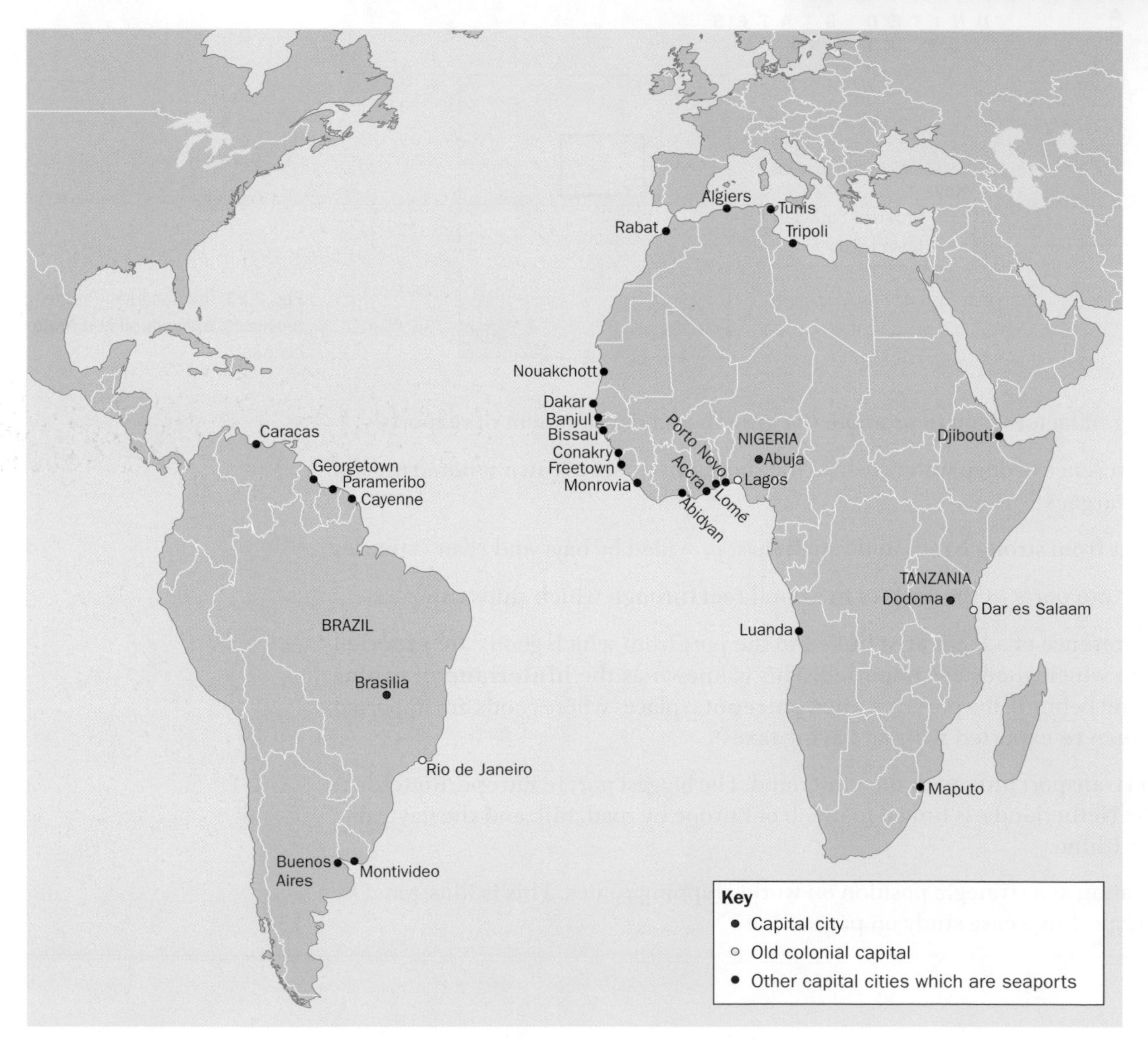

Fig. 2.24 Examples of seaport capitals, plus three examples of countries that changed their capital cities after the countries gained independence

Urban land use

Central Business District (CBD)

These are the features of the CBD:

- Government buildings
- High-order retail services, such as department stores, in the middle of the CBD and highly specialist shops on the outskirts
- Offices, including major company headquarters
- Theatres, hotels, and restaurants
- Old historic buildings
- Multi-storey buildings, developed in response to the high land values
- Concentration of public transport services, including buses and underground railways
- Few residents - the number of people in the CBD at night is low
- Zoning of different functions in different parts of the CBD. This is because certain shops, e.g. shoe shops, are better next to shops of the same type for comparison shopping. Businesses like banks and legal services also prefer to be next to each other for business contacts
- Vertical zoning, e.g. retail on the lower floors, offices on upper floors
- High numbers of pedestrians
- Pedestrianised areas

Fig. 2.25 The CBD of Cape Town in South Africa - the area of multi-storey buildings in the centre is the CBD, with lower buildings surrounding it

Fig. 2.26 The CBD often includes old historic buildings - like these in Tallinn, Estonia

Fig. 2.27 In smaller towns, the CBD will be more modest - like this shopping street in Barrow-in-Furness in the north of England

Reasons for the development of the CBD

- In many towns and cities, the CBD was the original 'core' of the settlement – so it contains the oldest buildings and the town expanded outwards from that point.
- The CBD was also the point where roads from the outskirts converged. This made the area the most accessible part of the town, i.e. the easiest place for all the people to get to.
- This, in turn, made it a very desirable place for services like retailing to locate themselves. As a result, land prices in the CBD became higher, and only certain services could afford to locate there. Buildings in the CBD began to be built taller to make the best use of the expensive land.

RESEARCH Certain areas of London's CBD are linked with certain functions or services. Use your own knowledge or the Internet to find out what function or service each of the following areas of London is associated with: Westminster, Oxford Street, The City, Soho, The West End, Harley Street.

Fig. 2.28 Harrods in London – a famous department store

Fig. 2.29 The British parliament – government buildings are usually found in the CBD

Residential areas

These are the areas where people live. Different styles of housing are found in different countries, so it is not easy to generalise about different types of housing.

High-density housing

This is where dwellings are relatively small and there is little or no open space between them. It is often found in the older parts of towns and cities, closer to the centre. In parts of northern Europe, long rows of houses joined together (called **terraces**) were built when the towns expanded rapidly during the Industrial Revolution. In other areas, flats (apartments) form the high-density housing. In some towns and cities, the high-density housing is several hundred years old (as shown in Fig. 2.30). The high-density housing in Fig. 2.31 is more modern.

Fig. 2.30 An old high-density residential area close to the centre of Granada in southern Spain

Fig. 2.31 A high-density residential area in Cartagena, Colombia

Low-density housing

In these areas there are fewer dwellings per km^2 and there is open space between the housing - usually in the form of garden plots for recreation. The areas are usually more modern and further away from the city or town centre.

11 **a** Describe the differences between the houses in Fig. 2.32.

b Which housing is the more expensive, do you think?

c Who is likely to live in each set of houses?

LEARNING TIP Question 11a asked you to write about differences. When this is the question, don't do two separate descriptions - compare the two things, giving pairs of points.

Fig. 2.32 Low-density housing areas in Mbabane, the capital of Swaziland

Flats (apartments)

These are multi-storey buildings containing a number of different units. In some European countries they are owned by local government, or a single owner who rents out the individual units. In theory, they allow more open space by building upwards.

This type of housing can be found in any part of the town or city. It can be very high quality and expensive, but it is not always suitable for families with children. Flats are also known as apartments and condominiums (condos), usually when the individual units are privately owned.

Shanty houses

Known in parts of South America as *favelas*, in parts of Asia as *bustees* - and by the United Nations as informal settlements - these are a type of slum housing. They are often built spontaneously out of any available materials, such as plastic sheets, metal sheets or even cardboard. They often lack normal services, such as a water supply and sewerage.

Shanty houses are associated with the outer areas of cities in LEDCs, although they may occur even in relatively small settlements (like the one in Fig. 2.34). They are described in detail on page 68, and in the case studies about Mumbai (pages 69-73) and Cape Town (81-4).

Fig. 2.34 Informal housing on the edge of Springbok, a small town in the Northern Cape of South Africa

Open spaces

Open spaces are found in most towns and cities - with few exceptions. In MEDCs, they are usually planned features like public parks or sports grounds. There may also be less attractive areas waiting for redevelopment. Planning regulations preserve these spaces and they are not allowed to be built on.

Fig. 2.33 Flats surrounded by public open space

In LEDCs, the open spaces have often developed much more informally and they are not necessarily well-kept areas. The case studies of New York (pages 75-80) and Cape Town (pages 81-4) provide good examples from an MEDC and an LEDC.

Industrial areas

Industrial areas are present in most towns and cities. They are of different types, including:

- older industries located close to railways or canals - and often quite close to the centres of older cities
- industries associated with sea or river ports, which involve imports or exports
- newer industrial areas closer to the outskirts of cities. These might be on **greenfield sites** and have good access to road transport.

Transport routes

Transport routes are a key feature of urban land use. They are dealt with in more detail on pages 63-6. There are big differences in transport between those cities which are very compact and have buildings located close together, and those cities with low-density housing spread over a large area. The high-density cities are easier to serve by public transport. There are great differences between cities in different parts of the world, as illustrated in Question 12 on the next page.

12 Look at Fig. 2.35.

a What type of relationship is shown – a positive correlation or a negative correlation? If you have not studied correlation it is explained in Chapter 12.

b Which country has the lowest urban population density and the highest petroleum consumption per person?

c Which region has the highest urban population density and the lowest petroleum consumption per person?

d There is little difference in population density between the cities of the USA and those of Australia and Canada. Suggest some reasons why much more petroleum is used per person in the USA.

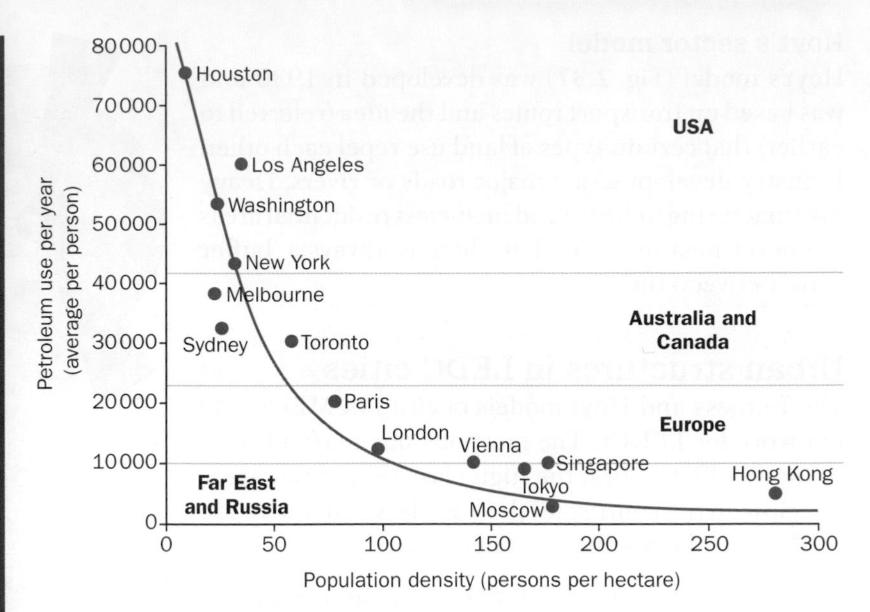

Fig. 2.35 The relationship between petroleum consumption per person and population density for selected world cities

Urban structures in MEDC cities

Urban structure, or urban morphology, is the distribution of different types of land use in a town or city.

In urban areas, the different types of land use (the CBD, different types of housing, industry, and open space) usually exist in separate zones. This is because:

- the value (cost) of the land is different, e.g. in the CBD rents are too high to allow much residential use
- certain types of land use do not mix, e.g. people in high-quality housing would not want to be next to ugly industry
- particular types of land use, once established, tend to continue that way.

Urban land use is often explained by models. A **model** is a simplified theory that attempts to explain how things work. No model works perfectly, but they do help to explain some of the features of urban structures.

Burgess's concentric zone model

This model (Fig. 2.36) was developed in 1925 and based on the structure of Chicago in the USA.

- The CBD develops at the original growth point, at the intersection of major roads. Here there is the greatest accessibility (due to public transport) and the highest land values.
- Beyond the CBD is a manufacturing zone.
- New immigrants moving into a city tend to move into inner-city areas with cheap housing, close to sources of employment.
- Housing quality and social class change with increasing distance from the centre. Increasing affluence and developing public transport allow people to live long distances from their places of work.
- As the city grows, all the circles push outwards. The area next to the expanding CBD is called the transition zone, where residential areas change to commercial use. This is like the modern core and frame concept (explained on page 61).

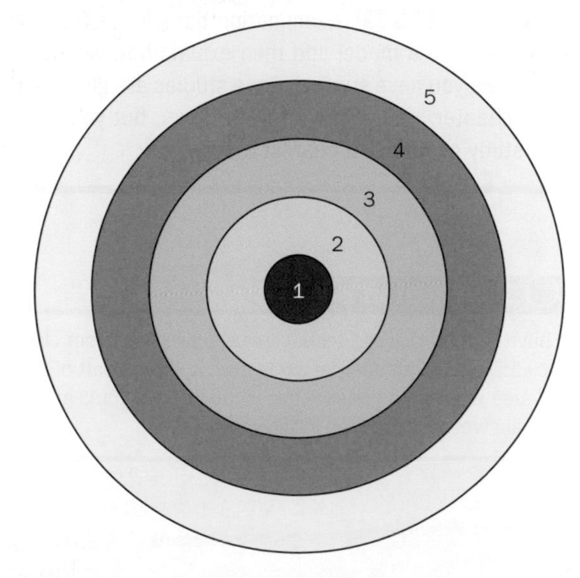

Fig. 2.36 Burgess's concentric zone model

Hoyt's sector model

Hoyt's model (Fig. 2.37) was developed in 1939 and was based on transport routes and the idea (referred to earlier) that certain types of land use repel each other. Industry develops along major roads or rivers. Heavy manufacturing industry and high-class residential areas are never next to each other; there is always a 'buffer zone' between them.

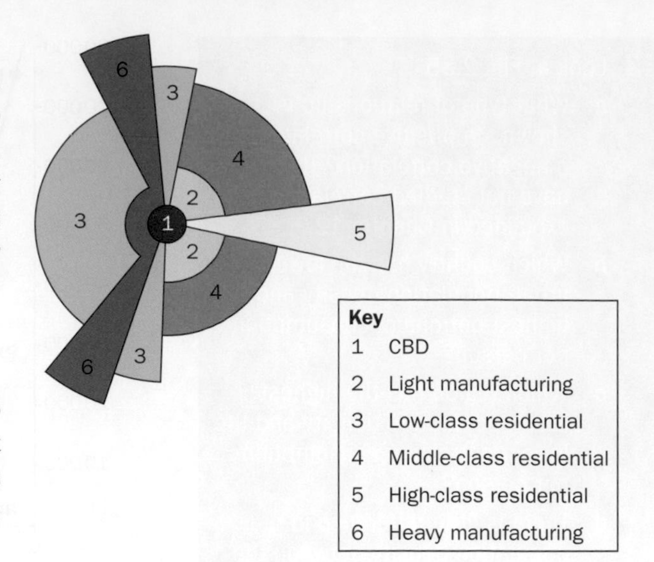

Fig. 2.37 Hoyt's sector model

Urban structures in LEDC cities

The Burgess and Hoyt models of cities in MEDCs do not work for LEDCs. The most obvious reason is that in many LEDC cities, the high-class residential areas are close to the centre and the low-class residences are on the periphery - the exact opposite of MEDCs.

The model illustrated in Fig. 2.38 is based on Latin American cities. The CBD is based on the old colonial centre, and has a sector of shops and offices leading from it along a major transport route. Either side of this are high-class residential sectors. These contain open areas, parks, homes for the upper and middle classes, and amenities such as good schools. The streets here are well maintained.

The other residential areas are based on concentric zones. Recent squatter settlements are found on the outskirts. Housing conditions closer to the centre are better, and there is older, more established squatter housing. It is easier for people living closer to the centre to find work. Manufacturing tends to be scattered throughout the city, although there may be industrial sectors along transport routes.

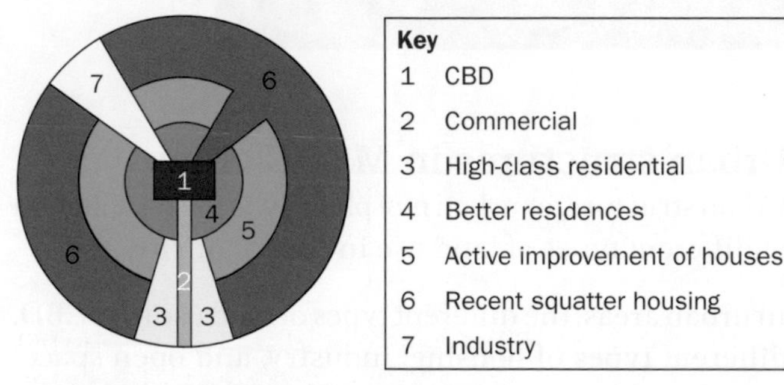

Fig. 2.38 A model of a Latin American city

LEARNING TIP Examination questions often ask you to describe a model and then explain how well it fits an example you have studied. Case studies are given later in this chapter, particularly of Cape Town, but you could use a study of your own nearest town or city.

13 What is meant by each of the following terms?

- **a** Greenfield site
- **b** Urban structure
- **c** Sector theory
- **d** Shanty
- **e** Accessibility

Discussion point

You have learned that residential areas sometimes occur close to the CBD and sometimes far away from it. Which part of an urban area would you prefer to live in, and why? Would an older person answer that question in the same way?

RESEARCH Get a map of your nearest urban area that you can use as a base map. Mark on your map: the CBD, industrial areas, residential areas (showing any different types), and open spaces. Describe how well your chosen area is like any of the models of urban structures (urban morphology). If you do this in detail, you could use your example to answer an examination question asking for a case study.

Problems associated with the growth of urban areas

Problems of the CBD

The CBD is a zone of constant change - whether it's in a large city or a small town. Problems arise because of this change, and also because there is a lack of space in the CBD and land there is expensive. In many countries, people are worried that the CBDs of towns and cities are in decline. Areas of the CBD in many towns and cities have been almost abandoned - with shops boarded up and properties in a poor state of repair.

The decline of retailing

One of the original reasons for the development of the CBD was that it was the most accessible place in the urban area for people to congregate. This is usually no longer the case. Increased car ownership has led to congestion - slowing down journey times, including bus journeys, and leading to a lack of parking space.

The first retailers to leave the CBD were the smaller independent shops, which could no longer afford the high rents. However, larger shops have now moved out as well:

- Shops such as DIY stores, furniture shops and carpet shops have often moved to **brownfield sites** in the inner-city area.
- Major department stores and hypermarkets have moved to greenfield sites on the outskirts of towns. Here the land is cheaper. People with cars can also drive easily to these locations, often along ring roads, and there is plenty of parking space.
- There has been a major growth in online shopping, particularly for clothing and electrical goods but also for retail services such as travel agents. In Asia online sales accounted for 12.1% of all retail sales in 2016 but only for 1.8% of retail sales in the Middle East and Africa. The loss of customers has forced some stores to close.

If more and more shops move:

- fewer people will go shopping in the CBD
- more shops will be forced to leave
- there will be a downward spiral of decline
- people's perception of the CBD might become one of a run-down area with empty derelict buildings, which is unsafe and dirty and suffers from litter and graffiti.

Decentralisation of companies and administration

The same issues of poor accessibility and car parking, plus high land values, have led many companies to think about whether the CBD is necessarily the best location for their offices. Modern electronic communication systems have also reduced the need to be near the offices of similar types of company. Now, when a company decides that its premises need renewing, it might decide instead to relocate to new purpose-built premises on the outskirts of the town, where land prices are lower and employees can drive to work and park more easily.

The CBD in the evening

The CBD can be very empty in the evening. If the only services open are bars, restaurants, and nightclubs, the CBD can become an unsafe place at night - with high crime rates.

The twilight zone

This name has been used to describe the problem areas in the transition zone on the edge of the CBD. Land uses at the centre and outsides of the CBD are different. They have been described as the **core and frame** (see Fig. 2.39).

The edge of the CBD is likely to be particularly affected by change. It might be improved if the CBD expands in that direction. However, it can also be an area of decline and suffer from:

- derelict land and buildings
- high rates of crime and social problems.

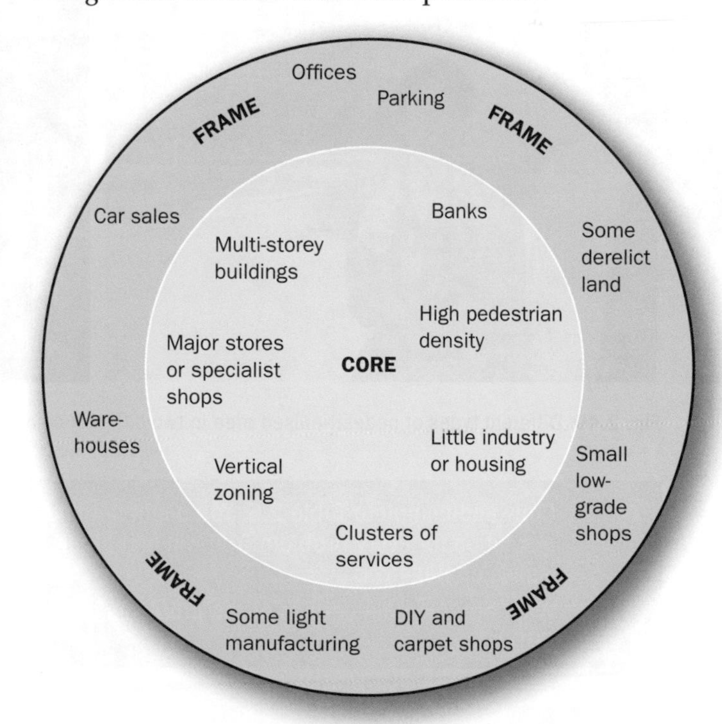

Fig. 2.39 The core and frame of the CBD

Solutions to the problems of the CBD

The following solutions have been proposed to address the problems of the CBD.

Pedestrianisation

Pedestrianised areas are traffic free, or only allow access to delivery vehicles at certain times of day. The aim of this policy is to create a safer, more relaxed environment, with less pollution from vehicles - especially air and noise pollution.

Shopping malls

These are undercover shopping areas. Shoppers can look around, compare goods and prices - and ignore the weather. They are air-conditioned in warm climates and heated in cold ones. Shopping malls are successful when they are new and purpose-built. They often try to have at least one major department store (an 'anchor store'), which encourages other shops to locate in the mall. The mall will also contain cafés and small restaurants.

Visual improvements

These can be achieved by providing flowerbeds, seated areas, trees and hanging baskets. Pavement cafés and bars are introduced. When a building becomes vacant, every effort is made to find a new tenant to occupy it quickly, rather than letting it stand empty. Street cleaning, litter and refuse collection are priorities. Many city centres have pieces of sculpture at important locations.

Security

Shoppers need to feel safe. This is helped by regular patrols by police or by private security firms. Closed-circuit TV is also a deterrent to pickpockets and shoplifters. This is especially relevant in the evenings.

Some of the more general problems of urban areas also apply to the CBD and these are discussed in this chapter on pages 63–8.

Fig. 2.40 Different types of pedestrianised area in two different cities

14 a Copy and complete Table 2.6. You might not find a solution for every problem. You will find other issues on page 63.

b What are the main causes of the problems of the CBD?

Problem	Solution

Table 2.6 Problems with the CBDs of cities today

CASE STUDY

London – improving the CBD's transport system

London has always had traffic jams. The first ones happened with horses and wagons long before motor vehicles were invented.

Underground railways, known as 'The Tube', were first developed in London in 1863 and have been extended ever since. The most recent extensions have been the Docklands Light Railway in 1995 and the Jubilee Line extension in 2001. The underground trains carry about 600 people each, run every few minutes, make frequent stops, and are not slowed down by traffic on the surface.

Bus lanes, which private cars cannot use, allow buses and taxis to move at their own speed – independent of other traffic.

The **London Congestion Charge** is a fee payable by motorists driving in central London. It aims to deter motorists from bringing their cars into central London, to reduce congestion, and to raise funds to improve London's public transport system. It was introduced in February 2003 and extended into parts of west London in February 2007. The congestion charge is payable each day for each vehicle travelling within central London between 7am and 6pm (Monday–Friday only). The system is mostly run on electronic technology, using automatic number plate recognition.

Electronic ticketing. This allows fares for buses, underground trains, and surface trains to be paid using a swipe card (the Oyster Card) or a contactless bank card, without the use of cash. This speeds up queues to pay for fares. It also allows cheap fares and bonuses for using public transport – encouraging people not to use their cars.

Crime and racial conflict

Cities in both MEDCs and LEDCs have areas where there are high levels of poverty. These areas usually have higher levels of crime and are less safe places to live than more affluent areas.

Cities have also always attracted high levels of immigration. Immigrants from a particular country or area tend to live in the same part of the host city. This might be because they are poor and need cheap housing, but also because they want to live near to people who share their culture and language.

An area of poverty, which also has a concentration of people from a particular immigrant community, is called a ghetto. Conflict can arise between immigrant communities and the native population, or between different immigrant communities. Both the Cape Town and New York case studies at the end of this chapter have examples of crime and racial conflict.

The solutions to these issues go beyond the scope of this book, but they include:

→ providing social facilities such as sports clubs

→ job-creation schemes to provide employment

→ special projects that bring communities together

→ a zero tolerance of crime

→ ensuring adequate policing on the streets

→ providing language lessons for immigrants.

Housing shortages

Housing shortages are a particular issue in larger cities in MEDCs, because:

→ older properties nearer to city centres require renovation or renewal

→ there has been a population increase through immigration and natural growth

→ property prices are too high for those who are unemployed or on low wages.

In the UK, between 1930 and 1970, there were major slum-clearance schemes in inner-city areas. The older housing was replaced by blocks of flats or new houses in the suburbs. These new properties were owned by the local government authorities. Several completely new towns were also built in the countryside. Many of the replacement properties are now owned by private landlords, and there is no longer a policy of government-owned housing.

In some Japanese cities similar problems are made worse by the shortage of flat land. As a result, the houses are sometimes extremely small and the population density is very high, as shown in Fig 2.19. In the urban area of Osaka-Kobe there are population densities of up to 10 000 people per square kilometre. The city has been forced to reclaim land from the sea to build more residences. Port Island and Rokko Island were reclaimed and new flats were built there. There was a great demand for this housing even though it was expensive, and the names of the first occupants were drawn from a hat.

Squatter settlements

As described earlier, these informal settlements are a type of slum housing - often located in the outer areas of cities and towns in LEDCs. The residents face the following difficulties:

- They do not own the land or have a legal right to occupy it, so they could be evicted at any time.
- The houses are not weatherproof and can be cold in winter.
- There is no proper sanitation and water supply, which can lead to outbreaks of diseases like cholera.
- There may be no refuse collection.
- In some cases they have no electricity supply, or they obtain it illegally.
- The location of their settlement on the outskirts of the urban area means that there is no local employment.
- They suffer from extreme poverty and high unemployment.
- There is extreme overcrowding, with whole families living in one or two rooms.
- There are high levels of crime and drug and alcohol abuse.
- The *favelas* of Rio de Janeiro, in Brazil, are built on steep slopes, and landslides and mudflows often destroy the housing (flash floods and mudslides killed more than 250 people in April and May 2010).
- The residents might be recent immigrants with no connection to the area.
- People face long journeys to the central areas of the city for work, and there may be little public transport.

Providing solutions to these problems is difficult, because the countries concerned are poor and the number of people living in these areas is so great. However, some solutions include:

- low-cost housing schemes that provide basic dwellings for families, with running water, electricity and proper sanitation. This approach often involves people who have some sort of employment and are able to pay a small amount of rent.
- self-help schemes which provide groups of people with the materials they need to build proper houses. These schemes can help produce a sense of community.
- the provision of basic services such as sewerage, piped water and refuse collection by the city authorities.

The case study on Mumbai (pages 69-73) describes some of these solutions.

15 Question 14 asked you about the problems of the CBD. Repeat the question but this time for problems facing other urban areas.

CASE STUDY

London – traffic congestion

(Also see the previous case study on transport in London's CBD.)

Traffic congestion occurs in London (population 8 million) because:

- it's an old city with an historic centre that was built long before the need for mass public transport
- the use of private cars has increased
- 2 million extra people travel into the city from elsewhere each day to work. This means that there is enormous movement inwards in the morning rush hour and outwards during the evening rush hour. Many of the trains and buses that carry these people are not needed during the rest of the day
- many people visit the CBD for sightseeing, shopping, or entertainment – estimated to be around 28 million people each year
- many people pass through the city on their way to other places, because London is the focal point of the UK's road and rail systems.

Integrated transport policy

London's underground railway (The Tube) and its bus services are co-ordinated by a single organisation – Transport for London. The overland train services are operated by different companies, but they must plan their services jointly to link them together.

London's road network

London has inner and outer **ring roads**. The motorway loop (the M25) is located at the edge of the built-up area, and allows traffic to avoid travelling through the city. This is important because the UK's national motorway network focuses on London. An inner ring road (the North and South Circular roads) performs a similar role for inner London. Traffic lights control the traffic. In places, traffic lights above the carriageways allow **tidal flow** – more lanes are available for traffic travelling inwards in the morning and outwards in the evening. **Roundabouts** (circles) are used on the major intersections. Smaller urban areas in Britain, for example Oxford and Cambridge, have **park-and-ride** schemes to increase the use of buses.

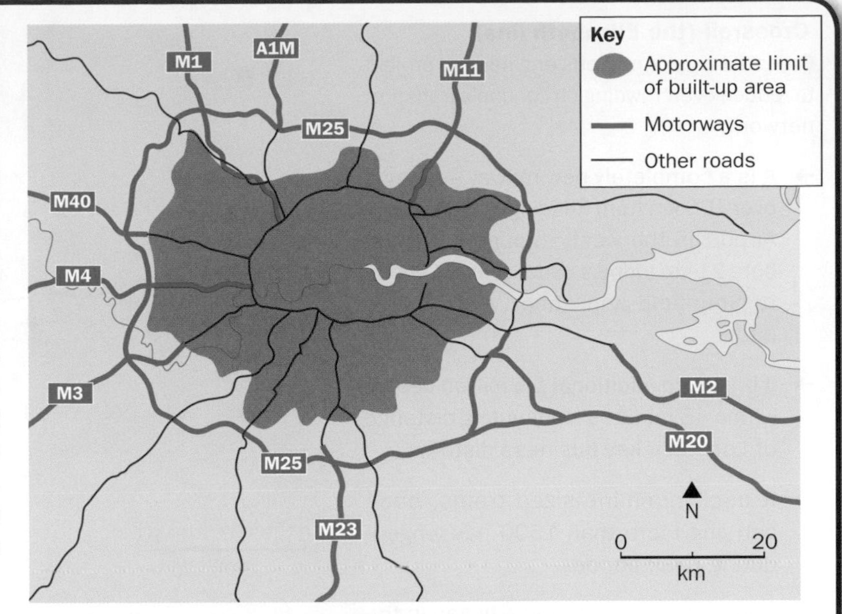

Fig. 2.41 London's motorway network and ring roads

London's rail network

The UK's national railway network focuses on London. Planning in the 1800s prevented the main national terminus railway stations from being built in the centre of the city, so they are located slightly further out. Suburban and national train services run from these stations.

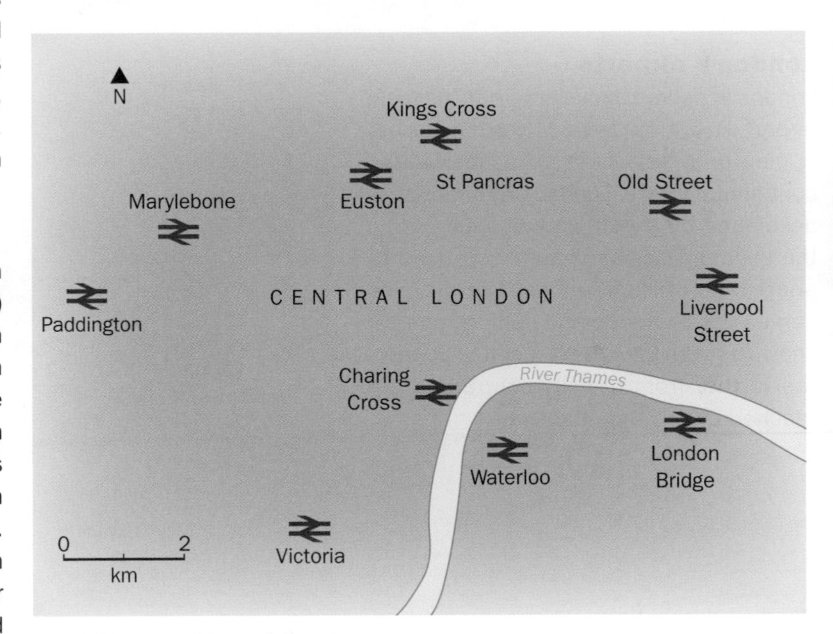

Fig. 2.42 London's main terminus railway stations

Eurostar

Eurostar is a high-speed passenger train service connecting London with Paris (France) and Brussels (Belgium). Great Britain is an island, so all of the trains travel through the 50-km-long Channel Tunnel – under the sea between Britain and France. The London terminus is St Pancras, with the main Paris terminus at Gare du Nord and the main Brussels terminus at Midi/Zuid station. The service uses 18 coach trains, which can reach speeds of up to 300 km per hour. Eurostar services began in 1994, and now carry more passengers than aircraft do between those three cities. Unlike the airport services, Eurostar transports international travellers from city centre to city centre.

Underground railway

The importance of the underground railway in the CBD has already been described. There is an extensive suburban network too, especially in north London. South of the River Thames, the soft rock makes tunnelling more difficult, so the underground network is not as extensive there. The river running through the centre of London causes road traffic congestion at the various bridges, but this is not a problem for the Tube service under the river.

Crossrail (the Elizabeth line)

Crossrail is an ambitious engineering project to reduce overcrowding on London's transport network.

- It is a completely new railway – running over 100 km from Reading and Heathrow Airport in the west, through new twin-bore 21-km tunnels under central London to Shenfield and Abbey Wood in the east.
- It brings an additional 1.5 million people within 45 minutes' commuting distance of London's key business districts.
- It uses mainline-sized trains, each carrying more than 1500 passengers during peak periods.

The main construction works began in the central section in 2010 with final completion due in 2019.

Fig. 2.43 The Crossrail Project

London's airports

London is served by five airports. The main airport, Heathrow, is said to be the busiest in the world. Gatwick also has many inter-continental flights. London City Airport was developed on a derelict industrial site in the London Docklands area. It is used by short take-off and landing aircraft, and serves the business and banking quarter known as 'The City'. The other four airports are further from the city centre and are reached by rail and motorway.

Fig. 2.44 London's airport connections

Discussion point

If you were in charge of improving the transport system of a town or city, what would you do? What might be the barriers that you would have to overcome?

Discussion point

This section has dealt with a range of problems faced by urban areas: crime and racial conflict, housing shortages, squatter settlements, various types of pollution, traffic congestion, and problems of the CBD. Not all these problems affect all urban areas. Which of the problems affect the urban area(s) that you know best? Can they be solved? How should these problems be approached?

The effects of urbanisation on the environment

These include the different types of pollution and the problems created by urban sprawl.

Air pollution

The major sources of air pollution are motor vehicles, industry, power stations, and open fires (all involving the burning of fossil fuels), so it's not surprising that urban areas have less clean air than the surrounding countryside. In MEDCs, the strict regulation of vehicle and industrial emissions has greatly reduced air pollution since the 1960s. Reducing traffic congestion has also helped. It's in the major cities of LEDCs and NICs that the highest levels of air pollution now occur. This is especially true in countries which are rapidly industrialising, such as China and India.

LEARNING TIP In examinations, don't just refer to 'pollution'. Make your answer specific to the types of pollution described on the following pages.

Pollutant	Source	Problem
Carbon monoxide	Vehicle exhausts. Concentrations will generally be highest close to busy roads.	At very high levels, it can lead to a significant reduction in the supply of oxygen to the heart, particularly in people suffering from heart disease.
Carbon dioxide	Vehicle exhausts, power stations, other industrial processes, and domestic heating.	No direct health effects. It's an important 'green-house gas', which contributes to global warming. Therefore the effects go beyond urban areas.
Nitrogen dioxide (one of a group of gases called nitrogen oxides)	The majority of nitrogen oxides emitted from a vehicle exhaust are in the form of nitric oxide, which is not harmful to health. However, this gas can react with other gases to form nitrogen dioxide. Power stations also release nitrogen oxides.	At very high levels, nitrogen dioxide gas irritates and inflames the airways of the lungs. This irritation causes worsening symptoms for those with lung or respiratory diseases. It's an important component in the formation of ozone (see below).
Ground-level ozone	Formed by a complex set of reactions involving nitrogen oxides and hydrocarbons (largely from vehicle exhausts) in the presence of sunlight. It's a particular problem in big cities with sunny climates.	While the naturally occurring ozone in the upper atmosphere (the ozone layer) protects the Earth, ground-level ozone is harmful to health. Ozone causes **photochemical smog**. Although often invisible, this can be extremely harmful – leading to irritations of the respiratory tract and eyes.
Particulate matter	Sand, sea spray, construction dust, soot from open fires, and fumes from diesel vehicles.	This is less of a problem for MEDCs today. But **smog** (fog + smoke) caused many deaths from respiratory diseases before the introduction of Clean Air Acts.
Sulfur dioxide	Coal- and oil-fired power stations (65% of the total emissions) and vehicle exhausts. The highest levels of sulfur dioxide are recorded in areas where coal is used extensively.	Short-term exposure to high levels of sulfur dioxide can cause coughing, tightening of the chest and irritation of the lungs. It increases the acidity of rain, which can be an issue outside urban areas.
Hydrocarbons (Including benzene)	Vehicle exhausts. Levels are therefore highest close to busy roads or in the vicinity of petrol filling stations. Levels are strictly monitored in MEDCs today.	Contributes to the formation of ground-level ozone. Long-term exposure to high levels of benzene and 1,3 butadiene has been linked to leukaemia and cancer.
Lead	Exhaust gases from leaded petrol.	Lead can harm the kidneys, liver, nervous system and other organs.

Table 2.7 The main air pollutants, where they come from, and the problems they cause

Water pollution

This is a particular problem in rivers that flow through urban areas, and also in the groundwater that might be used to supply drinking water from wells and boreholes.

Raw sewage is a health issue in both LEDCs and MEDCs. It can occur in areas such as shanty settlements, where there is no sanitation system at all and raw sewage is simply dumped or left in open drains to be collected by the 'night soil men'. In other areas, sanitation systems are in place but the sewage is not treated and is simply emptied straight into rivers or the sea. This leads to contamination of water supplies. It causes a variety of diseases, such as diarrhoea and dysentery, which can lead to death. The solution is for proper sewage pipes to be installed and for the sewage to be treated to make it safe before it's released into rivers.

Liquid industrial waste emptied into rivers from factories is also an issue for urban areas. In the 1950s, many of the urban rivers of Britain and Europe (e.g. the River Thames) were considered to be 'dead', because fish were unable to survive there. However, the strict regulation of industry has now improved this situation greatly and fish have returned to many rivers.

In some LEDCs solid domestic waste is also dumped into rivers.

Visual pollution

This means things in the urban area that look ugly or even offensive. There is a saying that 'beauty is in the eye of the beholder' - in other words different people have different ideas about what is and what is not beautiful. Do you think that the graffiti in Fig. 2.45 is ugly and should be considered visual pollution?

Noise pollution

Sources of noise pollution in urban areas include cars and lorries, trains, aircraft taking off and landing, factories, large congregations of people (e.g. football crowds), late-night noise from bars and nightclubs, and noise in residential areas (e.g. from house parties). The amount of noise pollution that each of these sources produces depends on their location in the urban area, the time of day and the day of the week. In some cases, noise can be seriously disturbing - especially to people trying to sleep. Noise pollution problems tend to be worst where cities have developed rapidly without proper planning (e.g. with industrial and residential areas right next to each other).

RESEARCH In groups, look at all the photographs in Chapter 2. Which ones do you think show visual pollution? You will not reach complete agreement, but draw up a list of what you think is visual pollution. Are there other things that should be included?

The problems of noise pollution have been addressed in some countries by:

- introducing laws which limit the noise from factories, bars, and homes
- separating noisy industrial areas from residential areas
- building solid fences along motorways and major roads to reduce the amount of traffic noise reaching residents
- restricting night flights from airports.

In many densely populated countries, there is simply nowhere you can go to experience complete silence.

Fig. 2.45 Graffiti

Discussion point

This chapter has described air, water, noise, and visual pollution. Which, if any, of these affect the area in which you live and in what ways?

CASE STUDY

Mumbai, India – the effects of urbanisation on the environment

Fig. 2.46 In Mumbai, rich and poor live side by side with very little spare space

Fig. 2.47 Dharavi, Mumbai

Sunita's story

I'm Sunita and this is my story. Two years ago, my parents brought my brother Rakesh and me to live in Mumbai – in an area called Dharavi. People say it's a slum. Maybe it is – we're all poor here – but my father says at least we have work. And one day maybe Rakesh or I will be rich.

Dharavi is very crowded and very noisy. Everyone is busy all the time. Just outside our house people wash laundry, sew clothes and bang the dents out of used oilcans, so they can be recycled. There are small workshops everywhere. Somebody told me there are 15 000 in Dharavi.

There are lots of open sewers. I like to walk down to the biscuit factory where it smells nicer!

I go to school every morning. The lessons are literacy and maths. In the afternoon I help my mother clean our house. It only has two rooms, but we do have electricity. Afterwards, I often go rag picking with my friends to earn a little money.

The informal economy

Many people in squatter settlements think that if you're old enough to walk and carry a bucket, you're old enough to work and earn your keep. Many people work for themselves in the **informal sector**.

People who work in the informal sector don't do a job that earns a regular wage. They make and sell goods and services unofficially – often on a 'cash-in-hand basis'. They don't have a contract, so there's no job security. They also don't have any health-and-safety protection, health insurance, or pension scheme to fall back on. If they can't work they don't earn anything. But they don't pay any taxes either!

Living and working in Dharavi

Dharavi lies between two railway lines in Mumbai. A lot of the homes there are pretty solid – made from brick, wood, and steel. And a lot of them have electricity (like Sunita's). Although people live there illegally, Dharavi has well-established communities that provide self-help clinics, food halls and meeting places – as well as thousands of small workshops like those in Fig. 2.48 and Fig. 2.49.

But average incomes in Dharavi are low. Rakesh Pol, a leather worker, earns about £40 a month. He can rent a room for about £12 a month. Gradually, families can acquire extra building materials to improve their homes, but few of them can afford to move out of Dharavi, because the rest of Mumbai is far too expensive.

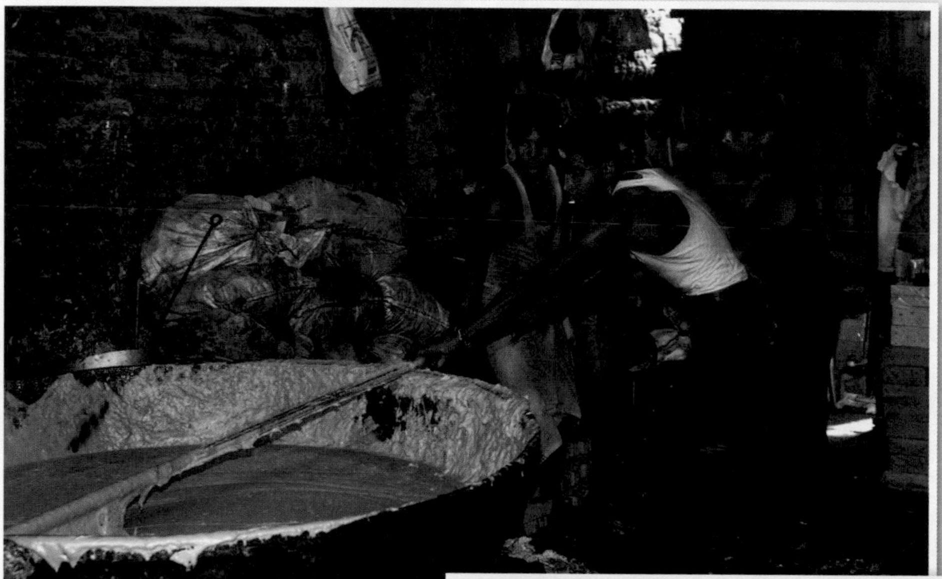

Fig. 2.48 Recycling soap in Dharavi

Fig. 2.49 One of the many small potteries operating in Dharavi

Improving squatter settlements

Around the world, people are trying to improve their quality of life. If you live in a squatter settlement, improving your home is a good place to start. Ways of doing this might involve the local authorities, or the residents themselves – or both.

The Dharavi Redevelopment Project

The city authorities in Mumbai have a big plan, called *Vision Mumbai*. Part of this plan is to try to tackle the poor quality of life of many Mumbai residents. Over the years, Mumbai's slums have multiplied and grown out of control, and pollution and water problems have rocketed. This situation has to be dealt with if Mumbai is to become more prosperous and successful in the future.

Dharavi's buildings might be poor quality, but the land they're built on is worth a fortune – US$10 billion! As part of *Vision Mumbai*, the plan is to demolish Dharavi's existing buildings and sell the land to property developers. As part of the deal, these developers will have to use some of the cleared land to build better homes for Dharavi's current residents. Up to 1.1 million low-cost, but higher-quality, homes could be built (many of them in high-rise tower blocks to fit in more homes in a smaller land area). This should cut the number of Mumbai residents living in slum housing by 90%. The water supply, sanitation, education, and health care would all be improved too.

But what's in it for the property developers? *Vision Mumbai* has encouraged the developers to get involved by offering them the land for less money than it's worth. Plus, as well as building high-rise tower blocks for Dharavi's existing residents, the developers will be able to use the land area saved by building upwards to build profitable shopping malls, office blocks and upmarket apartments for sale and rent to Mumbai's richer residents.

So, everyone's a winner? Not quite. Remember Sunita at the beginning of this case study? She says, '*Vision Mumbai* has a problem. What happens to the people who live in Dharavi now, while their new homes are being built? Where do they go? Some people have already been forced to leave their homes, so that they can be demolished to make way for the new buildings.' And Dharavi doesn't just provide homes – it provides jobs too. Where will all the small workshops and businesses, like those shown on page 70, go when the area has been redeveloped?

In 2009 it was announced that the plans to 'make over' Dharavi would be delayed because of the global economic crisis. Some of the organisations that had signed up for the Dharavi project had also dropped out. This may mean that Sunita will keep her home, at least for now, but that life won't get much better.

As you can see from the pictures of Mumbai, one of the problems facing this and many other shanty areas of the world is the size of the problem. It is very difficult to improve such large areas of settlement where so many people live. Even where improvements are going on, it is often difficult to make a difference, as more migrants move into the area.

Fig. 2.50 *Vision Mumbai* plans to replace Dharavi with buildings more like those in the background

Water pollution

Rapid urbanisation and industrialisation in poorer parts of the world have created big problems for the environment there. For example, getting rid of waste (of all kinds) has led to serious pollution in the Mithi River, which flows through Mumbai. For a long time, this river has been treated as a watery waste disposal unit – leading to pollution from a number of different sources:

- Big industries in Mumbai dump their untreated industrial waste straight into the river.
- The airport uses it to dump untreated oil.
- Every day, 800 million litres of untreated sewage go straight into the river.
- It's also used for dumping other waste – including food and cattle slurry, metals and old batteries – some of which is very toxic.

And in Dharavi, which sits right next to the river – apart from dumping human waste – the river is also used for things like washing out used oil drums.

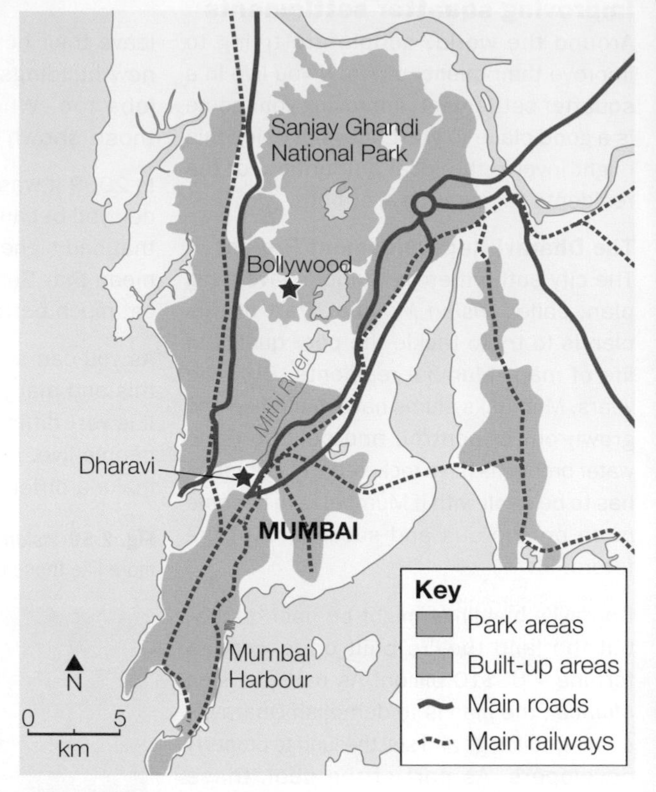

Fig. 2.51 Mumbai and the Mithi River

Flood risk

The solid waste dumped in the Mithi River (the metals and plastics) clogs it up and blocks the drains. Plants then grow on some of this waste, which helps to increase the risk of flooding.

In July 2005, the Mithi River flooded after a metre of rain fell in just 24 hours. Nearly a quarter of Mumbai was flooded. Roads and railway lines were under water for more than 24 hours. The airport was closed and many areas had no electricity for several days. People had to wade through water that was sometimes neck deep. The floods cost the city US$100 million, and 406 people died.

Fig. 2.52 The Mithi River is heavily polluted

What's being done? After the 2005 flood, the Mithi River Project was set up to try to prevent such a serious flood happening again:

- The river channel was dredged to make it deeper and increase its capacity to hold more water. It was also widened and obstacles were removed – and the banks were smoothed near bends in the river. All of this was designed to allow the water to flow more easily down to the sea.
- But none of those actions made the river any cleaner, so waste discharges from factories are now checked. More public toilets have also been built, to reduce the amount of raw sewage being dumped in the river.

Fig. 2.53 Dredging along the Mithi River helps the water to flow faster to the sea (along with all its waste)

Part of *Vision Mumbai* involves rebuilding homes in Dharavi and improving the area's water supply, sanitation, and drains. That should mean that less untreated sewage, ends up in the river.

Dharavi's workshops are also a source of pollution. But many are recycling materials that would otherwise be thrown away and add to Mumbai's waste disposal problem.

- Keeping the workshops going – but in a more environmentally friendly way – will therefore help to reduce the overall amount of waste.
- Education projects are also needed to help people understand why they shouldn't dump rubbish straight in the river.

Air pollution

Air pollution is a major problem in Mumbai. Exhaust gases from vehicles, and smoke from burning rubbish and factory chimneys, pollute the air. And, as the Indian economy grows, more and more electricity is needed – most of which is generated by burning fossil fuels like coal. As a result, large amounts of greenhouse gases, including carbon dioxide, are being released into the air.

Mumbai's residents, especially those who live in squatter settlements like Dharavi, suffer from very high rates of breathing problems. Illnesses like bronchitis are common.

Fig. 2.54 Serious air pollution in Mumbai

What's being done? Mumbai has concentrated its efforts to cut air pollution on transport. Vehicle exhausts are the biggest single source of air pollution there.

- A new metro system in the city aims to encourage people to use more public transport. By 2021, the planned metro system should have nine lines, and 32.5 of its 146.5 km of track will be underground.
- The city has also banned diesel as a fuel in all of its taxis. Many of Mumbai's 58 000 taxis now use compressed natural gas instead, which reduces greenhouse gas emissions.
- The main roads in and out of the city have been upgraded with 55 new flyovers. Smoother-flowing traffic should mean less congestion and less pollution.

Urban sprawl

Urban sprawl is the spreading outwards of a city and its suburbs – leading to changes in the surrounding rural area. It occurs in all areas of the world, but is particularly noticeable in countries like the USA, Canada and Australia – where urban areas tend to have low-density suburbs (single- or two-storey houses with large gardens).

In those countries, urban sprawl has caused problems such as:

- → high car dependence and increased vehicle emissions
- → inadequate facilities within the spreading suburbs, e.g. entertainment, shops, doctors, transport
- → higher costs to provide social facilities
- → high costs for public transport
- → lost work time spent commuting, and lower productivity
- → high levels of racial and socio-economic segregation
- → changing the character of the countryside and the loss of the rural way of life.

Fig. 2.55 Denver in the USA. Can you pick out the CBD, high-density inner suburbs and low-density outer suburbs?

In Europe, planning regulations have reduced urban sprawl. For example, in the UK, greenbelts (where development is restricted) were established around many cities. This was to protect the countryside and agriculture from urban sprawl and force developers to use brownfield sites (previously developed sites in urban areas) instead. Nevertheless many rural areas and villages have become urbanised and turned into dormitory villages for long-distance commuters.

In Europe, it has become common to refer to the **rural-urban fringe** (the transition zone where urban and rural land uses are mixed). The rural-urban fringe is characterised by agricultural land use alongside other types of land use linked to urban areas, for example:

- → roads, especially motorways and bypasses
- → recycling facilities and landfill waste sites
- → park-and-ride sites
- → airports
- → hospitals
- → sewage facilities
- → large out-of-town shopping facilities, e.g. large supermarkets
- → golf courses
- → parks and nature reserves.

Despite these urban uses, the fringe remains largely open, with the majority of the land used for agriculture or woodland.

CASE STUDY

New York, USA – an urban settlement

The area of New York was explored by the Dutch in 1609, and was originally called New Amsterdam. Two or three years after the first visit by Captain Henry Hudson in 1610, the Dutch occupied the southern end of Manhattan Island. The growth of the city was rapid. The population rose to 21 700 in 1700, 60 489 in 1800, 123 706 in 1820 – and 20 million today. By the 1950s, it was the most populous city in the world – although it is expected to fall to ninth place by 2020.

Fig. 2.56 The Statue of Liberty at the entrance to New York Harbour

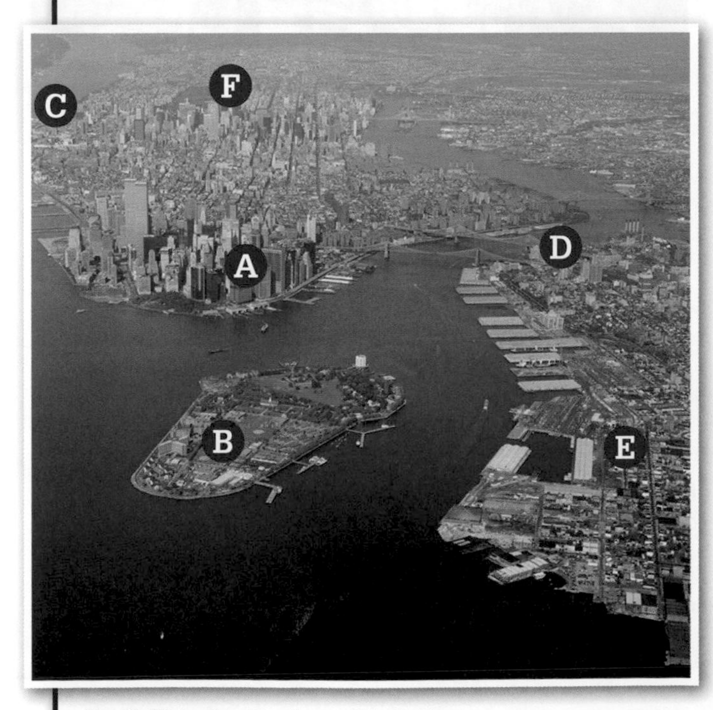

Fig. 2.57 Looking north over Manhattan

New York had great natural advantages that led to its growth.

Site

- The sheltered, natural harbour formed by the Hudson and East Rivers provided a safe, deep anchorage. There was an extensive waterfront for the development of docks.
- The Island of Manhattan was a rocky ridge extending north and south. The southern part was easy to defend and was the site of the first settlement.
- The country to the north was able to be cultivated.
- The solid bedrock which formed the ridge later provided good foundations when congestion on the island led to the development of skyscrapers.

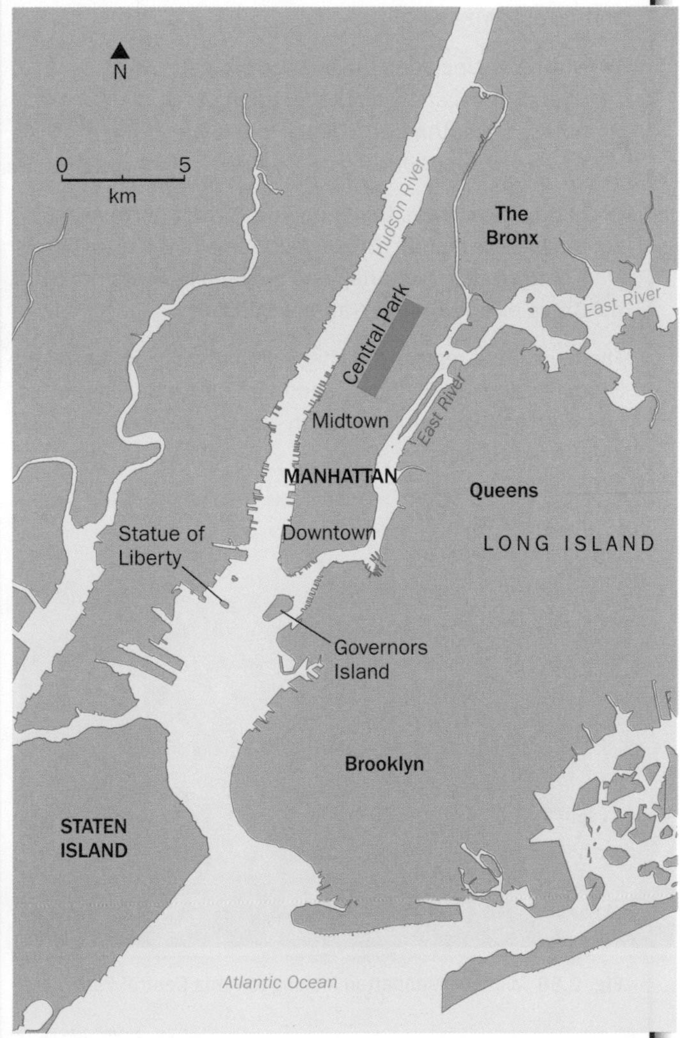

Fig. 2.58 The site of New York

16 Using Fig. 2.58, identify the areas shown by the letters on Fig. 2.57.

Situation

New York is a natural route centre. The navigable Hudson River provides communication with a large inland area. This was extended by the building of the Erie Canal (completed in 1825), which links the Hudson with the Great Lakes. Shortly afterwards, the construction of railways into the continent began. There are also the coastal routes to New England in the north and Philadelphia in the south.

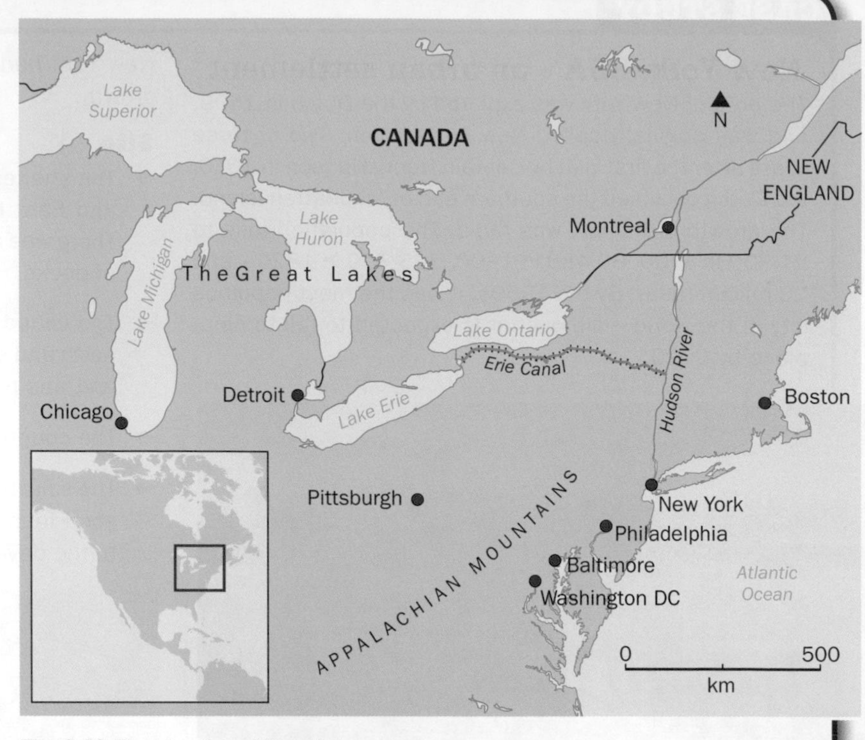

Fig. 2.59 The situation of New York

Urban structure

New York is made up of five boroughs: Manhattan, Brooklyn, Queens, the Bronx, and Staten Island. Brooklyn and Queens are on the western end of Long Island.

The CBD has two zones – Midtown Manhattan and Downtown Manhattan.

- Downtown includes the finance district (Wall Street) and was the location of – until 11 September 2001 – the Twin Towers of the World Trade Center.
- Midtown includes the shopping district (Fifth Avenue), the theatre district (Broadway), some of the main hotels (Plaza, Waldorf Astoria), and the Empire State Building, Chrysler and United Nations Buildings.

There are other regional business districts, e.g. in downtown Brooklyn and local ribbon shopping centres. Manhattan has a grid layout with 12 north–south Avenues and 219 east–west Streets.

Fig. 2.60 Midtown Manhattan looking towards Central Park

Fig. 2.61 Manhattan has a few older buildings. This is Macy's department store, built in 1902

Fig. 2.62 The docks and waterfront of Manhattan, with the Empire State Building in the background

Fig. 2.63 Central Park

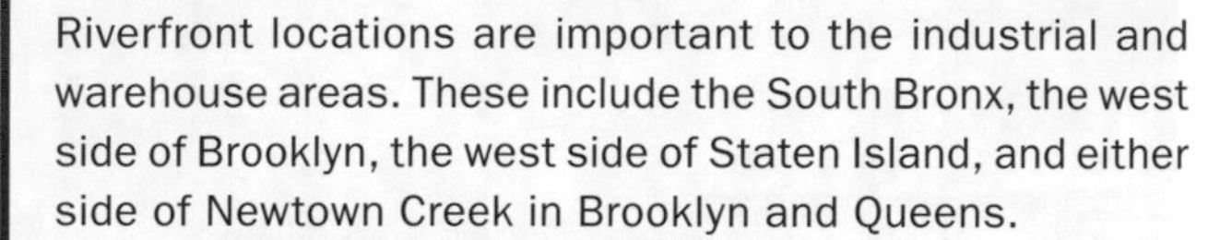

Riverfront locations are important to the industrial and warehouse areas. These include the South Bronx, the west side of Brooklyn, the west side of Staten Island, and either side of Newtown Creek in Brooklyn and Queens.

The highest-density housing – consisting of apartment blocks – occurs in Manhattan. Other high-density housing is found in areas of The Bronx, Brooklyn and Queens. Low-density housing occurs mostly in Staten Island, eastern Queens, southern Brooklyn and north–western and eastern areas of The Bronx.

Public open spaces make up 25% of the area. This is greatest in Staten Island and least in Manhattan – although New York's famous Central Park is in Manhattan.

Problems and solutions

Urban sprawl

There is a movement of the middle classes to the outer areas, and a movement of lower-income families to the inner city. As a result, the suburbs are expanding into the surrounding countryside. As well as residential land use, employment is also developing in the suburban areas. There has been growth outside the city in mid-Hudson Valley areas like Orange County. From 1990 to 2000, Staten Island (Richmond County) was the fastest-growing county in New York State. Over that decade, the borough grew by 65 000 people and added 24 000 new housing units – an increase of approximately 17%.

There are a number of reasons for this:

- The overall population is growing. The US Census Bureau reported that between 2000 and 2009 the population of New York increased by 383 195 people.
- People, especially the affluent middle classes, often leave the inner-city areas for a better lifestyle in the outer suburbs.
- Businesses have relocated to the suburbs because of cheaper land and better accessibility. In wealthy suburbs, like Westchester County, most people no longer commute to the city for work.
- Some suburban growth north of New York City has been attributed to worried city residents moving after the 11 September 2001 terrorist attacks.
- House hunters have been priced out of more expensive areas closer to the city.

In addition to the problems listed on pages 78–80, there are concerns about the new housing being too dense and not in character with the existing housing. This has led to the existing residents feeling that their quality of life has deteriorated.

Poverty and unemployment

Almost a million New Yorkers receive welfare payments. The situation is most serious in the inner areas. For example, in South Bronx about a third of the people receive welfare support, and the population has declined by about 50%. Many people in these areas are unemployed, underprivileged and poorly educated – which causes strains in society. The decline – in the 1980s – in the clothing industry and industries based around the port led to job losses. These lost jobs have not been replaced.

Urban decay and housing problems

In inner-city areas, like Harlem and the South Bronx, there are apartment blocks that have become empty, are in a poor state of repair, or are lacking modern amenities. Some schemes have now been introduced to tidy up these derelict houses. This creates jobs and also helps to reduce the number of homeless people.

Racial inequality and tension

The US population includes people whose families have originated from all over the world (for example, Chinatown and Little Italy are famous areas in Manhattan). At first, immigrants tend to occupy the poorest inner-city areas. Then, in time, those who become wealthy tend to move to the outer suburbs. However, many immigrants become trapped in poverty in the inner city and ghettos develop.

High cost of land

The high cost of land in Manhattan – which led to the development of its famous skyscrapers – also forced many smaller businesses to move to the outskirts, where land is cheaper. Land prices in Manhattan remain high.

Crime

In the 1980s, violent crime such as muggings and murder led to parts of New York becoming very dangerous. The Subway (underground railway) had a particularly bad reputation. Rudolf Giuliani, the Mayor of New York from 1994 to 2002, became famous for his 'zero tolerance' of crime. He increased the number of police on the streets, and minor offences and anti-social behaviour were not overlooked.

Fig. 2.64 Rundown apartments and a discarded fridge in the South Bronx

Fig. 2.65 A crowded street scene

Fig. 2.66 The New York Subway

Air pollution

The large number of vehicles in New York is the main source of the city's air pollution. The air contains high levels of ozone and particulates, and residents in some neighbourhoods have very high rates of asthma and respiratory conditions – especially in Manhattan.

Despite this, a greater use of public transport than in other areas of the USA means that greenhouse gas emissions in New York are lower than the national average. The city accounts for only 1% of the USA's greenhouse gas emissions, despite being home to 2.7% of its population.

New York has made efforts to reduce its particle pollution by adopting measures like:

- → fitting catalytic converters to the exhausts of diesel city buses
- → developing a biodiesel processing plant in Brooklyn to distribute biodiesel to filling stations in the city.

The city's government is also required to purchase only the most energy-efficient equipment for use in city offices and public housing. New York is also a leader in the construction of energy-efficient 'green' office buildings, including the Hearst Tower.

Water pollution

The Greenpoint neighbourhood of Brooklyn was once the location of many oil refineries. In 1950, the oil company which later became Exxon Mobil was alleged to have spilled a large quantity of oil into Newtown Creek. Although the oil industry has now moved elsewhere, oil still seeps into the creek from several on-land spills. This has damaged wildlife. The oil is now being pumped from the ground to remove it. In June 2006, the New York State Department of Environmental Conservation announced that it would sue Exxon Mobil to speed up the completion of the cleanup.

Visual pollution

Unattractive elements of a vista such as graffiti and derelict housing may be considered to be visual pollution.

Energy supply

Although the average New Yorker consumes less than half the electricity used by a resident of San Francisco, New York faces growing energy demands. This has led to a number of 'green projects', which include:

- → switching more than 11 000 traffic lights and 'Don't Walk' signals to energy-efficient light-emitting diodes (that use 90% less energy)
- → replacing 'cobra head' street lights with new energy-efficient types
- → powering the Statue of Liberty, Ellis Island and 22 other federal buildings in New York City using wind power
- → installing underwater turbines in the East River to take advantage of tidal currents
- → constructing windmills on a hill in the former Fresh Kills Landfill to power 5000 homes on Staten Island
- → introducing tax advantages for builders of energy-efficient buildings.

Fig. 2.67 Rush hour in Times Square

Water supply

Providing one of the world's biggest cities with water is not easy. New York City is supplied with drinking water from the Catskill Mountains, up to 200 km away. The area is in one of the largest protected wildernesses in the USA. The water from the Catskills is naturally filtered and pure enough to require only the addition of chlorine to ensure its purity for drinking. The water travels to the city downhill along aqueducts and does not require pumping. New York is now using 28% less water than it did in 1979. This is mostly the result of installing water-saving plumbing fittings, finding and fixing leaks in water supply pipes, and metering residential customers' water use.

Fig. 2.68 The Verrazano Narrows Bridge between Brooklyn and Staten Island

Waste disposal

New York's households produce 12 000 tonnes of waste a day. Until 2001, the refuse was disposed of at the Fresh Kills Landfill on Staten Island. After this closed, most of the city's waste was taken out of the city to landfill sites in other states. This led to a large amount of lorry traffic through low-income neighbourhoods. There is now a solid waste management plan, which uses barges and trains to export 90% of the city's waste.

Traffic congestion

Even though more than half of the households in New York do not own a car (the figure is even higher in Manhattan), its streets do become blocked with traffic – including yellow taxis. The daily commuter movement into and out of Manhattan is about 2 million people, which is similar to that of London. People travel by car, bus and subway. The bridges and tunnels linking the various New York islands also produce bottlenecks that add to the problem.

17 **a** Copy and complete Table 2.8 to summarise the urban problems of New York. You will not be able to find solutions to all the problems in the text.

b Which of the problems are environmental?

Discussion point

Do you think New York would be a good place to be a teenager? What would be the best and worst things?

Problem	Details	Solution
Urban sprawl		
Poverty and unemployment		
Urban decay and housing problems		
Racial tension		
High cost of land		
Crime		
Air pollution		
Water pollution		
Visual pollution		
Energy supply		
Traffic congestion		
Water supply		
Waste disposal		

Table 2.8 The urban problems of New York

CASE STUDY

Cape Town, South Africa – an urban settlement

Cape Town is a city of great contrasts:

- → Its CBD is modern, successful, and attracts tourists, but on its outskirts are some of the biggest shanty settlements in the world.
- → Its population is about 3.7 million, and consists of many races. However, the population was shaped by apartheid – the political system adopted in South Africa between 1950 and 1991. **Apartheid** segregated the races and discriminated in favour of the white population. There are still great social and economic differences between the white and black populations, despite the fact that apartheid came to an end in 1991.

Site

Cape Town developed because of its strategic location at the southern tip of Africa. The city originated around the sheltered natural harbour of Table Bay, which helped its development as a port.

Fig. 2.69 The sheltered harbour of Cape Town

The existing flat land next to the sea was extended by the original Dutch settlers through land reclamation, as they had done in the Netherlands.

Cape Town is built around Table Mountain, which is 1086 metres high. There are buildings on the gentle lower slopes but not on the steep and windy upper slopes.

Fig. 2.70 The lower slopes of Table Mountain are settled but not the upper slopes

Situation

Cape Town was established at a critical point on the shipping routes between Europe and Asia. Whoever controlled the Cape controlled these routes. The Dutch ruled Kaapstad, as it was then known, for 150 years before the British took over in 1814. There is still a naval base at Simon's Town, just outside Cape Town.

Gold and diamonds were discovered in the hinterland in the centre of South Africa in the 1870s and 1880s, and – as a major port – Cape Town benefited. That mineral wealth provided the basis for the development of the city.

There is also a rich agricultural hinterland, with wheat growing to the north, sheep farming in the Karoo area and vineyards (introduced by French settlers) immediately inland.

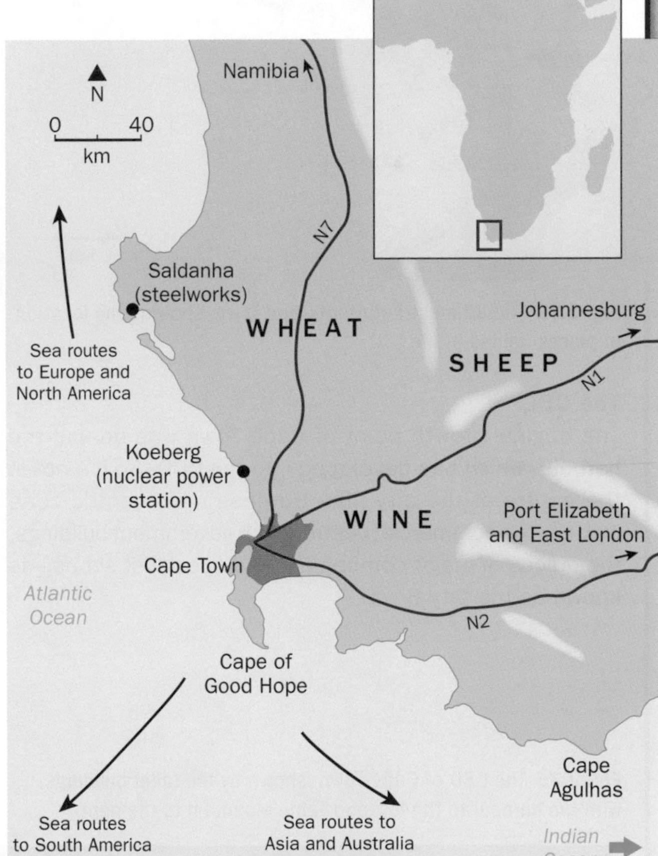

Fig. 2.71 The situation of Cape Town

LEARNING TIP Cape Town is a smaller city than New York or Mumbai. It provides a good example of how a city does and does not fit the models of urban structure described earlier in this chapter.

Urban structure

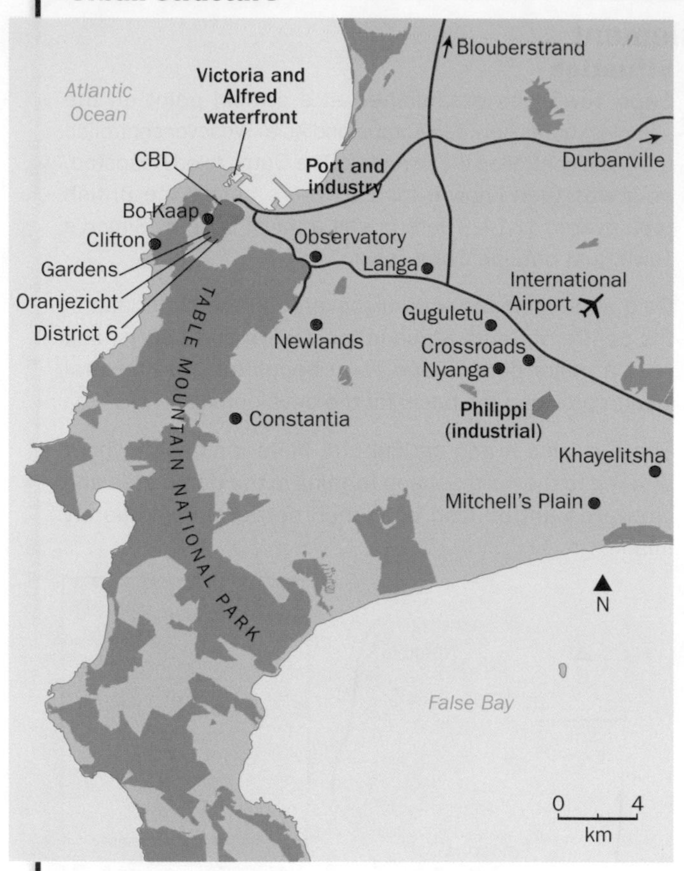

Fig. 2.72 The urban structure of Cape Town, showing the location of places named in the text

Fig. 2.73 The City Bowl

The CBD

The original growth point of Cape Town was around the harbour, which has developed into the CBD. So it's not in the centre of the city (like Burgess's concentric zone model). The commercial centre – with government buildings, the offices of major companies, and specialist shops – is known as the City Bowl.

The older part of the docks has been redeveloped through private finance. Warehouses have been converted into shops, restaurants, expensive loft-style apartments and hotels. This is the Victoria and Alfred Waterfront, which remains a working harbour for smaller boats.

The centre of Cape Town has not suffered the decline of other town and city centres, and property prices are booming. There are shopping districts and malls in the suburbs. The large Canal Walk Shopping Centre has been built on a brownfield site close to one of the main motorways that lead into the heart of the city.

Fig. 2.74 The Victoria and Alfred Waterfront, with Table Mountain in the background

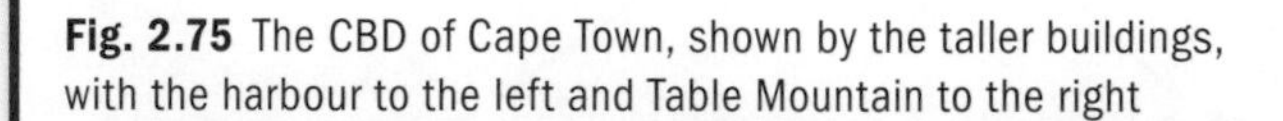

Fig. 2.75 The CBD of Cape Town, shown by the taller buildings, with the harbour to the left and Table Mountain to the right

Industry
There is little heavy industry within the city. The main iron and steel plant is at Saldanha Bay, 110 km to the north, and there is a nuclear power station at Koeberg, 30 km to the north. Other industry is close to the commercial port and at isolated locations in the suburbs, e.g. the Philippi industrial area. These may be closer to the areas of poorer housing.

Open space
There is a lot of open space within Cape Town's built-up area. Some of it is land that is too steep to build on, such as Table Mountain. The environment of this area – including its unique vegetation – is carefully protected. There are also well-managed public parks, like Company's Gardens in the City Bowl or Kirstenbosch Botanical Gardens in the inner suburbs. On the outskirts of the city in the Cape Flats area, close to the international airport, there is also a lot of open space.

District Six (Zonnebloem)
District Six is close to the City Bowl. It was originally a multiracial area, but in 1966 South Africa's apartheid government reclassified it as a whites-only district. Fifty thousand black residents were evicted to the Cape Flats area and their houses were demolished. The district remained empty for many years, but the land is now being handed back to the former residents under the District Six Beneficiaries Trust. Four thousand new homes are planned.

Inner suburbs
Many of the inner suburbs are affluent areas with luxurious houses. These include Tamboerskloof, Observatory, Gardens, Oranjezicht, Newlands and Constantia. Bo-Kaap is a more densely settled inner-city area, and home to Cape Town's Muslim community. There are also highly expensive suburbs along the coast to the south, such as Clifton.

Urban sprawl
Affluent areas have also developed on the outskirts of the city and beyond the built-up area. Examples are at Somerset West (40 km west of the CBD), Durbanville (to the north) and Bloubergstrand on the north coast. These areas have low-density housing with their own local shopping and business districts, such as Tyger Valley.

Migration to Cape Town and the development of squatter settlements
In Chapter 1, the push and pull factors of rural–urban population migration and its impacts in rural and urban areas were discussed. In this chapter, rural depopulation and urbanisation have already been discussed. How has this migration affected Cape Town?

About 15 to 20 km from the centre of Cape Town, on the western edge of the city next to the international airport, is the Cape Flats area, which is home to about a million people. It includes the townships of Khayelitsha, Crossroads, Mitchell's Plain, Guguletu, Nyanga, and Langa. These suburbs share the problems of other informal settlements described earlier in this chapter.

Fig. 2.76 District Six with the City Bowl beyond. The buildings in the foreground are some of the first new developments in the area

Rural–urban migration has led to people, mainly from the Eastern Cape (a province of South Africa), settling in the outskirts of Cape Town in search of work. People have also migrated to South Africa from other African countries, sometimes illegally, escaping wars or seeking a better standard of living. Some of these people have also ended up in Cape Town.

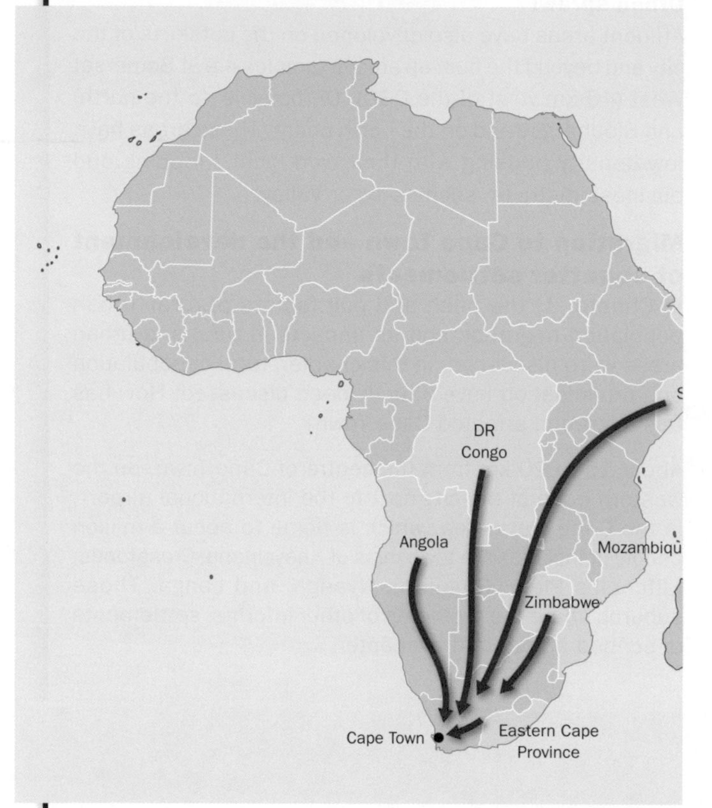

Fig. 2.77 Migration to South Africa and Cape Town

Fig. 2.78 Informal housing on the Cape Flats

When apartheid ended in 1994 there was a great influx of people in search of work and education. In 2011 around 62% of residents in the area of Khayelitsha were rural–urban migrants, most coming from the Eastern Cape.

Impact on the Eastern Cape

- Rural depopulation has occurred. Between 2001 and 2011 net migration was –278 000, the highest in South Africa.
- This has led to a shortage of labour, particularly in agriculture, as the migrants are usually of working age.
- The population has large numbers of young and elderly dependents.

Impact on the Cape Flats area of Cape Town

- Around 70% of the residents live in shacks, which are prone to fires.
- One in three people have to walk 200 m or further to access water.
- Only 53% of the working age population is employed.
- Almost 90% of households are either moderately or severely food insecure.
- Crime rates are high and often drug-related.
- With high unemployment levels among poorer South Africans, xenophobia is prevalent and many South Africans feel resentful of immigrants who are seen to be depriving the native population of jobs.

Changes in Cape Flats since 1994

The main improvements made to the area include:

- new brick housing
- new schools – there are now 37 primary schools and 21 secondary schools
- a new central business district in Khayelitsha, including the KCT mall, a 19 254m^2 shopping centre with 58 tenants
- social programmes such as AMANDLA EduFootball Safe-Hub, a youth-friendly space where young people can find safety, both physically and emotionally. This aims to prevent chronic youth-driven violence.

In addition, Metrorail now links six stations with the city centre. As in much of Africa, minibus taxis are important. Khayelitsha District Hospital opened in February 2012. Companies offer township tours, a form of social tourism.

3 Earthquakes and volcanoes

This chapter covers the following Cambridge IGCSE® and O Level topic:

- **2.1 Earthquakes and volcanoes**

- This photograph shows some of the devastation in the town of Ishinomaki, Japan, four days after the earthquake of 11 March 2011, which led to a powerful tsunami.
- People across the world were shocked to see images from Japan of the damage caused by this earthquake and tsunami.
- The pictures showed the power that exists inside the Earth. Earthquakes and volcanic eruptions can cost many thousands of lives and destroy whole towns and cities. Although we understand our planet much better than we did 50 years ago, we cannot control the forces beneath its surface.
- Why do these natural events occur?
- What causes them?
- Where do they occur and can we predict them?
- What can be done to limit the damage they cause?

In this chapter you will learn about:

- major features of the Earth's surface, such as mountain ranges, plates and plate margins, and volcanoes
- the major scientific breakthrough of plate tectonics, which changed our understanding of planet Earth
- volcanoes and their effects on human activity
- earthquakes and their effects on human activity.

Earth in motion

Earth is the name of the planet that we live on. Its surface constantly changes - sometimes leading to spectacular earthquakes and volcanic eruptions. But sometimes the changes occur very slowly over millions of years. The reason for this is that, deep within the Earth, heat is being produced by radioactivity. This makes the rocks very hot and, in some situations, they actually melt. This heat is the energy which produces volcanic eruptions and earthquakes, causes continents to move, and leads to the development of high mountain ranges.

Fold mountains

A glance at a world map, like Fig. 3.1, shows that the main mountain ranges are often long narrow belts, like the Andes in South America. Fold mountains form the highest of the world's mountain ranges. They are long, relatively narrow belts of mountains, with parallel ridges and valleys. The main range is made up of a series of ranges. Flatter areas form plateaux within the mountains. Active volcanoes form high conical mountains within the ranges. One of the world's highest volcanoes is Cotopaxi in the Andes. Its summit is 5897 m above sea level, 50 km from the capital of Ecuador, Quito. Fold mountain ranges, such as the Andes and the Alps, tend to be sparsely populated - but some communities are still found in these areas.

1 Name three ranges of fold mountains shown on Fig. 3.1.

Fig. 3.1 The global distribution of fold mountains

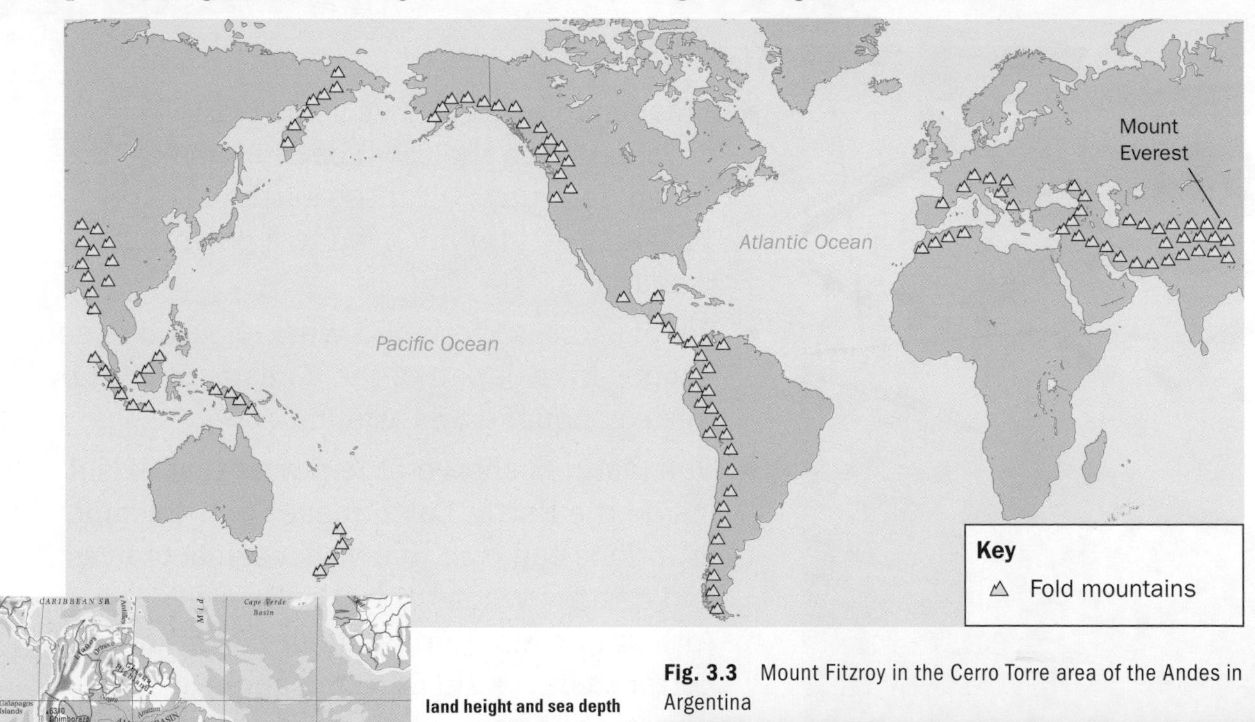

Fig. 3.2 The Andes in South America are an example of a fold mountain range

Fig. 3.3 Mount Fitzroy in the Cerro Torre area of the Andes in Argentina

Fold mountains form where the powerful **compression** of two tectonic plates (see page 91) squeezes the layers of rocks so that the **upfolds** form the ridges and the **downfolds** form the valleys. This could not happen close to the Earth's surface, because rock is too brittle and would snap. However, it does happen at great depths in the Earth, where the high temperatures and pressures make the rock behave as a plastic solid and allow it to be very slowly folded over long periods of time.

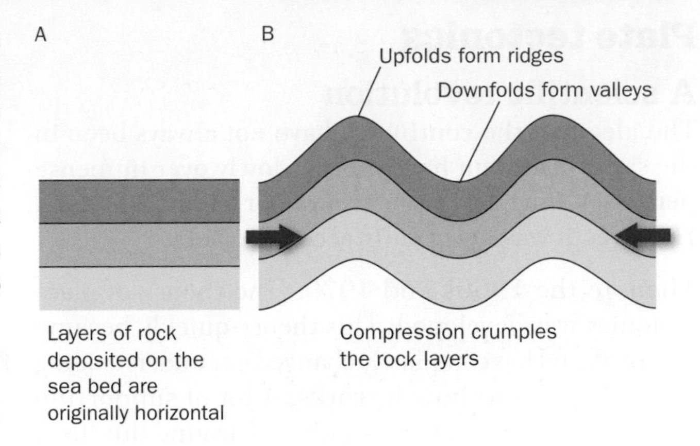

Fig. 3.4 The formation of fold mountains

Volcanoes and earthquakes

Fig. 3.5 shows the global distribution of **active volcanoes** (volcanoes which have erupted in the last 80 years). There are about 540 of these. Other volcanoes are described as **dormant** (resting, but which may erupt again in the future), or **extinct** (dead and will not erupt again). Notice how the active volcanoes form a series of relatively narrow belts around the planet. The Pacific Ring of Fire is the most famous of these belts, and it goes all the way round the rim of the Pacific Ocean. Fig. 3.6 shows the global distribution of earthquake activity. Earthquakes occur more frequently than volcanic eruptions.

2 Look at Fig. 3.5 and Fig. 3.6.

a What are the similarities between the two?

b List the areas that have earthquakes but not volcanoes.

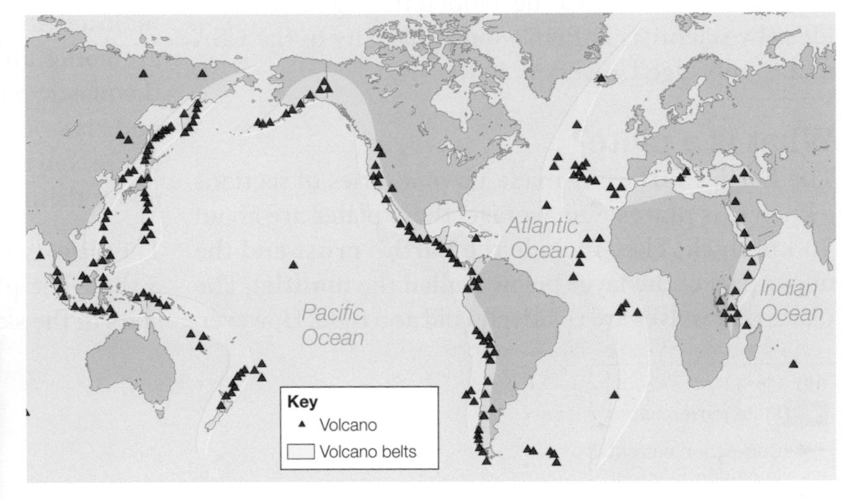

Fig. 3.5 The global distribution of active volcanoes

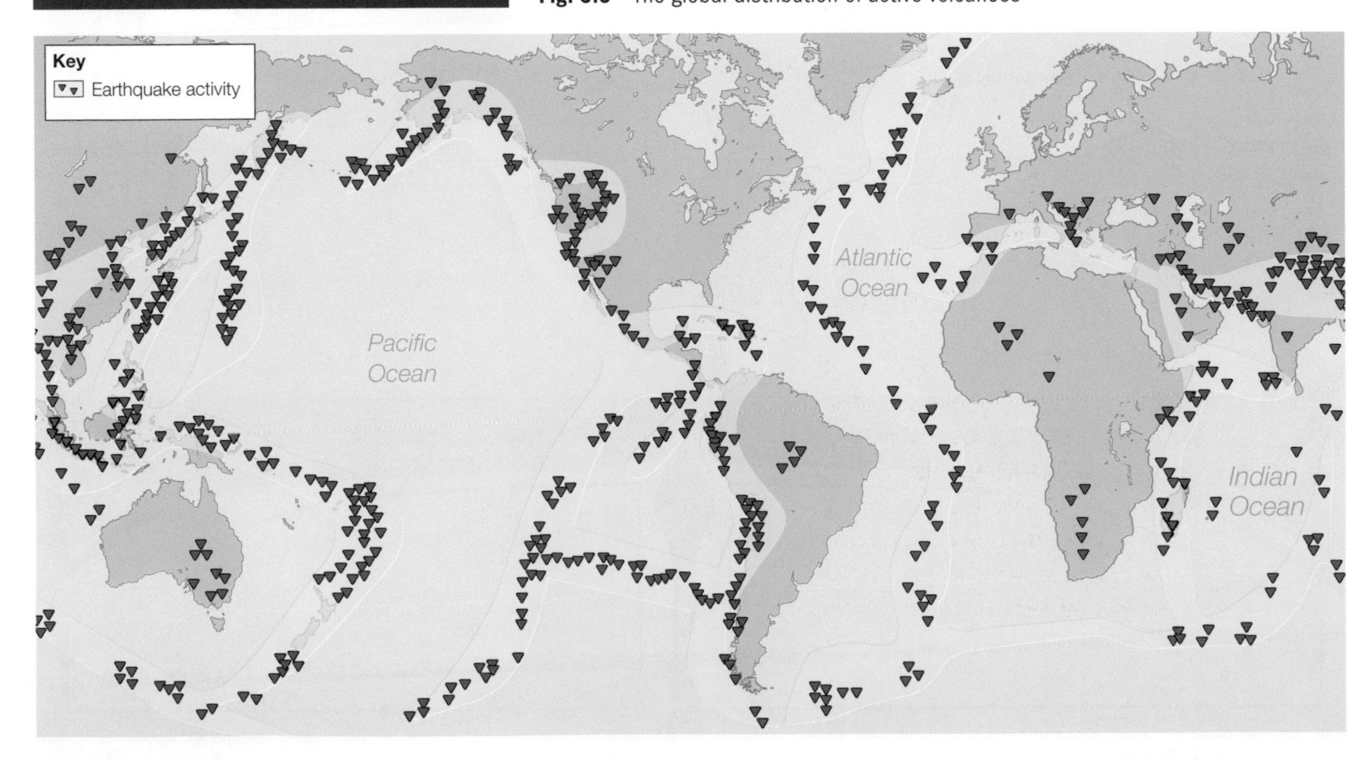

Fig. 3.6 Earthquake activity around the world

Plate tectonics

A scientific revolution

The idea that the continents have not always been in the same place, but have moved slowly over immense periods of time, has been around for about 100 years. However, it was never fully accepted.

Then, in the 1960s and 1970s, the theory of plate tectonics was developed. This theory quickly became accepted, and has completely changed our understanding of the Earth and how it works. A lot of supporting evidence for the theory was gathered around this time, as a result of the first surveys of the deep ocean floors. This evidence was very important in convincing other scientists about the new ideas.

No single scientist can claim to have invented the theory of plate tectonics. It was developed through a series of ideas by scientists at Princeton University in the USA and Cambridge University in the UK.

What is a plate?

The Earth's surface is made up of a series of sections - known as **plates**. On average, these plates are about 50 km thick. They include the Earth's **crust** and the upper part of the layer below (called the **mantle**). The plates themselves are relatively cold and rigid. However, the rocks underneath them reach temperatures of more than 1300 °C and behave plastically (in other words, they can flow rather like a jelly). The plates interlock - a bit like the panels that make up the outside of a football. However, unlike the panels on a football, the plates can move relative to each other - flowing over the hotter, more plastic, rocks below (that act like a lubricating layer).

LEARNING TIP Remember that 'the plates' and 'the crust' are not exactly the same thing. Make sure that you know the difference between the two.

Why do the plates move?

Deep within the Earth, heat is being produced by radioactivity. This heat is not evenly distributed, so there are hotter areas and colder areas. At the hotter areas, the plastic rocks in the Earth's mantle become lighter and rise - causing convection currents (see Fig. 3.7). These convection currents drag at the rigid plates sitting above them - causing them to move.

The plates typically move at rates of between 1 and 10 cm a year. The plate movements can be measured using stars in the sky as reference points.

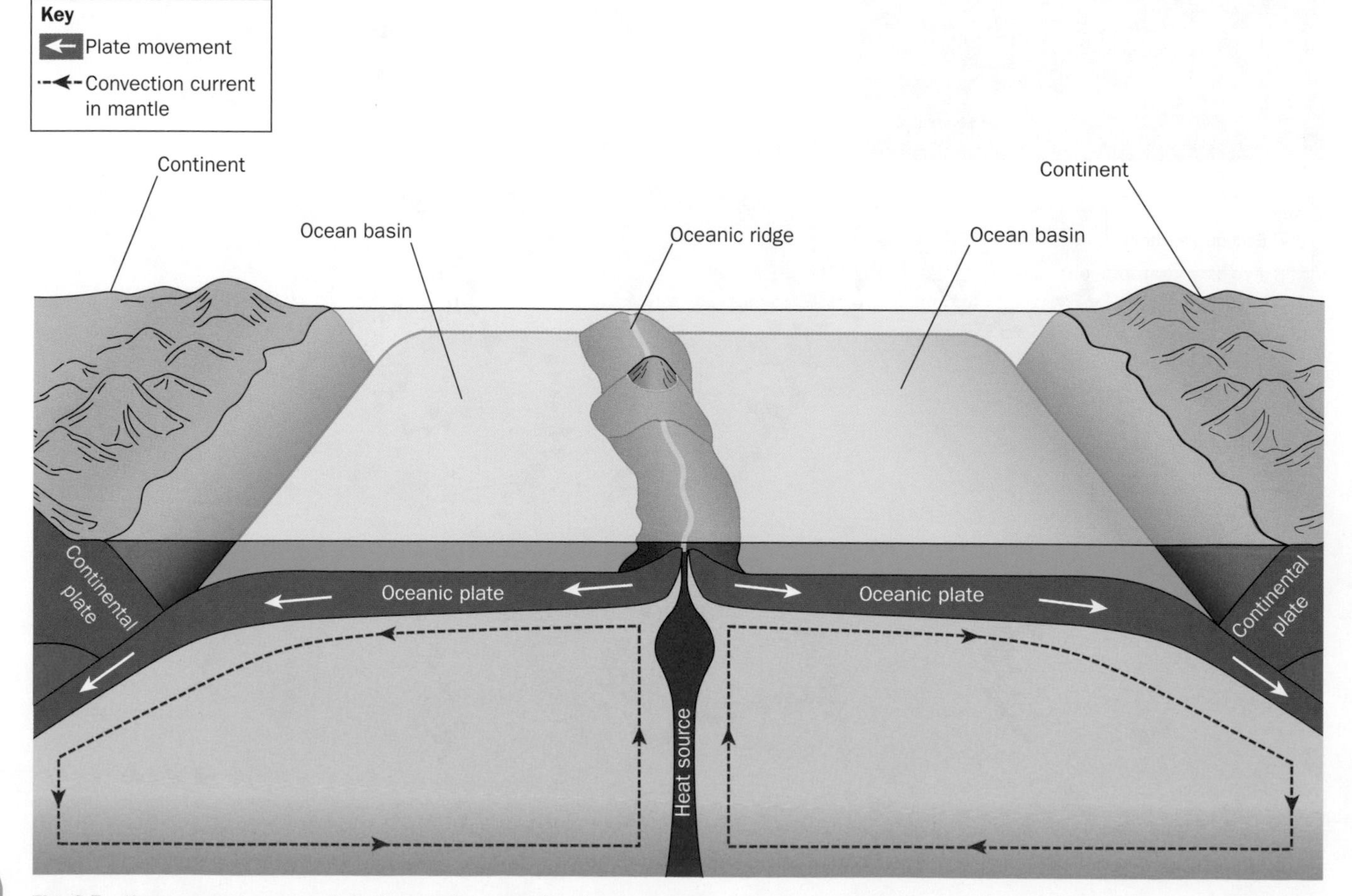

Fig. 3.7 How convection currents in the mantle cause the plates to move

Plate boundaries

Do you live on a plate margin (plate boundary)? If you do, you will certainly experience large earthquakes from time to time. You might also be on a plate boundary where there is volcanic activity. And you might live in, or close to, high fold mountains.

If you live closer to the centre of a plate, any earthquakes near your home will be relatively small. There will also be no active volcanoes, and any mountain ranges will be smaller.

Discussion point

Living near to a plate margin can be hazardous. There are examples in both MEDCs (e.g. New Zealand and the West Coast of the USA) and in LEDCs (e.g. the Philippines and Peru). Why do people live in these areas?

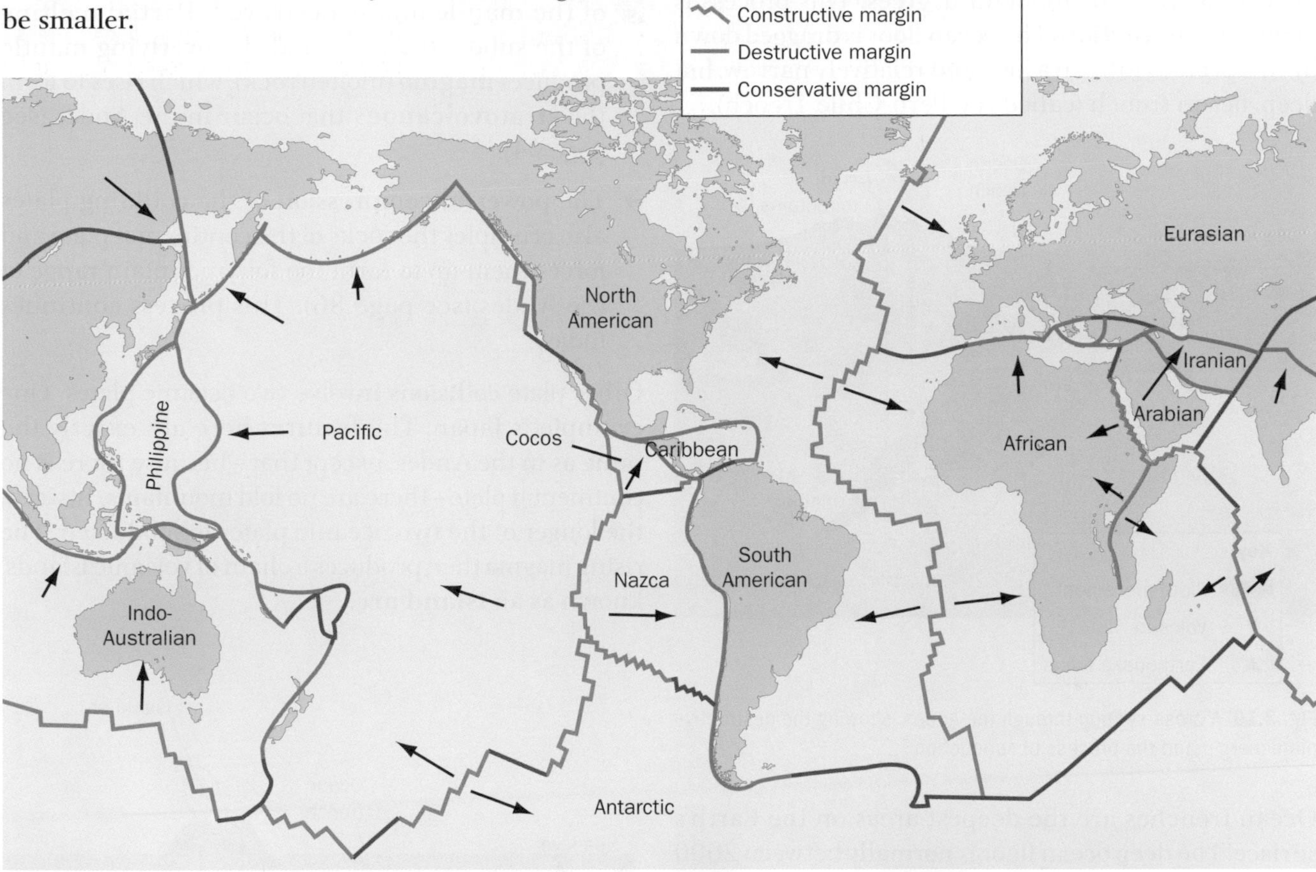

Fig. 3.8 Plates, plate boundaries and plate movements (the *Oxford International Students' Atlas* has a more detailed map on pages 8 and 9)

3 Look at Fig. 3.8.

- **a** Name a plate which includes both part of a continent and part of an ocean.
- **b** Name a plate which contains just ocean floor and no part of a continent. (There are no very large plates which are just made up of a continent.)
- **c** Name a place where two plates are moving towards each other.
- **d** Name a place where two plates are moving away from each other.

Fig. 3.9 The 2011 earthquake in Christchurch, New Zealand, brought down many buildings and killed 65 people

Destructive (convergent) plate margins

Destructive margins are places where plates move towards each other (collide). They are called destructive because oceanic plate is destroyed. A good example is along the west coast of South America, beneath the Andes (see Fig. 3.10). Here, an oceanic plate (the Nazca Plate) collides with a continental plate (the South American Plate). Because the oceanic plate is denser, it is forced beneath the less dense and more buoyant continental plate at an angle of about 45 degrees. This process is known as **subduction**. The ocean floor is dragged down by this process to form a long and relatively narrow, but deep, ocean trench (called the Peru-Chile Trench).

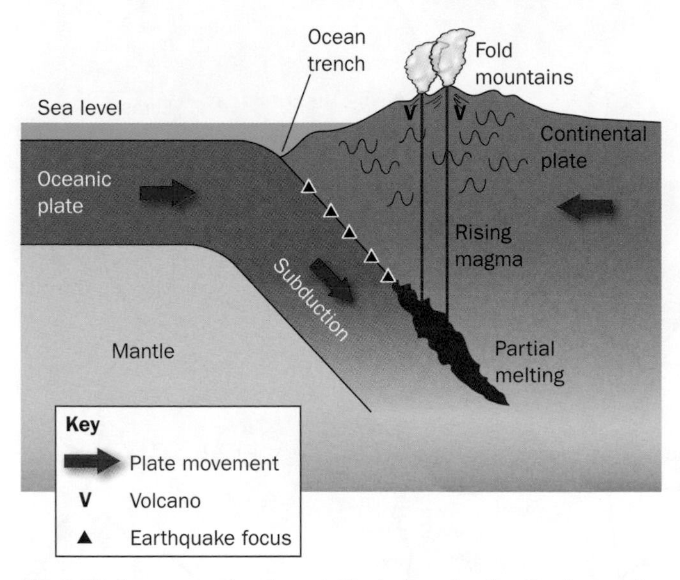

Fig. 3.10 A cross-section through the Andes, showing the destructive plate margin and the process of subduction

Ocean trenches are the deepest areas on the Earth's surface. The deep ocean floor is normally between 2000 and 5000 m below sea level, but the deepest point on the Earth's surface (the Mariana Trench in the Pacific Ocean) is 11 034 m deep. This is deeper than the highest mountain (Mount Everest) is high, as Fig. 3.11 shows.

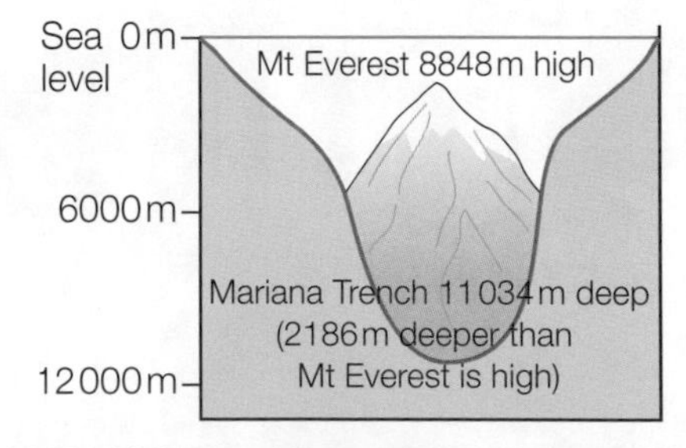

Fig. 3.11 The highest and lowest points on the surface of the Earth

→ As the oceanic plate is subducted beneath the continental plate, friction between the two plates causes earthquakes. The earthquakes get deeper and deeper the further down the subduction zone they are. Shallow earthquakes occur beneath the Peru-Chile Trench but deeper earthquakes occur below the Andes.

→ Eventually, the oceanic plate is forced so deep into the Earth that it is partially melted, becomes part of the mantle and is destroyed. Partial melting of the subducted plate and the overlying mantle produces **magma** (molten rock), which rises to form the **stratovolcanoes** that occur in the Andes (see page 94).

→ The powerful compression of the colliding plates also crumples the rocks of the continental plate and forces them up to form the fold mountain range of the Andes (see page 86). This process continues today.

Other plate collisions involve two oceanic plates. One example is Japan. The features here are exactly the same as in the Andes, except that - because there is no continental plate - there are no fold mountains. Instead, the longer of the two oceanic plates is subducted. The rising magma then produces a chain of volcanic islands, known as an **island arc**.

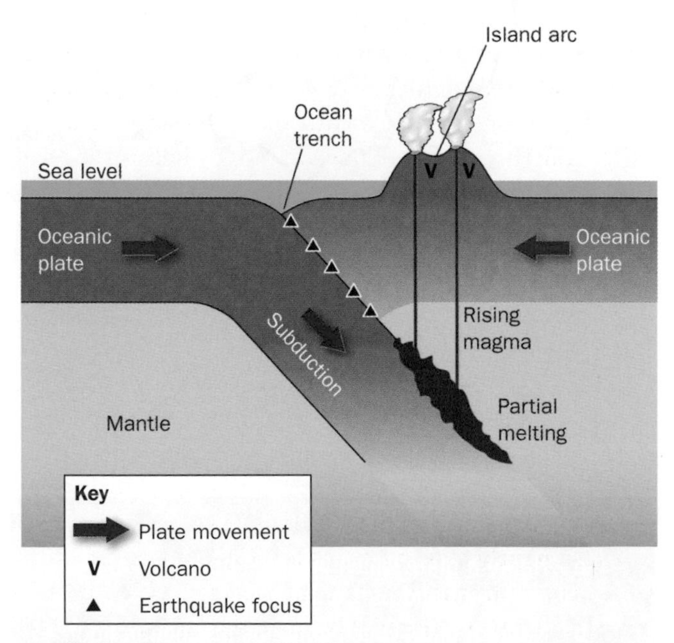

Fig. 3.12 A cross-section through Japan, showing a destructive margin involving two oceanic plates

In the case of the Himalayas (the highest mountains in the world), there is a collision between two continental plates (see Fig. 3.13). Fold mountains form and earthquakes occur – but there is no subduction of an oceanic plate, so there are no volcanoes in the Himalayas.

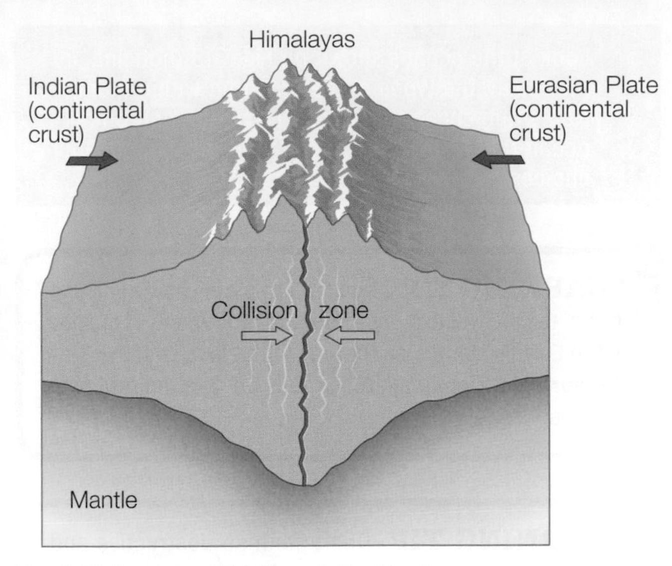

Fig. 3.13 A cross-section through the Himalayas

> **4** Explain the meaning of the terms: subduction, magma, plate, plate margin, compression.

Constructive (divergent) plate margins

Constructive plate margins are the places where new oceanic plate is created. Continental plate is neither created in the same way nor destroyed by subduction. At constructive plate margins, the stresses within the Earth are of tension (stretching) rather than compression (see Table 3.1).

Tension	Compression	Shearing
Plates diverge.	Plates converge.	Plates slide past each other sideways.
The Earth's surface is stretched and gets longer.	The Earth's surface is squashed and gets shorter.	The Earth's surface does not change in length.

Table 3.1 Types of stress

The only constructive plate margins on the planet are the great **ocean ridge** systems (e.g. the Mid-Atlantic Ridge, the East Pacific Rise, and the Carlsberg Ridge in the Indian Ocean). These ridges are huge submarine mountain ranges, which form some of the largest features on the Earth's surface. They are mostly below sea level, but occasionally rise above it. One example of this is the volcanic island of Iceland in the North Atlantic Ocean (see pages 99–100).

Beneath these great ocean ridges, deep in the Earth's mantle, there is a concentration of heat, which causes partial melting (see Fig. 3.14). Small pockets of magma slowly collect and rise towards the Earth's surface – where they cool and solidify to form new oceanic crust. This new rock forms below as well as on the surface. The lava often flows out from long cracks (fissures). As a result, gently sloping areas are built up, rather than conical mountains. The oceanic crust cracks and diverges – pushed apart by the newly formed crust and dragged by the convection currents in the mantle (see page 88).

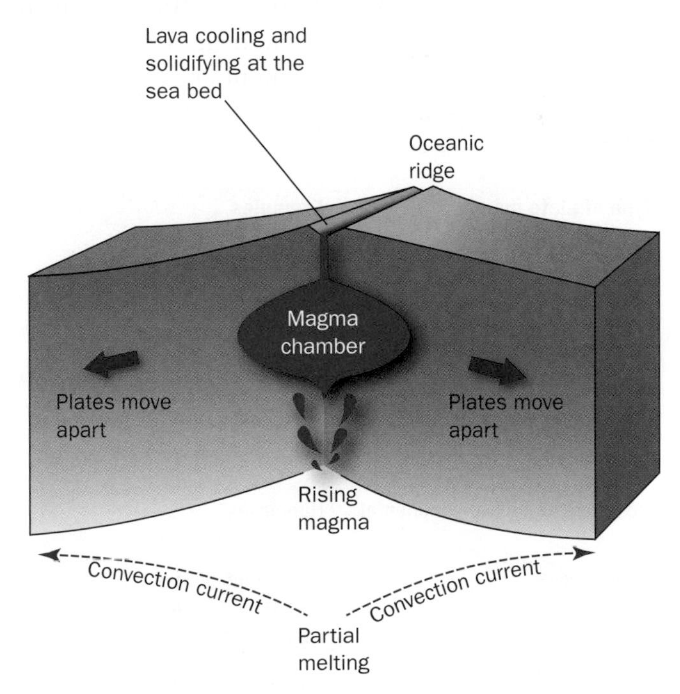

Fig. 3.14 A cross-section through a constructive plate margin

5 Look at the oceans shown on Fig. 3.8 (noticing, in particular, the types of plate margin within them). Name one ocean that is getting bigger and one ocean that is getting smaller. Give reasons for your choices.

LEARNING TIP If you are explaining how a volcano forms, tell the whole story – from the production of magma in the Earth's mantle, all the way to the lava solidifying on the surface. Simple sketches can be very helpful with explanations.

LEARNING TIP Don't confuse destructive and constructive margins! Some candidates get mixed up because fold mountains form at destructive margins. These plate margins are called destructive because plate is destroyed.

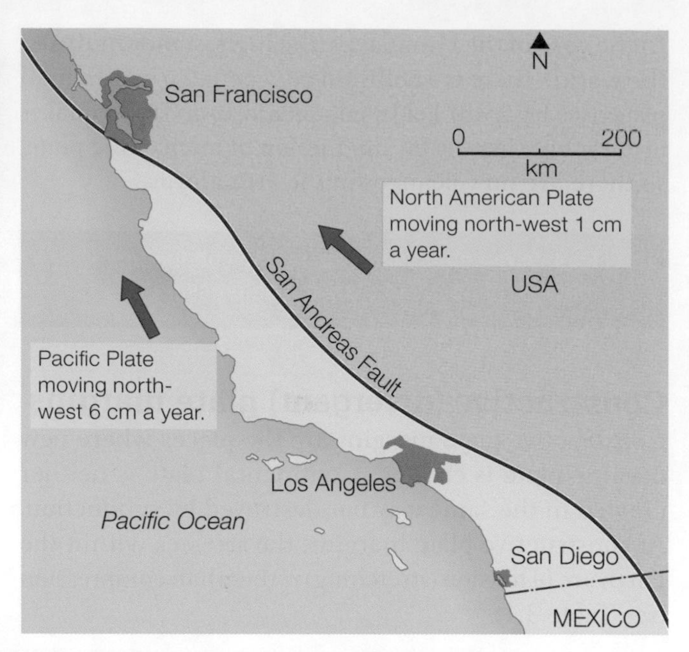

Fig. 3.15 California and the San Andreas Fault system

Conservative plate margins

Conservative plate margins are called that because the plates are being conserved. In other words, they are neither being created nor destroyed. Shearing stress occurs at these margins (see Table 3.1), and the plates slide past each other sideways. Friction between the two plates causes earthquakes but volcanoes do not occur.

The San Andreas Fault system in California (in the USA) is an example of a conservative plate margin. Although Fig. 3.15 shows both plates moving in the same direction, their speeds of movement are different. As a result, the Pacific Plate is moving past the North American Plate and shearing occurs.

Type of plate margin	Examples	Type of stress	Features
Destructive with an oceanic plate and a continental plate	Andes	Compression	Earthquakes Fold mountains Volcanoes Ocean trenches
Destructive with two oceanic plates	Japan Philippines West Indies	Compression	Earthquakes Island arcs Volcanoes Ocean trenches
Destructive with two continental plates	Himalayas	Compression	Earthquakes Fold mountains
Constructive	Mid-Atlantic Ridge East Pacific Rise Carlsberg Ridge	Tension	Earthquakes Ocean ridges Volcanoes
Conservative	San Andreas Fault	Shearing	Earthquakes

Table 3.2 A summary of the features of plate margins

Volcanoes

A volcano is a hole or crack in the ground through which gases, **lava** (liquid), and **pyroclastic** material (solid) are erupted. The vent is connected to a **magma chamber** beneath the ground.

What comes out of a volcano?

Gases

The main gas to be emitted by volcanoes is water vapour (50–80%), but there may also be emissions of sulfur dioxide, hydrogen sulfide, nitrogen, hydrogen, and carbon dioxide. Some of these gases are poisonous.

The gases can become trapped in the viscous lava - causing pressure to build up and leading to frothing of the magma and explosive eruptions.

Liquids

Magma is molten rock material below the Earth's surface. Lava is the flows of molten rock material which have erupted on to the Earth's surface.

Solids

These are known as pyroclastic material.

- Ash is made up of the smallest particles (less than 4 mm in size). However, blocks of the coarsest material are much larger.
- The smallest particles can be held in suspension in the air, as clouds, for months or even years.
- The particles get finer the further away they are from the volcanic vent. Because of its weight and size, the largest material is dropped nearest to the vent. More material is therefore found close to the vent than further away.

Types of volcano

Shield volcanoes

Shield volcanoes are formed in the oceans - often at constructive plate margins. They occur on the Hawaiian Islands (e.g. the volcanoes Mauna Loa and Kilauea), as well as on Iceland. They:

- rise from the deep ocean floor
- have gentle upper slopes (at an angle of about 5 degrees), and steeper lower slopes (at an angle of about 10 degrees)
- usually have a roughly circular or oval shape in map view, and cover a wide area
- are composed almost entirely of long thin lava flows, built up over a central vent
- have very little pyroclastic material associated with them
- are mostly formed by runny lava that flows easily down the slope away from the summit vent. This lava is dark in colour and has a low silica content. (It is sometimes called basic lava in older textbooks.) The low viscosity of the magma allows the lava to flow quickly down a gentle slope. But - as it cools and gets less runny - its thickness builds up on the lower slopes, which explains their steeper profile.

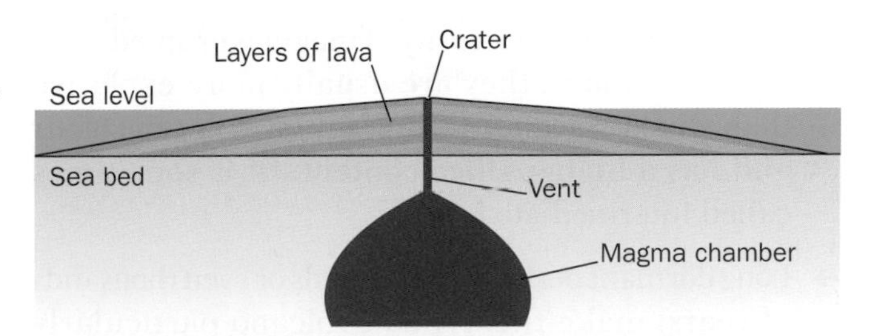

Fig. 3.16 A cross-section through a shield volcano

Fig. 3.17 The shield volcanoes of the Hawaiian Islands, as seen from outer space. These volcanoes are some of the biggest features on the Earth's surface.

Stratovolcanoes

→ They have steeper slopes and narrower bases than shields – with slope angles of 6–10 degrees at the bottom of the volcano and up to 30 degrees near the summit.

→ The steep slopes result from short, wide and very viscous lava flows that don't travel very far from the vent.

→ There are alternating layers of lava and pyroclastic material. Pyroclastic material can make up over half the volume of a stratovolcano.

→ Due to the higher viscosity of magmas erupted from stratovolcanoes, they are usually more explosive than shield volcanoes. Their lava is paler in colour and has a higher silica content. (It is sometimes called intermediate lava.)

→ Long dormant periods (of hundreds or even thousands of years) make this type of volcano particularly dangerous, because people are reluctant to heed warnings about possible eruptions. Sometimes the eruptions have two phases – first an explosive phase that unblocks the vent and produces pyroclastic material, and then a second phase which produces lava.

Fig. 3.18 Stromboli, Italy, a stratovolcano in the Mediterranean Sea

Discussion point

Which of the two main types of volcano is the most dangerous, and how is it related to the chemistry of the molten rock?

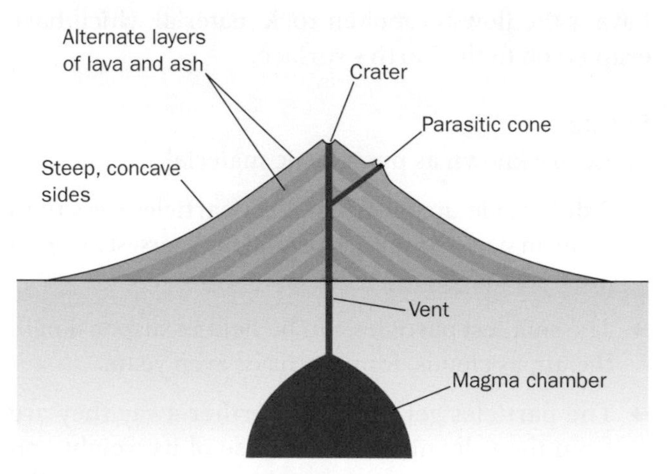

Fig. 3.19 A cross-section through a stratovolcano

Features of volcanoes

Craters

Craters are circular depressions that are usually less than 1 km in diameter. Both types of volcano sometimes have a crater at the summit. It is formed by the explosive ejection of material from a central vent.

Calderas

A caldera is a huge crater caused when a volcanic cone collapses into a partly empty magma chamber after a powerful eruption. Yellowstone National Park in the USA and the Ngorongoro Crater in Tanzania are well-known examples.

Fig. 3.20 The crater of Mount St Helens in the USA

Fig. 3.21 A crater in a small volcanic cone in Timanfaya National Park, in Lanzarote, Canary Islands

Parasitic cones

An example of a parasitic cone is shown on Fig 3.19. These are smaller cones which develop on the sides of a bigger volcano. They form when the main vent becomes blocked and the magma finds another outlet. Fig 3.22 shows part of Mount Etna in Sicily, which has many parasitic cones covering its slopes.

Fig. 3.22 Parasitic cones on Mount Etna, Sicily, Italy

Lava domes

Lava domes are features that often grow on the sides of stratovolcanoes. They form from very viscous lava that is pale in colour and has a high silica content. (It is sometimes called acid lava in older textbooks.) This lava cannot flow very far before solidifying, so the cones produced have steep convex sides. Lava domes often collapse – leading to explosive eruptions and pyroclastic flows, like those seen on the Soufrière Hills volcano on the island of Montserrat in the Caribbean.

The dangers of volcanic eruptions

Ash falls	Fine ash is blasted into the atmosphere, where it can stay in suspension for many months – affecting areas far away from the volcano. It mostly damages property by burying buildings; people are not usually harmed directly. Ash can also be a hazard to aircraft and lead to the cancellation of flights. Sometimes ash clouds can block the sun, causing the weather to be cooler and affecting crops.
Pyroclastic flows	Very hot solid material can travel rapidly down valleys and slopes. It is impossible for people to escape, so pyroclastic flows can be responsible for many deaths. A famous example was the eruption of Mt Pelée in Martinique in the West Indies in 1902, when a white-hot glowing ash cloud killed 40 000 people.
Lateral blasts	Sometimes a volcano can explode sideways, which can be very destructive for areas within 40 km of the volcano. It can destroy houses and property.
Mudflows (lahars)	These form when ash mixes with water and travels down river valleys. Because mud is much denser than water, mudflows are very destructive – washing away buildings, roads, bridges, and people.
Volcanic gases	Carbon dioxide is a dense, non-toxic gas that can flow downhill, causing suffocation. Other gases that are poisonous can burn or cause lung diseases.
Acid rain	Because of the sulfur dioxide and hydrogen sulfide released, very large eruptions cause acid rainfall. This can damage buildings, and may have had very serious effects on plant and animal species in the past.
Post-eruption famine and disease	The disruption to homes, roads and services caused by the effects described above can result in famine and disease, especially in LEDCs.
Tsunamis	The collapse of volcanoes into the sea can result in a tsunami (see page 102).
Lava flows	Although lava flows can destroy buildings, they rarely result in a direct loss of life – they travel slowly enough for you to walk away!

Table 3.3 The many hazards of volcanic eruptions

Fig. 3.23 The ash cloud emitted during the eruption of Mount Pinatubo in the Philippines in 1991

Discussion point

Which of the volcanic hazards described in Table 3.3 pose the greatest threat to people, and why?

LEARNING TIP Examination questions will often ask you not just to explain what the hazard is but to explain its effect on people's lives, such as their homes or their jobs.

RESEARCH The Internet has a wealth of material about active volcanoes around the world. Sites such as that of the United States Geological Survey (http://volcano.si.edu/reports/usgs/) give up-to-date information about volcanic activity. Choose a volcano and keep a log of its activity. Sometimes these sites have webcams which show the current activity. Choose a volcano from close to your time zone, otherwise it might be night time at the volcano when it's daytime where you live!

6 Look at Fig. 3.24 and Fig. 3.25.

a List the main hazards and explain what risks they could bring to people living in the area.

b Suggest reasons for the pattern of the hazards on the map.

Fig. 3.24 Mount Rainier, a stratovolcano in north-west USA

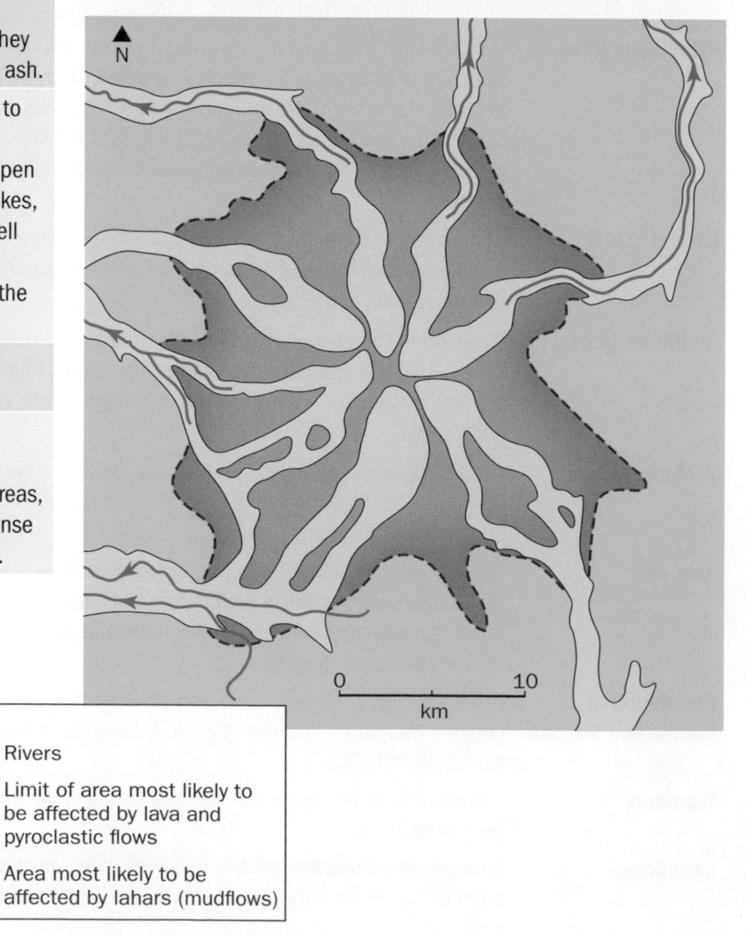

Fig. 3.25 A hazard map for Mount Rainier

What can be done to reduce the risk?

Lava flow diversion	Mechanical excavators can be used to channel lava flows away from buildings. Lava flows can also be sprayed with water to cool them down and make them solidify and stop flowing.
Mudflow barriers	These are walls built across valleys to trap mudflows and protect settlements further down the valley.
Building design	Although little can be done to stop a violent volcano, stronger roofs can be built so that they are less likely to collapse when covered with ash.
Volcano monitoring	Recently active volcanoes can be monitored to give early warning of future eruptions. This usually involves measuring features that happen before an eruption – such as small earthquakes, ground deformation (the ground tends to swell before an eruption) and gas emissions (the mixture and amount of gases released from the vent changes before an eruption).
Remote sensing	Monitoring the location of ash clouds from satellites is useful for warning aircraft.
Hazard mapping and planning	This involves looking at the pattern of past eruptions in order to predict future eruptions. It can lead to a ban on building in high-risk areas, or simply the preparation of emergency response plans such as constructing evacuation routes.

Table 3.4 Ways to reduce the hazards of volcanic eruptions

Advantages brought by volcanoes

Geothermal power	In some volcanic countries, such as Iceland, great use is made of the fact that the rocks beneath the surface are very hot and water in the ground is also hot. Electricity is generated, either directly from steam in volcanically active areas or by water pumped down and heated from hot rocks. Hot water from the ground can be used directly in central-heating systems and even in swimming pools.
Fertile soils	Some types of lava and ash weather rapidly in tropical conditions and form a rich, thick soil layer, abundant in trace elements. This soil can be extremely fertile and produce high crop yields.
Creation of new landmass	Volcanoes produce new islands and enlarge existing landmasses. This is shown in the Iceland case study on pages 99–100.
Tourism	When safe, volcanoes tend to attract tourists. This has helped the economy in places such as Iceland and the Canary Islands (Tenerife and Lanzarote) and creates jobs, e.g. for tour guides and hotel workers.
Minerals and mining	Much of the sulfur mined is from around active volcanoes. Other mineral deposits were formed by volcanoes that are now extinct.
And long in the past ...	Volcanoes supply large volumes of gases to the atmosphere, which initially created the Earth's atmosphere. All the water now in the oceans originated as volcanic gas in the form of water vapour.

Table 3.5 Volcanoes – not all bad!

7 Look at the two photographs of Mount Etna in Fig. 3.27. Describe the features of tourism shown, which bring employment to the area.

Fig. 3.26 Geothermal power in Iceland – how hot is the water in the pipeline?

LEARNING TIP Examination questions often ask you to use case studies of chosen volcanoes (and earthquakes). Make sure that you know about the different plate tectonic settings of your chosen volcanoes, how they formed, their features, and their effects on people's lives when they erupt.

Discussion point

If the Earth's internal heat died away, there would be no more mountains being formed, no more earthquakes and no more volcanoes. Would this bring any disadvantages to people? If so, what are they and over what timescale?

Fig. 3.27 The slopes of Mount Etna, an active volcano in Sicily, Italy

How volcanoes form

Earlier in this chapter, you learned that volcanoes form at constructive plate margins (like the Mid-Atlantic Ridge), destructive plate margins (like the Andes), and occasionally away from plate margins (like in Hawaii).

Magma is produced deep within the Earth, in areas that are hotter than the melting point of the rocks. The magma rises because it is less dense than the surrounding solid rocks.

An effect can then take place that is like taking the top off a bottle of fizzy drink that has been shaken:

→ Magma often contains water dissolved within it as gas. As the magma rises, it may reach a depth where the pressure is lower. The dissolved gas can no longer be held in solution in the magma and it begins to form bubbles, which expand.

→ In runnier magmas, the gas is able to escape. But in thick viscous magmas, the gas is released explosively at the surface - producing very violent eruptions that spray lava high into the air.

→ Bubbles of liquid lava burst explosively in the air and then the material cools and solidifies and falls to the ground. This is how the pyroclastic material (solid) is produced.

→ The build-up of this material leads to the formation of the volcano.

Table 3.6 A summary of volcano types and eruptions

	Shield volcano	Stratovolcano
Shape		
Slope angles	Gentle	Steep
Plate tectonic setting	Constructive margins Mid-plate volcanoes	Destructive margins with an oceanic plate
Products	Mostly lava	Lava and pyroclastics (ash)
Type of eruption	Continuous and non-violent	Explosive with dormant phases

8 Copy and complete Table 3.7. Show which features occur at which types of plate margin by writing either yes or no in each box.

Type of plate margin	Earthquakes	Volcanoes	Fold mountains	Ocean trenches	Ocean ridges
Constructive					
Destructive - two oceanic plates					
Destructive - two continental plates					
Destructive - an oceanic plate and a continental plate					
Conservative					

Table 3.7 Features of plate margins

CASE STUDY

The Eldfell volcano, Iceland

Iceland is an island in the North Atlantic Ocean, on the constructive plate margin of the Mid-Atlantic Ridge. The country has experienced many volcanic eruptions over the centuries, including the eruptions of Eyjafjallajokull in March and April 2010, which caused an ash cloud that prevented all air travel over much of Europe for many days. However, this case study concentrates on the eruptions of Eldfell on Heimaey (the largest of the Vestmannaeyjar islands, off the south coast of Iceland) in 1973.

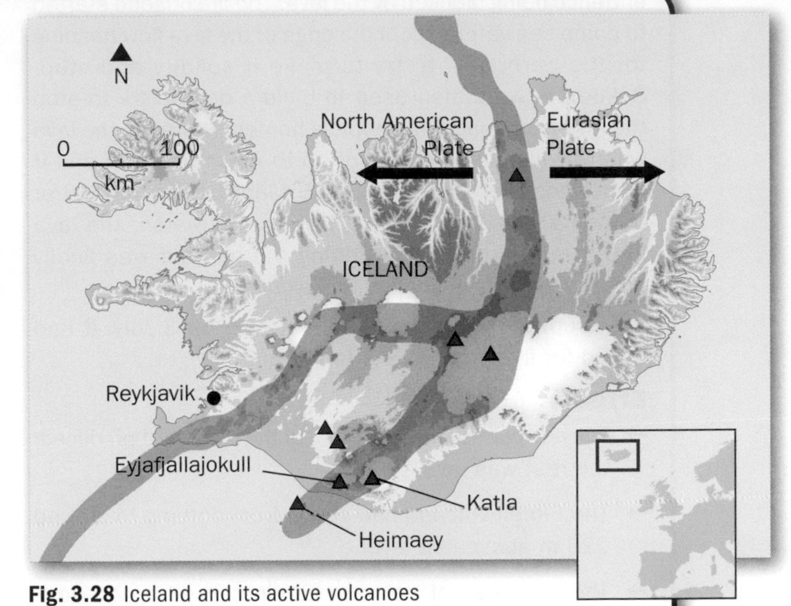

Fig. 3.28 Iceland and its active volcanoes

The 1973 eruption

On 22 January 1973, all of Heimaey's fishing boats (plus some others from other ports that had taken shelter there) had to stay in the island's harbour all day, because of a storm and very rough seas. That night the islanders went to bed as normal, but from 10pm onwards they began to experience a series of weak earthquakes.

Then, at 2am the following morning, the police station in Heimaey town centre received a report that a volcanic eruption had started near Kirkjubær, on the eastern edge of the town. The eruption had begun from a 1600-m-long fissure (crack) in the ground. When the police officers drove to the area, they found that the fissure had opened up right along its length as far as they could see – and was erupting with a row of lava fountains like a wall of fire. A new volcano, later known as Eldfell, was beginning to form.

Responses

The fire alarm was sounded to wake everybody up, and police cars patrolled the streets with their sirens going to make sure. The islanders put on warm clothing, gathered a few belongings and began to meet down at the harbour. The town council immediately decided to evacuate the whole population of the island – apart from those employed on essential work. The first fishing boat left for the mainland at 2.30am, followed by many others.

The airport was in danger if the fissure were to open up southwards. But, despite this, aircraft from the mainland landed on Heimaey and 300 people – mostly the sick and aged – were taken to the mainland by air during that first night. Altogether, about 5000 people were evacuated from Heimaey on the night of 22–3 January.

The houses on the eastern side of the town began disappearing under lava and ash up to 4 m thick. Many houses collapsed under the weight. As the eruption continued, a dark cloud often hung over the town.

Fig. 3.29 A map of Heimaey showing the effects of the 1973 eruptions. Changes to the area of the main town are not shown.

By 6 February, there was a real danger of the harbour entrance being blocked by the lava. The fire brigade started to pump seawater to cool the edge of the lava flow heading for the harbour – to try to make it solidify and stop. Bulldozers were also used to build a dam to try to stop the lava from reaching people's homes. However, the lava was travelling at a speed of up to 40 m per hour and it went right over the dam. By 25 March, it had buried or burned about 111 houses. But, on 26 March, the lava flow that had been threatening the harbour was finally halted – the water cooling had worked.

The eruption was declared finally over on 3 July. It had lasted for five months and ten days.

Effects

- Ash and pumice made up 10% of the material it produced; the rest was lava.
- The Eldfell volcano had grown from nothing to around 225 m above sea level.
- The land area of Heimaey had increased by 2.5 km^2. The island had previously been 12 km^2 in area, but that was extended to 14.5 km^2 by the end of the eruption. The new land that was created gave the harbour extra shelter from rough seas.
- One third of the houses in the town were destroyed and another third damaged.

By 1975, most of the ash had been cleaned away and moved to the new lava field, where it was used to make roads, enlarge the airstrip and make foundations for new houses. 3500 out of the 5300 islanders came back after the eruption. The total population is now about 5000.

9
- **a** What was the first warning of the Heimaey eruption?
- **b** From what feature did much of the lava flow?
- **c** Why was the storm on 22 January lucky for the islanders?
- **d** How many people were evacuated from the island?
- **e** What were the effects of the ash from the eruptions?
- **f** What were the effects of the lava from the eruptions?
- **g** The eruptions had some advantages for the island. What were they?
- **h** Eldfell, a new volcano, formed during the eruptions. How high was it?

Fig. 3.30 A photograph of Heimaey in 2009. The enlarged harbour can be seen with the lava flows from the 1973 eruptions in the foreground.

Earthquakes

Fig. 3.8 shows the location and type of plate margins around the world. Strong earthquakes occur at all of these different plate margins, as shown in Fig. 3.6.

What causes earthquakes?

Earthquakes are caused by plate movements - either towards each other, away from each other or sliding past each other. The plates don't always move at a constant rate; they are often 'stuck' in one position. Stress and pressure builds up as the plates try to move. Then there is a sudden release of pressure when the plates break free (along a crack in the Earth called a fault). Huge amounts of energy are released and the shock waves or vibrations travel through the Earth as an earthquake wave or seismic wave. The point within the Earth where the earthquake originates is called the **focus**. The point on the Earth's surface directly above the focus is called the **epicentre**.

Assessing earthquakes

The effects of an earthquake can be assessed on a 12-point scale - named after the seismologist Guiseppi Mercalli (see Table 3.8). These effects can also be shown on a map using the **Mercalli Scale** (see Fig. 3.32).

Earthquakes can also be assessed using the **Richter Scale** of magnitude, which measures the total amount of energy released by an earthquake. Powerful earthquakes have Richter values between 5 and 9. An increase of 1 on the scale means that the energy released increases by about 30 times.

Value	Intensity	Description
1	Instrumental	Not normally felt. Animals uneasy.
2	Feeble	Felt only by a few people at rest.
3	Slight	Vibrations like a lorry passing. Felt by people at rest.
4	Moderate	Felt indoors by many. Cars rock.
5	Rather strong	Sleepers awakened. Some windows broken.
6	Strong	Bells ring. Trees sway. Loose objects fall.
7	Very strong	Difficult to stand up. People run outdoors. Walls crack.
8	Destructive	Collapse of some buildings. Trees fall.
9	Ruinous	Ground cracks. Pipes break.
10	Disastrous	Landslides. Many buildings destroyed.
11	Very disastrous	Few buildings left standing.
12	Catastrophic	Total damage. Ground surface rises and falls in waves. Objects thrown into the air.

Table 3.8 The Mercalli Scale of earthquake intensity

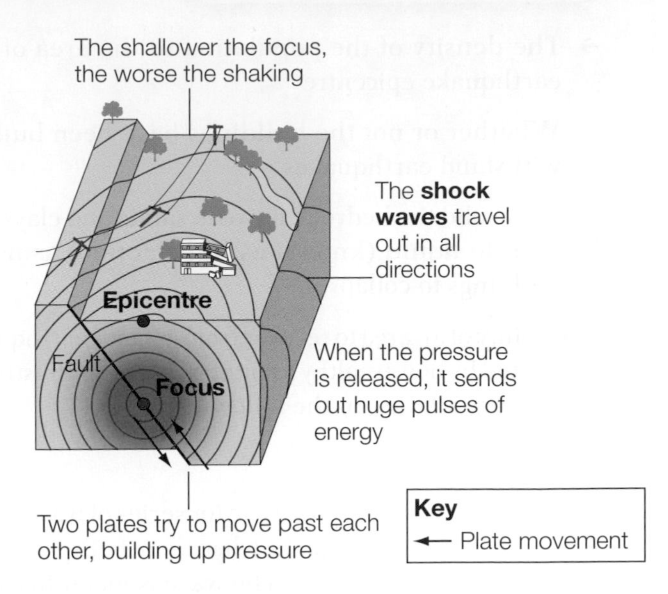

Fig. 3.31 How earthquakes occur

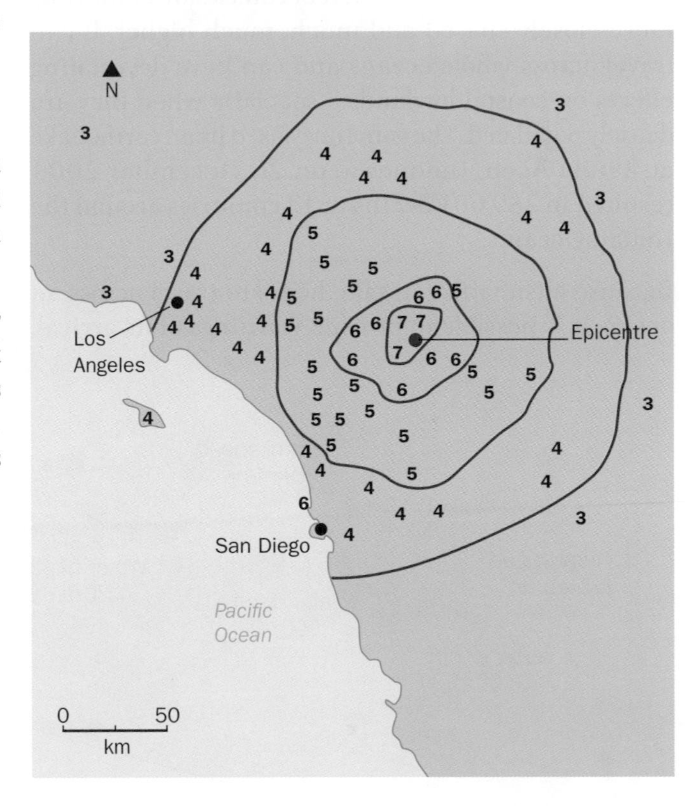

Fig. 3.32 Lines of equal intensity around the epicentre of an earthquake

LEARNING TIP Notice how the lines on Fig. 3.32 go around the numbers not through them. To draw lines of equal intensity on a map don't try to join the points.

The amount of damage that an earthquake causes will be affected by the following factors:

→ The amount of energy released (as measured by the Richter Scale)

→ The depth of the focus beneath the surface (shallower earthquakes have a greater effect)

- The density of the population in the area of the earthquake epicentre
- Whether or not the buildings have been built to withstand earthquakes
- How solid the bedrock is; weak sands and clays can turn to liquid (known as liquefaction), causing buildings to collapse.

The ability of an area to recover from a major earthquake is affected by how wealthy a country is. This is illustrated by the case studies on the following pages.

Tsunamis

A tsunami is a giant ocean wave (or series of waves) that is generated by an earthquake when there is displacement (movement) of the seabed. The wave is magnified as it travels into shallower water. It becomes slower moving, more closely spaced and much, much higher. It can travel across whole oceans and can have devastating effects on coastal lowlands, especially when they are densely populated. The tsunami caused by an earthquake at Banda Aceh, Indonesia, on 26 December 2004 resulted in 289 601 deaths in 12 countries around the Indian Ocean.

Because a tsunami can take hours to travel across an ocean, it is possible to provide warnings of its arrival.

Tsunami speed 50 km/hr
Tsunami speed 340 km/hr
Tsunami speed 835 km/hr
Sea level
Water depth 20 m
Water depth 900 m
Water depth 5500 m
Displacement

Fig. 3.33 How a tsunami forms

Discussion point

If a tsunami warning is given, what can be done to protect people and property? What are the barriers which might prevent this happening?

LEARNING TIP Do not refer to a tsunami as a 'tidal wave'. It is nothing to do with the tides.

The Pacific Ocean has had a tsunami warning system in place for many years. After the terrible tsunami in 2004, the Indian Ocean countries also introduced a tsunami warning system, together with emergency drills and procedures to keep casualties to a minimum.

Monitoring to predict earthquakes	Like volcano monitoring, before a major earthquake, there may be small earthquakes (foreshocks) and ground deformation detected by seismometers and lasers.
Tsunami warning systems	These use buoys and sensors in the water. A system was put in place in the Indian Ocean after the devastating tsunami of December 2004.
Evacuation of people	People can escape to higher ground if warned in time, which has proved successful in saving lives. However, evacuating a major city would be very difficult.
Earthquake drills	Schools in countries such as Japan have regular earthquake drills. They teach people how to react to an earthquake, e.g. not to run out of buildings because they might be hit by falling objects.
Hazard mapping	Building development can be prevented in high-risk areas.
Building design	Design features to minimise earthquake damage include: • strong foundations, e.g. liquid cement pumped into micropiles drilled in rock • reinforced structures, e.g. diagonal bracing, reinforced concrete (safer than brick) • flexible structural supports that absorb sway via computer-controlled hydraulics • ground isolation systems: Teflon pads, rubber pads, rollers – ground moves, building doesn't

Table 3.9 What can be done to reduce the risk of tsunamis?

Fig. 3.34 A tsunami warning sign on the island of Grenada in the Caribbean

CASE STUDY

The Haiti earthquake of 2010

Haiti is an LEDC in the Caribbean, near Cuba. Before 2010, there had been no major earthquakes there in living memory, so its buildings were not constructed to withstand them. People in the capital, Port-au-Prince, were housed in crowded conditions. The government was not well organised and the country was very poor. Haiti was 145th out of 169 countries on the UN HDI index and 86% of the population were living in poorly built slums.

Then, on 21 January 2010, a 7.0 magnitude earthquake occurred 25 km from Port-au-Prince (with a focus depth of 13 km). The cause of this earthquake was movement along the destructive plate margin between the Caribbean Plate and the North Atlantic Plate (see Fig. 3.8).

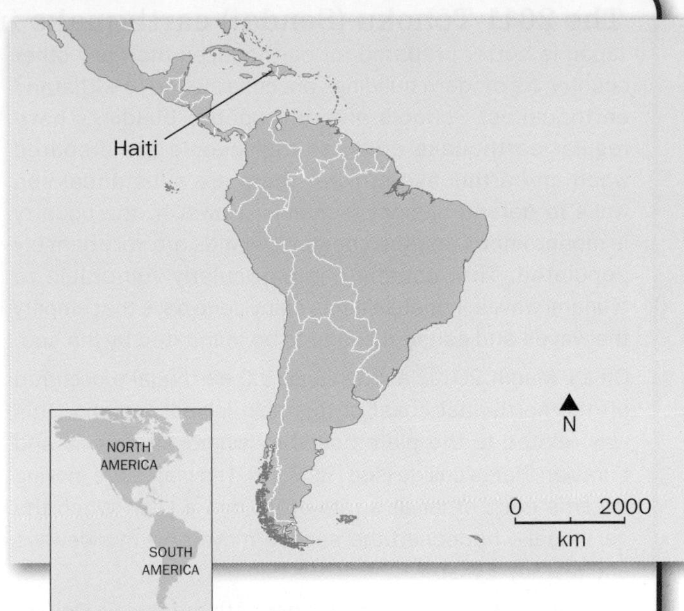

Fig. 3.35 The location of Haiti

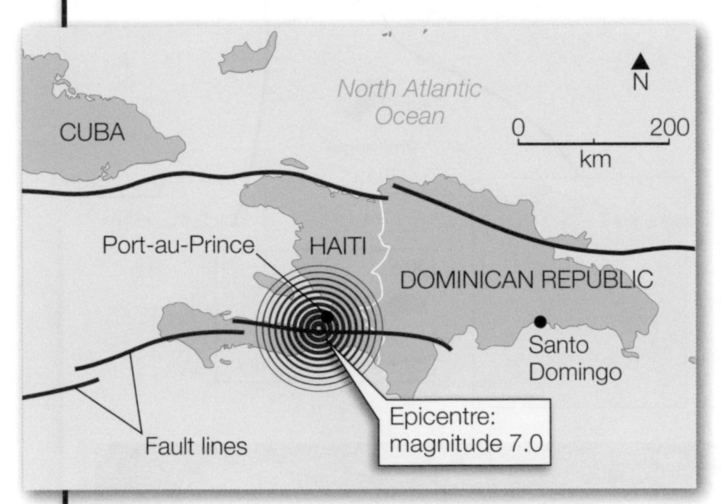

Fig. 3.36 The location of the Haiti earthquake of 2010

Haiti is part of the long chain of the Caribbean Islands (the West Indies). This is another example of an island arc. The plate margin has a collision between two oceanic plates. As well as earthquakes like that in Haiti in 2010, many of the islands have active volcanoes which sometimes erupt violently. The eruption of Mt Pelée in Martinique in 1902 was very destructive. More recent eruptions occurred on Montserrat. You can research these eruptions on the Internet.

Effects

- More than 220 000 people were killed and about 300 000 were injured.
- Many buildings were destroyed and 1.3 million people were made homeless.
- Hospitals and government buildings were destroyed.
- The port was destroyed and many roads were blocked.
- Looting became a problem.
- Over 2 million people were left short of food and water. Power supplies were cut.
- There were outbreaks of cholera, a disease caused by poor sanitation in the temporary camps, affecting 720 000 people of whom 9000 died.

Responses

- Damage to the port and roads meant that aid supplies were difficult to deliver. The airport couldn't handle the number of relief aircraft required.
- American engineers helped to clear the port and airport.
- The USA sent 10 000 troops and police.
- About 1.5 million people were living in temporary tented camps and 200 000 moved to less damaged cities.
- Bottled water and water purification kits were provided.
- Field hospitals were set up.
- Because of its poverty and other problems, Haiti is relying on overseas aid to help it recover – but this will take time. Five years after the earthquake the rubble had been removed from the streets but people were still living in tents and temporary accommodation.

CASE STUDY

The 2011 Tōhoku (Sendai) earthquake in Japan (or the Fukushima earthquake)

Japan is better prepared for earthquakes than any other country. Its modern buildings are constructed to withstand earthquakes. Schools and other public buildings have regular earthquake drills, so that people are prepared when an earthquake strikes. There are substantial sea walls to defend against tsunamis. However, the country is mountainous and the coastal lowlands are very densely populated. That coastline is particularly vulnerable to tsunami waves, because it has many deep bays that amplify the waves and cause the land to be inundated by the sea.

On 11 March 2011, a magnitude 9.0 earthquake occurred off the north-east coast of the main island, Honshu. This was related to the plate boundary where the Pacific and Eurasian Plates collide (see Fig. 3.12). The plates are moving towards each other at a rate of 83 mm a year. When the earthquake happened the seabed moved 50 m sideways and rose by 7 m.

The earthquake epicentre was beneath the Pacific Ocean, 129 km east of the port of Sendai and 373 km north of the capital city, Tokyo. The depth of the focus was 32 km. There were a series of large foreshocks over the previous two days and major aftershocks.

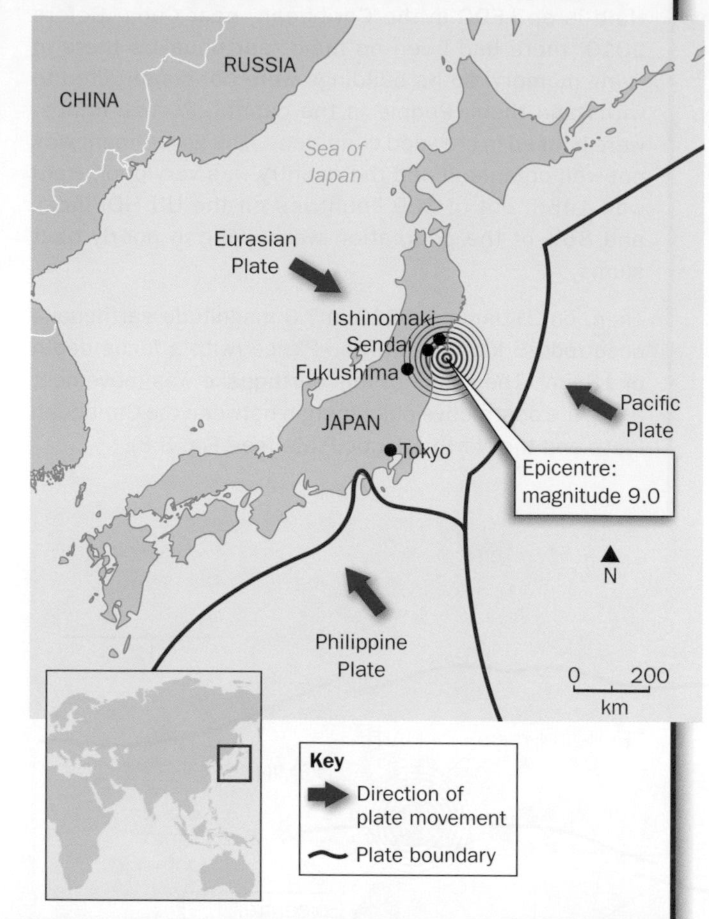

Fig. 3.37 The epicentre of the Sendai earthquake and the plates in the area

Fig. 3.38 The town of Ishinomaki four days after the tsunami struck

Effects

- The earthquake triggered a tsunami which was up to 10 m high. It passed over the defensive sea wall and caused great destruction inland.
- As a result, 15 894 people died (92% by drowning), 6152 were injured, and 2562 people went missing.
- Police said that 215 000 people had fled their homes; 127 290 buildings totally collapsed, a further 272 788 partly collapsed, and another 747 989 were damaged.
- Whole villages were simply washed away, and the town of Sendai was badly hit. The town of Rikuzentakada was mostly under water – with barely a trace of any buildings left. About 1800 homes were reported destroyed in the city of Minamisoma. One third of the city of Kesennuma was also under water.
- A muddy torrent of water swept cars and homes far inland, turning residential areas and paddy fields into a lagoon of debris-filled seawater.
- A dam burst in north-eastern Fukushima prefecture, sweeping away homes.
- Japan Railways said it could not trace four trains along the north-eastern coast, and a ship carrying 100 people was also reported missing.
- In central Tokyo, 373 km from the epicentre, a number of office workers spent the night in their offices because the lifts stopped working. Millions of commuters were stranded overnight and others walked home after train services were suspended. Food supplies ran short in the shops.
- About four million homes in and around Tokyo suffered power cuts.
- Soil liquefaction occurred in areas around Tokyo.
- The tsunami caused meltdown at reactors at the Fukushima Daiichi nuclear plant. It is considered the second worst nuclear accident ever. An evacuation zone affecting 200 000 people was set up.

Responses

- The Earthquake Early Warning system was activated one minute before the earthquake and this saved many lives. The tsunami warning was sent out and 58% of people in coastal areas headed for higher ground, saving many lives.
- A huge relief mission swung into action in north-eastern Japan, the day after the devastating earthquake.
- The country's military mobilised thousands of troops, 300 planes and 40 ships for the relief effort.
- Prime Minister Naoto Kan visited the disaster zone by helicopter, including the Fukushima nuclear plant.
- Rescue teams from South Korea, Australia, New Zealand, and Singapore arrived.
- An American aircraft carrier was already in Japan and another was sent for.
- Japan's recovery from the earthquake has been rapid. However, the problem of the Fukushima nuclear plant will take much longer to solve. The plant will take 40 years to decommission and contaminated soil may need to be removed.

10 **a** Describe what causes an earthquake. Use the words fault and energy in your answer.

b Explain the difference between the Mercalli Scale and the Richter Scale.

11 Explain how the Haiti and Japan earthquakes are linked to plate tectonics.

12 **a** Make a table comparing the effects of the two earthquakes.

b How well prepared were both countries and why were more people killed in the Haiti earthquake?

c How good were the responses to the earthquakes?

Discussion point

How should the world respond to natural disasters such as earthquakes and violent volcanic eruptions? Who should be responsible for paying for and organising relief efforts? What happens today?

RESEARCH Keep monitoring the United States Geological survey website (http://earthquakes.usgs.gov/earthquakes/) for recent earthquake activity. This shows earthquakes that have occurred in the last week, day and hour. You will find reports of earthquakes before they appear on the TV news programmes.

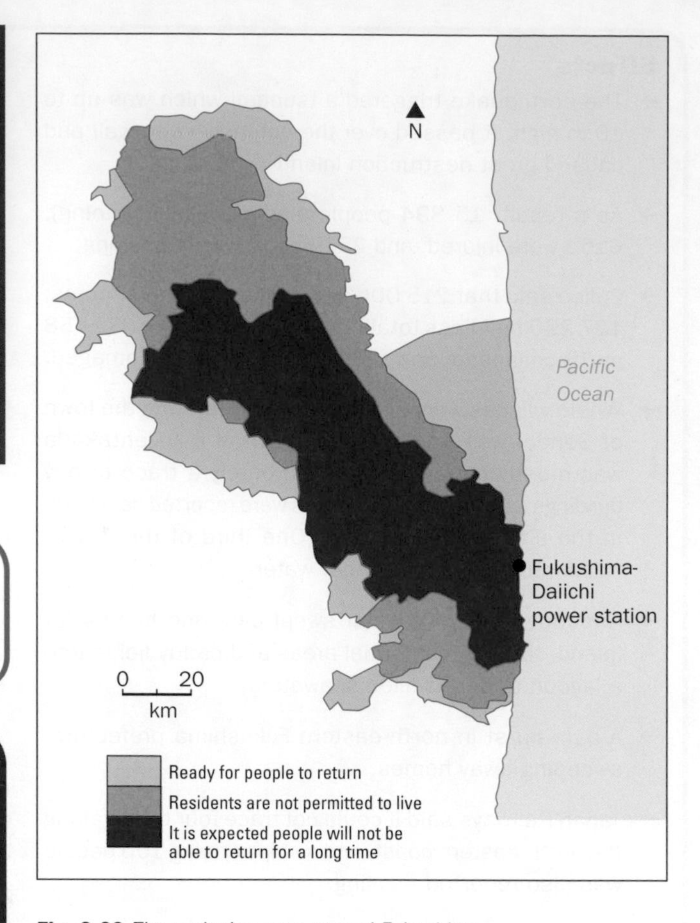

Fig. 3.39 The exclusion zone around Fukushima

4 Rivers

This chapter covers the following Cambridge IGCSE® and O Level topic:

- **2.2 Rivers**

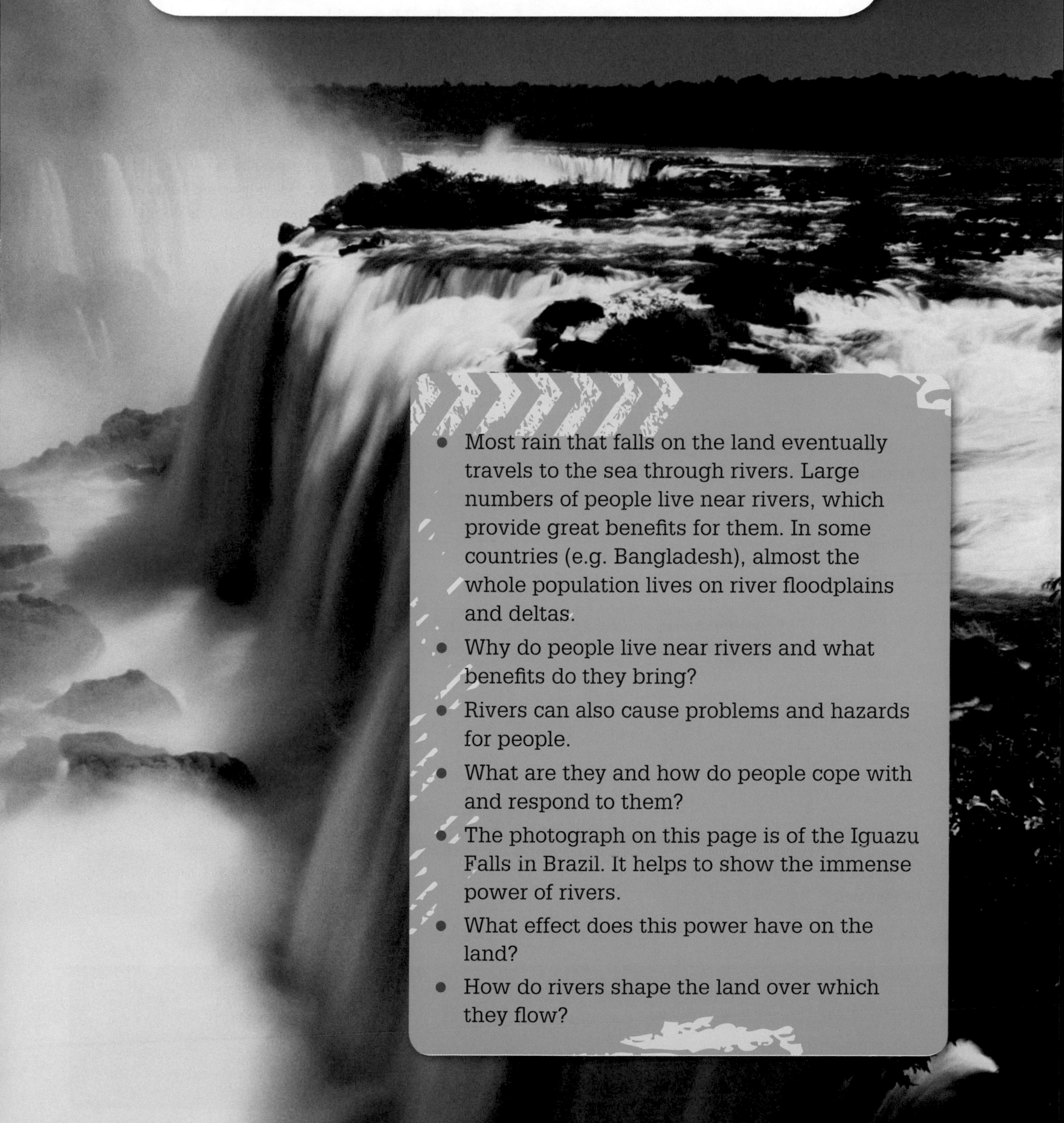

- Most rain that falls on the land eventually travels to the sea through rivers. Large numbers of people live near rivers, which provide great benefits for them. In some countries (e.g. Bangladesh), almost the whole population lives on river floodplains and deltas.
- Why do people live near rivers and what benefits do they bring?
- Rivers can also cause problems and hazards for people.
- What are they and how do people cope with and respond to them?
- The photograph on this page is of the Iguazu Falls in Brazil. It helps to show the immense power of rivers.
- What effect does this power have on the land?
- How do rivers shape the land over which they flow?

In this chapter you will learn about:

- the work carried out by a river – erosion, transportation, deposition
- the landforms that occur in a river valley
- how flooding can be a hazard in river valleys.

(The use of rivers for water supply will be described in Chapter 11.)

River processes – the work carried out by a river

- A river gradually wears away and removes material from its channel (the river bed and banks). This is called **erosion** and it can make the river channel deeper and/or wider.
- The boulders, pebbles, sand, silt, and mud eroded by the river are carried downstream - a process called **transportation.** The material being transported is called the river's load.
- When the river no longer has enough energy to carry its load, it gradually drops it on the river bed - a process called **deposition**. The largest and heaviest material (like boulders) is deposited first, while the lightest material (like silt and mud) is deposited last.

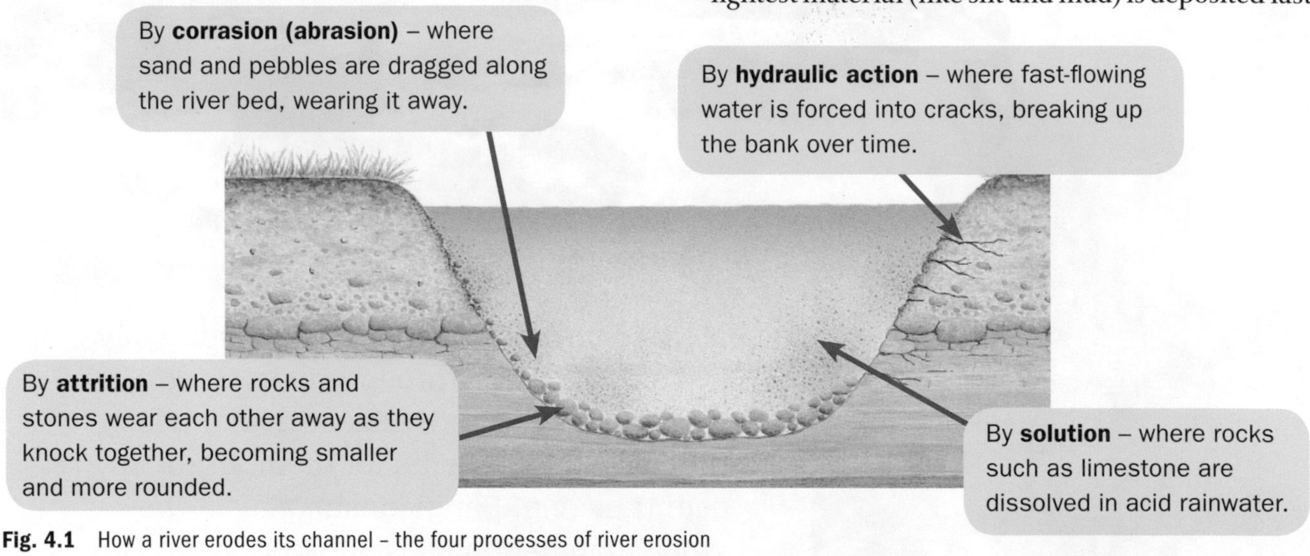

By **corrasion (abrasion)** – where sand and pebbles are dragged along the river bed, wearing it away.

By **hydraulic action** – where fast-flowing water is forced into cracks, breaking up the bank over time.

By **attrition** – where rocks and stones wear each other away as they knock together, becoming smaller and more rounded.

By **solution** – where rocks such as limestone are dissolved in acid rainwater.

Fig. 4.1 How a river erodes its channel - the four processes of river erosion

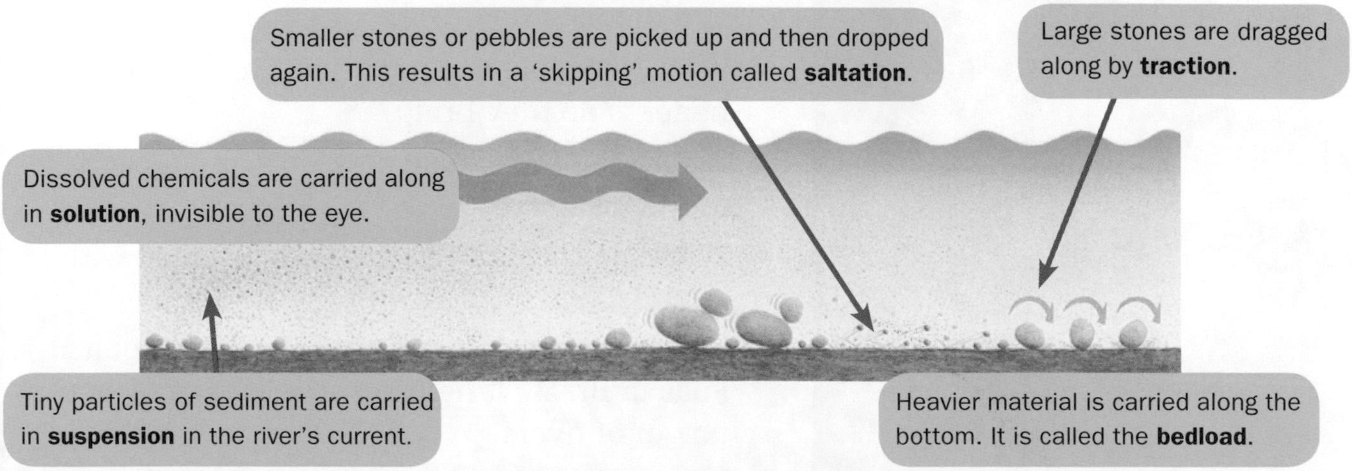

Smaller stones or pebbles are picked up and then dropped again. This results in a 'skipping' motion called **saltation**.

Large stones are dragged along by **traction**.

Dissolved chemicals are carried along in **solution**, invisible to the eye.

Tiny particles of sediment are carried in **suspension** in the river's current.

Heavier material is carried along the bottom. It is called the **bedload**.

Fig. 4.2 How a river transports its load - the four processes of river transport

1 **a** Work in pairs and test each other.

  i One partner should try to describe the four types of erosion.

  ii The other partner should try to describe the four ways in which a river transports its load.

b For each set of words below, decide which is the odd one out and explain your choice.

- solution, attrition, valley, abrasion
- suspension, channel, traction, load
- hydraulic action, saltation, suspension, traction
- corrasion, hydraulic action, attrition, traction

Factors affecting the work carried out by a river

The different processes that a river might be carrying out at any one time vary according to the following factors:

- The **velocity** of the running water. This is sometimes called the **energy** of the river. The faster the river is flowing, the larger the material will be that can be transported in the load. Large boulders can be transported only by very fast flows. After heavy rain, rivers often look brown because of the suspension load.
- The **volume** of the running water. The more water there is in the river, the greater the volume of the load.
- The **bedrock** along the course of the river. Hard rocks (like granite) are very slow to erode. Soft rocks (like shale) erode easily. Some rock types (like limestone) are soluble and can be dissolved by the river water.

Fig. 4.3 The River Rhine in Germany after heavy rainfall

2 Look at the colour of the water in the main river in Fig. 4.3.

- **a** What is the cause of this? What type of transport is taking place? What size of particle is being transported?
- **b** Suggest why the tributary's water is clear.

Why deposition happens

Deposition occurs when a river loses velocity (energy). This can be caused by:

- a decrease in gradient
- a decrease in river flow (discharge) as water drains away after heavy rain
- the river meeting the sea or a lake, often forming a delta
- the river flowing more slowly on the inside of bends.

Fig. 4.4 A river channel in the Northern Cape of South Africa during the dry season

Discharge

This is the volume of water flowing down the river at any one time. It is measured in cubic metres per second (often referred to as 'cumecs'). In climates with wet and dry seasons, or those that are affected by melting snow in spring, the discharge can vary considerably (see Fig. 4.4).

Discussion point

For a river or rivers that you have seen in your area, what is the evidence that erosion, transportation, and deposition are going on? Are the banks being eroded? Are there areas of deposition? Is the water brown at any time? When do these processes happen?

The river channel

The river channel seen on a map or from the air has three main forms, as shown in Table 4.1. The cross-section of the river channel is described on page 128.

> **3** Draw a diagram of a meandering river that enters a lake. On your diagram, label two different places where deposition is taking place.

Meandering	Straight	Braided
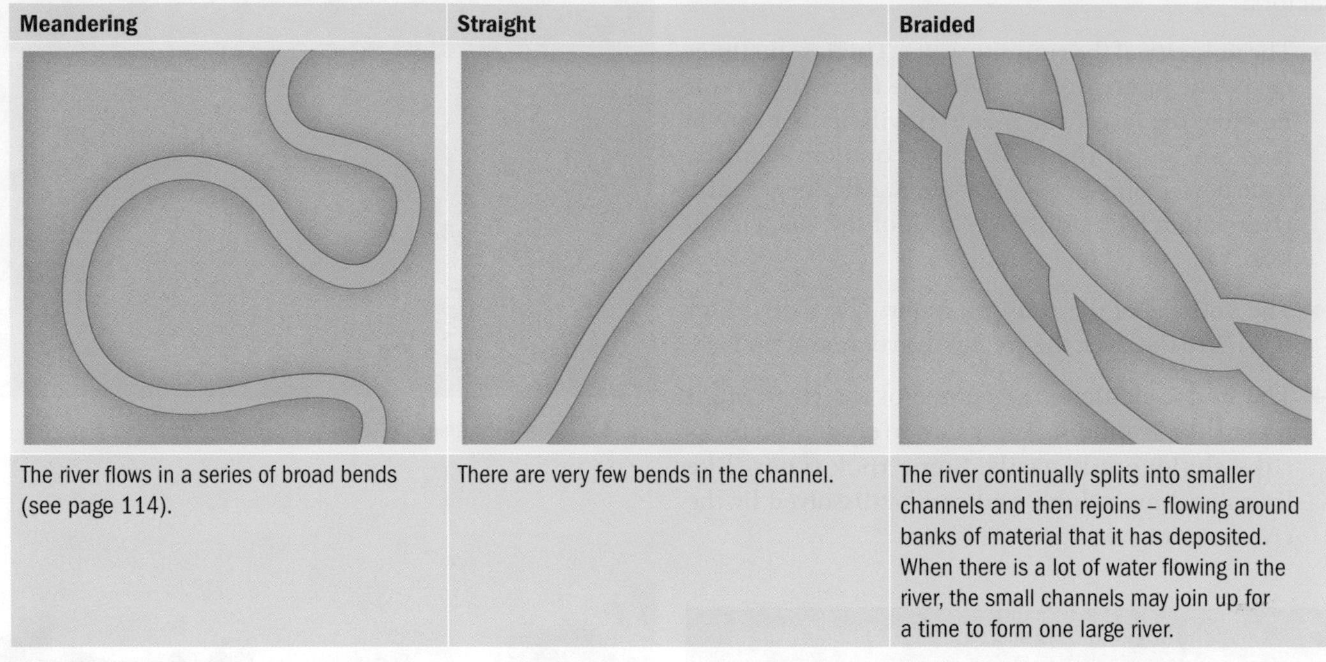		
The river flows in a series of broad bends (see page 114).	There are very few bends in the channel.	The river continually splits into smaller channels and then rejoins – flowing around banks of material that it has deposited. When there is a lot of water flowing in the river, the small channels may join up for a time to form one large river.

Table 4.1 Different river channel patterns

The upper course of a river

The place where a river starts its course is known as the **source**. This might be in an upland area or mountains, and the gradient of the river could be quite steep. In cross-section, the river valley at this point is a V-shape (see Fig. 4.5). The valley floor is very narrow and the river channel may occupy the whole of it. The valley sides are steep. The river also winds its way around interlocking spurs (see page 113). Large boulders in the river – the bedload – are only moved after heavy rainfall, when the river becomes a powerful torrent.

> **LEARNING TIP** Because the gradient is steep, people often think that the river in its upper course is flowing quickly. In fact this is not true, because there is a lot of friction between the relatively small river and its channel, which makes the water flow more slowly. Remember that a steep gradient does not always mean a faster flow.

Fig. 4.5 An upper-course river valley in the Sierra Nevada mountains of Spain, showing a typical V-shaped cross-section

The river is carrying out **vertical erosion**. Occasionally, the vertical erosion can be much greater than any sideways (lateral) erosion. This might lead to the formation of gorges, or even canyons like the one in Fig. 4.6.

Potholes are smooth, rounded hollows formed in the bedrock of the river bed by vertical erosion. They are often about 30 cm across, and are formed by stones trapped in hollows on the river bed. Eddies in the water swirl the trapped stones around, causing corrasion (see page 108), which drills down into the rock. The hollows become deeper and wider and eventually join together.

Rapids are common features in the upper course of a river. They form at places where the water is shallow and the river bed is rocky and irregular - making the water rough (see Fig. 4.8). The gradient here is often steeper than at other points in the river's course. Rapids are often a barrier to river navigation (unless you are a white-water rafter or a kayaker). They can be caused by a band, or bands, of hard rock in the river bed.

Fig. 4.6 The Fish River Canyon in Namibia

Fig. 4.7 Potholes in the Cederberg Mountains, South Africa

Fig. 4.8 Rapids on the Orange River at Augrabies in South Africa

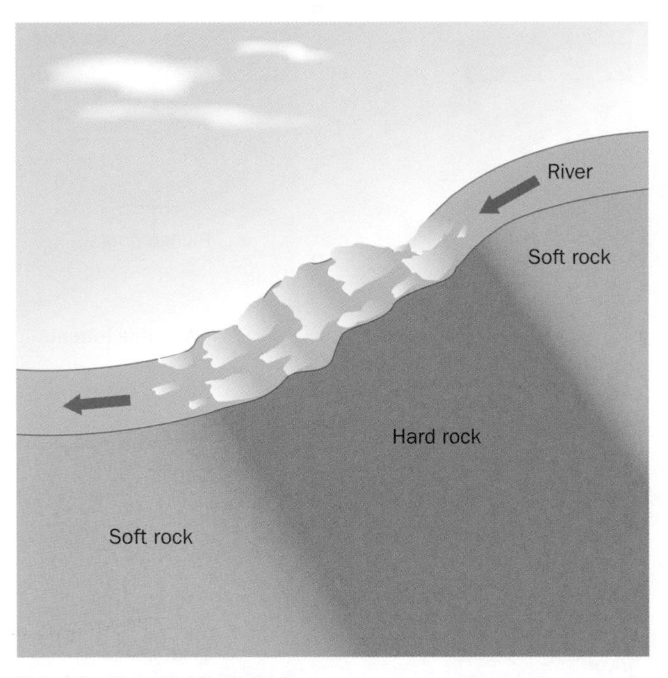

Fig. 4.9 How rapids might form

Waterfalls and gorges

- Waterfalls form where a horizontal layer of hard rock lies on top of a layer of soft rock in a river valley.
- The soft rock underneath is eroded more quickly by the river, and gradually a **plunge pool** develops.
- The splashing water and eddy currents in the plunge pool undercut the hard rock layer above. The hard rock eventually creates an unsupported overhang and collapses.
- If the processes of undercutting and collapse are repeated over a long period of time, the waterfall will retreat upstream - forming a deep, steep-sided valley called a **gorge**.

Waterfalls, especially famous waterfalls like Victoria Falls on the Zambezi River (on the border between Zambia and Zimbabwe), can bring major economic benefits to an area. They create beautiful scenery, which encourages the development of the local tourist industry. This can lead to employment for local people in hotels, transport and tour companies, as well as providing a market for local produce and handicrafts.

The massive power of large waterfalls can also be used to create hydroelectric power (see Chapter 11). These HEP schemes then attract even more development (a regional multiplier effect). Economic development then often becomes concentrated at the location of a waterfall (e.g. at Niagara Falls on the US-Canadian border).

However, waterfalls do have disadvantages as well. In particular, they cause problems with navigation - and crossing the river may also be difficult.

Fig. 4.10 Skógarfoss, Iceland

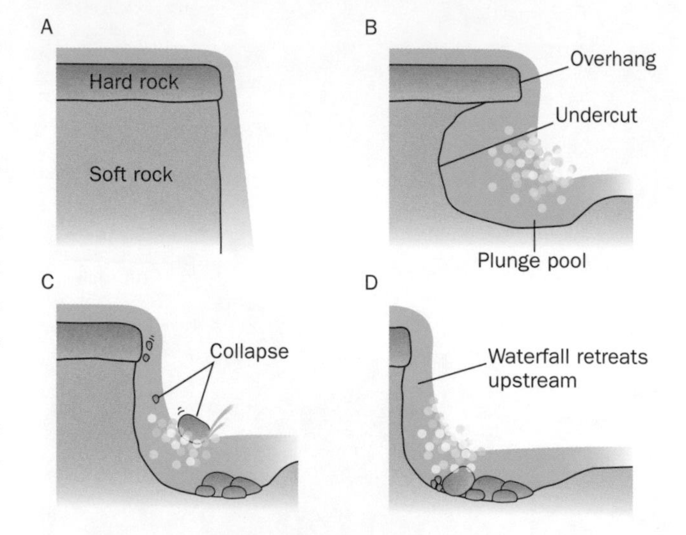

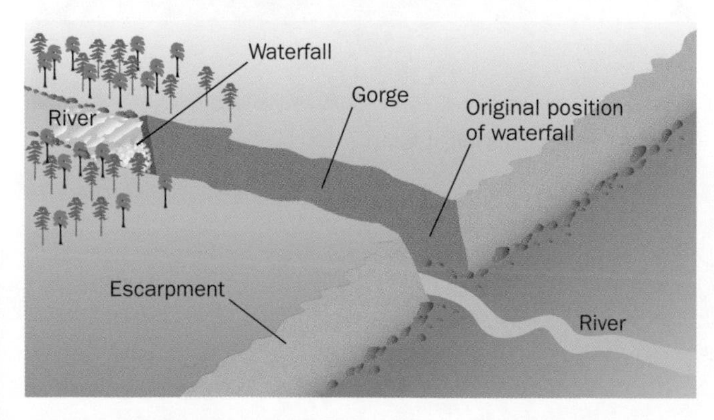

Fig. 4.11 The formation of a waterfall (A–D) and the formation of a gorge by waterfall retreat

Discussion point

Think about a river or rivers that you know from your own area (particularly if they have waterfalls). What benefits and disadvantages do they bring?

RESEARCH Use books or the Internet to find an example of a large waterfall on one of the world's major rivers.

1. Explain, with diagrams, how the waterfall formed.
2. Describe the effect of the waterfall on the economy of the surrounding area.

Fig. 4.12 Interlocking spurs on the upper course of a river in the Western Cape, South Africa. Notice how the spurs prevent a view along the valley floor. Contrast this with a river valley in its middle course (Fig. 4.13).

Interlocking spurs

In the upper course of a river, where the valley is narrow, the river flows around spurs of land on either side of the valley.

4 Draw a large simple sketch of Fig. 4.12. Add labels identifying the features of the valley.

The middle and lower courses of a river

Further downstream, the river valley begins to change shape. Usually, the whole valley widens - the slopes become gentler and the valley floor becomes wider and flatter. The gradient of the river also lessens. Fig. 4.13 is a typical example of the valley of a river in its middle course.

Vertical erosion becomes less important in the middle course of a river, and **lateral** (sideways) **erosion** and deposition start to take over. In the lower course, vertical erosion may stop altogether.

Fig. 4.13 The valley of a river in its middle course in Cumbria, northern England

Fig. 4.14 The lower course of a river in the Northern Territory, Australia

Meanders and oxbow lakes

Meanders are sweeping bends in a river, which tend to occur in its middle and lower courses, as shown in Fig. 4.14. They are natural features, which form as a result of both erosion and deposition.

1 The water in a river flows naturally in a corkscrew pattern. This is called **helical flow**.

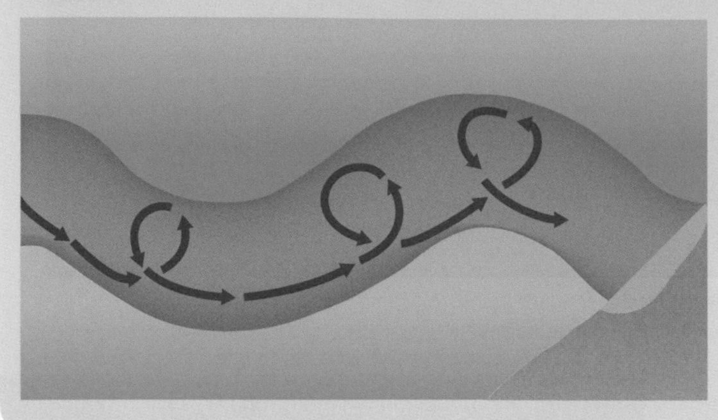

2 Helical flow sends the river's energy to the sides (laterally). The fastest current is forced to the outer bend (A), where it undercuts and erodes the bank to form a **river cliff**.

The helical flow then transports sediment from (A) across the channel to the inner bank (B), or **slip-off slope**, where the slower-moving water deposits it to form a **point bar.**

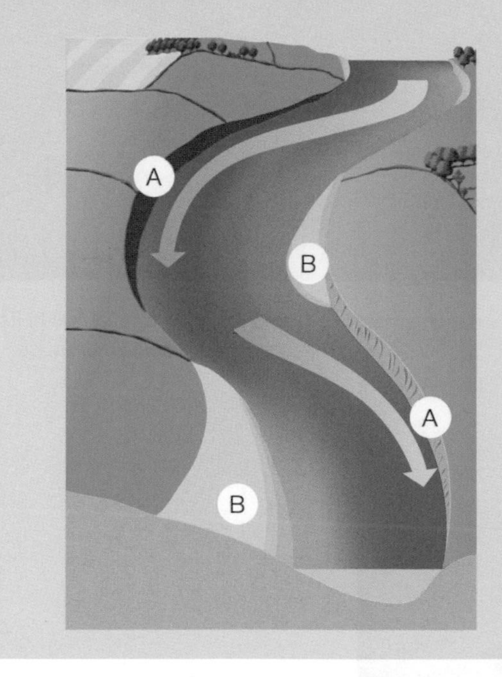

3 Continued erosion sometimes creates a narrow neck between two meanders (X). Eventually, the neck is cut through at Y, and the river creates a new channel for itself across the neck of the meander (an easier route for the water). The old meander then becomes an **oxbow lake** (Z) when deposition seals the ends – completely separating it from the river.

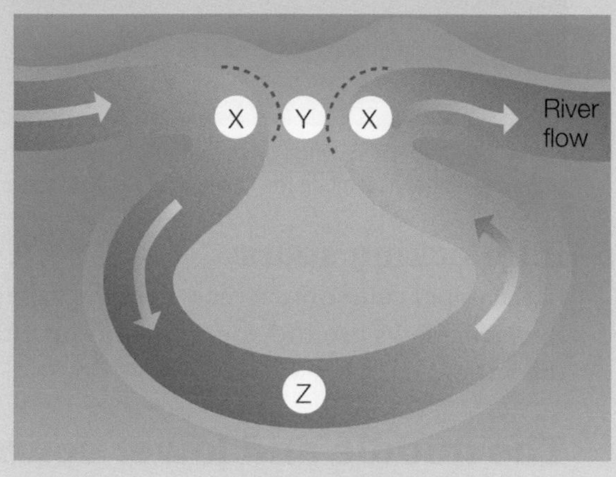

Fig. 4.15 The formation of a meander, slip-off slope and river cliff

Fig. 4.16 A river meander. Notice the gravel deposited on the inner bank of the river (left side of the photograph). Erosion will occur at the outer bank (right side of the photograph).

Floodplains and levees

A floodplain is the flat land next to a river which is liable to flood - sometimes every year, sometimes once in 10 years, sometimes once in 100 years. The floodplain could be less than 100 m across in the case of smaller rivers, but more than 100 km across in the world's largest river valleys. Floodplains are often marshy and poorly drained. Occasionally, the river flows above the level of the surrounding floodplain but is enclosed by raised embankments, or levees.

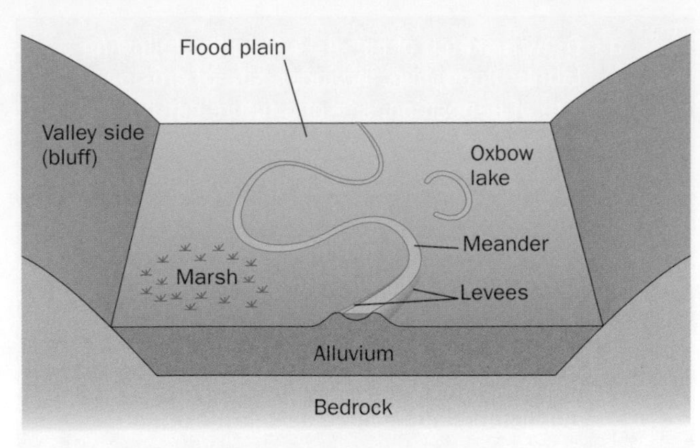

Fig. 4.17 The valley of a river in its lower course

Three types of deposition help to build up the flood-plain:

- → The deposition of point bars on the insides of meanders (see page 114). These deposits are spread across the valley as the meanders migrate - both sideways and downstream
- → The deposition of gravel on the river bed (part of the bed load)
- → The deposition of fine silt and mud (part of the suspension load) on the floodplain itself, when the river overflows its banks during floods.

Levees are formed naturally when a river floods. When the water overflows the river channel, it slows down (loses energy). As a result, the coarsest part of the river's load is deposited close to the channel - making the banks naturally higher. However, during normal flows, the river deposits material from its load on the river bed within the channel - causing the bed to be higher than the surrounding floodplain (see Fig. 4.18).

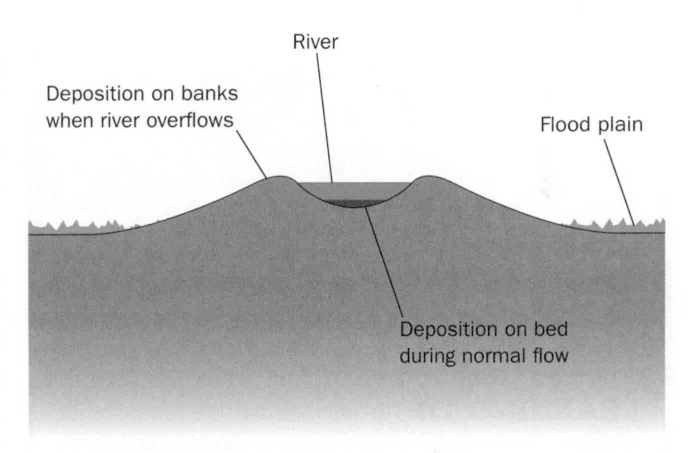

Fig. 4.18 How levees form

Often, natural levees are raised or strengthened in an attempt to stop flooding. Sometimes, artificial embankments are built, but they should not be confused with naturally formed levees.

Deltas

A delta is an area of low-lying, flat, marshy land where a river meets the sea or a lake. There are often lakes or lagoons within the delta. The river channel divides into **distributaries**, which rejoin to form a **braided** drainage pattern. The Greek letter D is delta, which describes the shape of a river delta.

Deltas form when a river carrying its load (mostly mud and silt by this stage) meets the still water of a sea or lake. The loss of velocity (energy) leads to the deposition of the river's remaining load, which builds up gradually to form the delta. Over time, the delta expands out into the sea or lake. Continued deposition blocks the river's main channel, which then leads to the formation of distributaries.

Fig. 4.19 A lake delta in the right foreground of the photograph. Notice how low and flat the land is.

5 **a** Draw a sketch of Fig. 4.14. Add the following labels: floodplain, meander, site of erosion, site of deposition, site of a future cut-off, future oxbow lake.

b Draw a cross-section of a river channel in its middle or lower course. Label on it fast flow, slow flow, slip-off slope, river cliff.

c Explain the meaning of the terms levee, floodplain, delta, distributary.

d Copy and complete Table 4.2. Decide whether the features listed are formed by erosion or deposition and place a tick in the correct column.

Landform	Formed by erosion	Formed by deposition
Pothole		
Levee		
Rapids		
Floodplain		
Waterfall		
Delta		
Gorge		
Interlocking spurs		

Table 4.2 Features formed by erosion or deposition

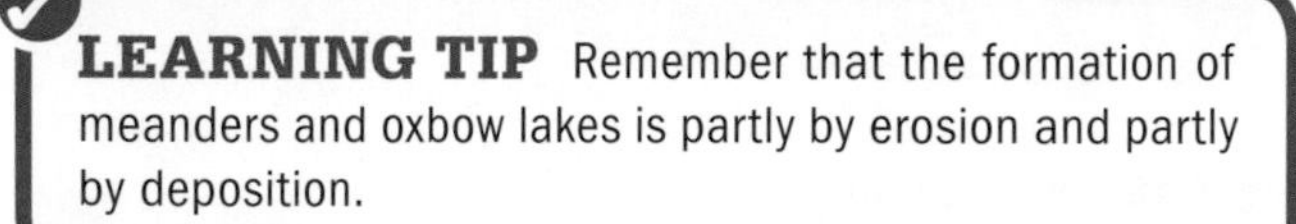

LEARNING TIP Remember that the formation of meanders and oxbow lakes is partly by erosion and partly by deposition.

The long profile of a river

The long profile of a river is a line drawn from the source of the river to the mouth, showing how the gradient changes. A typical long profile is steep in the upper course with waterfalls, rapids, and lakes, and more gentle and smooth in the lower course. Over time, erosion and deposition remove irregularities in the profile, making it smooth and concave.

6 **a** List the landforms which are common in the upper course of a river.

b List the landforms which are common in the lower course of a river.

Discussion point

You have learned a lot about the landforms that result from erosion and deposition by rivers. How long do you think these features take to form? Do they form slowly and gradually, or do occasional major events play a part?

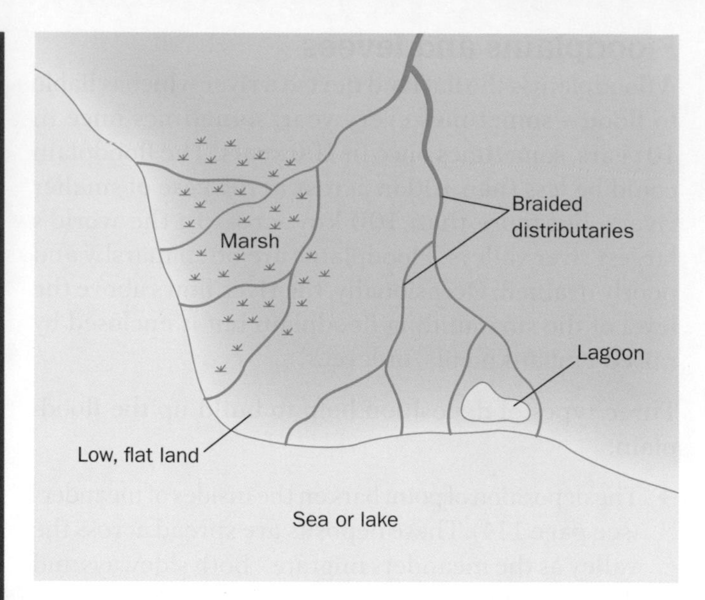

Fig. 4.20 A river delta

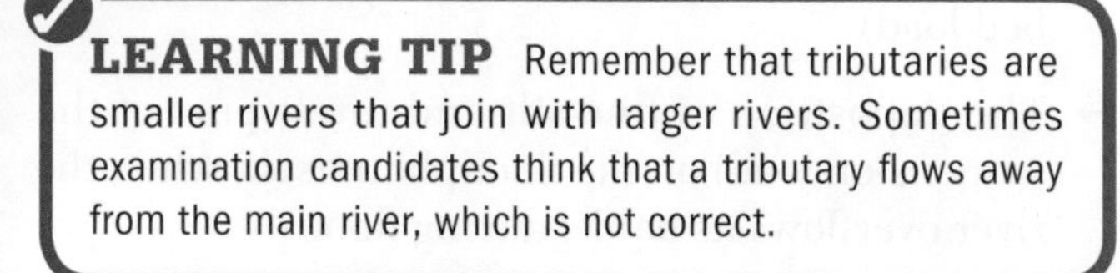

LEARNING TIP Remember that tributaries are smaller rivers that join with larger rivers. Sometimes examination candidates think that a tributary flows away from the main river, which is not correct.

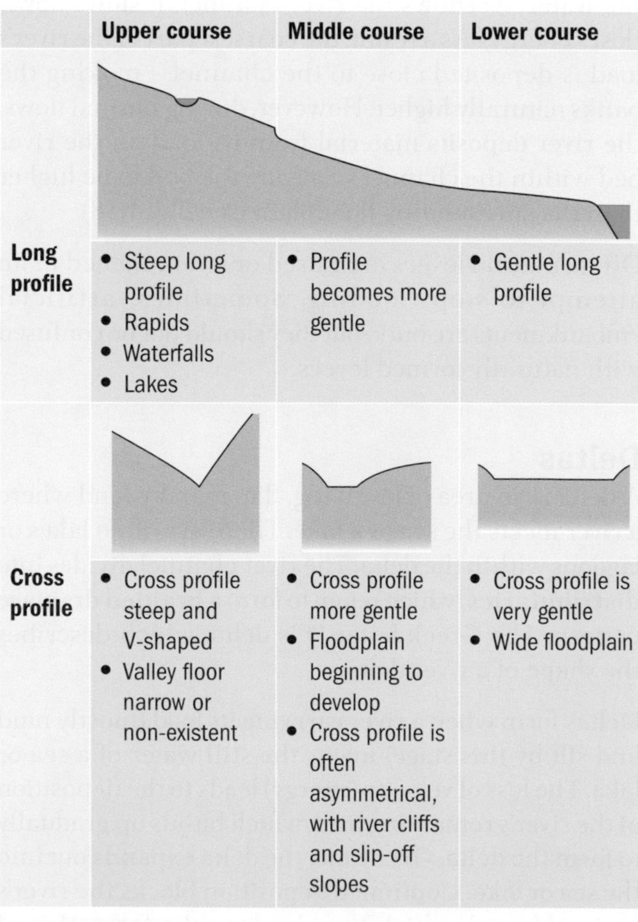

	Upper course	Middle course	Lower course
Long profile	• Steep long profile • Rapids • Waterfalls • Lakes	• Profile becomes more gentle	• Gentle long profile
Cross profile	• Cross profile steep and V-shaped • Valley floor narrow or non-existent	• Cross profile more gentle • Floodplain beginning to develop • Cross profile is often asymmetrical, with river cliffs and slip-off slopes	• Cross profile is very gentle • Wide floodplain

Table 4.3 The long and cross profiles of a river valley

Opportunities presented by rivers, the associated hazards and their management

Floodplains and deltas are often some of the most densely populated areas on the planet. This is because they offer great advantages for settlement, including:

- flat land which makes it easy to build roads and settlements and carry out agriculture
- soils that are often mineral-rich and fertile, due to the silt and mud deposited by the river during floods (alluvium) - so agriculture is profitable
- river valleys that are often natural route ways
- rivers which may be navigable (allowing transport and trade), and provide water for drinking and other uses such as irrigation, as well as fish for food
- the potential to produce hydro-electricity in some cases.

There is a case study about Bangladesh in Chapter 1 that discusses these issues.

However, floodplains and deltas also present problems for human activity:

- In the Tropics, these areas often suffer from diseases carried by insect pests, such as malaria (mosquitoes) and sleeping sickness (tsetse flies).
- In all climatic zones, floodplains and deltas are liable to flooding. The dense population can make the effects of the flooding particularly severe.
- Erosion on the outer banks of meanders can destroy houses, roads and farmland.

Flooding

The discharge of a river is the volume of water flowing down the river at any one time. It is measured in cubic metres per second. When the discharge can no longer be contained within the channel, the river overflows and floods the surrounding area.

From rain to river

How does the water get into a river in the first place? When it rains, very little falls directly into the river itself - most of it falls elsewhere. Fig. 4.21 shows what happens next.

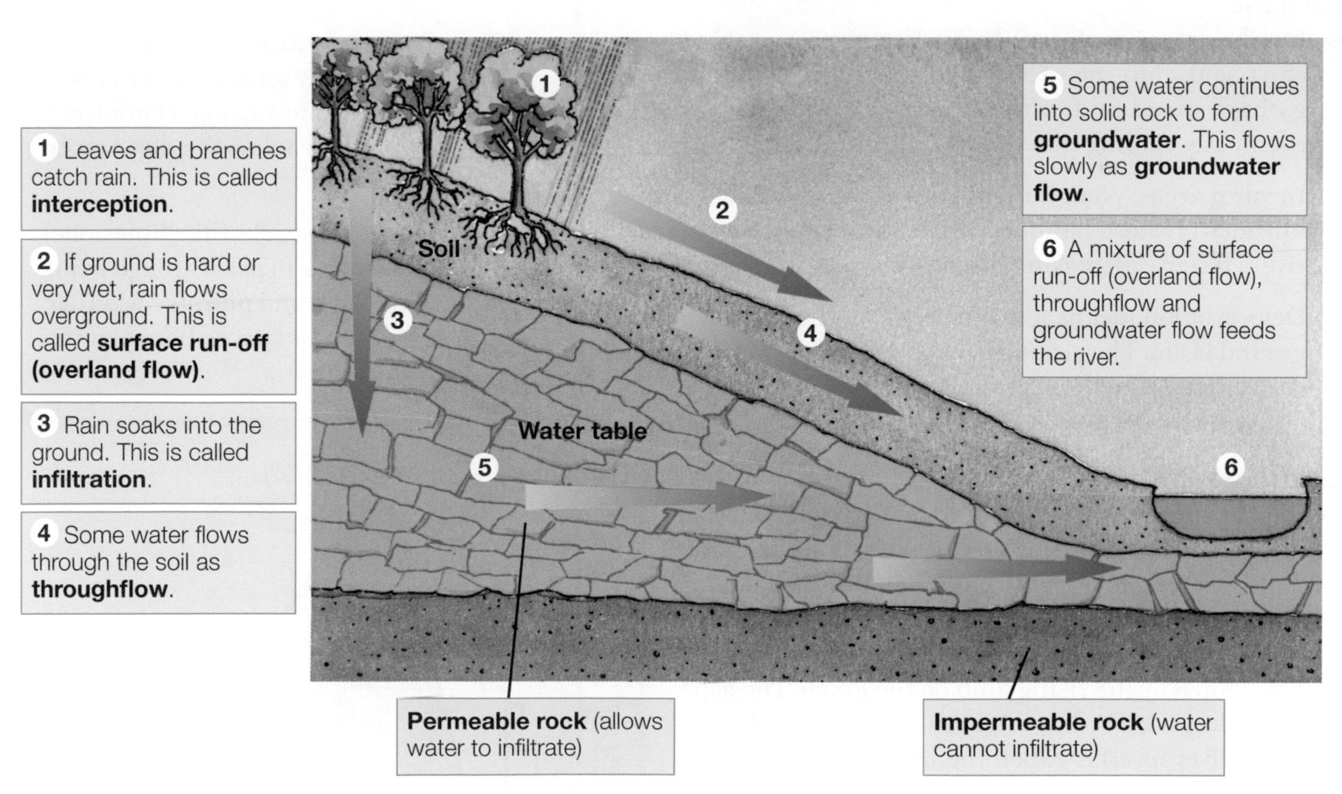

Fig. 4.21 How water reaches a river

Factors affecting discharge

The volume of water in a river can change due to the factors outlined here.

Rainfall

The amount and type of rainfall will affect a river's discharge:

- **Antecedent rainfall** is rain that has already happened. It can mean that the ground has become saturated. Further rain will then flow as surface run-off towards the river.
- Heavy continual rain, or melting snow, means more water flowing into the river.

Weather conditions

- High temperatures increase evaporation rates from water surfaces, and transpiration from plants - reducing discharge.
- Long periods of extreme cold weather can lead to frozen ground, so that water can't soak in.

Discharge

Relief

Steep slopes mean that rainwater is likely to run straight over the surface before it can infiltrate. On more gentle slopes, infiltration is more likely.

Rock and soil type

- **Permeable** rocks and soils (such as sandy soils) absorb water easily, so surface run-off is rare.
- **Impermeable** rocks and soils (such as clay soils) are more closely packed. Rainwater can't infiltrate, so water reaches the river more quickly.
- Permeable rocks (like limestone) allow water to pass through joints in the rock, and **porous** rocks (like chalk) have spaces between the rock particles.

Land use

- In urban areas, surfaces like roads are impermeable - water can't soak into the ground. Instead, it runs into drains, gathers speed and joins rainwater from other drains - eventually spilling into the river.
- In rural areas, ploughing up and down (instead of across) hillsides creates channels which allow rainwater to reach rivers faster - increasing discharge.
- Deforestation means less interception, so rain reaches the ground faster. The ground is likely to become saturated and surface run-off will increase.

Hydrographs

A flood or storm hydrograph is a graph showing how a river responds to a particular storm. It shows rainfall and discharge (see Fig. 4. 22). When it rains, it takes a while for water to reach the river (as shown in Fig. 4.21). Once water does enter the river, discharge increases. This is shown by the rising limb on the graph. The gap between the peak (maximum) rainfall and peak discharge (highest river level) is called the lag time. Changes to land use can mean that water gets into the river faster. This makes the lag time shorter, the rising limb steeper, and the peak discharge higher.

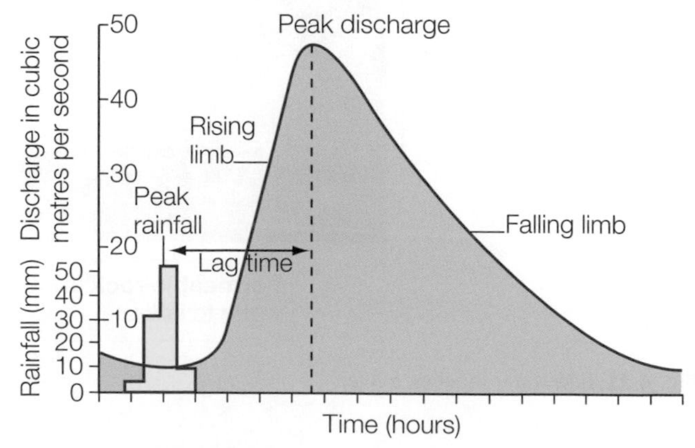

Fig. 4.22 A storm hydrograph

Factors affecting discharge

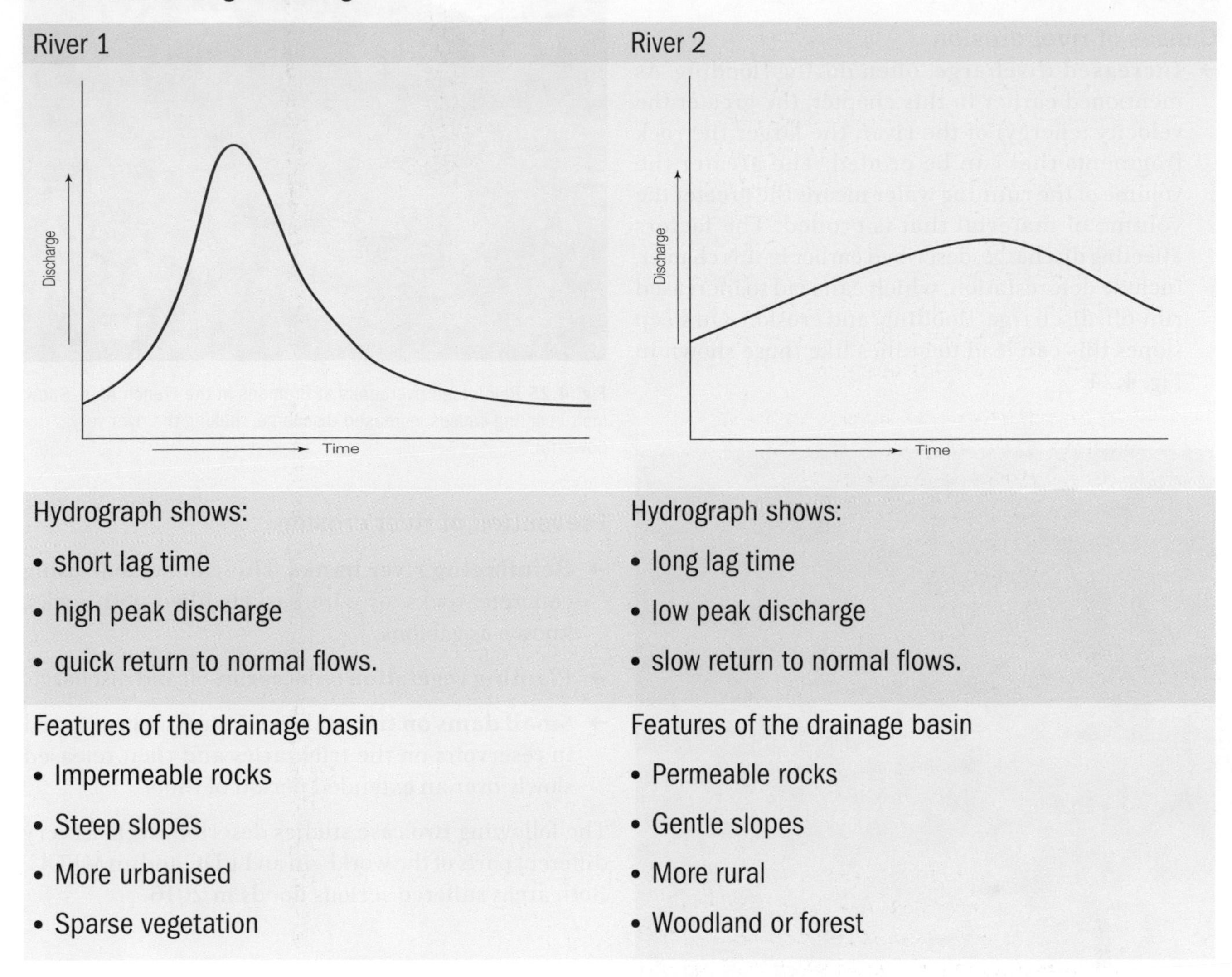

Fig. 4.23 The hydrographs of two rivers, affected by the same rainstorm. River 1 is more likely to flood.

Managing river flooding and its impacts

Planting vegetation

Planting vegetation, such as trees, in the area drained by a river works in two ways. The trees take in rainwater through their roots and lose it by transpiration, so it does not reach the river. Vegetation also acts like a sponge, intercepting the rainwater and releasing it slowly, so that flood peaks are reduced.

Reservoirs

Water can be trapped in reservoirs on the tributaries and then released slowly over a longer period of time.

Straightening the channel

This shortens the river and gets the water away faster.

Dredging the channel

Making the channel deeper increases its capacity and makes it less likely to overflow.

Artificial levees

These increase the capacity of the channel, just like dredging. Usually the banks are strengthened, for example with concrete or stone, so that they are less likely to break.

Bridge design

Bridges with wide pillars and walls on top act like dams, which hold back the water and allow it to spread. Modern bridges are slim and streamlined to prevent this happening.

Wash lands and building planning

Because floods cannot be completely prevented, land use on floodplains should be carefully controlled. Ideally, residential land use should be prevented and these areas used for things such as recreation. The flood waters can then be channelled into areas where they will cause less damage.

Erosion

Causes of river erosion

- **Increased discharge**, often during flooding. As mentioned earlier in this chapter, the greater the velocity (energy) of the river, the larger the rock fragments that can be eroded. The greater the volume of the running water means the greater the volume of material that is eroded. The factors affecting discharge, described earlier in this chapter, include deforestation, which can lead to increased run-off, discharge, flooding, and erosion. On steep slopes this can lead to gullies like those shown in Fig. 4.24.

Fig. 4.24 Gully erosion in Kwazulu Natal, South Africa

- **Soft rocks**. This can be seen on floodplains when the soft clays and sands of the alluvium are easily eroded.
- **Channel size and shape**. Erosion may be concentrated on the outside of meanders where velocity is greater.

Fig. 4.25 Reinforced riverbanks at Bramans in the French Alps. Snow melt in spring causes increased discharge, making the river very powerful.

Prevention of river erosion

- **Reinforcing river banks**. This can be done using concrete, rocks, or wire baskets filled with rocks, known as gabions.
- **Planting vegetation** reduces run-off and discharge.
- **Small dams on tributaries.** Water can be trapped in reservoirs on the tributaries and then released slowly over an extended period of time.

The following two case studies describe rivers in very different parts of the world - in an LEDC and an MEDC. Both areas suffered serious floods in 2016.

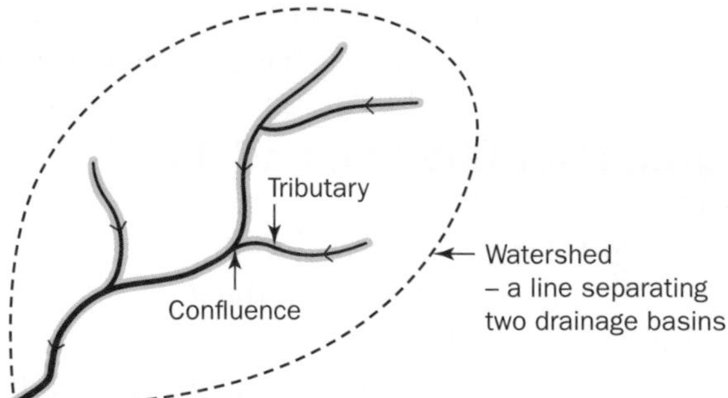

Fig. 4.26 A drainage basin – the area drained by a river and its tributaries

CASE STUDY

The Indus River Valley, Pakistan (an LEDC)

The Indus is one of the world's great rivers. It rises in the Himalayas and Hindu Kush mountains and flows to the Arabian Sea. It drains most of Pakistan (an LEDC) but also parts of Afghanistan, China, and India. Flooding occurs each summer.

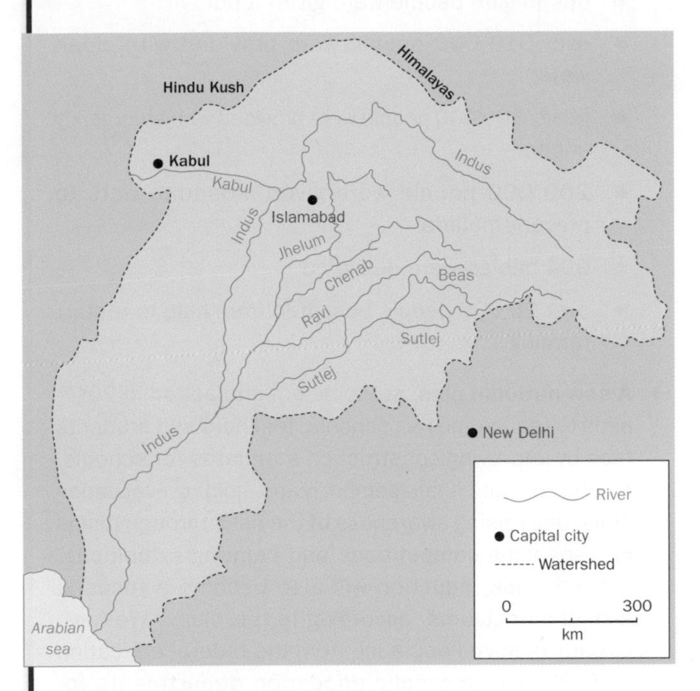

Fig. 4.27 The Indus drainage basin

Opportunities

- **Water supply**. The Indus and its tributaries are the main water supply for Pakistan's population of over 170 million. There are treaties between India and Pakistan about use of water in the Indus tributaries.
- **Irrigation and agriculture**. Rainfall in the southern parts of the country is less than 250 mm per year but there are rich alluvial soils. There is a complex irrigation network of dams and canals, including the 1350-m-long Guddu Barrage. Pakistan is one of the world's top ten producers of both wheat and cotton.
- **Hydroelectricity**. For example, the Taunsa Barrage near Dera Ghazi Khan produces 100 000 kilowatts of electricity. This is vital for urban centres and heavy industry.
- **Flat land for building**.
- **Fishing**. This is particularly important in the Sindh area. Palla fish is a delicacy for people living along the river with Sukkur, Thatta, and Kotri being the major fishing centres. Fish farming of pomfret and prawns is important in the delta.

Hazards

- **Deforestation and industrial pollution are** affecting the vegetation and wildlife of the delta.
- There are concerns that the **Indus River may be shifting its course** westwards, although the change may take centuries.
- **Sediment is clogging irrigation canals**, affecting agricultural production and vegetation.
- Extreme heat has caused water to evaporate, leaving **salt deposits** that make land useless for cultivation.
- There is **conflict** between India and Pakistan over water use.
- Climate change could increase **flooding** (see below).
- Back in 1960, Pakistan had a population of just under 50 million. According to the census of 2017, Pakistan now has a population of around 200 million. This means that the pressures on the food and water supplies are naturally greater than in the past. Flooding can now have a greater impact on water supply, food supply, sanitation, and sewage plants.

The causes of flooding

- Torrential summer **monsoon rains**. In August 2010, more than half the normal monsoon rain fell in only one week. Normally it is spread over three months. Hyderabad recorded 77 mm of rain in 24 hours on 7 August 2016.
- **Melting snow and glaciers** in the Himalayas.
- **Global warming** could mean more glacial melting and more monsoon rains.
- In some countries flood risk has been reduced by building **levees** along rivers. These barriers prevent the rivers from bursting their banks in extreme floods. However, because the Indus is choked with sediment from erosion in the Himalayas, building levees has caused the river channel to silt up, causing even bigger floods when the levees break.
- **Deforestation** in the upper parts of the tributaries has caused erosion and more sediment has been transported down the rivers, blocking the channel.

The effects of flooding

- **Loss of life**. In 2010 1700 people died. In 2016 434 people died. People were killed by drowning or electrocution, or they were crushed when walls or buildings collapsed on them.
- **Displacement**. In 2010 14 million people were displaced from their homes. In 2016 5.3 million people were displaced.
- **Loss of homes**. In 2010 over 700 000 homes were damaged or destroyed, causing widespread homelessness.
- In 2010 the floodwaters covered roughly one-fifth of Pakistan's land area. In 2016 at least 690 000 ha of arable land were inundated. Many people suffered from **malnutrition** and a **lack of clean water**. In 2016 six million needed food aid.
- **Infrastructure damage**. Damage occurred to roads, railways, bridges, the electricity network, and the irrigation system.
- **Loss of school buildings**. In 2010 10 000 schools were damaged or destroyed. Since then another 10 000 have been affected in the same way.
- **Loss of livestock**. In 2010 about 450 000 livestock were lost.

Responses to the floods

- **Appeals** were immediately launched by international organisations, like the UK's Disasters Emergency Committee, the UN, the Red Crescent and Médecins Sans Frontières.
- **Pakistan's government also tried to raise money** to help the huge number of people affected.
- **Foreign governments donated millions of dollars**, and Saudi Arabia and the USA promised $600 million in **flood aid**.
- The UN's World Food Programme provided crucial food aid.
- Following the 2010 floods, the DEC reported that:
  - one million people were given food
  - over 510 000 people were provided with clean water
  - nearly 300 000 people were provided with emergency shelter
  - 200 000 people were given mosquito nets to prevent malaria
  - 994 babies were delivered
  - over 26 000 people benefited from help to restart farming.
- **A new national plan**, set to be put into action in 2017, aims to reduce the risk schools, teachers and students face by improving construction standards for schools, creating disaster management plans, holding evacuation drills, and raising awareness of the risks through things like speaking competitions and painting exhibitions. Disaster risk reduction will also become a focus in school curriculums, according to the plan. "We have already received applause from the federal education ministry and provincial education departments for framing the plan, and we feel really proud of it," said Major-General Asghar Nawaz, chairman of the National Disaster Management Authority (NDMA).

Fig. 4.28 People walking through floodwaters after their houses were flooded at UnarPur, Pakistan

CASE STUDY

The Eden and Derwent Valleys, UK (an MEDC)

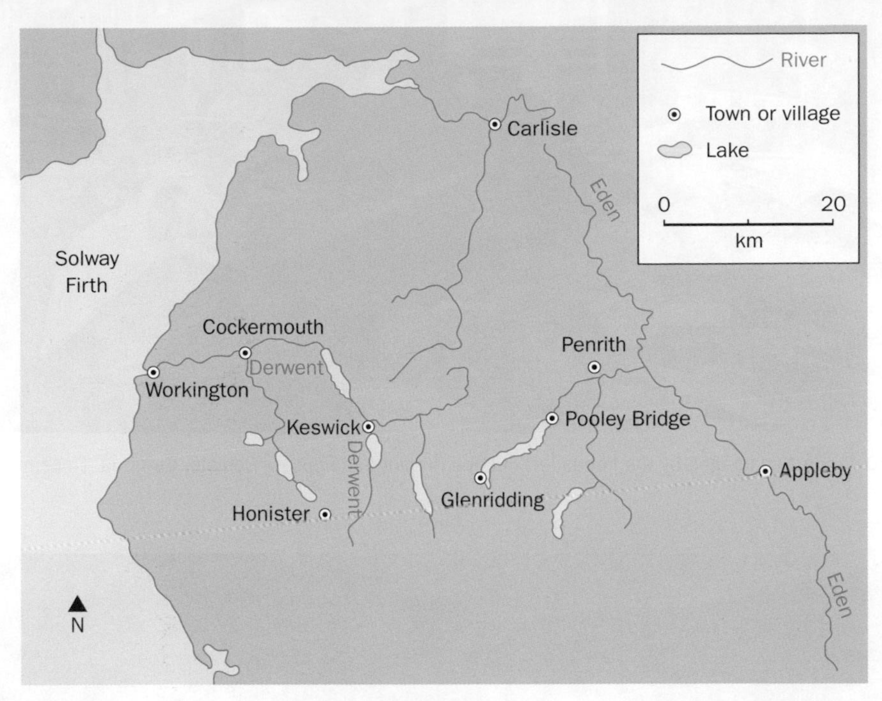

Fig. 4.29 The Eden and Derwent drainage basins

Opportunities

- The scenic beauty of the Lake District has made **tourism** a significant industry, centred on the town of Keswick.
- The fertile alluvial plains are important for **agriculture**, including dairy cattle, and wheat and barley.
- The focus of valleys at Carlisle (population 107 000) has led to its development as a **route centre**, railway junction, and administrative centre.

Hazards

- **Flooding** is the only significant hazard, with significant floods in 2005, 2009, and 2015.

The causes of flooding

- **Heavy rainfall**. In December 2015 Storm Desmond brought heavy rainfall and flooding to Cumbria. 150 to 200mm were expected but Storm Desmond broke the UK's 24-hour rainfall record, with 341.4 mm of rain falling at Honister Pass, on 5 December.
- **Steep mountain slopes** leading to rapid run-off.
- **Impermeable rocks** such as volcanics lead to rapid run-off.
- **Deforestation**. The natural forest in the mountains was felled centuries ago to allow sheep farming to take place. It has been claimed that the sparse grassland vegetation contributes to rapid run-off.

The effects of flooding

- **Loss of life** has been restricted to isolated incidents and the main damage has been to property. The 2015 floods were estimated to have caused £500 million worth of damage.
- **Damage to property**. More than 5000 households are estimated to have been affected by the 2015 floods. These were in Carlisle, Cockermouth, Appleby, and Keswick.

Fig. 4.30 Property damaged by the floods left outside the flooded houses, Carlisle, Cumbria, December 2015

Fig. 4.31 The A591 main road through Cumbria collapsed and washed away, December 2015

- **Damage to roads and bridges**. Roads were washed away and bridges were destroyed at Workington and Pooley Bridge. They took many months to rebuild, with long-term traffic disruption. Landslides damaged roads, notably the A591 through the centre of the Lake District tourist area (see Fig. 4.31).
- **Loss of tourist income**, claimed to be £1 million a day.
- **Industries** were forced to close, for days, weeks or months, for example the McVitie's biscuit factory in Carlisle, which employs 640 people. The Glenridding Hotel was flooded four times in the winter of 2015–16 and had to close for six months for repairs.
- **Agricultural** land was flooded and covered in deposits of gravel. Livestock were swept away. In 2015, one of Gordon Tweedie's cows was found alive, 30 km away from his farm in the village of Lazonby.

Fig. 4.32 Building new flood defences in Carlisle, Cumbria

Responses to the floods

- Temporary **evacuation** and relief were provided by the army and fire service. Some flooded property was insured. The 2005 floods in Carlisle cost insurance companies £272 million. Many houses were not reoccupied for up to a year. Future insurance costs for policy holders are likely to be higher.
- Damage to roads and bridges is not covered by insurance. **Rebuilding** the A591 took five months.
- The government said it would give **temporary relief** on business and council tax for homes and businesses.
- After the 2016 floods, **embankments** were built to protect houses in Carlisle.

Fig. 4.33 New food defences being built, Glenridding, Cumbria

5 Coasts

This chapter covers the following Cambridge IGCSE® and O Level topic:

- **2.3 Coasts**

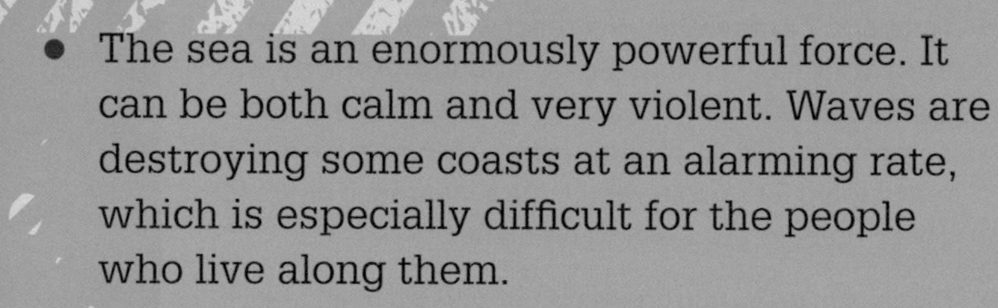

- The sea is an enormously powerful force. It can be both calm and very violent. Waves are destroying some coasts at an alarming rate, which is especially difficult for the people who live along them.
- Did you know that, in some areas with weak rocks, the coast is retreating at a rate of more than 6 metres a year? Will we ever be able to stop this destruction? Yet, in other places, the sea is building new land.
- How do the waves do these things? The shape of the coastline is constantly changing. Beautiful landforms are being made and destroyed.
- Coasts attract many tourists, who spend much of their time on beaches made from grains of sand – the end product of weathering and erosion. Tourist resorts rely on their beaches, but a single storm can remove them.
- Coral reefs are some of the most amazing and beautiful places on the planet. But why are they found only in tropical seas?

In this chapter you will learn about:

- the work of the sea in eroding, transporting, and depositing
- the landforms that result from the interaction of marine processes with the rocks along coastlines
- the work of the wind in forming coastal sand dunes
- coral reefs and mangrove swamps and the conditions necessary for their development
- how coastal erosion and tropical storms cause hazards for people
- the opportunities coasts provide for people.

Introducing the coast

- The **coast** is a zone where the land meets the sea.
- The **coastline** (the outline of the edge of the land on a map) follows the mean **high water mark** on a lowland coast, and the foot of **cliffs** on a steeply sloping coast.
- The sea rises to high **tide** (high water mark) and falls to low tide (low water mark) normally twice a day. This results in the high and low water marks. This tidal range varies both from coast to coast and with the time of year.
- The area between the lowest tide level and the highest point reached by storm waves is known as the **shore.**
- The tide's only importance in the formation of coastal landforms is in controlling how high and low the waves can work. It is the action of waves that produces the coastal features.

Explaining waves

Freak waves can be caused by movements of the seabed during earthquakes, or by very large ships passing too quickly near the coast, but most waves result from friction between the wind and the surface of the sea. This causes part of the sea to rise at right angles to the wind. Seen from above, the resulting wave is long and narrow. The wave form increases in height as it is driven forward by the wind.

The size of a wave depends on:

- the wind speed
- the length of time during which the wind blows in the same direction
- the length of sea over which the wind blows (the **fetch** of the wave). The greater these factors are, the bigger the wave.

Hawaii is famous for its big waves - some are more than 20 m high. Sometimes they are formed by storms that occur far away from where the waves eventually break on to the shore. An example is the North Pacific Swell. The wind in Hawaii might be very light when these large waves arrive at its coast.

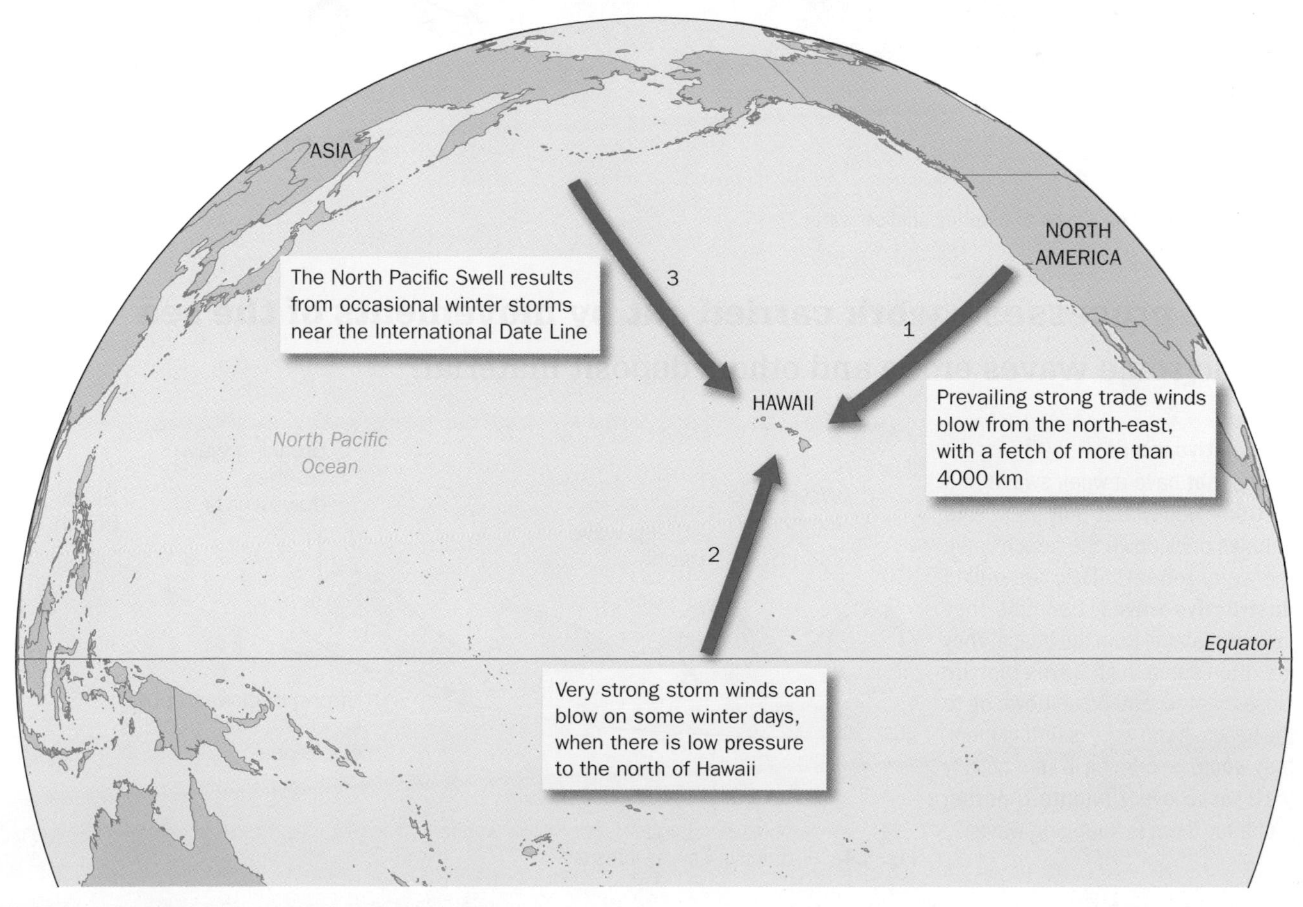

Fig. 5.1 Influences on the waves that reach Hawaii

A boat with its engine switched off bobs up and down as a wave passes beneath it – but it does not move towards the shore. This is because only the wave *form* moves beneath it (see Fig. 5.2):

→ Each water particle moves in a vertical circle to form the wave.

→ As water rises, it forms the **wave crest**. As it falls, it forms the **wave trough**.

→ The **wave length** is the distance between two wave crests.

→ The **wave height** is the difference between the height of a wave crest and the adjacent wave trough.

Water particles can no longer move in a circular manner when the wave reaches shallow water, so the wave breaks.

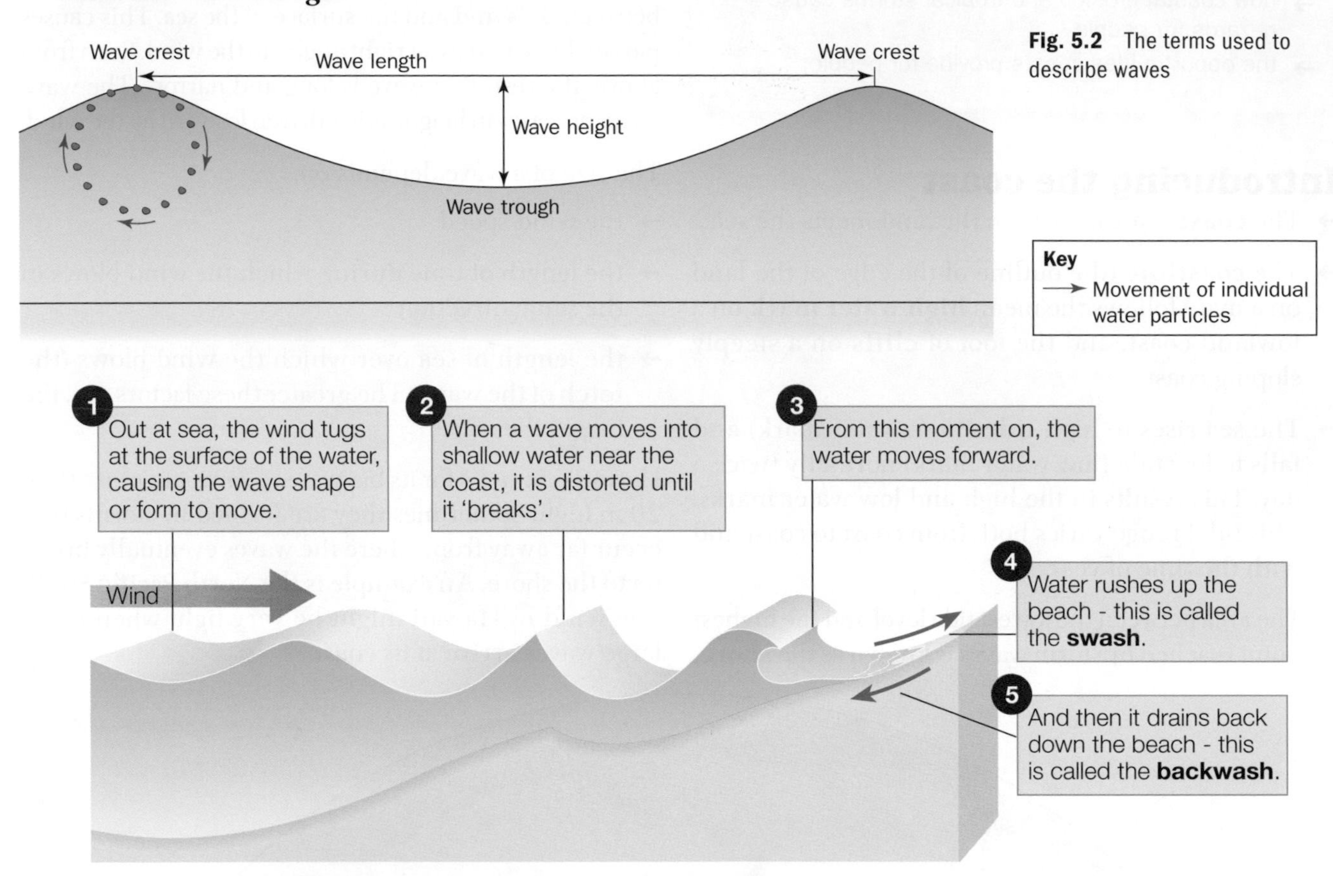

Fig. 5.2 The terms used to describe waves

Fig. 5.3 The effect on waves of entering shallow water

Marine processes – work carried out by movements of the sea

Why do some waves erode and others deposit material?

Destructive waves

Waves that have a weak swash and a strong backwash pull sand and pebbles back down the beach when the water retreats. They are called **destructive waves**, because they remove material from the beach. They are often steep, high waves that are close together and crash down on to the beach. If you were counting them, they would be coming in very quickly – 10 to 15 every minute. Another name for them is 'plunging waves'.

Waves close together

Steep wave front

Breaking wave plunges downwards

Steep beach

Strong backwash pulls sand and even pebbles out to sea

Fig. 5.4a Destructive waves – erosion

Constructive waves

Waves that have a very strong swash and a weak backwash are known as **constructive waves**, because they build up the beach. They push sand and pebbles up the beach and leave them behind when the water retreats, because the backwash is not strong enough to remove them. They are often low waves with a longer time between them. As they break, they spill up the beach, so they're also known as 'spilling waves'. They come in at a rate of 6–8 every minute.

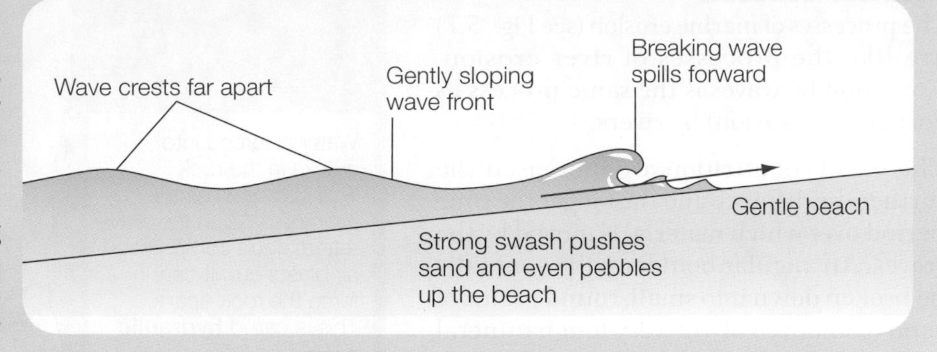

Fig. 5.4b Constructive waves – deposition

Discussion point

Some students have never been able to stand in the sea and feel the power of waves washing in and out. If you have experienced the force of waves at the coast, tell your classmates what they were like and how powerful they felt.

Why should people relaxing at the coast always keep an eye on the sea?

Oamaru is located on an exposed straight stretch of the east coast of New Zealand's South Island. The headland in the background of the photo can give shelter only when the winds come from the south and south-east. Destructive waves are being driven by strong winds approaching from the east, over a very long fetch. The waves are plunging and digging down at the sand. The swash is small and the backwash powerful.

Fig. 5.5 Destructive waves attacking the coast of New Zealand just south of Oamaru

The waves in this photo are in a sheltered **bay**. Note that the wave crests are a greater distance apart than those shown in Fig. 5.5. The swash has moved a long way up the beach and the backwash is sinking into the porous sand, so it is weak. These waves have a much lower height and the beach has a very gentle gradient.

Fig. 5.6 Constructive waves on Zihuatanejo Beach on the west coast of Mexico

Marine erosion

The processes of marine crosion (see Fig. 5.7) are like the processes of river erosion. Corrasion by waves is the same process as corrasion (abrasion) by rivers.

The effects of attrition are increased the further the distance and the longer the time period over which material is moved by the waves. An angular boulder will eventually be broken down into small, round grains of sand (composed of the very tough mineral quartz), which are resistant to further breakdown. Rounded beach material of intermediate size between boulders and sand is known as **shingle**.

LEARNING TIP When writing the terms 'corrasion' and 'corrosion', be sure to make the 'a' and 'o' clear, so that your meaning is clear.

There is another process of erosion – called **corrosion**. This happens when seawater dissolves material from the rock. It happens along limestone and chalk coasts, when calcium carbonate is dissolved.

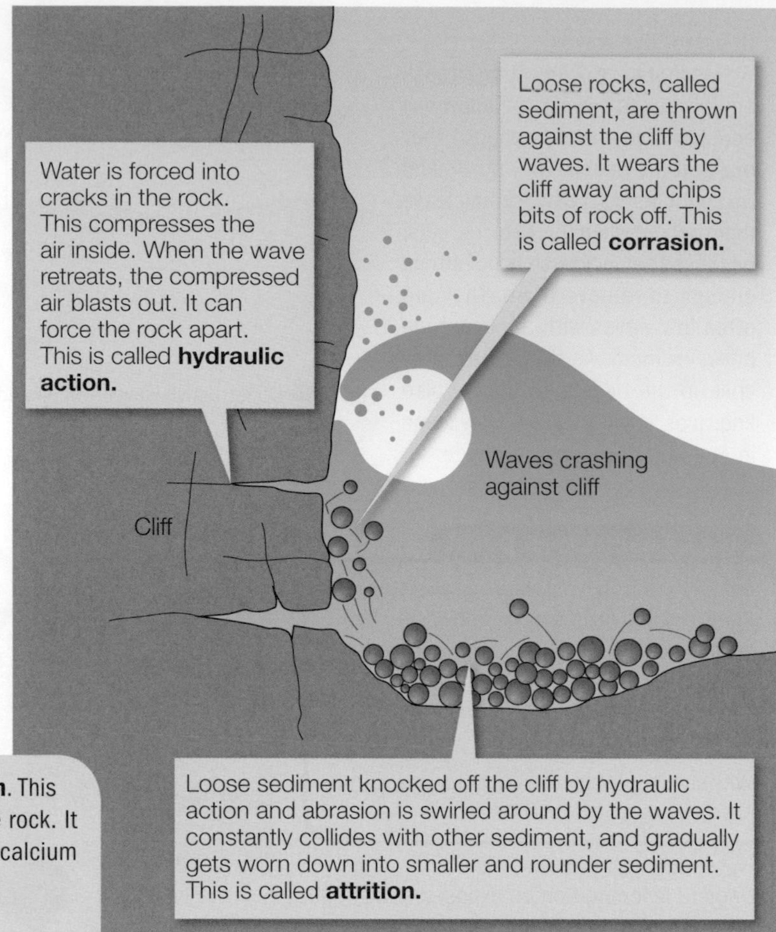

Fig. 5.7 The processes of marine erosion

Marine transportation

The sea transports the load or sediment that it obtains by erosion in the same ways that a river does.

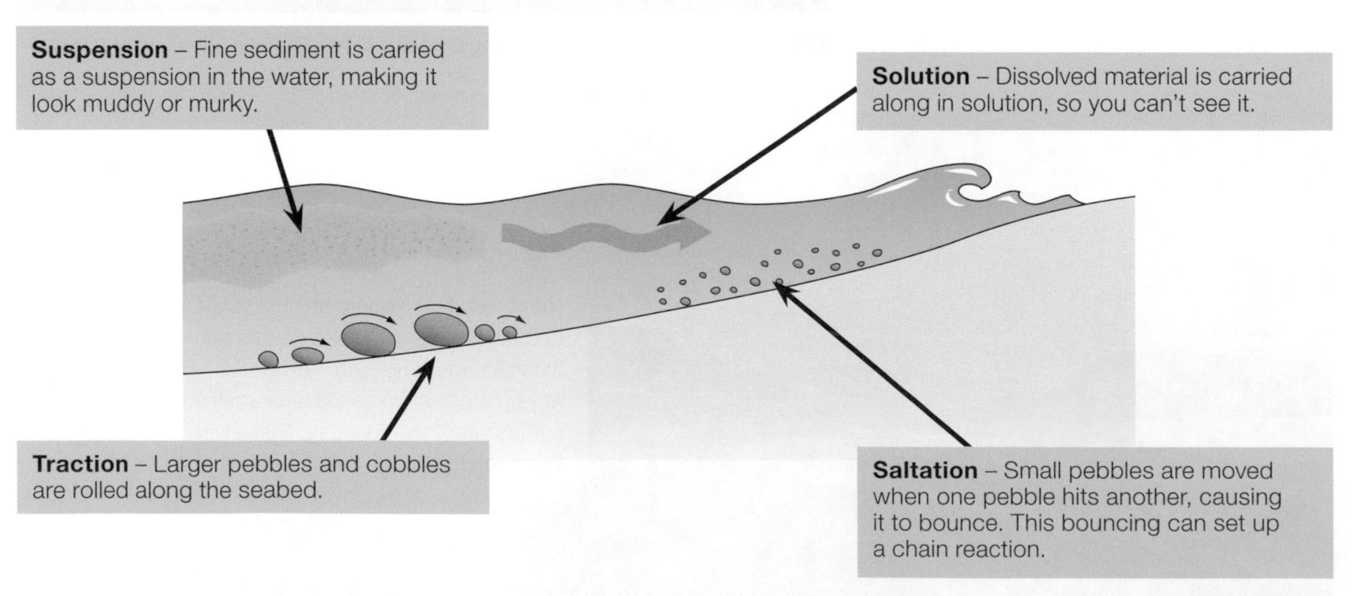

Fig. 5.8 The processes of marine transportation

As well as being moved up and down a beach, sediment can also be moved along it - if the incoming waves are driven by onshore winds at an oblique angle to the coast. This movement of sediment along a beach is called **longshore drift** (see Fig. 5.9).

On coasts where longshore drift occurs mainly in one direction, beach sediment is transported further down the coast. If an obstruction prevents its replacement from further up the coast, the beach will be depleted. This causes two problems for the local authorities:

- The smaller beach is less attractive to tourists, causing a loss of income.
- It removes the protection from erosion that the beach provides for cliffs.

To counter this, some local authorities put barriers called **groynes** at right angles to the beach - to trap the sediment and reduce longshore drift (see Fig. 5.11).

Material may also be moved along some coasts in the offshore zone by longshore currents. These are like rivers of water moving through the sea along the coast.

LEARNING TIP The prevailing wind is not always responsible for the direction of longshore drift. In some locations, the strongest winds - known as dominant winds - approach from a different direction. Many spits are recurved, or hooked, as a result of material being moved in two directions.

LEARNING TIP Do not confuse longshore currents offshore with longshore drift, which has a much greater influence on coasts. Longshore currents are also different from the major ocean currents that circulate in the oceans.

LEARNING TIP Longshore drift does not *form* landforms. It supplies the sediment for the process of deposition to form them.

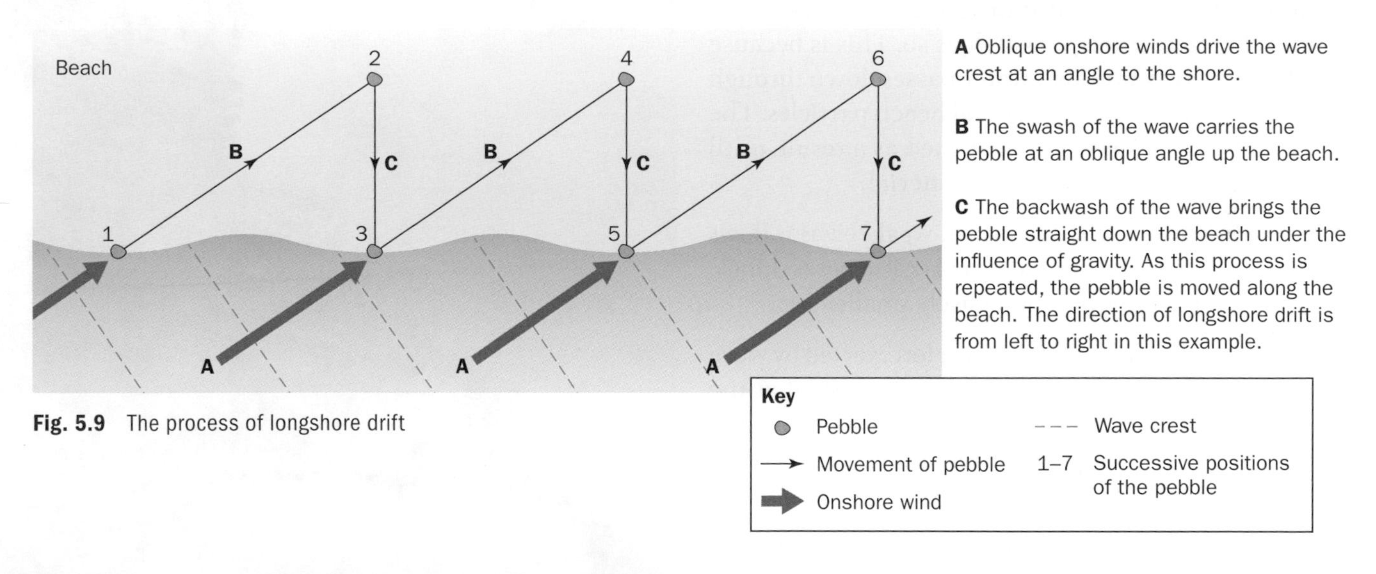

Fig. 5.9 The process of longshore drift

Fig. 5.10 Waves breaking obliquely on a shingle beach

Fig. 5.11 Groynes on a sand and shingle beach

1 **a** Was longshore drift moving material to the left or the right of the photograph when Fig. 5.10 was taken? What is the evidence for your answer?

b Describe the waves in Fig. 5.11.

c What evidence is there in Fig. 5.11 that the groynes are being successful in reducing longshore drift?

d Why might a local authority further down the coast be concerned about these groynes?

e What evidence is there in both photographs that, as rivers do, the sea sorts material during deposition?

f How is the slope of a beach related to the size of its material?

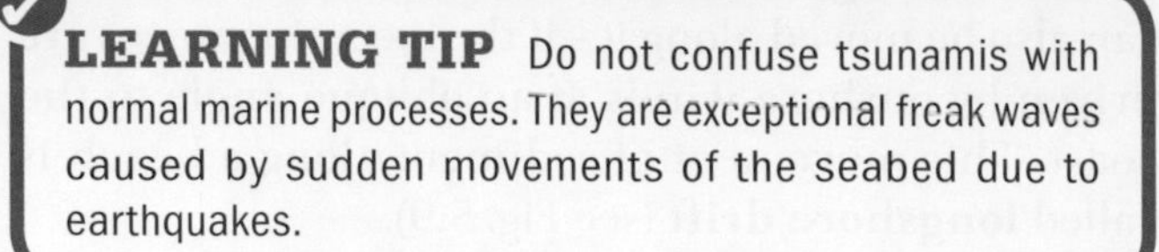

LEARNING TIP Do not confuse tsunamis with normal marine processes. They are exceptional freak waves caused by sudden movements of the seabed due to earthquakes.

Marine deposition

- When the strong swash of a constructive wave moves up a beach, it carries sand or shingle with it.
- The largest material is deposited at the upper limit reached by the swash.
- The backwash then carries smaller material back down the beach - but it progressively loses water, and therefore energy, as it does so. This is because a beach is very porous - water passes down through spaces between the individual beach particles. The flow of the backwash is weakened as a result, until it can carry only the lightest material.
- Consequently, as the backwash weakens as it flows back towards the sea and gets weaker, it deposits shingle and sand particles of progressively smaller size.
- The material on a beach is, therefore, sorted by wave deposition - the largest shingle is deposited at the top of the beach and the finest sand is deposited near the sea. The smallest mud particles settle in the low-energy environment offshore.
- When a storm occurs at the time of the highest tides, large shingle is tossed above the usual high tide level to form a ridge at the top of the beach.

Landforms formed by marine processes

Every coast is unique, because its features depend on the nature of the local rock, the processes acting upon it, and how long they have been operating. However, there are relatively few different types of coastal landform, but they occur frequently.

Landforms formed by marine erosion

Cliffs and wave-cut platforms

Cliffs are vertical or steeply sloping rocks. The angle of their slope depends on the nature of the rocks that form them, and also on the amount and ferocity of wave attack at their bases. Many cliffs have an indentation - called a **wave-cut notch** - at about the high-tide level, where wave attack has undercut the rock. Undercutting keeps the cliff steep.

Fig. 5.12 An undercut cliff made from layers of sedimentary rock, with fallen blocks of rock at its base. In this cliff (at Burton Bradstock, UK), less resistant sandstone layers have been worn back more than more resistant layers cemented with calcite.

Waves continue to attack the base of the cliff until the rock above becomes unsupported and collapses. This process continues and the cliff is steadily worn back. The retreat of the cliff leaves a **wave-cut platform** where the cliff once stood. Fig. 5.13 shows a low cliff at Cape Town in South Africa, with a wave-cut platform in front of it.

Look at an atlas map of South Africa and note how exposed the **headland** called the Cape of Good Hope is. Fig 5.15 shows the wave-cut platform nearest to the cliff littered with boulders broken from the cliff face by destructive waves. The boulders have been shaped by different amounts of attrition.

Fig. 5.13 A low cliff and wave-cut platform at Cape Town, South Africa

Fig. 5.15 The cliff and wave-cut platform at the Cape of Good Hope

Wave-cut platforms extend between the low-water mark and the cliff base at the high-water mark. They slope gently towards the sea and may have rock pools eroded into them at weak points. These solid rock platforms are often covered with debris eroded from the cliffs.

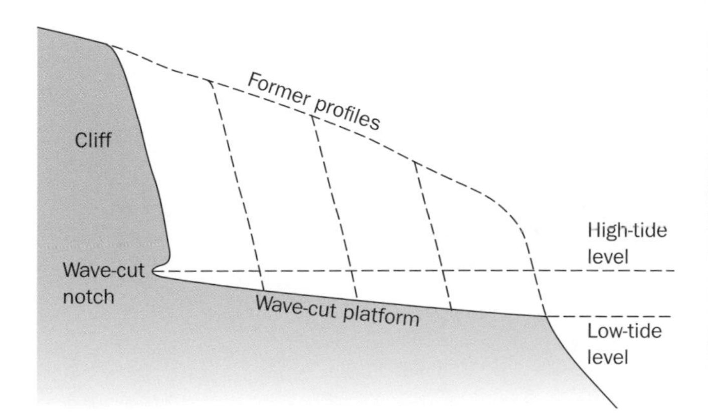

Fig. 5.14 The formation of a wave-cut platform

Fig. 5.16 The outer edge of the wave-cut platform at the Cape of Good Hope

The profile of a cliff varies according to the nature and dip of the rocks which form it (see Figs. 5.12, 5.15, 5.17 and 5.18).

2 **a** Compare the cliff profile at the Cape of Good Hope (Fig. 5.15) with the one in Fig. 5.17. Suggest why they are different.
b What was the state of the tide when Fig. 5.16 and Fig. 5.18 were taken? Give a reason for your answer.
c What type of wave is breaking in Fig. 5.16?
d Describe the wave-cut platform at the Cape of Good Hope.

Fig. 5.17 The base of the cliff below the red rock is a soft rock. The rock layers are horizontal, so the cliff profile is vertical.

Fig. 5.18 Here the rock layers dip steeply away from the sea, so the cliff profile slopes more gently. Rock pools are visible in the extensive wave-cut platform.

Headlands and bays

Coastlines with alternate hard and soft rocks consist of a series of headlands and bays.

A headland:

- projects out into the sea.
- is usually longer than its breadth
- has sides which form cliffs.

A bay usually has:

- an approximately semicircular shape of sea extending into the land
- a wide, open entrance from the sea
- land behind it that is lower than the headlands on either side.

Headlands and bays form most readily on **discordant coasts**, where different types of rock lie at right angles to the sea and are subjected to differential erosion. The soft rocks are more easily eroded than the hard rocks, so they are worn back more quickly to form bays. The hard rocks resist erosion and form headlands, which protrude out into the sea between the bays.

Active undercutting at the base of a cliff keeps it vertical. When this ceases, the slope becomes less steep as weathering and sub-aerial erosion become dominant.

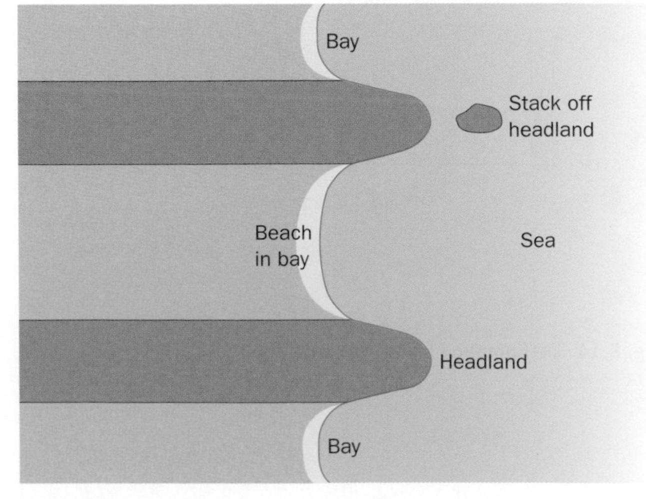

Fig. 5.19 A discordant coast

Caves, arches, stacks, and stumps

As oblique waves enter shallow water, they tend to turn so that their crests are parallel to the coast. This is known as **wave refraction**. Waves can be seen refracting around the headland and bay in Fig. 5.20. This refraction concentrates wave attack on all sides of the headland. Any line of weakness in the rocks is then subjected to hydraulic action and corrasion - forming **caves** and narrow inlets and, eventually, **arches**, **stacks**, and **stumps**.

If the rock contains soluble minerals, corrosion will also weaken the rock.

Arches, stacks and stumps are most common on discordant coasts. They are rarer on concordant coasts, where the rock layers are parallel to the sea.

Stacks are tall, steep-sided rocks, surrounded by sea. They wear down over time to short stumps, the tops of which are smooth, low and only visible at low tide, as the tops lie below high-tide level.

Fig. 5.20 Wave refraction on a headland and bay coast on the island of St Kitts in the Caribbean Sea

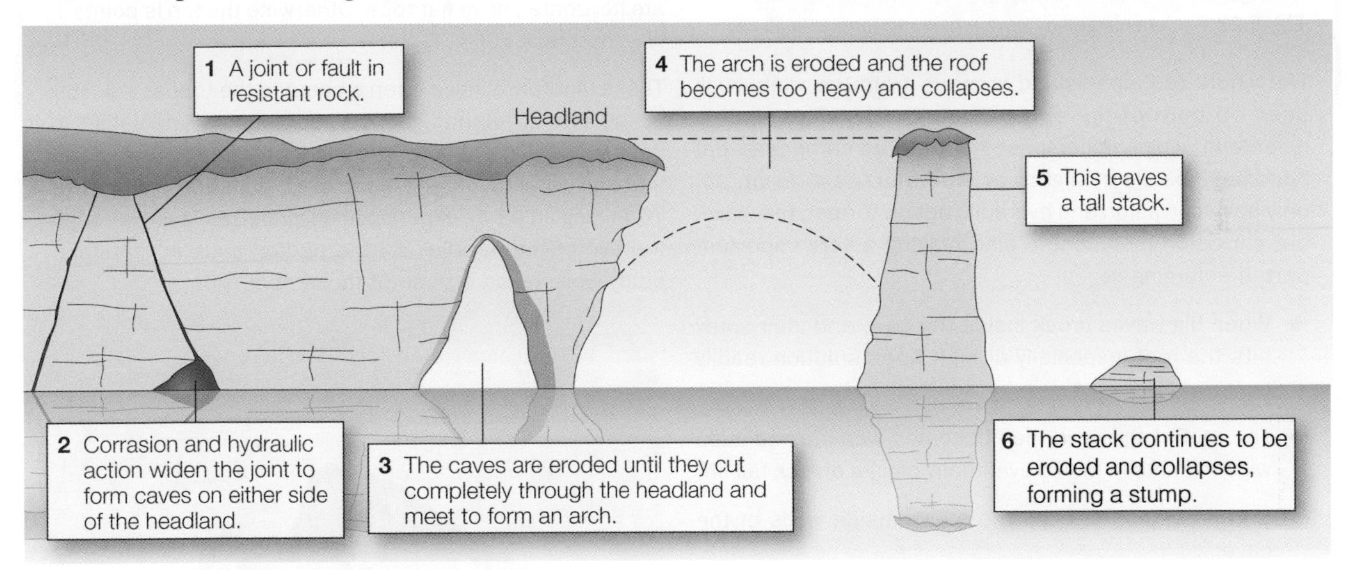

Fig. 5.21 How caves, arches, stacks and stumps form

CASE STUDY

The Bay of Islands, New Zealand

The Bay of Islands is on the north-east coast of New Zealand's North Island. Cape Brett is the bounding headland on the south side.

All the photographs in this case study were taken on one rocky island – Piercy Island – in this bay with many islands. Piercy Island, which lies at the seaward end of the bay (off Cape Brett), is too small to be included on most atlas maps. Although now located a considerable distance from the headland, it is in fact a stack.

Fig. 5.22 shows a major joint being widened at and above sea level by corrasion and hydraulic action. Fig. 5.23, taken nearby, shows a later stage in the erosion process. A sizeable cave, known as the Grand Cathedral Cave, has been formed.

Fig. 5.22 A joint enlarged by wave erosion to form a small cave

Fig. 5.23 A larger cave has formed

The whole of Piercy Island is made from the white rock seen on both of the above photographs. This rock is limestone, which is composed of calcium carbonate that becomes soluble in slightly acidic water. As a result, not only have corrasion and hydraulic action formed this cave, but corrosion (solution) is also playing a very important part in enlarging it:

- When big waves break inside the cave and their spray hits the roof (especially at high tide), solution readily occurs.
- The roof of the cave then becomes weak (as solution widens joints in it) and eventually blocks of rock fall off.
- Corrosion also attacks the steep inside walls of the cave.
- As Fig 5.23 shows, the cave is now much wider inside than it is at the entrance – as a result of corrosion.

Two caves previously formed on opposite sides of the island – along a major line of weakness. These caves eventually met up to form the arch in Fig. 5.24. A stack can also be seen to the right of this arch. It was once connected to the larger island by a second arch. However, further erosion caused the roof of this arch to collapse – leaving the stack as an isolated pillar of rock, surrounded by sea.

Both islands were once part of a long headland, which is now eroded back to a position to the right (off the photograph). Stacks have steep or vertical sides and, if the rock strata are horizontal, have flat tops. Otherwise the top is pointed, like the stack in Fig. 5.24.

These landforms have been a boost to the tourist industry of the Bay of Islands. Several companies specialise in boat trips taking people to circle Piercy Island. If the tide and waves allow it, the boat even passes through the 70-m-long arch – an experience described as 'going through the eye of the needle'. All the photographs in this case study were taken on one of these boat trips.

Fig. 5.24 An arch and stack formed at Piercy Island

Fig. 5.25 About to go through the 'eye of the needle'

Fig. 5.26 The arch is 70 m long and the indentation in its roof suggests that a vertical line of weakness is located there

Trips to Piercy Island are often combined with dolphin watching. The companies running these boat excursions are very concerned to help sustain the local **ecosystem** and protect the **biodiversity** of the bay. Among the measures taken are placing bins on board for litter disposal and also discharging effluent at controlled sites on land, instead of at sea. They also use the most economical cruise speed to reduce fuel consumption and help conserve the world's fuel resources.

LEARNING TIP When answering questions that require case study information, make sure that you name and locate the study and also use place names to locate features within it where possible.

The stacks and arch shown in Fig. 5.28 have formed at Cabo San Lucas at the tip of the long peninsula known as Baja California, in western Mexico. This is another exposed location where wave attack is strong, especially during tropical storms. The end of the headland is shown on the right of the photograph. Steep lines of weakness are clearly visible in its rocks. At one time, the three stacks would have been part of this headland. Now they remain as the detached but most resistant parts of it.

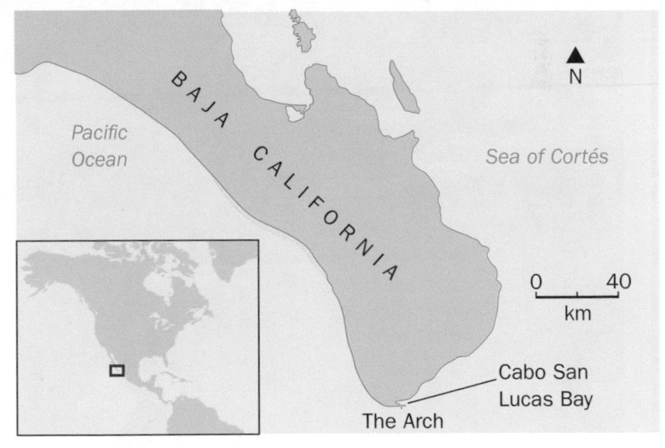

Fig. 5.27 The location of the arch and stacks at Cabo San Lucas Bay in western Mexico

LEARNING TIP When explaining the formation of caves, arches or stacks, you must refer to a line of weakness – not a layer of soft rock – as the cause of them.

Fig. 5.28 The arch and stacks at Cabo San Lucas

CASE STUDY

Coastal hazards

Hazards at coasts include erosion and tropical storms.

Can cliff erosion be stopped?

Pacifica is a settlement on the coast just south of San Francisco in the USA. It is situated on top of a sandstone cliff, 27 m above the Pacific Ocean. In the winter, powerful waves approach from the south-west – driven by prevailing south-westerly winds that blow over an enormous fetch. They attack and undercut the base of the cliff. After a time, the unsupported cliff falls into the sea.

- Between January and April 2010, an apartment block on the edge of the cliff was evacuated, after cliff falls resulted from undercutting of the cliff.
- Some attempts had already been made to slow down the erosion of the cliff. Large boulders had been placed at its base, to try to break the energy of the waves. These boulders are known as **rip-rap**. Engineers also tried to strengthen the cliff by reinforcing it and surfacing it.
- By the end of December 2010, even more of the cliff had fallen during storms – and a third of the apartment building was hanging over the ocean! The cliff had been eroded back by almost 7 m in just a year.
- More severe storms from the south-west removed the cliff from beneath another apartment block in January 2016. Undermining is likely to continue as grains in the sandstone are only weakly bonded together and earthquakes have caused cracks in the rock.

The owner of apartments further down the road wanted a very large sea wall to be built to deflect the incoming waves' energy and prevent the cliff from continuing to erode back towards his apartments. However, this would have been a costly project. Planners had to weigh this cost against the value of what would have been lost if it was not done.

Another way in which wave energy can be reduced is by putting **gabions** at the base of the cliff. These are metal baskets filled with stones. They can be used in a single layer or piled up to form a wall. Gabions are a cheaper alternative to concrete sea walls.

Discussion point

Should owners of houses destroyed by cliff falls be able to claim compensation from their government, because it failed to protect the base of the cliff from erosion?

Fig. 5.29 A concrete sea wall at the head of a shingle beach with an unstable cliff behind it. The sea wall's purpose is to deflect waves, so its top is curved outwards.

Managing the impact of coastal erosion

The methods used at Pacifica to reduce cliff erosion are, together with **groynes** (page 141), examples of **hard engineering** methods which use solid structures to absorb or reflect wave energy. Another method is the building of sloping walls, known as **revetments**, along the coast parallel to the sea to absorb the energy of the waves.

Soft engineering methods, such as **beach nourishment**, try to act in the same way as natural, protective, coastal processes. Another example is drainage pipes in cliffs which remove water to reduce the stress of its weight and the danger of cliff collapse.

Sea walls are very expensive to construct but walkways can be put on top of them. Gabions and rip-rap are quite cheap.

3 **a** Make copies of the diagrams and label each one with the type of engineering method it shows.

b Which two methods improve the coast for tourists? Explain how they improve it.

c Name a method that can be considered to ruin the appearance of the coast.

d List the methods that make access to the beach more difficult.

e Which method will cause difficulties further along the coast? Explain your answer.

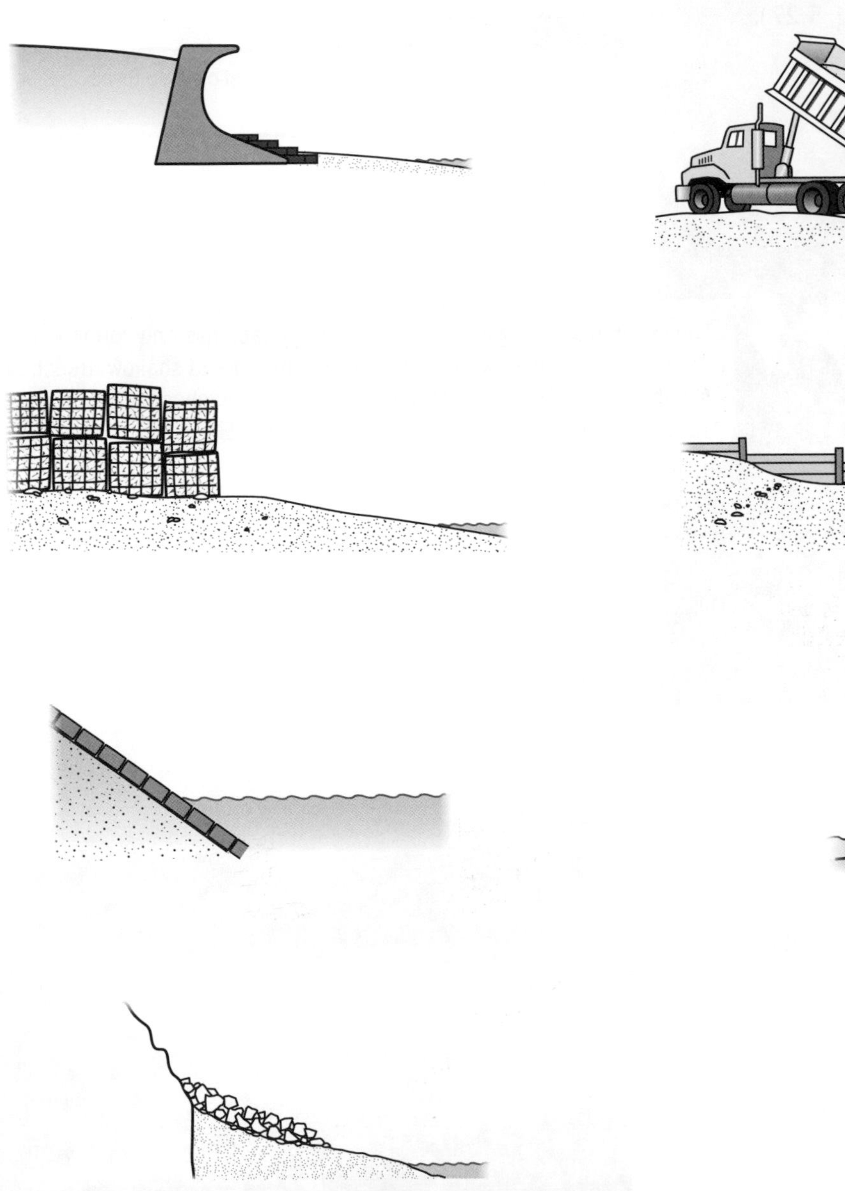

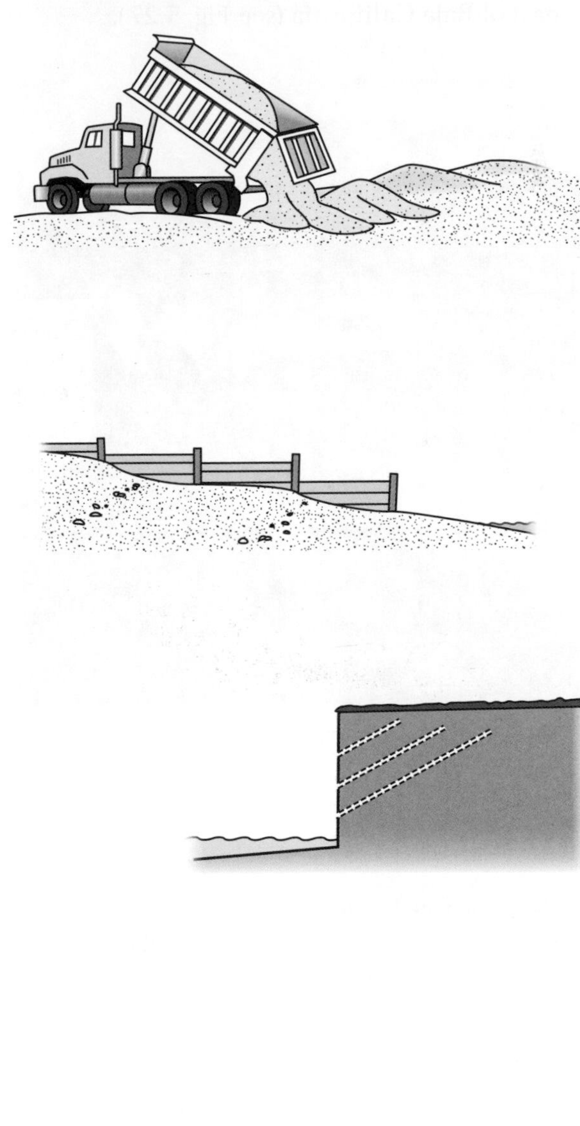

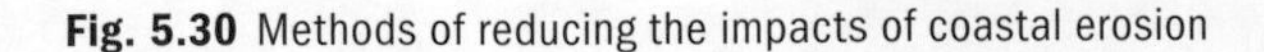

Fig. 5.30 Methods of reducing the impacts of coastal erosion

Landforms of coastal deposition

Beaches

Beaches are composed of sand or shingle, or both. When both sediments are present, the shingle forms a steep slope at the highest parts of the beach and the sand a gentle slope on the lower parts (as shown in Fig. 5.11). On a purely sand beach, the coarse sand will be at the top and the finest particles will be by the sea.

All beach material has been eroded from further along the coast and then transported by longshore drift. Often, sand is deposited in the relatively calm waters of a bay, or at the head of an inlet (see Fig. 5.31).

Some beaches are remarkably straight. One example is the beach that runs for much of the length of the Pacific coast of Baja California (see Fig. 5.27).

RESEARCH If you have access to Google Earth on a computer, look closely at some coasts. You may choose those mentioned in this chapter or others. Try to look at the Twelve Apostles on the coast of Victoria, Australia. One of the stacks is 45 m tall. Erosion has reduced their number from 12 to 8. The Acapulco area of Mexico is also worth looking at.

Discussion point

Consider the different ways in which countries with coasts have advantages over landlocked countries. What are the disadvantages of coastlines?

Discussion point

Are coastal depositional processes beneficial or a disadvantage for human activity, or both?

The water in this long inlet is deep, so the beach has only formed at the head of the inlet, where it is most sheltered and shallow. Beaches like this tend to be semi-circular in plan, while beaches formed in bays tend to have a crescent shape (like the one in Fig. 5.32).

Fig. 5.31 Cala 'n Porter Beach in Minorca

On this photograph, half of the crescent is hidden beneath the wall, but the wet patch on the sand shows the extent covered by the long swash. It is also possible to see how the waves fan out in the bay to approach parallel to the shore. The backwash is moving down the beach and two wave crests are about to push more sand up the beach.

Fig. 5.32 Hout Bay beach, a crescent-shaped beach south of Cape Town

RESEARCH Choose a country with a coastline. Make a note of place-names which locate the advantages and disadvantages of the coast.

CASE STUDY

The Cape coast of South Africa

The features referred to above (and shown in Figs. 5.13, 5.15, 5.16 and 5.32) are all located on Fig. 5.33. This coast is under almost constant wave attack. For most of the year, winds are onshore from a westerly direction, and the fetch from their origin is more than 2000 km.

CASE STUDY

Can anything be done about beach erosion?

Groynes are often used to try to reduce beach erosion. The other method is to add sand to the beach from an area where sand is plentiful. This has been successful at Teluk Chempedak, a bay beach on the east coast of Malaysia. The beach in front of two five-star hotels there was replenished with nearly 180 000 cubic metres of sand in 2004. The natural supply of sand to the beach is prevented by dredging the port at nearby Kuantan.

Discussion point

Should artificial alterations to a beach be subject to permission from neighbouring coastal authorities?

4 Identify as many coastal features as possible on Fig. 5.34. Try to identify more than five.

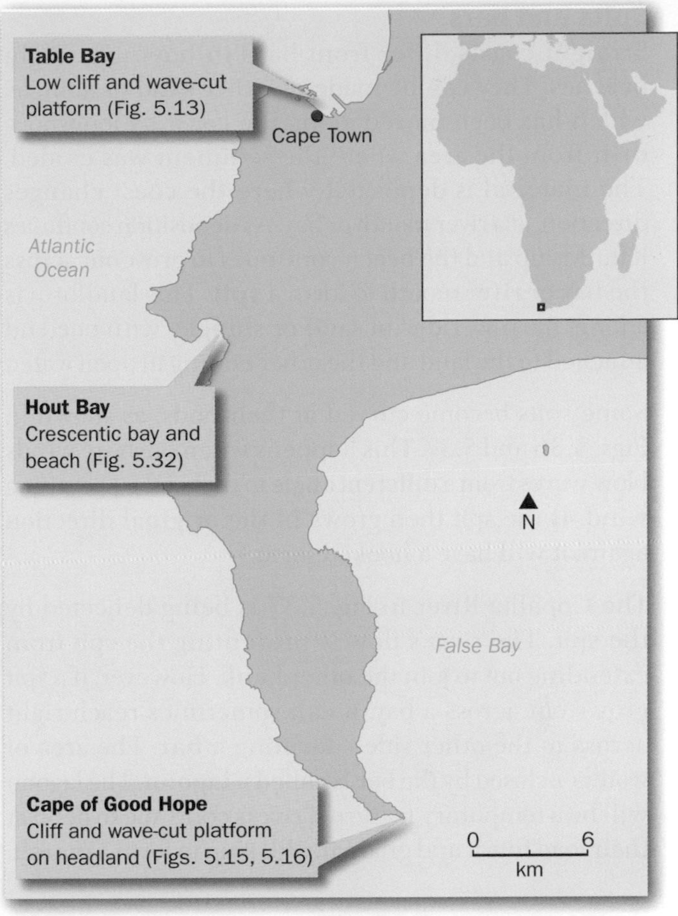

Fig. 5.33 The Cape coast of South Africa

Fig. 5.34 Landforms at a coast

Spits and bars

Straight coasts differ from bays in having straight beaches. They can be made of either sand or shingle, which has been moved along the coast by longshore drift from the area where the sediment was eroded. The material is deposited where the coast changes direction, at a river mouth or bay. As deposition continues it builds up and the beach continues to grow out across the bay or river mouth to form a **spit**. This landform is a long, narrow, ridge of sand or shingle, with one end attached to the land and the other ending in open water.

Some spits become curved at their ends, as shown in Figs. 5.36 and 5.37. This happens when onshore winds blow waves from a different angle to that of the prevailing wind. If the spit then grows in the original direction again, it will have a hooked form.

The Copalita River in Fig. 5.37 is being deflected by the spit. The river's flow is preventing the spit from extending out to join the other bank. However, if a spit grows out across a bay, it can sometimes reach right across to the other side – forming a **bar**. The area of water enclosed by the bar is called a **lagoon**. The lagoon will be a temporary feature if rivers continue to deposit their load into it and gradually fill it up to form a **marsh**.

Discussion point

Articles removed by the tsunami, in March 2011, from coastal areas of Japan (on the west side of the Pacific Ocean) arrived on beaches in Oregon (on the east side of the Pacific Ocean) in June 2012. Use atlas maps and your knowledge to discuss how this could occur.

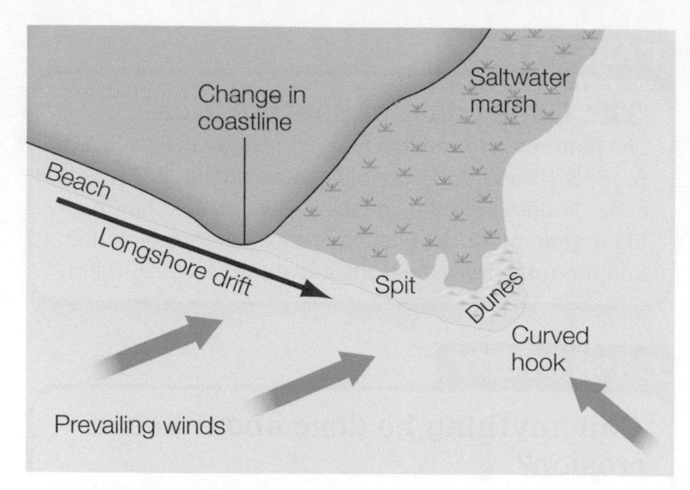

Fig. 5.36 How a spit forms

Fig. 5.37 A recurved spit across the mouth of the Copalita River in south-west Mexico

Fig. 5.35 A sand spit with small **sand dunes** on top of it at Poole in the UK

Fig. 5.38 A bar and lagoon at Slapton Ley, UK

CASE STUDY

The site of Kingston's international airport

The Palisadoes is a long, narrow spit near the capital of Jamaica. It is so large that the international airport's runway fits into it with room to spare. This spit was flattened easily and cheaply to provide a suitable site for the airport. It has sand dunes on its south side and mangrove swamp at its northern edge, which could be reclaimed if there were a need to expand the airport.

Discussion point

What are the advantages and disadvantages of the site and location of Kingston's international airport?

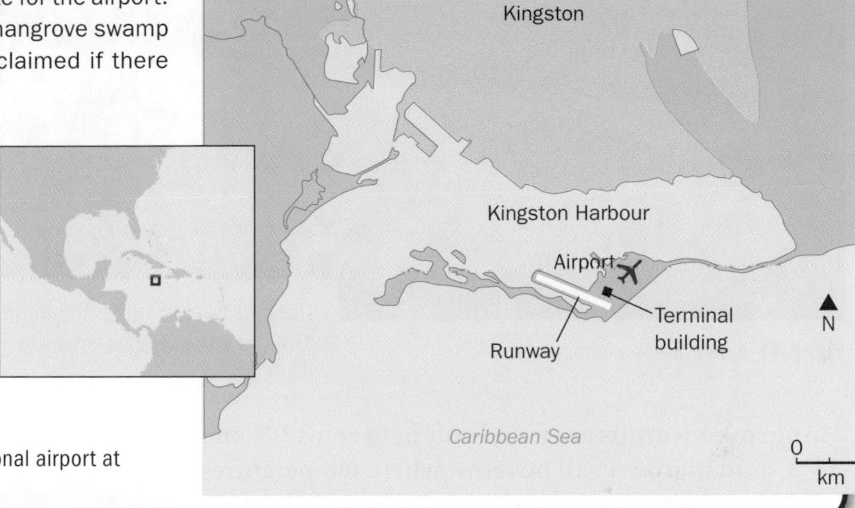

Fig. 5.39 The site and location of the international airport at Kingston, Jamaica

LEARNING TIP Deltas have extensive areas of marsh (some of which are salt marsh) at their seaward edges, but deltas are classified as landforms of ***river*** deposition. You must not choose deltas when you are asked to write about features of ***coastal*** deposition, even though they are found at the coast.

LEARNING TIP Mangroves and marshes are vegetation. The mud flats on which they grow are landforms. You may use this case study as an example of the threats to the environment where economic development is taking place.

Salt marshes and mangrove swamps

The water behind a spit is a low-energy environment in which small particles of mud can settle. The mud brought in when the tide rises sinks in the sheltered water. Gradually it accumulates to form a mudflat on which plants tolerant of saltwater - and of being covered by water twice a day - begin to grow. These plants encourage more deposition by slowing the water movement further. Their roots help to trap more mud and hold it firmly in place. As the surface level of the mud rises, vegetation increases and the area becomes a salt marsh or mangrove swamp forest in sheltered waters such as estuaries. About 65% of tropical and subtropical coastlines are fringed with mangroves, which provide important protection against erosion.

Fig. 5.40 A tidal channel crossing a salt marsh

Fig. 5.41 A salt marsh nearer the sea

The spit behind which this marsh formed can be seen in the background. Most of the marsh is only just above the low-tide level.

Mangrove swamps are found only between 32°N and 38°S, as mangroves will not grow where temperatures fall below 20°C or where the coolest season is more than 5°C cooler than the warmest season. Sea surface temperature should never be below 16°C. Where there is a wide tidal area (between low- and high-tide marks), a dense vegetation of mangrove swamp forest - consisting of mangrove trees and other plants - develops provided there are no strong waves or tidal currents.

Mangroves are able to withstand brackish water and can grow where other plants cannot because some species have roots that filter salt out and others have leaves that excrete salt. Their conical breathing roots (aerial roots) stick vertically up out of the water at low tide and enable them to live in soils without oxygen. All have lateral spreading roots which rapidly extend out from seedlings and anchor them in the soft mud. Prop or stilt roots extend down from the trunk to give them stability and strength as tides rise and fall and when waves attack or storm surges occur.

The swamps have many tidal channels which are swept clear of mud. Many different species inhabit mangrove swamps, including crocodiles, frogs, snakes, fish, birds, and even some terrestrial mammals like monkeys.

Mangrove swamps are invaluable for protecting the coast from flooding. As mangroves can grow to a height of 15 m, they also offer some protection from the strong winds of hurricanes. The importance of mangrove swamps as a coastal ecosystem was made very clear during the devastating Indian Ocean tsunami in December 2004. There were more than 6000 deaths in one coastal village in Sri Lanka that had cut down its mangroves, compared with only two deaths in a neighbouring village that had kept its mangroves.

Fig. 5.42 Mangroves with stilt roots and, on the right, breathing roots, Bahamas

Fig. 5.43 Mangrove forest by the side of a creek on Grand Cayman island

Discussion point

Should the coast be left in its natural state or exploited for human enjoyment and gain?

CASE STUDY

Mangroves vs development on Grand Cayman

Originally, 66% of Grand Cayman in the Caribbean was occupied by mangroves. The largest area was in the centre of the island. Because it is located in a tropical storm zone, this very low-lying island has always needed the protection that mangroves provide. It's also at risk from tsunamis caused by tectonic activity along the Caribbean Plate boundary.

However, since 1980 there has been a lot of development on the island, such as the building of new hotels and condominiums.

- By 2009 at least 10% of the island's mangroves that existed in 1997 had been lost to various developments.
- By 2010, 66% of the mangroves that existed on the western peninsula had been removed (Fig. 5.46).

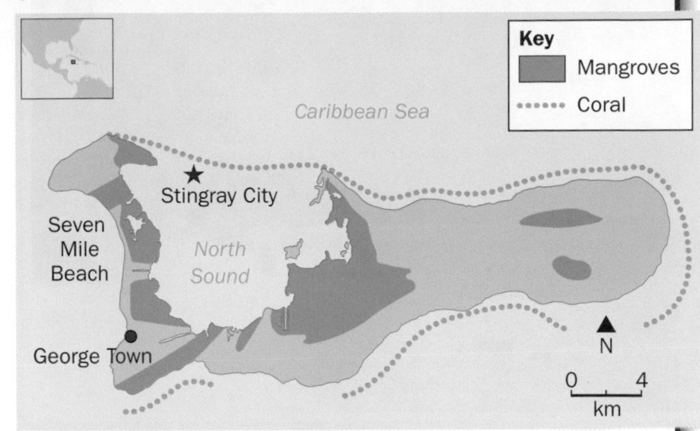

Fig. 5.44 The area covered by mangroves on Grand Cayman in 1960

Fig. 5.45 Mangroves on the coast of Grand Cayman in 1997

Fig. 5.46 The extent of the remaining mangroves on Grand Cayman's western peninsula in 1997 and 2009

Compare Figs. 5.47 and 5.48, which show the same part of Seven Mile Beach in 1997 and 2009. Mangroves and other trees can be seen as green areas lining the beach in the background of the 1997 photograph. The 2009 photograph shows that much of the green area has been replaced by the white of new buildings.

In 2010, a further 83 hectares of mangroves were removed to make way for a development at Dragon Bay, which stretches from Seven Mile Beach to the North Sound (see Fig. 5.46). Some of the properties in the Dragon Bay development have private beaches. There is also a golf course, tennis courts, and a marina, among other facilities. This project caused controversy. The developer maintained that the area of mangroves removed for it had been badly damaged by Hurricane Ivan in 2004. Others, supported by aerial photographs, argued that the mangroves had largely recovered from the hurricane.

Tourists started to visit Grand Cayman in the 1960s, and tourism now provides 75% of the GNP (a measure of wealth) of the country. Many areas of mangrove swamp have been removed for road building, golf courses, tourist accommodation, marinas, and housing.

Local people have mixed opinions about this issue and objections are increasing as the desire to conserve what remains of the natural environment is growing. Some describe the western side of the island as a concrete jungle. There is also considerable anger about the development because mangroves are valued as protection against high winds and storm surges. The effects of high winds are known, because 75% of homes and other buildings on the island were severely damaged by Hurricane Ivan in 2004. The damage would have been much worse without the protection of the mangrove buffer zone.

Fig. 5.47 Seven Mile Beach in 1997, looking north

Fig. 5.48 Seven Mile Beach in 2009, looking north

Mangroves in Grand Cayman have other uses too:

- Their roots help to stabilise the coast against erosion caused by large waves and storm surges that batter the coast during storms.
- They are a source of firewood.
- Mangrove swamps absorb inorganic nutrients that drain into them in water from farmland and urban areas. This prevents them from being deposited in the sea and harming marine life.
- Mangrove swamps catch sediment washed towards the sea after heavy rains, preventing it from harming the coral reefs.
- Mangrove leaves decay and add organic nutrients to the water, which provides food for the small fish that hatch and shelter in the mangrove area.
- Mangrove swamps provide a nursery for fish and shellfish, such as young barracuda and spiny lobster which can hide from predators among the roots, along with baby turtles and some iguanas. By acting as a place of safety for young marine species, the mangrove helps maintain the biodiversity of the coral reefs offshore.
- They are important wildlife habitats and breeding and feeding grounds for birds. Parrots nest in hollow trunks of dead mangrove trees. Heron, frigate birds and Whistling Duck shelter in the mangrove to breed, and feed.
- They are also important for recreation – especially for fishing, bird watching, wildlife photography, and boating.

However, people in favour of development argue that more employment will be provided directly in the tourist industry. There will be an increase in people working in hotels, running boat trips to the famous Stingray City (where tourists can swim in the North Sound with stingrays) and diving expeditions down the amazing Cayman Wall. Indirect employment will also result, as shopkeepers and restaurants will gain more revenue and the overall economy will be better.

If all unprotected areas of mangrove are developed, only 12 km^2 will remain. The increasing population puts the mangrove under greater threat. What will be the cost to the environment of continuing to develop homes and tourist accommodation? Will jobs and the economy take precedence over wildlife and their habitats?

On such a small island, caution is needed regarding further development. When an area becomes overdeveloped, it loses its attractiveness and tourists find other places to go.

5 Copy this table and expand on it, summarising the importance of mangrove swamps under the headings shown.

How mangrove swamps provide ...	
... protection for the coast	
... protection for the land	
... protection for the sea	
... an important habitat	
... a recreational resource	

Table 5.1 The importance of mangrove swamps

Coastal sand dunes

Sand dunes are ridges of sand which form at the back of beaches and on spits. The conditions necessary for their formation are shown in Fig. 5.49.

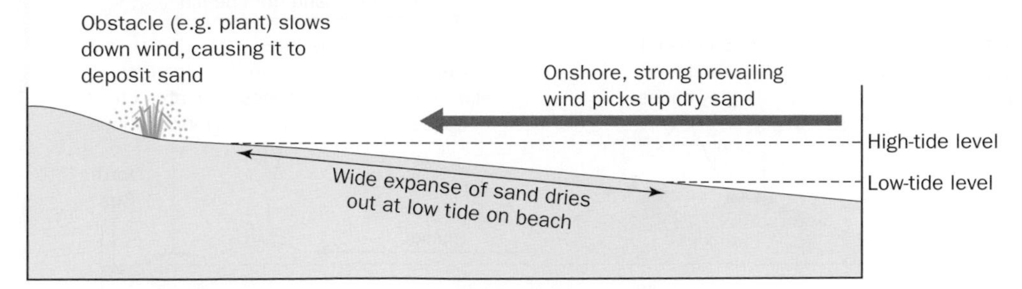

Fig. 5.49 Conditions necessary for the formation of coastal sand dunes

The obstacle needed to begin the process of dune formation can be any material or plant on the beach. Friction with the obstacle slows the onshore wind so that it loses energy, can no longer carry the sand in suspension or saltation, and deposits it around the obstacle. This increases the size of the obstacle, so the deposition process continues and the dune grows. While it is still within reach of spray from the waves, only salt-tolerant plants can grow on it.

Over time, the embryo dunes grow and join together to form a line - known as a fore dune. This moves in an inland direction because the wind picks up sand from the seaward face of the dune and deposits it on the leeward side. Meanwhile, a new embryo dune can be forming nearer the sea. Eventually, lines of dunes are formed parallel to the sea (see Fig. 5.51).

As the dunes are moved away from the sea and grow higher, they are colonised by marram grass, a plant resistant to the drought conditions that prevail. Marram grass plays an important part in the growth of the dunes, because it grows upwards rapidly after being covered in windblown sand. It also has a network of very long roots that help to anchor the sand. The leaves and stems slow down the wind and protect the sand from erosion. The yellow mobile dune ridge is often the highest. Marram grass is the dominant plant on the seaward slope, but other plants grow on the leeward slope.

In time, the dune becomes a semi-fixed grey dune, as soil develops on it. Finally, the dune becomes a fixed dune with an almost total cover of vegetation. The fixed dunes furthest inland are the oldest.

Fig. 5.50 Young or embryo sand dunes on the coast of Scotland

CASE STUDY

The sand dune system at Gibraltar Point

These dunes have formed just south of Skegness on the North Sea coast of England. The beach has a very gentle gradient, so the onshore wind blows over a wide expanse of sand – picking up and carrying dry sand for dune building.

Wind carries sand from beach

Trees on grey fixed dunes

Coast in 1824

Shrubs on semi-fixed grey dunes

Marram grass on yellow mobile dunes

Marram grass on fore dunes

Salt-tolerant plants on embryo dunes

North Sea

Sand

Slacks

Older rocks

West

East

Yellow dunes

Increased age of dune ridges

Fig. 5.51 Sand dunes at Gibraltar Point, England

Long, marshy depressions, some with strips of water, lie between the dune ridges. These depressions are known as **slacks**. They contain water-loving plant species.

Sand dunes are important as they protect inland areas behind them from erosion and flooding.

Discussion point

A defence post built in the Second World War to defend the beach at Gibraltar Point is now almost 400 m inland from the beach, has no view of the beach and is partly covered by sand. How could this have happened?

6 Read the information in Fig. 5.53. On the south-east coast of Florida, sea oats are planted on the dunes instead of marram grass. On the notice 6" is equivalent to 15 cm and 5' is equivalent to 1.5 m. Explain why sea oats are used.

Fig. 5.52 A slack between dune ridges

Fig. 5.53 A notice on Florida's east coast

7 Study Fig. 5.54 and describe the features of the area between the line of huts and the sea to the right.

Fig. 5.54 A coastal landscape

Young dunes are very fragile, because marram grass cannot tolerate trampling. When the grass dies, the wind easily removes the exposed sand. This leaves a depression or a valley-like shape cutting right through the dune - called a **blow-out**. The dunes in Fig. 5.55 have been badly eroded, as shown by the patches of bare sand. Fencing has been used to keep people and animals out to give the vegetation time to recover and stabilise the dunes.

An example of a very large area of sandy beaches and sand dunes is Les Landes in south-west France, where the prevailing winds are strong onshore westerlies. Here, on the older dunes, the soils are deep enough and contain sufficient water and nutrients for coniferous forest to grow. As Les Landes is an important tourist area, information boards and leaflets have been used to educate the public about the fragility of the dunes. Wooden boardwalks have been provided for people to walk on, and access to the most fragile areas has been prevented by fencing them off.

Fig. 5.55 An area of eroded dunes

Discussion point

Should further development along coasts be stopped so that their remaining beauty can be kept for future generations to enjoy?

CASE STUDY

Sand dune and beach destruction on the Moroccan coast

Sand extraction is one of the economic opportunities coasts offer to people. The removal of sand from dunes and beaches on the Moroccan coast, for use by the construction industry, has been done on a massive scale. The sand is cheap, easy to remove, and very accessible. It is reputed to be the world's largest excavation of coastal sand.

The areas of coastal destruction near the major cities of Casablanca, Rabat, and Tangiers have been described as looking like an ugly moonscape. Such enormous lengths of sand dune have been removed that the area is now believed to be totally useless for tourist development – leading to a big loss of potential earnings. The removal of the soft sand has left the area with a hard rocky surface, which is difficult to walk on.

Another potential economic impact is the loss of the protection given by the beaches and dunes to the coastal road and buildings along it. These are now vulnerable to the effects of storms and a rising sea level. The loss is not just economic but also environmental, because marshes inland have been threatened and dune ecosystems destroyed.

Coral reefs

Corals are tiny, marine animals called polyps that form reefs when they live in colonies in their millions. Their skeletons are calcareous cup-like structures which are joined with others in the colony to form a hard, stony mass. As one generation dies, the next grows on top of it, so the reef grows upwards and outwards. There has to be a solid surface from which the growth starts. This is usually rock but could be another feature, such as a shipwreck in shallow water.

Reef-building corals cannot grow just anywhere. They have specific needs, which must be met if they are to grow healthily. They are generally only found between the latitudes 30° North and 30° South, because they need sea temperatures which are higher than 18 °C. They grow best where the mean temperature of the water at the surface is 22–25 °C. If sea temperatures rise because of global warming, existing corals may die but other waters may become warm enough for reefs to develop.

Corals also need oxygen and food, both of which are brought by breaking waves. They cannot grow much above low tide level, because they cannot survive long periods exposed above the water. The top of the reef is usually level with low tide level and the highest part of the reef is the outer edge where oxygen and food are most abundant.

Fig. 5.56 The coral itself is not coloured, but the algae that live with it are.

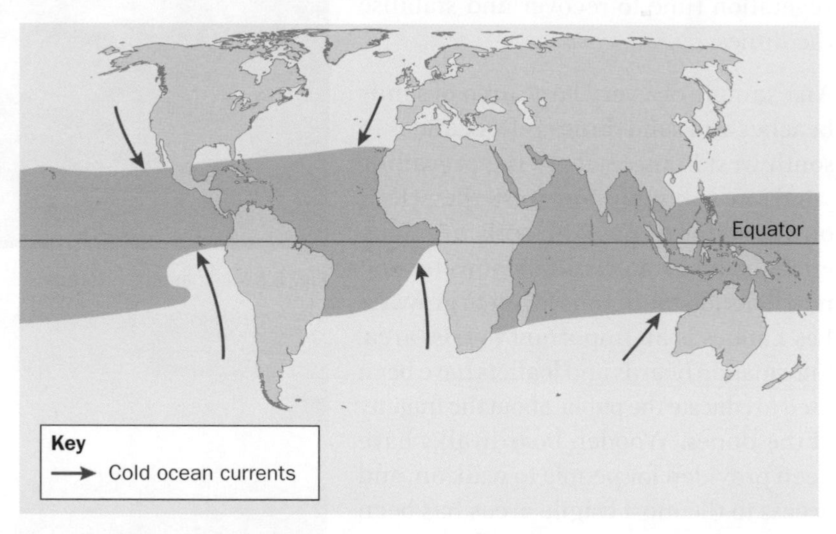

Fig. 5.57 The dark shading shows areas with sufficiently warm temperatures for coral reefs to grow

RESEARCH Find the island of Bermuda in an atlas. It has coral reefs. What is surprising about its latitude? Use a map of ocean currents to explain this anomaly.

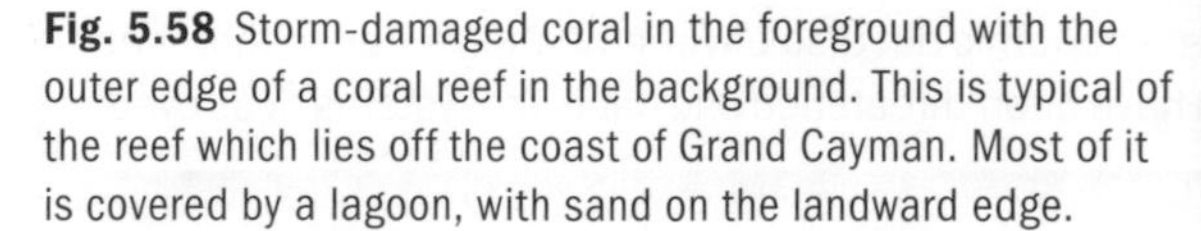

Fig. 5.58 Storm-damaged coral in the foreground with the outer edge of a coral reef in the background. This is typical of the reef which lies off the coast of Grand Cayman. Most of it is covered by a lagoon, with sand on the landward edge.

Coral reefs run parallel with coasts but are not continuous. They have breaks in them, which are usually at river mouths. This is because polyps need clean, clear water in which to live, so they cannot live where sediments enter the sea at the mouths of rivers:

- If the water is cloudy with sediment in suspension, sunlight (essential for the life of the plankton on which polyps feed) cannot penetrate far enough.
- If sediment settles on polyps, they are unable to feed.
- Sediment also provides the ammunition for corrasion during storms - causing reefs to break up.

Another reason why breaks in reefs occur at river mouths is that the influx of freshwater lowers the salinity of the sea - and corals grow best in conditions of normal salinity. Freshwater is less dense than seawater, so it tends to affect the surface zone where the living corals are.

Rivers can also bring pollutants into the sea, especially after heavy rain. Sewage and phosphorous from fertilisers can cause a population explosion of Crown of Thorns, a species of starfish which feeds on coral.

There is a break in the coral reef off George Town - the port and capital of Grand Cayman. Originally the reef was probably continuous around the whole island, because Grand Cayman has no rivers. However, the movements of ships in and out of the port have undoubtedly destroyed any coral reef that was there. The island is now a port of call for many cruise ships, so the reef off George Town is not likely to regenerate.

Corals are not found at great depths. This is because they cannot live without algae - and algae need abundant sunlight to survive. Light decreases with depth, so living colonies of corals are not found below depths of 50 m and few exist below 30 m. For the reasons explained earlier, shallow, agitated waters to about 10 m depth are best.

Corals and algae live in a symbiotic relationship. The algae use waste products from the polyps, together with sunlight, to make food and oxygen that the polyps then live on.

8 **a** Study Fig. 5.57. Explain why there are no coral reefs off the western coasts of continents in subtropical latitudes.

b Study Fig. 5.58 and describe the visible reef and the water between it and the mainland.

c Make a list to summarise the conditions required for the growth of coral reefs.

Discussion point

What human activities are suggested by the evidence in Fig. 5.60?

CASE STUDY

Opportunities for people provided by the coasts of the Cayman Islands

In addition to the many uses of mangrove swamps already described, the Cayman coasts present a variety of opportunities.

- The 300 000 **tourists** who stay on the main island each year are attracted by the beautiful coast with its sandy beaches, such as Seven Mile Beach, coral reefs, warm sea for swimming, and hot climate.
- Tourism provides a lot of **employment** in hotels, restaurants, and transporting tourists to a diverse range of activities, including sea fishing and diving along the outer edge of the coral reef, known as the Cayman Wall.
- Some Caymanians earn their living by **fishing**, as the coastal waters and coral reefs are home to many species of fish and shellfish.
- A gentle indentation on the sheltered west coast provides a suitable location for the **port** of George Town. It imports most of the consumer goods and food needed by its population and visitors. Many of the 1.7 million cruise ship passengers who step ashore at the port every year rush to its duty-free shops for souvenirs.
- High-quality **sea salt** produced on the island from evaporated seawater earns export income (Fig. 5.70).
- The sea also supplies desalinated **water**, a vital necessity on an island without rivers.

Fig. 5.59 Containers on the quay at the port of George Town, on the leeward side of the island (sheltered from the prevailing NE trade winds)

Fig. 5.60 The coast just north of George Town

Other opportunities of coasts for people

Coasts provide essential raw materials for the construction industry, concrete manufacture and road aggregate. In addition to the extraction of sand from beaches and sand dunes (see the case study on page 149), gravel is extracted from shingle beaches (Fig. 5.10), and stone for crushing from beaches with larger cobbles and boulders (Fig. 5.15).

9 Look through all the case studies in this chapter and make a list of the different opportunities that coasts provide for people. For each one, note the location where it occurs and the main ways in which it is an advantage.

Types of coral reef

Coral reefs are classified into three types, the most common of which is the fringing reef.

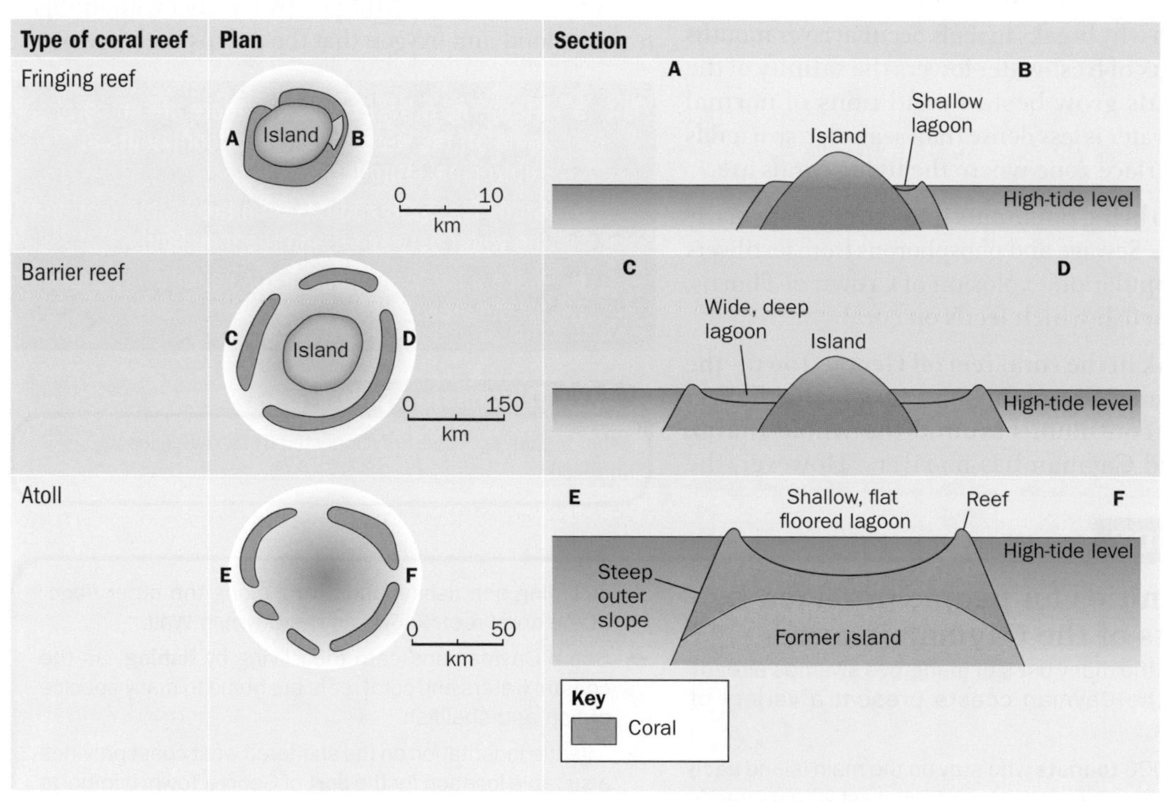

Fig. 5.61 The three types of coral reef

Theories suggest that the formation of the three different types of reef is linked. As the fringing reef grows upwards and outwards, it could develop into a barrier reef, if the sea level were rising or its island base sinking (which would have the same effect as a rise in sea level). If this process continued, an atoll would be formed as the coral grew upwards to keep access to light, oxygen, and food as the sea level rose.

10 a Use the map of Grand Cayman (Fig. 5.43) to estimate to the nearest kilometre the average distance from the land to the outer edge of the reef. What type of reef is this?

b Find the Great Barrier Reef on an atlas map of Australia. Describe its location, distance from the mainland, and features.

CASE STUDY

Australia's Great Barrier Reef

This is the world's largest reef system. It is so large that it can be seen from space. It stretches parallel to the coast of Queensland for more than 2600 km, and is nearly 350 000 km^2 in area. It contains almost 3000 reefs and is a very important tourist attraction.

A large part of it has been protected by the Great Barrier Reef Marine Park, in which fishing and tourism are limited by the Park Authority. However, its greatest threat is partly natural. Every few years, Queensland experiences severe flooding from heavy rains. The pollutants in the run-off from such events usually cause a population explosion of Crown of Thorns starfish, which eat the polyps and do a lot of damage. The Great Barrier Reef is also threatened by warming of the waters caused by an El Niño event (see page 153). More coral died in 2016 than in any previous year as a result of bleaching caused by temperatures one degree higher than normal between February and April. Global warming could have the same effect.

CASE STUDY

Sustainable management to conserve Fiji's coral reefs

Since 2002 there has been an enormous effort to conserve coral reefs in Fiji – a group of tropical islands in the South Pacific Ocean. All the islands are surrounded by coral reefs, two-thirds of which are thought to be at risk.

The reefs are vitally important to Fiji for both environmental and economic reasons. They are very rich ecosystems with great biodiversity. Fiji's 300 species of coral are home to almost 2000 species of fish, including sharks and tuna and many shellfish (such as giant clams). Endangered turtle species also live there. The reef ecosystem has links with bigger ecosystems through the food chain. A simple example of this is shown in Fig. 5.62.

plankton ➔ coral ➔ parrot fish ➔ larger fish (e.g. snapper) ➔ humans

Fig. 5.62 A simple coral reef food chain

As everywhere, Fiji's reefs are important barriers to storm waves – reducing their impacts on the land. The plants and animals in the reef ecosystem are also being researched as possible future sources of medicines.

The reefs are of enormous importance to Fiji's economy. Both subsistence and commercial fishing provide good livings. The colourful reefs and their beautiful marine life also attract many tourists.

In Fiji the reefs are also of cultural importance and the local people hold ceremonies to thank them. They have long been aware of the importance of the reefs to their lives and have traditionally regulated their use. For example, they have banned fishing in certain areas when stocks have been threatened by over-fishing.

Why are reef conservation measures needed?

There are both human and natural causes of coral destruction.

Human and economic causes

Corals are extremely sensitive animals that can die if touched. They have been affected by boat and anchor damage. Tourists walking on the coral and diving and snorkelling among the reefs have also caused damage. The population of Fiji is rising rapidly and most people live around the coasts. The development of settlements has led to harmful sediment, sewage and other pollutants being washed out into the reefs. The natural forest has also been replaced with sugar cane fields in places, leading to **eutrophication** from the nitrates and phosphorous in fertilisers which drain into the sea. Sewage also causes eutrophication: the addition of large amounts of nutrients results in rapid plant growth, which uses up oxygen and harms other organisms.

Natural causes

Every three or four years, Fiji is battered by tropical storms which generate destructive waves that break the coral. They also lead to more nitrates and phosphates reaching the reefs, which causes the number of Crown of Thorns starfish (which feed on the polyps) to increase.

Every so often, an increase in sea temperature causes bleaching of the reefs, because it leads to the deaths of the colourful algae that live with the polyps and leaves the bleached white coral behind. In time the polyps, deprived of food, also die. This happens when there is an **El Niño** event. This reversal of the equatorial ocean currents leads to warm water from Indonesia moving east and warming the sea around Fiji. A record amount of coral died in warm water in February 2016 and when two tropical storms passed through in March and April that year.

Efforts by a Fijian resort to conserve the reef

Many hotels and communities in Fiji are trying to conserve the coral reefs on which their livelihoods depend. Fiji's Coral Coast is fringed with the world's second largest reef. The Hide-away Resort in the middle of the Coral Coast uses a world ecotourism, award-winning coral reef conservation programme, known as 'Integrated Coastal Management'.

The resort fronts directly on to the fringing reef and makes visitors aware of its great importance. The area has been made into a protected zone. Guests can snorkel at high tide, but putting feet down on to the coral is not allowed. Guided walks are organised along a specially provided reef path. At high tide, guests can view the marine life on the reef from the resort's glass-bottomed boat. In the lagoon there is a coral nursery where coral is planted and protected. Notices remind visitors of the importance of the reef (see Fig. 5.63).

THE P SHAPE ON THE TOP OF THE REEF WALK MARKERS INDICATES THE SIDES OF THE PATH TO WALK ON

CORAL REEFS

Coral reefs are the largest living structures on earth that are visible from space. They are among the most productive ecosystems on the planet – comprising roughly 0.2% of the world's oceans, but 11% of the world's fish harvest.

UNDER PRESSURE

Coral reefs around the world are under pressure from:

- coastal developments
- pollution
- sedimentation
- over-fishing
- walking on the reef
- natural disasters
- eutrophication

HOW DOES EUTROPHICATION AFFECT THE REEF?

High nutrient levels (especially nitrogenous and phosphorus compounds, such as fertilisers and detergents) lead to high seaweed populations.

The relatively slow-growing corals can't compete with the rapidly growing phytoplankton (seaweed) when there are raised nutrient levels in the water

HOW YOU CAN MAKE A DIFFERENCE

1. Do not walk on the reef other than on the reef walk path.
2. Sponsor and plant a piece of coral on the reef.
3. Do not take anything from the sea.
4. Do not buy shells.

Fig. 5.63 Notice to hotel guests

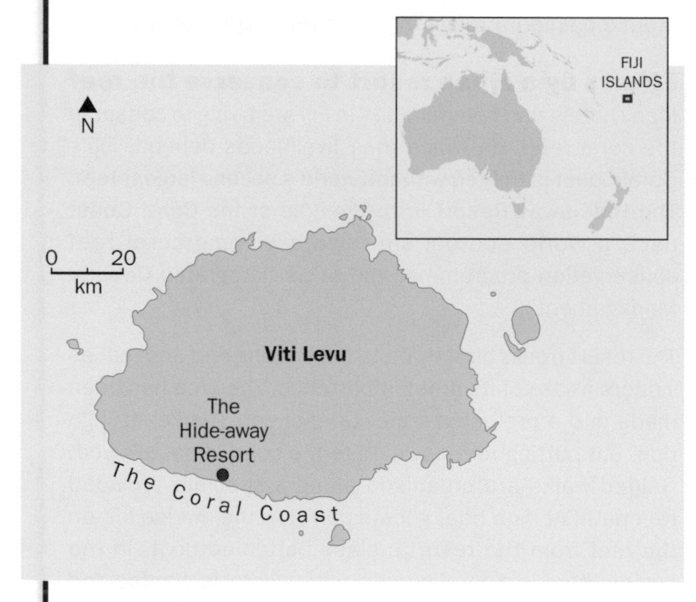

Fig. 5.64 Fiji and the location of the Hide-away Resort on the main island of Viti Levu

Fig. 5.65 The glass-bottomed boat on the lagoon. The outer edge of the reef can be seen in the background

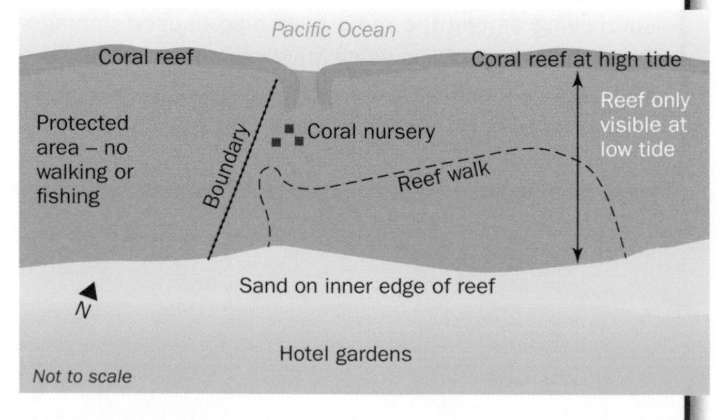

Fig. 5.66 Conservation measures in the hotel grounds

Discussion point

Are coral reefs so valuable that humans should not be allowed to touch them?

How tropical storms affect coasts

CASE STUDY

Tropical cyclone Winston, Fiji, February 2016

The islands of Fiji are often in the path of **tropical storms** (see Chapter 6) during the southern hemisphere's summer and autumn. In February 2016 Fiji experienced what is believed to be the most ferocious storm ever in the South Pacific – the category 5 **tropical cyclone** Winston. It tracked across central Fiji from east to west, battering its islands with sustained wind speeds of 230 km per hour and gusts as high as 325 km per hour. The winds caused very destructive waves which affected a wide area at the coast because of the **storm surge**. During a storm surge the sea surface is raised because atmospheric pressure is low and the weight of the atmosphere is lessened. Strong onshore winds drove large storm waves on to the coast, flooding large areas of lowland and causing destruction, especially where the water was funnelled up shallow estuaries and bays.

- Most people live on the coasts of the islands, mainly on the large island, Viti Levu, where the interior is steep hills. Whole villages were destroyed, including many schools and health centres, as well as homes.
- More than 40 people lost their lives and 40 000 people needed immediate help for serious injuries.
- The main coast road was cut, trees were blown down and 80% of the population lost their electricity supply. Water supplies and sanitation were damaged. Communication systems between the islands were destroyed.

Effects of Winston on the coral reefs and mangrove forests

There was extensive damage.

- **Corals were broken off** and many fish, shellfish and plants killed. The costs of restoring the coral reefs are very high.
- The **livelihoods** of coastal villagers were ruined, including the inhabitants of Namatala in Ra Province whose fishing grounds were totally destroyed.
- The **mangrove trees blown down** and stripped of their foliage will take more than ten years to recover.

Restoration

The debris had to be removed. Mangroves had to be replanted and new coral grown, transplanted to the reef and protected. Damaged boats and fishing gear had to be mended. Seawalls had to be repaired and undercut cliffs protected.

Damage to Fiji's agriculture

Tourism is a major source of income but agriculture is the main occupation, with 70% of the population working in the sector. Most croplands are near the coast on small patches of coastal plain and deltas. Sugar cane and coconuts are the main cash crops and important exports.

- On the south coast of the second largest island, Vanau Levu, the storm surge waves reached 200 m inland, destroying the crops and leaving the soil too salty for crops to grow for some time.

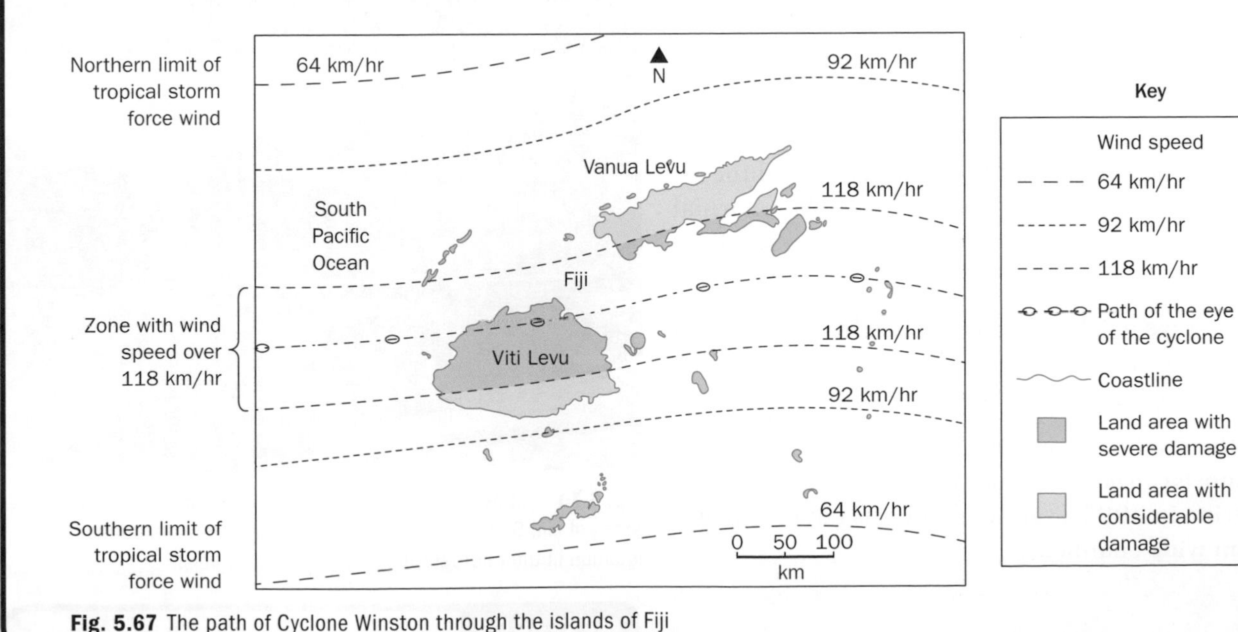

Fig. 5.67 The path of Cyclone Winston through the islands of Fiji

- Most of the sugar cane is grown on the north-east side of the islands, where it was exposed to the full force of the storm. Most of the crop was uprooted. One of the country's four sugar mills was badly damaged.
- Coconut grows well on sandy shores, so the coconut palms were also in the path of the storm surge.
- Estimates put the cost of damage to crops and livestock together as over US $100 million.

Help provided after the tropical storm

The government organised the distribution of food rations because people had lost both their subsistence crops and income from fishing. They could not afford to buy even staple foods.

The FAO and the Red Cross helped to provide food aid. Many countries gave financial donations to help Fiji import large quantities of food as three years' crop production would be lost and the fisheries would take more than six years to recover. Fiji's balance of payments was badly affected. Exports were down, especially of sugar and coconut oil, and imports were up as foodstuffs, machinery, and materials needed for reconstruction had to be brought in. The total cost of the damage was nearly 10% of Fiji's GDP.

Fig. 5.68 Coconut palms growing along Fiji's coast on Viti Levu

Fig. 5.69 A strip of coconut palms, bananas and fruit trees on the side of the coast road

The effects of tropical storms on low sandy coasts

The low sandy coast of the southern USA is also vulnerable to erosion as a result of tropical storms:

- As beach material is eroded and deposited out at sea, the sandy beaches have been lowered by as much as 1.5 m and the coast moved inland by 15 m.
- The seaward face of sand dunes is carved out into a very steep slope and the height of the dunes is reduced as waves attack at a higher than normal level during a storm surge and move the sand out to sea.
- If the wave height exceeds the height of the dune it is destroyed as its sand is washed inland, burying everything under a deep layer of deposited sand.
- As the sea floods over and then retreats, seawater channels can be cut through the sand. After Hurricane Katrina in 2005 a sandy island was cut in two by a 2 km wide channel.

11 Look through this chapter and make a list of the different hazards for people living in coastal environments that must be included in an answer to a question set on the topic. Include example locations.

Fig. 5.70 Salt extraction is an important economic activity in coastal areas of Fiji, SE USA and other subtropical coastal locations. The location in the photograph is in the Canary Islands.

RESEARCH Find out why and how how salt is extracted from sea water.

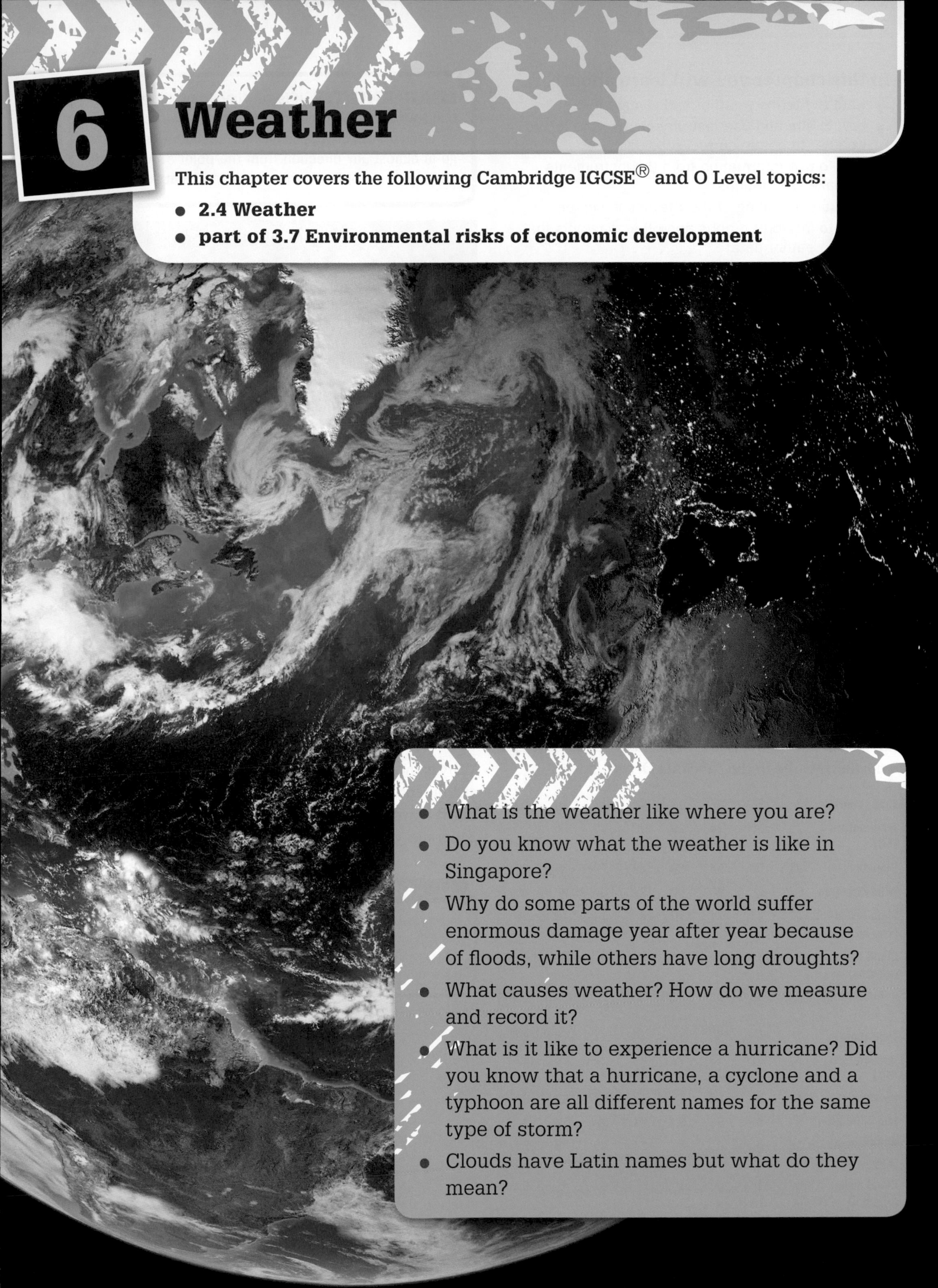

6 Weather

This chapter covers the following Cambridge IGCSE® and O Level topics:

- **2.4 Weather**
- **part of 3.7 Environmental risks of economic development**

- What is the weather like where you are?
- Do you know what the weather is like in Singapore?
- Why do some parts of the world suffer enormous damage year after year because of floods, while others have long droughts?
- What causes weather? How do we measure and record it?
- What is it like to experience a hurricane? Did you know that a hurricane, a cyclone and a typhoon are all different names for the same type of storm?
- Clouds have Latin names but what do they mean?

In this chapter you will learn about:

- what the terms weather and climate mean
- how to site and use instruments used to record weather to obtain correct readings
- how to use information from the instruments to make calculations
- the use and siting of the Stevenson screen
- how to interpret graphs and other diagrams that show weather data
- the main types of cloud, and how to estimate the amount of cloud cover
- the formation of weather hazards (*not* required syllabus knowledge), to aid your understanding of tropical storms, droughts, and desertification.

What is meant by the term weather?

Weather is the state of the **atmosphere** at a particular time. The atmosphere is a mixture of gases that encircles the Earth. The lower part is called the **troposphere** and is the zone of weather. This zone's vertical extent is least at the poles (8 km) and greatest at the equator (16 km). Airplanes ascend and descend through this zone of weather, and many cruise above it in the lower **stratosphere**. The top of the troposphere is called the **tropopause**.

Weather varies from place to place and at different times. Table 6.1 shows the weather forecasted for 1 March 2011 at Balikpapan - near the equator in Kalimantan (the Indonesian part of the island of Borneo) - and at Iquique in northern Chile in the Atacama Desert.

Time (hours)	2	5	8	11	14	17	20	23
Temperature (°C)	25	25	26	29	29	28	26	26
Humidity (%)	93	93	98	80	79	77	84	89
Wind direction	SW	W	W	SW	SW	SW	W	W
Wind speed (km/hour)	22	11	14	10	6	14	10	18
Cloud cover (eighths)	6	7	5	8	5	6	6	8
Rainfall (mm)	0.76	5.08	3.56	3.30	0	2.29	2.79	6.86

Time (hours)	2	5	8	11	14	17	20	23
Temperature (°C)	22	22	19	22	24	25	25	22
Humidity (%)	59	64	68	60	57	53	53	68
Wind direction	S	S	S	S	SW	SW	SW	S
Wind speed (km/hour)	18	6	6	5	24	35	38	26
Cloud cover (eighths)	6	1	0	1	3	0	4	0
Rainfall (mm)	0	0	0	0	0	0	0	0

Table 6.1 Weather forecasted for 1 March 2011 at Balikpapan (top) and Iquique (bottom)

LEARNING TIP It is very important to remember that a wind is named according to where it has come *from*. So, a south-west wind has come from the south-west. It could go in almost any direction from the point at which it is measured - according to the circumstances at the time.

1 Use the data in Table 6.1.

a Form groups of 8 to 12 and then subdivide each one into two groups. One subgroup should write a weather forecast in words for Balikpapan, and the other for Iquique.

i Divide the time up into sections, to emphasise changes in the weather.

ii Mention any obvious relationships that are evident from the table, e.g. between temperature and humidity.

iii Use descriptive words and refer to night and day.

iv Read each forecast to the whole group and discuss and note down the main differences between them.

b Find both places in your atlas. Can the differences in weather be explained by the distances of the two places from the sea? Explain your answer.

Weather forecasts contain references to **weather elements**, like temperature and pressure. There are also weather events, such as droughts, tropical storms, and thunderstorms.

Why is the weather so different at different places and at different times? This will be explored fully in Chapter 7, but it is worth noting now that all of the energy that generates weather is provided by the sun.

Discussion point

How typical of your location's normal weather was the weather you experienced yesterday? In what ways was it usual, and in what ways rather different, from the weather usually expected there at this time of year?

How weather data is collected

Some weather elements are measured by instruments, and some by eye.

Measured by instrument	Measured by eye
precipitation	cloud type
temperature	cloud amount
humidity	(visibility, fog, and mist)
pressure	
wind speed	
wind direction	
sunshine	

Table 6.2 Ways of collecting weather data (note that the CAIE IGCSE® and O Level syllabuses do not require you to know how the bracketed elements are measured)

Discussion point

As a class, form an opinion about what the temperature, rainfall, relative humidity, cloud cover, wind direction, and wind speed are at the moment in your location. Check with a weather information service to see how right or wrong your perceptions were.

2 The rain gauge must be correctly sited to enable a correct reading to be obtained. Write an account of how the rain gauge must be sited – and why – by pairing up the different site requirements in the left-hand column of Table 6.3 with the correct reasons from the right-hand column. Some reasons can be used twice. Afterwards, make sure that you check your answers and correct any errors.

Site requirement	Reason
• On grass and not on a hard surface ...	• ... to avoid surface run-off entering
• Part buried (two reasons) ...	• ... to avoid drips entering the gauge
• Rim 30 cm above the ground surface (two reasons) ...	• ... to avoid shelter and too little rain entering
• No trees or roofs overhanging ...	• ... to avoid splashes entering the gauge
• Away from buildings, trees, or other objects ...	• ... to prevent evaporation
• Standing vertically upright ...	• ... for stability
	• ... to keep the correct gauge diameter

Table 6.3 Factors to consider when choosing the site for a rain gauge

Precipitation

Amounts of **rain**, **drizzle**, **snow**, **hail** and **sleet** are measured using a **rain gauge**.

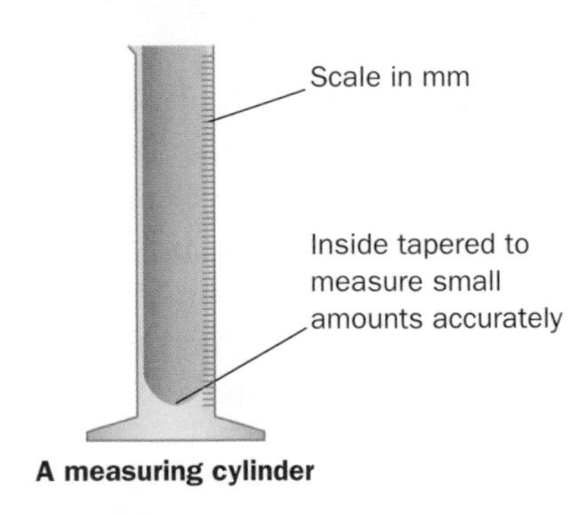

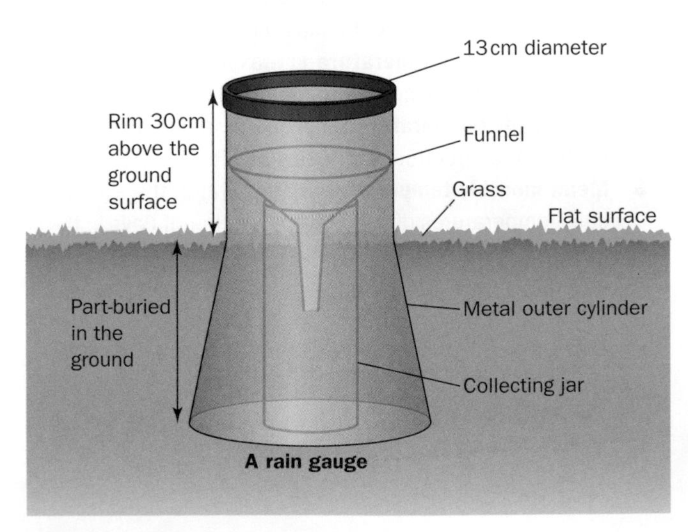

Fig. 6.1 A rain gauge and how it is sited

The rain gauge must be in an open space (the distance from the nearest object should be twice its height). In countries with long hot dry seasons, surfaces are hard and lack grass, so the rain gauge should be raised on a tripod.

→ The instrument must be read at the same time each day (generally at about 9 am).

→ Any snow or frost present must be melted first.

→ The reader should pour the water from the collecting jar into the tapered measuring cylinder.

→ The measuring cylinder should then be placed on a flat surface.

→ The water level should be read at eye level – to avoid parallax error. If the eye is not at the same level as the lowest part of the concave **meniscus** (curved upper surface), the angled reading will be inaccurate.

→ Any amount too small to measure should be recorded as a **trace**. This is probably the result of drizzle.

Recordings can be totalled to give monthly and annual rainfalls. The mean (average) monthly and annual rainfalls can also be calculated (usually over a minimum of 30 years).

LEARNING TIP Unlike analysis of temperature, it is not conventional to refer to ranges of rainfall.

Temperature

Air **temperature** is the degree of warmth in the air. It is measured in degrees Celsius (Centigrade), abbreviated as °C. The words used to describe temperature can have different meanings in different **latitudes**. For example, a person living in cool temperate latitudes might consider a temperature of over 30 °C to be very hot, 20-29 °C as hot, 10-19 °C as warm, 5-9 °C as mild in winter and cool in summer, 5 °C to -5 °C as cold - and anything below that to be very cold! On the other hand, a person living in the tropics might consider a temperature of 19 or 20 °C to be chilly.

A **maximum-minimum thermometer** is used to measure the warmest and coldest temperatures of the day. The air temperature at any time can also be read from it.

There are two versions of this instrument:

→ a minimum thermometer and a separate maximum thermometer

→ a **Six's thermometer** (which combines the two).

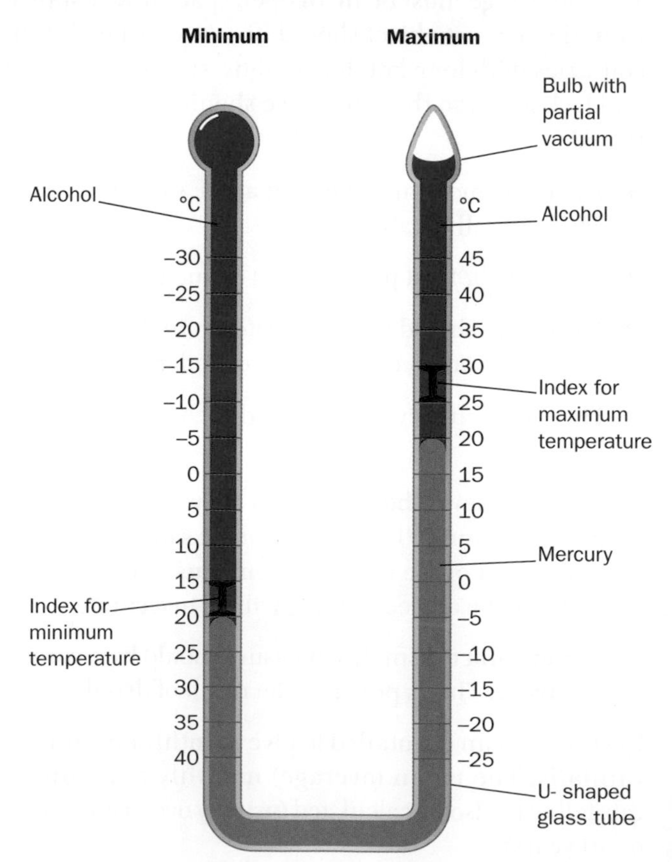

Fig. 6.2 The Six's thermometer

These instruments are read at the same time of day, every day, to give maximum and minimum temperatures for each 24-hour period. It is important to avoid parallax error by making sure that the reading is taken with the eye at the same level as the correct part of the index.

Fig. 6.2 shows maximum and minimum thermometers combined into one curved glass tube that contains both alcohol and mercury. This is the Six's thermometer.

In a Six's thermometer, the glass tube contains mercury with alcohol on either side of it.

- When the temperature rises, the alcohol in the left arm expands, while some of the alcohol in the right arm evaporates into the space in the bulb.
- The expanding alcohol on the left is then able to push the mercury up the right arm.
- This pushes an index, which is left at the maximum temperature reached.
- When the temperature cools, the alcohol in the left arm contracts. Some of the alcohol vapour in the bulb turns back to liquid.
- The mercury then moves up the left arm - pushing a metal index as it does so - to indicate the minimum temperature reached.

The instrument is read at eye level from the lower end of each index. It is reset using a magnet to draw each index back to the mercury.

In order to measure the air temperature, thermometers have to be shielded from direct sunlight and other sources of heat. They also have to be raised so that their bulbs are 1.25 m from the ground - to avoid heat being both radiated from the ground and being chilled by contact with cold ground. They must also be away from buildings that might radiate heat.

Temperature recordings can be used to calculate various useful statistics. These are the most common:

- **Daily range of temperature** is maximum temperature minus minimum temperature for one day.
- **Mean daily temperature** is maximum temperature plus minimum temperature of the day divided by two.
- **Mean monthly temperature** is the total of the average daily temperatures divided by the number of days in the month.
- **Annual range** is mean temperature of the warmest month minus mean temperature of the coldest month.
- **Mean annual temperature** is the total of the monthly means divided by 12.

LEARNING TIP Mean (average) figures are of little use when there is a large range of values and there are some very extreme figures at one end of the range. These statistics are most useful for comparison purposes when they are averaged over a period of at least 30 years.

3 Study Fig. 6.2.

a State the three temperature readings possible on Fig. 6.2.

b For the readings, calculate:

i the daily temperature range

ii the mean temperature of the day.

Sunshine

Sunshine hours are measured by a Campbell-Stokes sunshine recorder, which consists of a small glass sphere about 10 cm in diameter with a curved metal housing around its back and sides. A recording card on which the hours of one day are marked is placed in a holder on the metal housing. In bright sunshine, the glass sphere concentrates the sun's rays on to a point on the card, which burns. The burn extends along the card as the earth's position relative to the sun changes during its daily rotation. It records only the hours of sunshine bright enough to burn a hole through the card.

The sunshine recorder should be in an open space without any shade. In temperate and subtropical latitudes the instrument is placed to face south in the northern hemisphere and north in the southern hemisphere. In the tropics two instruments are used, one facing north and the other south.

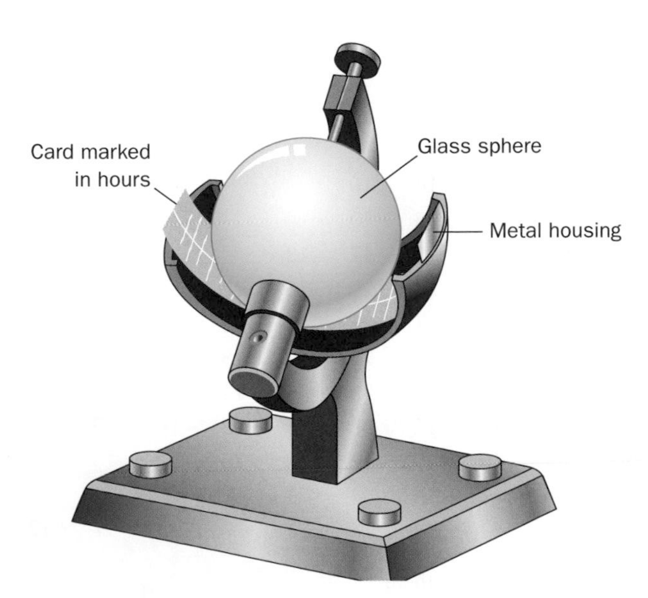

Fig. 6.3 The Campbell-Stokes sunshine recorder

The user has to insert a new card and leave it for a certain period (e.g. each day), then calculate the hours of daily sunshine from the length of the burns on the card. However, weak sunshine, such as when the rays are low in the sky near dawn and dusk, are not recorded.

Humidity

Humidity is the amount of **water vapour** in a given volume of air. Water vapour is an invisible gas, and even dry air - such as that over deserts and in classrooms - contains some water vapour. The amount of water vapour that any air can hold depends on its temperature. Warm air can hold more water vapour than cold air:

- When the air temperature increases, it **evaporates** water from water surfaces, such as seas, lakes, rivers, and vegetation. This increases the air's humidity.
- When the air cools, it can hold less water vapour, which reduces its humidity. If the air is cooled sufficiently, it will reach the temperature at which it is holding the maximum amount of water vapour that can be held at that temperature. That temperature is called the **dew point** and the air is said to be saturated (in a state of **saturation**). Any further cooling will result in **condensation** - a process by which the excess water vapour changes into water droplets (a change from water vapour to ice, known as deposition, occurs in very cold air).

LEARNING TIP Refer to water *vapour* condensing, not water condensing.

Knowing how near the air is to being saturated is vital for accurately forecasting, whether or not precipitation will occur. The **relative humidity** of the air is expressed as a percentage. The actual amount of water vapour in the air is divided by the maximum amount of water vapour that the air can hold at that temperature, and then multiplied by 100 to give the percentage. When the air is saturated, its relative humidity is 100%.

Relative humidity is measured using **wet- and dry-bulb thermometers**, otherwise known together as a **hygrometer**.

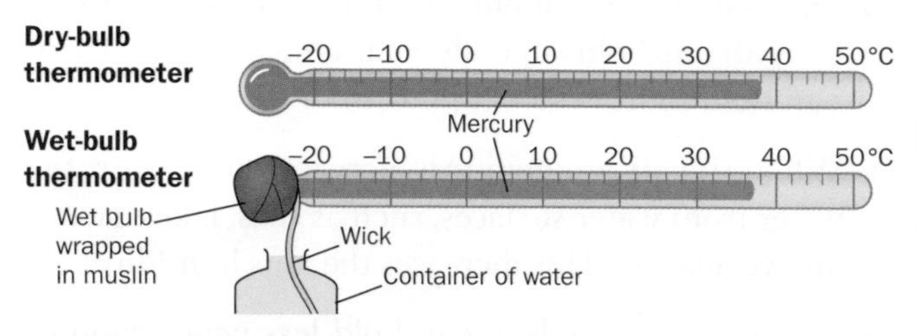

Fig. 6.4 Wet- and dry-bulb thermometers

- → The dry-bulb thermometer is an ordinary thermometer giving the air temperature (the dew point temperature of that air is known).
- → The bulb of the other thermometer has a fine muslin cloth wrapped around it. Beneath it is a container of water from which a wick of cloth leads to the muslin round the bulb - keeping the bulb constantly moist.
- → If the air is not saturated, water will evaporate from the muslin. **Evaporation** causes cooling of the bulb, which causes the mercury to contract and register a lower temperature.
- → The difference between the wet-bulb temperature and the dry-bulb temperature is called the wet-bulb depression. The larger this is, the lower the relative humidity will be.
- → If the air is already saturated, evaporation will not be possible, so the thermometers will show the same temperature. If the air is not saturated, the relative humidity can be found using a table (see Table 6.4).

LEARNING TIP If you use the term *hygrometer*, be careful to spell it correctly because there is a different instrument known as a hydrometer.

4 **a** Work out the relative humidity for the time the wet- and dry-bulb thermometers had the readings shown on Fig. 6.4.

b What was the temperature reading of the wet-bulb thermometer when the dry-bulb reading was 18 °C and the relative humidity was 73%?

c What would the relative humidity be if the wet- and dry-bulb thermometer readings were the same?

	Depression of wet bulb (°C)																		
Dry bulb (°C)	1	2	3	4	5	6	7	8	9	10	11	12	13	14	15	16	17	18	19
8	87	74	62	50	39	28	17												
10	88	76	65	54	44	34	23	14											
12	89	78	68	58	48	38	30	21	12	4									
14	90	79	69	60	51	41	33	24	16	10									
16	90	81	71	62	54	45	37	29	21	14	7								
18	91	82	73	65	57	49	42	35	27	20	13	6							
20	91	82	74	66	58	51	44	36	30	23	17	11							
22	92	83	75	68	60	53	46	40	34	27	21	16	11						
24	92	84	77	70	63	56	49	43	37	31	26	21	14	10					
26	92	85	77	71	64	57	51	45	39	34	28	23	18	13					
28	92	85	78	72	65	59	53	47	42	37	31	26	21	17	13				
30	93	86	79	73	67	61	55	50	44	39	35	30	24	20	16	12			
32	93	86	79	73	68	62	57	52	46	41	37	32	27	23	19	15			
34	93	87	80	74	69	63	58	53	48	43	38	34	30	26	22	18	10		
36	93	87	81	75	70	64	59	54	50	45	41	36	32	28	24	21	13		
38	94	87	81	76	70	65	60	56	51	46	42	38	34	30	26	23	16	10	
40	94	88	82	76	71	66	61	57	52	48	44	40	36	32	29	25	19	13	
42	94	88	82	77	72	67	62	58	53	49	45	41	38	34	31	27	21	15	
44	94	89	83	78	73	68	63	59	54	50	47	43	39	36	32	29	23	17	12

Table 6.4 Part of a relative humidity table (note: as percentages are rounded to whole numbers, there may be slight differences between published relative humidity tables). You do not need to learn this table.

Pressure

There is a column of atmosphere above every part of the Earth's surface. The atmosphere exerts **pressure** on the surface, because it has a weight. This weight varies from place to place and from time to time.

- Pressure is measured in **millibars** (mb).
- The mean (average) pressure at sea level is considered to be 1013 millibars, so pressures below that are usually described as low and those above it as high. However, this is not always the case, because a higher-pressure area surrounded by lower pressures would be described as a high-pressure system.
- As altitude increases, pressure decreases - because there is less air above the ground surface, so recordings are converted into the sea-level equivalents when plotted on isoline maps. If this were not done, the pressure map would look like an inverted relief map.
- Isolines on a map showing pressure are known as **isobars**.

Three kinds of instrument are used to measure atmospheric pressure: a **mercury barometer**, in which the pressure of the air forces mercury up a glass tube; an **aneroid barometer;** and a **barograph**.

Fig. 6.5 An aneroid barometer

An aneroid barometer has a corrugated metal box inside. This box expands when pressure is low, and is compressed when pressure is high because the greater pressure of the denser air pushes down on it. It can change in this way because it is partly evacuated of air.

The changes in the top of the box are transmitted, by means of a series of levers, to a pointer that moves on the face of the dial. Some dials also have an external pointer, which can be used to mark the previous position of the internal pointer. Then, when the reader gently taps on the glass and the inner pointer moves to the present pressure reading, it is immediately obvious whether pressure has fallen or risen.

A scale is marked on the dial, and often weather conditions associated with different pressures are also shown. The scale on Fig. 6.5 is in inches, but many modern aneroid barometers have a scale in millibars (ranging from 950 on the left to 1060 on the right).

Unlike temperature and rainfall, humidity and pressure are not totalled for a week, month or year. However, their ranges and averages can be used to help explain certain weather occurrences. Pressure maps show pressures reduced to sea level to eliminate the influence of altitude and allow high- and low-pressure systems to be seen.

The only site requirements for barometers and barographs are that they should be kept away from strong air movements, direct sunlight and heat sources. The barometer is kept in a Stevenson screen (see page 167) at a weather station, but it can also be kept indoors.

Fig. 6.6 A barograph

A barograph has an aneroid barometer, and a revolving drum covered in graph paper, inside a case. Changes in the level of the top of the metal box in the aneroid barometer are transmitted by levers to a pen that marks a continuous trace of the pressure on to the rotating graph paper. The paper has vertical lines marked in hours and parts of hours, with millibars on the horizontal scale.

Wind speed

Wind is air moving over the surface of the ground. Its velocity is measured using an **anemometer**.

Fig. 6.7 An anemometer

The three cups are supported on a tall shaft, so that the slowing effect on the wind of friction with the ground is minimised. The speed at which the cups rotate when driven by the wind is measured and shown on a meter or dial (in metres per second). Of course, wind speed changes, so it's important to look for a sufficient time to be able to estimate an average velocity or to note any stronger gusts of wind. This problem is often overcome by transmitting a continuous signal of the speed to a moving chart.

Wind can be very destructive. A scale known as the **Beaufort Scale** has been devised to classify different wind speeds according to what effect they have (see Table 6.5). This table is for reference only - an examiner would not expect you to memorise it!

Wind direction

A **wind vane** has the points of the compass fixed and sited, so that they point in the correct directions. Above them is an arrow that can be moved by very light winds. When the wind blows, it catches the thickened end of the arrow's shaft and swings it round so that the arrow is pointing to the direction from which the wind has come.

It is important to know where the wind has come from, because it brings with it some of the characteristics of temperature and moisture of the areas over which it has passed. It can therefore influence the weather at its new location. The most frequently occurring wind in an area is known as the **prevailing wind**.

Fig. 6.8 A wind vane

Anemometers and wind vanes should both be sited on poles 10 m above the ground (to avoid gusts and ground friction). They should also be in an open space, at least three times the height of the nearest obstacle away from it (e.g. if a tree is 10 m high, the instrument pole should be sited at least 30 m away from it). In urban areas, they often have to be placed on top of buildings.

LEARNING TIP When describing wind direction, do not state that it is moving in a certain direction. This statement is ambiguous. Always express the direction clearly. For example, 'it is a westerly wind' or 'the wind is from the west' cannot be misinterpreted. It is also correct to state for a west wind that 'the wind blows west' but not all will be familiar with that expression.

LEARNING TIP It is not correct to state that these instruments are sited high in order to have contact with the strongest wind speeds. The strongest gusts in urban areas occur where the wind is funnelled down narrow passages between buildings, but the instrument recording wind speed should be placed on top of the buildings to avoid the gusts.

5 **a** Explain why pressure readings taken on a mountain top will always be lower than at the foot of the mountain.

b What wind direction is shown in Fig. 6.8?

c How many cups does the anemometer have?

Beaufort Scale number	Wind speed (km/h)	Wind description	Visible effect
0	less than 1	Calm	Smoke rises vertically.
1	1-5	Light air	Smoke moves, but not a wind vane.
2	6-11	Light breeze	Wind is felt on your face, leaves move slightly, a wind vane moves.
3	12-19	Gentle breeze	Leaves, twigs and a flag move constantly.
4	20-28	Moderate breeze	Small branches move.
5	29-38	Fresh breeze	Small trees sway.
6	39-49	Strong breeze	Large branches move.
7	50-62	Moderate gale	Whole trees move.
8	63-73	Fresh gale	Twigs break off trees.
9	74-86	Strong gale	Roofs are slightly damaged.
10	87-100	Whole gale	There is considerable damage to structures, and trees are uprooted.
11	101-118	Storm	There is widespread damage.
12	119 and above	Hurricane	There is widespread devastation.

Table 6.5 The Beaufort Scale

Digital weather recording instruments

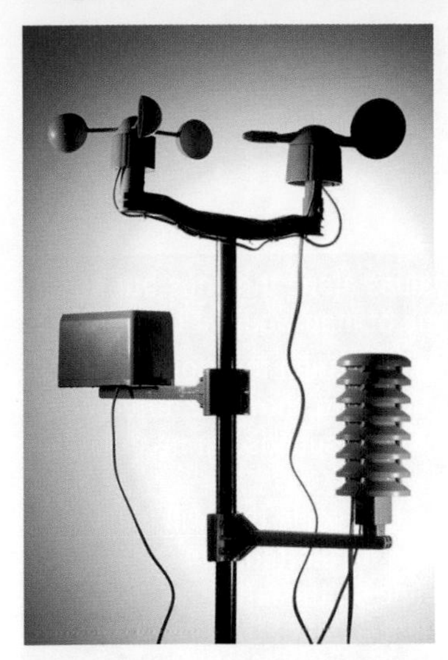

Fig. 6.9 Digital instruments at a weather station

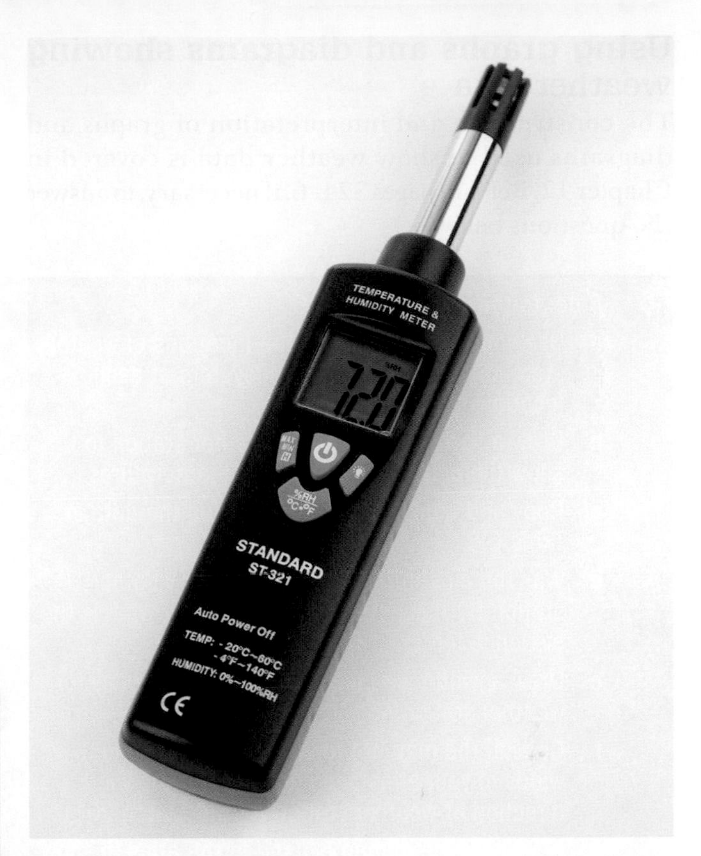

Fig. 6.10 A digital hygrometer

The three cups of the anemometer can be seen on the top left of the photograph. To the right of that is the digital wind vane, with its prominent paddle to catch the wind (its current position indicates a wind blowing from the left). Below that is the corrugated metal box of the anemometer. Cables are clearly visible, along which the instruments send their results to computers.

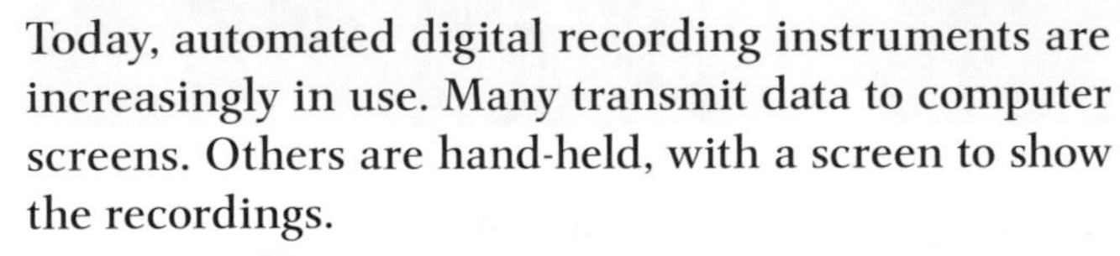

Today, automated digital recording instruments are increasingly in use. Many transmit data to computer screens. Others are hand-held, with a screen to show the recordings.

The digital hand-held hygrometer in Fig. 6.10 indicates a relative humidity of 72%. The reading is obtained quickly and easily.

Digital maximum-minimum thermometers are also used. They measure to a 10th of a degree - a far finer measure than is possible on a non-digital instrument. The biggest problem is that data may be lost if they get wet. The thermometer also has to be kept out of direct sunlight.

Tipping bucket rain gauges use a bucket with a known volume that tips over when it is full. The total volume of water collected is calculated by multiplying the number of times that the bucket tips over by the volume of a single bucket. The total result for a specific period of time (e.g. 30 minutes) is then sent automatically to a computer. The distribution of rainfall throughout the day can then be recorded.

Sunshine sensors are also used to log duration and intensity of the sunshine and transmit the data to weather station computers.

The advantages of digital recording instruments

There are several advantages to using digital weather recording instruments compared with conventional ones:

- → They are more quickly and easily read as the data is displayed in clear figures.
- → The reading is more likely to be accurate as it cuts out the chance of misreading by parallax errors, reading the wrong end of the index and other errors.
- → The portable ones are also more useful for fieldwork because they can be used at any site.
- → Readings from instruments at digital weather stations download directly to a computer, saving the time and inconvenience of having to go to the instrument.
- → Digital readings are also very frequent, so more analysis can be done, such as calculating hourly averages.

Using graphs and diagrams showing weather data

The construction and interpretation of graphs and diagrams used to show weather data is covered in Chapter 12. Refer to pages 324–6, if necessary, to answer the questions below.

6 Refer to the data in Table 6.6.

a Use your atlas to identify the locations of weather stations A and B and the ecosystem at each location.

b Draw a line graph for each weather station to show the actual and dew-point temperatures.

c Copy and complete Table 6.7, using the data in Table 6.6.

d Was the air saturated at any time at weather station B? If so, state when – and say what is the evidence for it.

e Draw a wind rose (see page 326) for the day for weather station A.

f Consult the map of prevailing (most common) winds in your atlas and comment on the wind direction at weather station A on 4 March 2011.

g **i** What type of map is used in the atlas to show the distribution of pressure, temperature, and precipitation?

ii What convention is used to give a good visual impression by the shading of increasing amounts of precipitation?

iii If the map had values along the lines but no shading, what type of map would it be?

h **Dispersion diagrams** are useful for showing distributions. For example, to show the difference in the distributions of wind speeds during the day at each location:

- Draw a vertical line to a scale that will allow the strongest wind to be plotted.
- Number the scale line on the left.
- On the right of it, plot a dot at the level of all the values for weather station A. If you have more than one plot of the same value, place each one to the right of the previous ones (side by side), keeping them at the correct level.
- Next, starting a new column, plot the values for weather station B (using a cross).
- Add a key.
- Comment on the differences in wind strength between the weather stations.

i What type of graph should be used to show daily, monthly, or annual precipitation totals? Refer to types of data in Chapter 13 (page 336) and explain your answer.

Weather Station A (30° 08´ N, 31° 24´ E)

Time (hours)	Temp. (°C)	Dew point (°C)	Pressure (mb)	Wind direction	Wind speed (km/h)	Rainfall (mm)	Weather event
midnight	15	12	1020	NE	29	0	-
1.00	15	13	1020	NE	29	0	-
2.00	15	13	1019	NE	29	0	-
3.00	15	13	1019	NE	27	0	-
4.00	15	13	1019	NE	22.5	0	-
5.00	15	13	1019	NE	22.5	0	-
6.00	14	14	1020	NE	21	0	-
7.00	14	14	1020	NE	24	0	-
8.00	15	14	1021	NE	26	0	-
9.00	18	13	1021	NE	32	0	-
10.00	20	13	1021	NE	27	0	-
11.00	21	13	1021	NE	24	0	-
12.00	23	8	1019	NE	27	0	-
13.00	24	7	1018	NE	34	0	-
14.00	25	9	1018	NE	22.5	0	-
15.00	25	11	1018	N	27	0	-
16.00	24	10	1018	NE	32	0	-
17.00	23	10	1018	N	34	0	-
18.00	22	7	1018	N	32	0	-
19.00	20	9	1019	NE	24	0	-
20.00	18	11	1019	NE	24	0	-
21.00	18	11	1019	NE	21	0	-
22.00	17	11	1020	NE	22.5	0	-
23.00	16	11	1019	NE	14.5	0	-
24.00	15	11	1019	NE	13	0	-

Weather Station B (1° 22´ N, 103° 59´ E)

Time (hours)	Temp. (°C)	Dew point (°C)	Pressure (mb)	Wind direction	Wind speed (km/h)	Rainfall (mm)	Weather event
midnight	26	24	1010	NE	11	0	-
1.00	26	24	1010	NE	8	0	-
2.00	26	24	1009	Variable	3	0	-
3.00	25	24	1008	NE	8	0	-
4.00	25	24	1008	Variable	3	0	-
5.00	25	24	1008	Variable	3	0	-
6.00	25	24	1008	Variable	3	0	-
7.00	25	24	1009	N	5	0	-
8.00	26	24	1010	Variable	5	0	-
9.00	26	25	1011	Variable	5	0	-
10.00	29	24	1011	N	11	0	-
11.00	30	24	1010	NE	14.5	0	-
12.00	32	24	1009	Variable	16	0	-
13.00	31	24	1008	NE	13	3	-
14.00	27	24	1008	Variable	11	5	-
15.00	25	24	1007	Variable	3	9	thunder
16.00	25	24	1007	Variable	3	9	thunder
17.00	25	25	1006	Variable	3	1	-
18.00	26	25	1007	Variable	1.5	0	-
19.00	25	24	1007	Variable	3	0	-
20.00	25	25	1007	N	5	0	-
21.00	26	24	1008	NE	9.5	0	-
22.00	26	25	1009	NE	9.5	0	-
23.00	26	24	1009	NE	9.5	0	-
24.00	26	24	1009	NE	10	0	-

Table 6.6 Weather recordings taken at two weather stations (A and B) on 4 March 2011

Date: **4 March 2011**	**Weather station A**	**Weather station B**
City/town		
Ecosystem		
Maximum temperature		
Minimum temperature		
Daily range of temperature		
Mean daily temperature		
Pressure (high or low)		
Description of the humidity and reason for your description		
Most frequent wind direction		
Description of strongest wind (using the Beaufort Scale)		
Precipitation (daily total)		
Precipitation (distribution)		
Weather event		

Table 6.7 A comparison of weather on one day at two weather stations

Lines on isoline maps have different names, according to the weather element being shown:

→ **Isohyets** are lines joining places with the same rainfall.
→ **Isotherms** are lines joining places with the same temperature.
→ **Isobars** are lines joining places with the same pressure.

A synoptic chart (another type of map showing weather) is described later in the chapter on page 175.

The Stevenson screen

The maximum-minimum thermometers and wet- and dry-bulb thermometers are housed in a wooden box called a **Stevenson screen**. The barometer can also be put in it.

Fig. 6.11 The inside of a Stevenson screen

This box is designed to ensure that the instruments give the correct readings.

Air temperatures have to be measured in the shade out of direct sunlight and at the same height above the ground at all weather stations, so that data is standardised and can be compared. The screen has to be positioned so that the door opens away from the sun, facing north in the northern hemisphere and south in the southern hemisphere. Slatted sides called louvres allow the free flow of air into and out of the screen; this is essential for measurement of the outside air temperature and to allow evaporation from the wet-bulb thermometer.

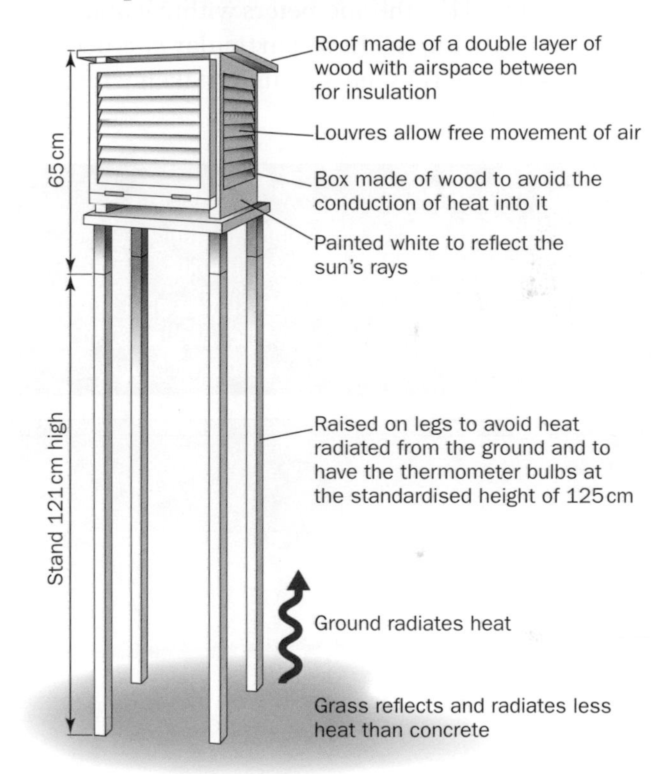

Fig. 6.12 Features of the Stevenson screen

The screen is painted white to reflect the sun's rays. The roof has double layers of wood so that the airspace between them insulates the screen below. Wood and air are bad conductors of heat. The slanting roof sheds rain.

Stevenson screens should be sited on grass that is kept short – to standardise the influence of the type of ground surface and record what is considered to be the real air temperature. If the screen has to be sited on concrete, it will have to be placed much higher (because concrete heats up and radiates the heat, as well as reflecting some of the sun's rays upwards).

LEARNING TIP Know the difference between insolation (the incoming radiation from the sun) and insulation (protection from incoming heat or cold).

7 **a** Explain why the Stevenson screen is raised on a stand with legs about 121 cm long.

b Identify the other weather recording equipment at the weather stations in Figs. 6.13 and 6.14.

Discussion point

Where would be the best places to site weather instruments in your school grounds?

Weather station layout

The site of the Stevenson screen is determined by the requirements of the thermometers within it and, as all the other instruments have particular site needs, a weather station layout must be undertaken with all of them in mind.

8 Look at Fig. 6.15.

a What errors have been made in siting the instruments at this weather station?

b Make a copy of the weather station plan and move the instruments to better positions, so that they can make accurate weather recordings.

Fig. 6.13 A Stevenson screen located at a weather station in Death Valley, USA (a hot desert area)

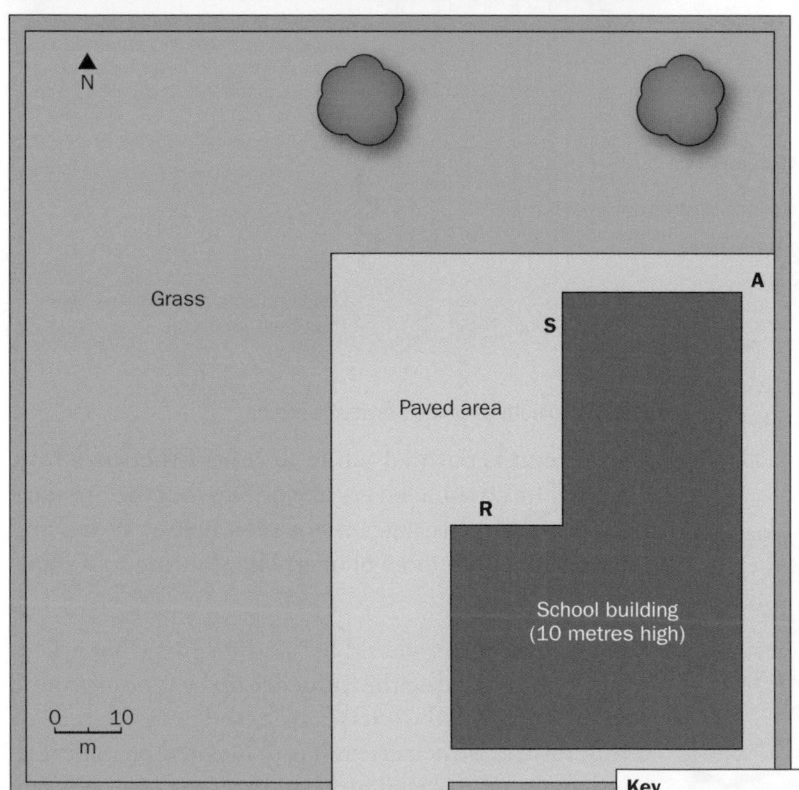

Fig. 6.15 A weather station with badly sited instruments

Fig. 6.14 A Stevenson screen located at a weather station in the Austrian mountains

Cloud types and extent

Clouds consist of tiny water droplets, or ice particles, that are too light to fall to Earth. They are formed when air containing water vapour is forced to rise. As it rises into lower pressure, the air expands and that expansion causes cooling. If it is uplifted enough to cool to below dew-point temperature, water vapour then condenses into water droplets or, if it is sufficiently cold, ice crystals form by the deposition process.

LEARNING TIP Rising air does not cool because it is rising into cooler layers of the atmosphere. It cools because it expands and uses energy in doing so.

Air will continue to rise while it is warmer, and therefore lighter, than the air into which it is rising. The tallest clouds form in the tropical zone, where the tropopause is at its highest.

Only three types of cloud produce precipitation. For this to happen, there have to be a lot of water or ice particles, plus sufficient air movement for the particles to be moved so that they collide and join together. If they grow large and heavy enough to fall through the rising air currents that form the cloud they are in, they might fall to the ground as precipitation. However, some droplets evaporate before they reach the ground.

The density and vertical extent of a cloud determine whether it looks white, grey, or black from below. If it is thin enough for sunlight to pass through it, it looks white. However, a cloud with a very large vertical extent will prevent sunlight from penetrating to its base, so it will appear to be black.

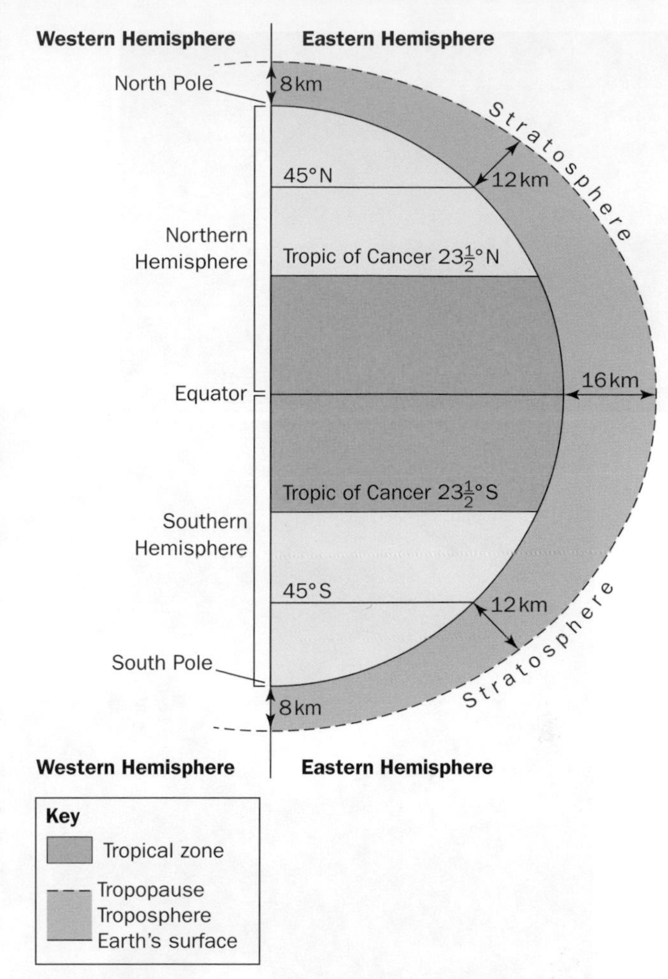

Fig. 6.16 The tropical zone and zone of weather for the Earth's Eastern Hemisphere

The main types of cloud

Clouds are classified into three main types, according to their shapes. These, in turn, depend partly on how far the air has been made to rise - and over what horizontal area it has been pushed up:

- Where there is little vertical uplift, but the uplift is over a wide area, clouds form in layers and are known as **stratus**.
- More vertical, but localised, uplift results in heaped **cumulus** clouds with flat bases and globular upper surfaces.
- Where condensation occurs at very high levels, wispy **cirrus** clouds are formed (made of ice crystals). Cirrus is often described as feather-like (see Fig. 6.17). It is likely that the cloud in Fig. 6.17 formed from the condensation of a water-vapour trail from an aircraft and that high-level air movement then expanded it into a feather shape.

Fig. 6.17 A feather-shaped cirrus cloud

Fig. 6.18 Stratus often forms Table Mountain's famous 'tablecloth' in Cape Town
Fig. 6.19 Cumulus cloud in the Canadian Rocky Mountains
Fig. 6.20 The globular upper surface of cumulus cloud seen from the air
Fig. 6.21 This view of cirrus in New Zealand illustrates why the cloud is often described as thread-like
Fig. 6.22 Wispy cirrus cloud
Fig. 6.23 Very low-level stratus in Alaska

The three main types of cloud can be further subdivided:

- Very high clouds could be in layers called **cirrostratus**, or globules called **cirrocumulus**.
- Middle-level cloud is prefixed by 'alto'. The clouds formed here could be **altostratus** or **altocumulus**.
- Low-level cloud is stratus or cumulus. If it has the characteristics of both, it is called **stratocumulus**.

Discussion point

What types of cloud can you see in the sky now? If there are no clouds, why is this so? If there are, what does it tell you about the state of the atmosphere and the processes occurring?

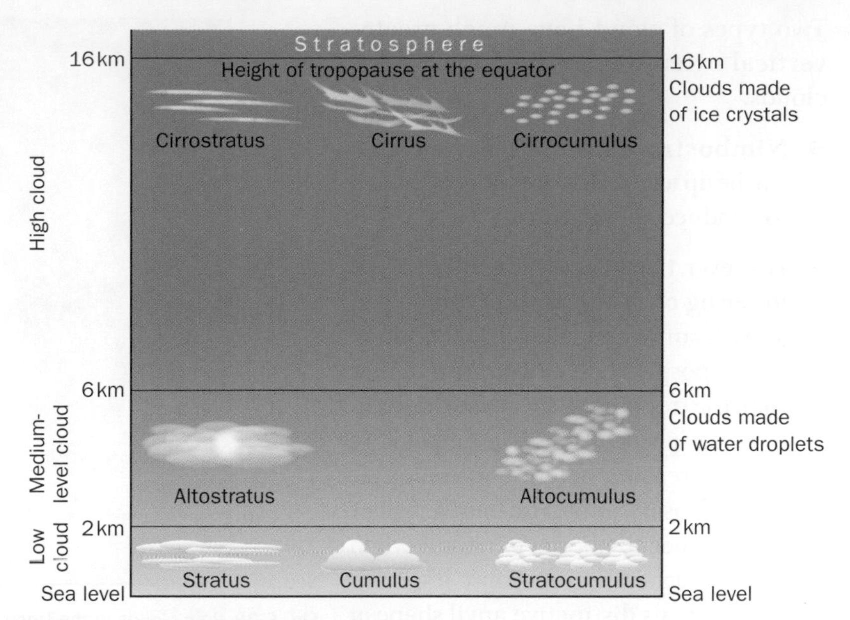

Fig. 6.24 Fair weather clouds in the Tropics

Fig. 6.25 Cirrostratus and cirrocumulus above airplane trails, seen from ground level

Fig. 6.26 Altocumulus off the coast of California from the air

Fig. 6.27 Layers of altostratus cloud in Monument Valley, USA

Fig. 6.28 A layer of stratocumulus cloud just above the sea in Fiji

Two types of cloud have much greater vertical extents than the fair weather clouds:

- **Nimbostratus** is a layer cloud that can be up to 5000 m deep (deep enough to produce steady rain).
- However, the biggest cloud by far is the towering **cumulonimbus**, which can grow from near sea level to the top of the troposphere. It cannot rise higher than the tropopause, because the air in the stratosphere increases in temperature with increasing height - so cooler air (being denser) cannot rise into the lighter and warmer air above. As a result, when this air reaches the tropopause, it spreads out to form a distinctive anvil shape at the top of the cumulonimbus cloud.

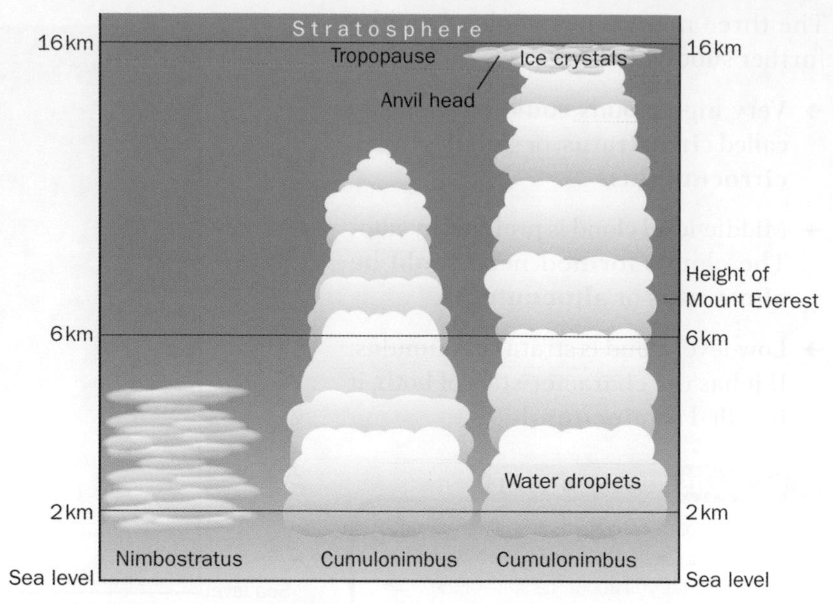

Fig. 6.29 Rain clouds in the Tropics

Fig. 6.30 Nimbostratus over salt flats in New Zealand

Fig. 6.31 Cumulonimbus cloud developing over the Canadian Rocky Mountains, with cirrocumulus at the top of the photograph

Fig. 6.32 Cumulonimbus cloud with an anvil top

Fig. 6.33 Cumulonimbus cloud with the dark base of one cloud over a harbour in Mexico

Cloud	Level	Description	Associated weather
Cirrus	High (above 6 km)	This cloud is thin, white, and made of ice crystals. It forms narrow wisps or threads (cirrus means hair-like). It can also be feather-like in shape.	Fine
Cirrostratus	High (above 6 km)	This cloud is a thin, white layer made of ice crystals, with a wide horizontal extent. It often covers the whole sky.	Fine
Cirrocumulus	High (above 6 km)	This cloud is thin, white, made of ice crystals and slightly heaped.	Fine
Altostratus	Medium (2-6 km)	This cloud is a layer of water droplets, which can be thin and white or thick and grey.	Fine
Altocumulus	Medium (2-6 km)	This is a heaped cloud of water droplets, which can be white or thick enough to look light grey.	Fine
Stratus	Low (0-2 km)	This cloud is a thin, uniform, grey layer or sheet of small water droplets, with a fairly flat base.	It may be thick enough to produce drizzle
Cumulus	Low (0-2 km)	This cloud is white with a darker, flat base and globular upper surface. It is made of water droplets. It may have a small or considerable vertical extent.	Sunny by day, fine weather
Stratocumulus	Low (0-2 km)	This is a layer of cloud with some heaped sections, giving white and grey parts. It is made of water droplets.	Fine
Nimbostratus	The base can be low, or above 2 km	This cloud is a thick, dark grey layer of water droplets.	Steady rain or drizzle
Cumulonimbus	A low base, but the cloud extends up to high levels	A dense, dark grey cloud with a great vertical extent. It grows from a cumulus cloud to have a high, billowy head (or a flat top if it reaches the tropopause). If it then spreads out, it has an anvil top. It is composed of ice crystals at the top and water droplets at lower levels.	Very heavy rain, or snow showers, often with hail and thunder and lightning

Table 6.8 Cloud characteristics and their associated weather

Discussion point

Would it be dangerous for a parachutist to jump out of an aircraft in some or all types of cloud? Which types would be the most dangerous?

How cloud extent is measured

The extent of cloud cover is estimated by eye and expressed in the number of oktas (eighths) of the sky covered with cloud. For example, full cloud cover is expressed as eight oktas, no cloud cover as zero oktas and half the sky covered as four oktas.

9 **a** Identify the main cloud types in Figs. 6.34–6.38 on the next page.

b There are several different cloud types in Fig. 6.39. Locate, describe, and identify three of them.

c Fig. 6.40 was taken from an aircraft. It shows cloud cover over a mountain range. Describe the characteristics of the types of cloud shown.

d Estimate the extent of the cloud cover in each of Figs. 6.34, 6.35, 6.36, 6.37 and 6.40.

Discussion point

How could flying through different types of cloud affect airplanes?

Discussion point

What determines the shape of a cloud?

Discussion point

Do clouds have any advantages?

On a synoptic chart (see page 175) cloud extent is shown by the following symbols

Key

CLOUD

Symbol	Cloud amount (oktas)
○	0
⦶	1 or less
◔	2
◔	3
◑	4
◕	5
◕	6
●	7
●	8

Figs. 6.34–6.40 Different cloud types

Readings taken at weather stations are plotted on synoptic charts (Fig. 6.41 is an example).

- ➔ Pressure over the area is shown by isobars.
- ➔ Circles are used to represent each weather station. The shading inside each circle shows the extent of the cloud cover there.
- ➔ An arrow into the station circle shows the wind direction.
- ➔ The number of feathers on the arrow represents the wind speed.
- ➔ A symbol by the circle shows the type of precipitation, if any.
- ➔ Temperature is written in numbers above the station circle.

10 Look at Fig. 6.41.

a What is the wind direction at (i) Cairo, (ii) Khartoum and (iii) Kisangani?

b What is the cloud cover at the four weather stations?

c Describe the location and other features of the weather at (i) Kisangani and (ii) Tamanrasset.

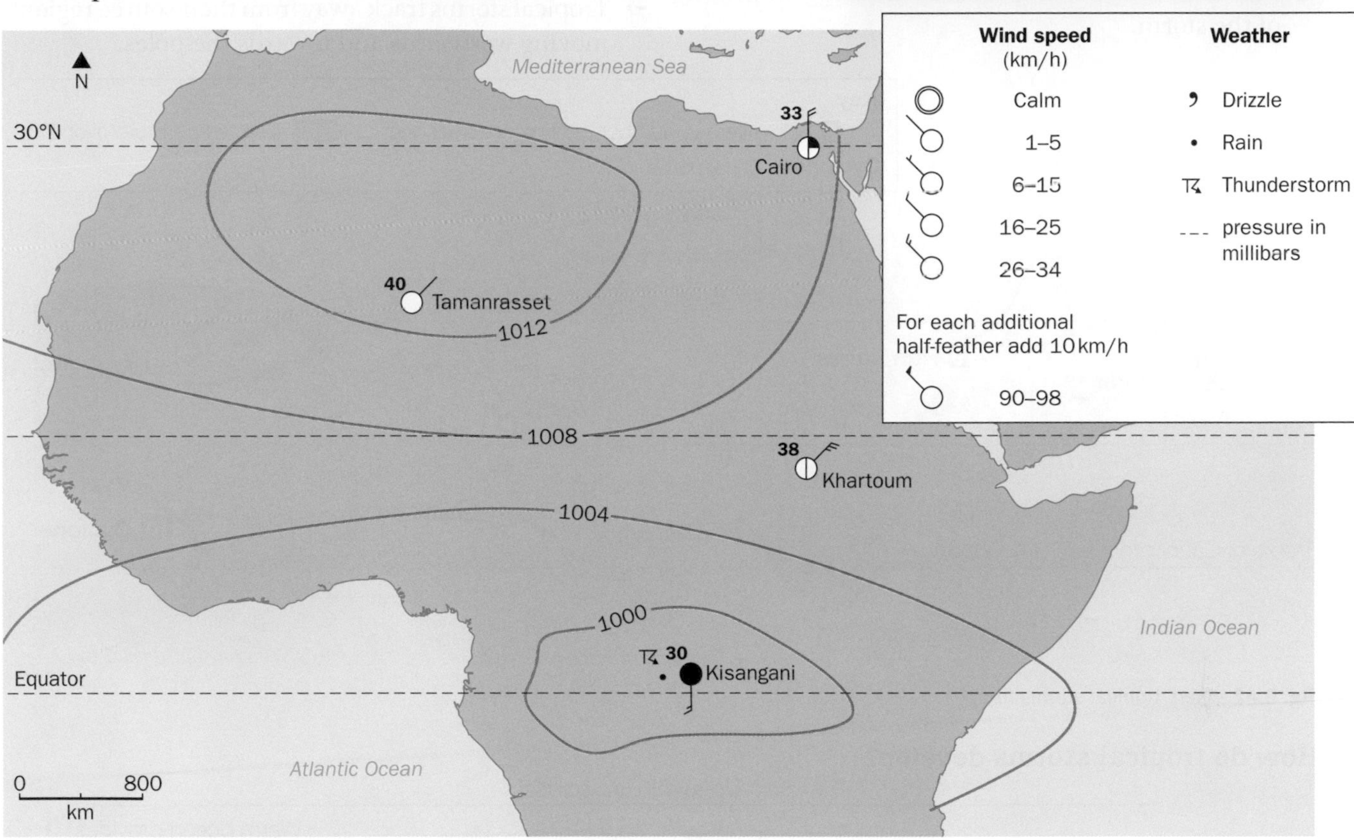

Fig. 6.41 A synoptic chart for part of Africa for 2 pm on a day in March. The key to cloud amounts is on page 173.

Weather hazards

Weather brings many benefits: warm temperatures and steady rainfall allow farmers to earn a living by producing food to support people; rainfall can also be stored to provide the water needed for people's survival, as well as for many other activities; wind provides power for renewable and clean energy. However, extreme weather can also be incredibly destructive. For example, flooding can result from long periods of intense rainfall or rapid thawing of heavy snowfall (see Chapter 4 for the effects of this).

Information on tropical storms and droughts is included here to aid your understanding of the syllabus sections on hazards affecting coasts, food production and desertification. Knowledge of the formation of tropical storms is not required by the syllabus.

Tropical storms

A tropical storm is equivalent on the Beaufort Scale (Table 6.5) to gale and storm categories 8-11 (where the wind speed is 63-118 km per hour). However, reference in this syllabus to tropical storms includes storms with even stronger winds. Tropical storms sometimes intensify into deep low-pressure systems, with ferocious whirlpools of air. These are known as **hurricanes** in the Caribbean Sea, Gulf of Mexico, and west coast of Mexico, as **cyclones** in the Indian Ocean, Bay of Bengal, and northern Australia, and as **typhoons** in the South China Sea and western Pacific Ocean. But, whatever their name, with wind speeds of 119 km per hour or more, they are greatly feared by people living in their tracks.

The average lifespan of a tropical storm is 7-14 days. Each one is given a name. A list of names is used each year, beginning with A and alternating between men and women's names.

Where, when, and why do tropical storms develop?

- They form over oceans between May and November in the Northern Hemisphere and between November and May in the Southern Hemisphere.
- Sea surface temperatures have to be a minimum of 27 °C, so they form only in the summer or autumn after a long period of intense heating. The hot sea surface warms the air in contact with it so that large amounts of water vapour are evaporated into the air. This moisture is important for fuelling the growth of the storm.
- Tropical storms originate in the tropics between 5° and 20° North and South. They do not form nearer the equator, because the effect of the Earth's rotation is needed to make the rising air spin. They do not form in higher latitudes, because the sea is not warm enough.
- With the exception of those affecting the west coast of Mexico, tropical storms form towards the west sides of the oceans.
- Tropical storms track away from their source regions, moving westwards and towards the poles.

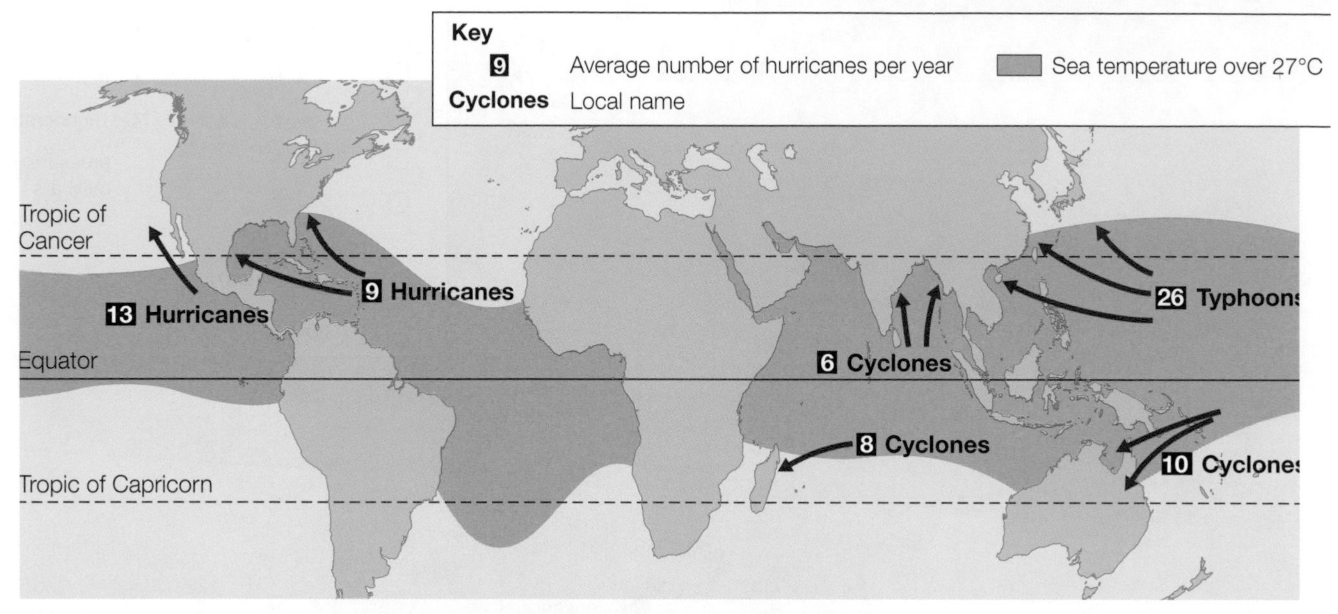

Fig. 6.42 Where tropical storms form

How do tropical storms develop?

Up to 480 km across

1. Heated, very moist air rising from the ocean creates low pressure at the surface.
2. More warm, moist air moves in over the ocean to replace the air that has risen.
3. The spiralling mass rises rapidly.
4. Thick cumulonimbus clouds are produced, from which very heavy rain falls.
5. The air cools at high levels and sinks, forming the eye in the centre of the storm.

Eye of the storm

Warm ocean provides heat and moisture to power the storm

Up to 15 km in height

Very low pressure

Area of strongest winds, heaviest rain, and thunder and lightning

Surface winds rotate around the eye

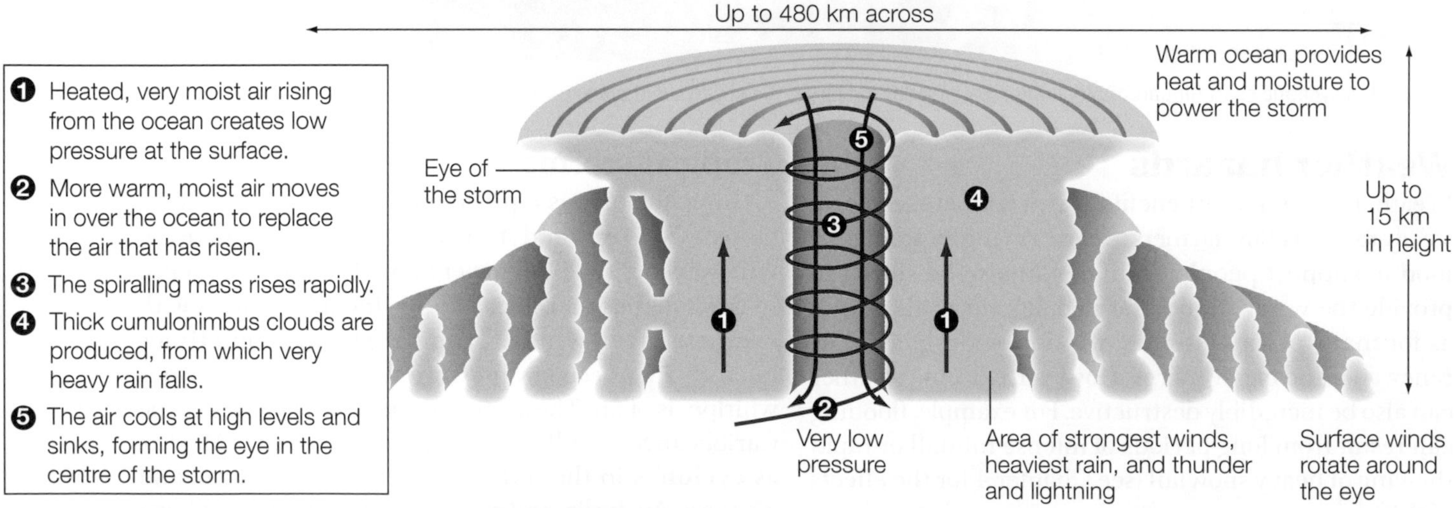

Fig. 6.43 Processes in the formation of tropical storms

While the system remains over warm water, these processes continue and intensify. This is because, as the rising air cools to below dew point and the moisture condenses into water droplets, condensation releases heat. This causes the air to rise even faster, the low pressure to lower even more and the wind to rush into the low-pressure centre even faster. The eye is a calm, sunny area in the centre of the storm. Over land the power of the storm lessens as it is cut off from its fuel: water.

The weather in a tropical storm

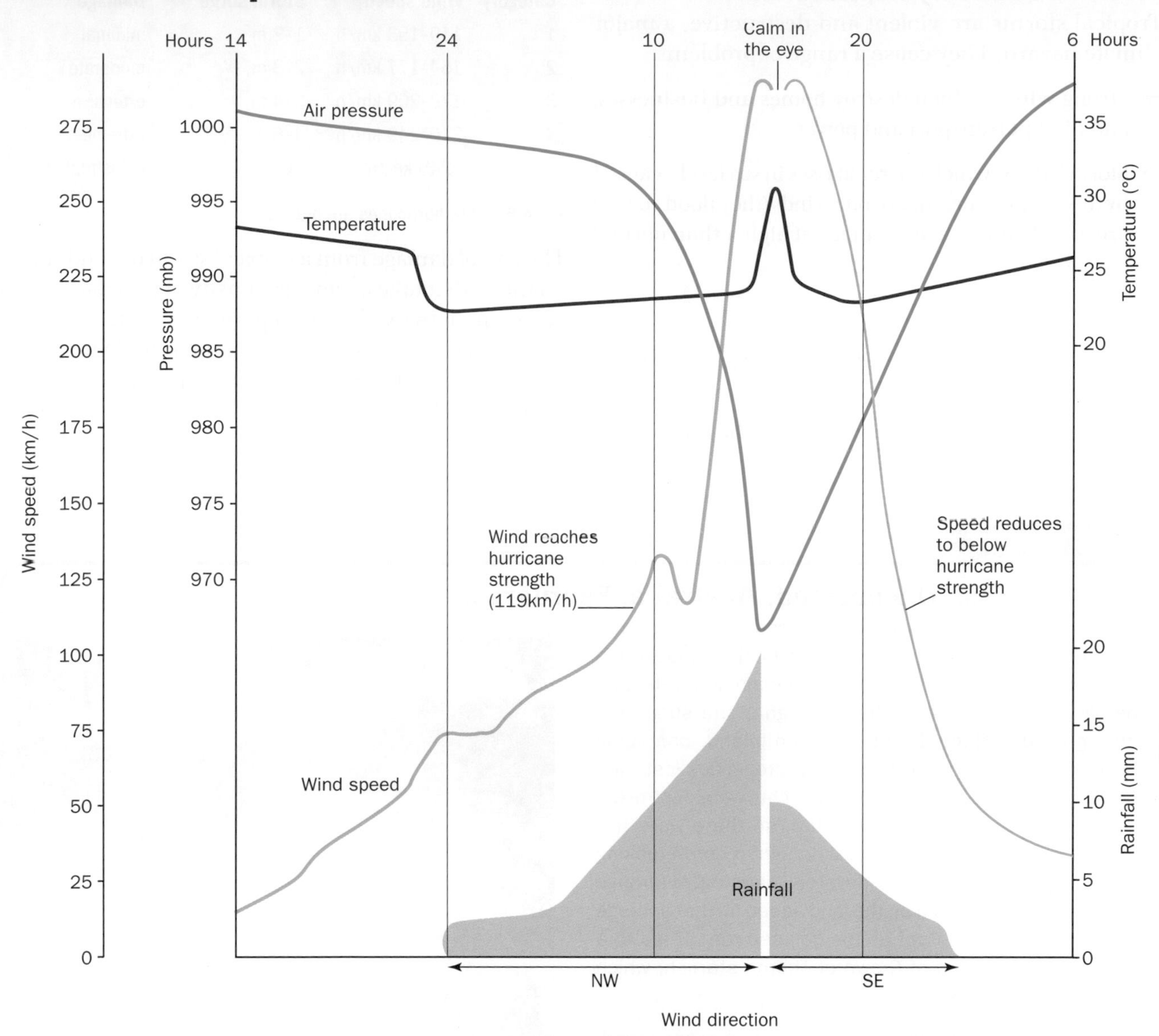

Weather ...	... before the eye	... in the eye	... after the eye
Air pressure	Falls gently, then rapidly	Very low (can be below 970 mb)	Rises rapidly
Wind strength	Gusty, then increases rapidly near the eye	Calm	Very strong again
Wind direction	Opposite to after the eye	No wind	Opposite to before the eye
Cloud	Heavy cumulonimbus	No cloud	Heavy cumulonimbus
Precipitation	Torrential rain starts suddenly, with thunderstorms	Dry	Intense rainfall, but less than before the eye
Temperature	Drops as cloud increases	Sudden rise in the sunny eye	Drops again as cloud cover recurs

Fig. 6.44 Weather recordings during the passage of a typical tropical storm

11 Imagine that you live in the path of the tropical storm above. Write an account of the weather during the day. Divide it into sections: before the storm, the first part of the storm, the eye, the second part of the storm, and after the storm has passed.

Discussion point

Would you expect people in LEDCs or MEDCs to suffer more as a result of tropical storms? What do you think are the reasons for this?

Problems caused by tropical storms

Tropical storms are violent and destructive, a major climate hazard. They cause a range of problems:

→ Strong winds, which destroy homes and businesses, and disrupt transport and power

→ Storm surges, which are rapid rises in sea level - caused by the low pressure and strong winds (they flood coastal areas and allow waves to attack at higher than normal levels)

→ Torrential rain

→ Inland flooding and landslides, caused by the torrential rain.

Hurricanes are measured according to the hurricane's strength, or intensity.

Category	Wind speed	Storm surge	Damage
1	119–153 km/h	1–2 m	minimal
2	154–177 km/h	2–3 m	moderate
3	178–209 km/h	3–4 m	extensive
4	210–249 km/h	4–6 m	extreme
5	>249 km/h	>6 m	catastrophic

Table 6.9 How hurricanes are measured

The cost of damage from a tropical storm depends more on the path it takes, and the density of population and economic activity along that path, than on the category of the storm. Compare the consequences of Storm Yasi for an MEDC, described below, with that of Winston for Fiji, an LEDC (see Chapter 5).

CASE STUDY

Cyclone Yasi, Queensland, Australia, February 2011

Cyclone Yasi was Category 5 as it crossed the Great Barrier Reef off the coast of Queensland. It badly damaged 6% of the coral before moving on to the mainland, affecting nearly 550 km of coast. The 7 m high storm surge and strong winds affected up to 300 km inland, damaging buildings, roads, and railways. Many properties lost their power supply. Expensive boats and yachts were destroyed. Fortunately, people had evacuated, so there were few injuries and only one death. Yasi brought record-breaking daily rainfall totals to a wide area, causing extensive flooding. The heavy rain on the land added further damage to the Great Barrier Reef as the polluted run-off led to a population explosion of Crown of Thorns starfish, which ate the coral polyps.

Yasi was the costliest cyclone Australia had experienced; the state's main exports were lost, including 75% of Australia's banana crop. Large areas of sugarcane fields were also extensively damaged. Losses to agriculture, mining, and local government were estimated at US$2 billion, with a further US$1 billion cost to the tourist industry.

Most of the infrastructure was repaired quickly and, one year after Cyclone Yasi, farmers were already growing bananas and sugar cane again. Bank loans and insurance pay-outs enabled people to rebuild their homes and restart their businesses.

Fig. 6.45 The enormous size of Cyclone Yasi – 1450 km wide; note the small size of the eye in comparison

How the USA prepares for hurricanes

Meteorologists at the National Hurricane Center in Florida monitor satellite images to track storms. Their aim is to provide an early warning for people in the USA and surrounding countries, to enable them to protect their properties and evacuate their families to a safer area.

However, predictions of where hurricanes will hit are not always accurate. This is because tropical storms can suddenly change course. Exactly where they will make landfall is not known until very near the time it actually happens. If a state governor orders an evacuation, and the storm does not hit the expected settlement, there will have been a considerable economic cost and disruption to people's lives for nothing.

The National Hurricane Center also uses various means to educate people in advance about the dangers they face, and what to do if they are caught in a hurricane. These include Family Disaster Plans.

The Federal Emergency Management Agency was set up to make sure that places are prepared for disasters of whatever kind, and to give help in an emergency and aid recovery after it.

12 Explain why:

a the social and economic impacts of tropical storm hazards on coastal communities in LEDCs, such as Fiji, are likely to be greater than those in MEDCs, such as Australia.

b LEDCs are likely to recover more slowly than MEDCs after the hazard.

13 Explain why it is very difficult to prevent damage by a tropical storm in:

a an agricultural area

b an area with a lowland sandy coastline.

Drought

What is drought?

Drought is a longer than usual period of dry weather. Many people live in areas where the climate has a dry season, and some live in deserts where it is dry all year round. But they adjust to those conditions and learn to cope with the difficulties of the environment. Droughts occur when rain fails to fall when it is expected - causing problems for vegetation and for human activities.

Where do droughts occur?

Droughts can occur almost everywhere, but in certain regions of the world they are particularly severe and frequent (see Fig. 6.46). The effect of drought on the vegetation and land in some of these areas has been so severe that it has led to **desertification** and **soil degradation**.

LEARNING TIP Hazards exist only because people live in locations prone to storms, volcanic eruptions and other damaging forces. Population increase is resulting in more people living in hazardous areas, such as floodplains and on the slopes of volcanoes.

LEARNING TIP Researching the numbers of deaths and costs of the damage from different sources gives a variety of figures. In disasters, no one can be absolutely certain that the statistics are correct, because many people remain missing and others are thought incorrectly to be missing.

Discussion point

What would your feelings be if you were forced to evacuate your home because it was thought that a hurricane might hit the area, but then it changed course at the last minute? Why would the local authorities feel that they were correct to make you evacuate?

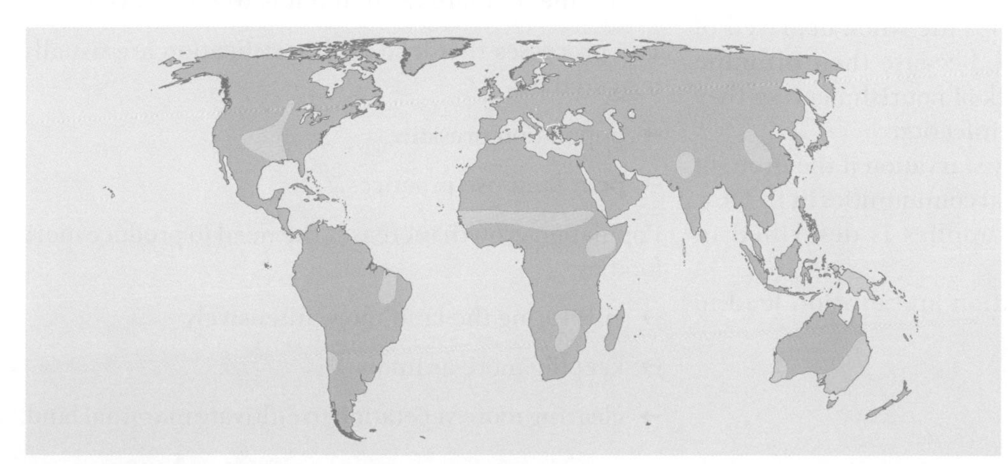

Fig. 6.46 Regions (in yellow) where frequent droughts occur

RESEARCH You could compare the tropical storms in the case studies with one you research yourself. The website www.nasa.gov has useful images from space and www.hurricanezone.net shows any storm that is active at the time you access the site. www.noaa.gov is also a useful source of information.

Impacts of drought

The photograph shows wildlife after four years of drought. The animals have gathered at a borehole that has little water left in it. The grass has been overgrazed and killed by the drought, and the soil structure has been ruined – leaving the sandy conditions seen in the foreground. There are still leaves on the trees and bushes that browsing animals like giraffes can eat, but the grazers will find little food.

Fig. 6.47 Wildlife in drought conditions

→ Water sources dry up if water levels in the rocks fall too low to keep water flowing at the surface.
→ The concentration of harmful chemicals in the water increases as water levels fall, reducing its quality and safety for drinking.
→ Air is polluted with fine dust as dry soils blow away.
→ Dry vegetation can easily be ignited by lightning, causing large-scale wildfires which are common in Australia, Africa, and western USA. These fires pollute the air with smoke and increase soil erosion as the protective vegetation is lost.
→ Animals, plants, and crops die when deprived of sufficient water and also because their immune systems weaken from lack of nourishment, so they cannot fight disease and infection.
→ Malnutrition, followed by starvation if the drought lasts for years, affects rural communities in LEDCs. (The impact on food supplies is described in Chapter 9.)
→ Increased soil degradation and erosion lead to desertification.

14 a i Draw a dispersion diagram to show total rainfalls at Masinga Dam for the months March to June in three years:

1997 = 305 mm, 2006 = 260 mm, 2009 = 110 mm (compared with the mean of 240 mm)

ii How would these variations in annual rainfall affect hydroelectricity production at Masinga Dam?

b Draw a compound bar graph to show the cost to Kenya of drought. Use these figures:

58% loss of industrial production, 26% loss of hydroelectric power production, 10% loss of crop production, 6% loss of livestock.

Desertification

What is desertification?

Desertification refers to processes that cause vegetation and crops to be unable to grow where they used to be present, so the area begins to *resemble* a desert, even though it does not have a desert climate. The United Nations Environment Program defined desertification as 'land degradation caused by adverse human impact'. This sets in motion a chain of physical processes which lead to the degradation of what used to be productive land. The term desertification has now been widened to include situations in which natural processes, such as frequent or prolonged droughts and climate change, lead to land degradation.

Many semi-arid regions, such as the Sahel, are undergoing desertification, but areas well away from deserts can also be affected. Many of the areas shown have:

→ fragile ecosystems
→ soils which are loose and easily eroded because they lack humus and structure
→ a dry season or frequent droughts – but these do not always *cause* desertification, as the land can recover in times of more rainfall if it is well managed.

The processes that lead to desertification are usually triggered by:

→ population pressure
→ poor land-use practices.

Population growth increases the need to produce more food by:

→ cultivating the land more intensively
→ keeping more animals
→ clearing more vegetation to cultivate marginal land.

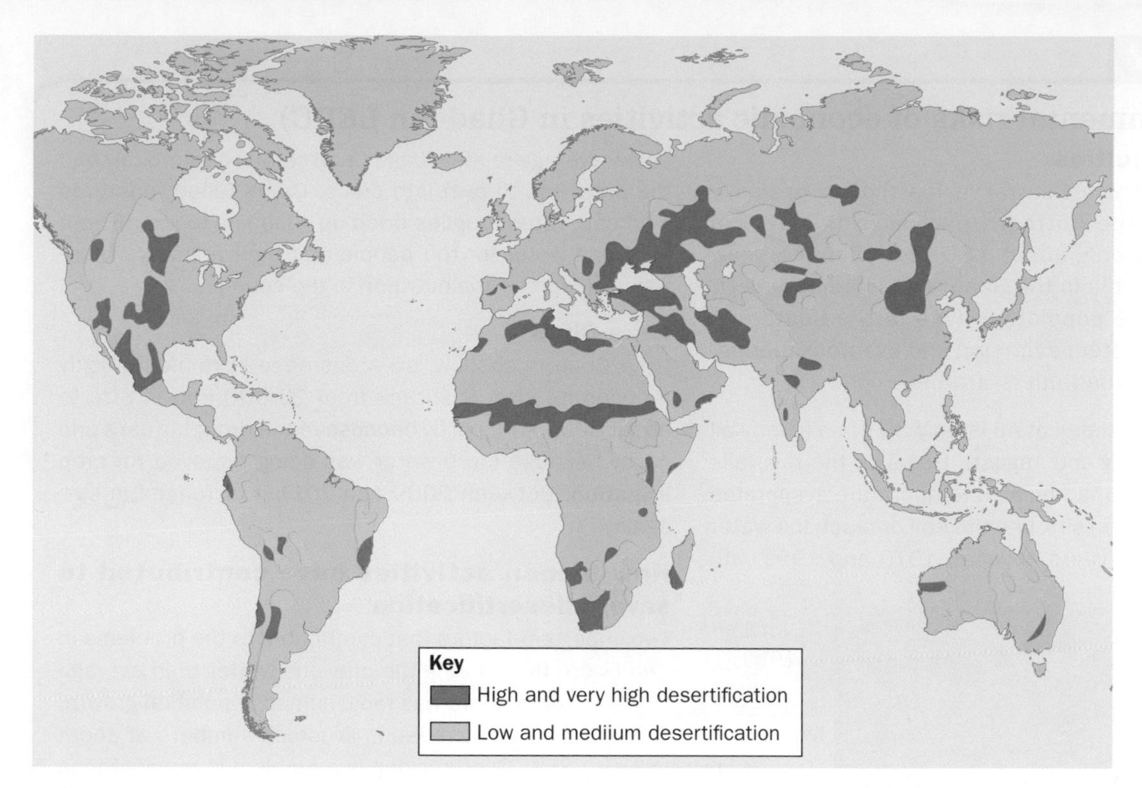

Fig. 6.48 Areas affected by desertification

Poor land-use practices result from lack of knowledge. They include ploughing up and down slopes, monoculture, over-cultivation, over-grazing and excessive gathering of fuelwood. All these activities leave the soil bare and unprotected from strong wind or heavy rainfall, so that it is easily eroded. These processes are explained in Chapter 9.

Between 1970 and 1993 the Sahara Desert was thought to be advancing southwards when drought was extremely severe on its southern fringe - the semi-arid area known as the Sahel. Then the desert edge retreated north again as higher than average rainfall fell in 11 of the following 20 years to 2013.

Long-term average rainfall is not a very useful indicator of the expected rainfall in areas like the Sahel, because rainfall is so variable. The Sahel is typical of areas with a semi-desert climate, because they are transitional areas between a dry climate and one that has a marked wet season. In such areas, the median is usually a better indicator of the rainfall than the mean.

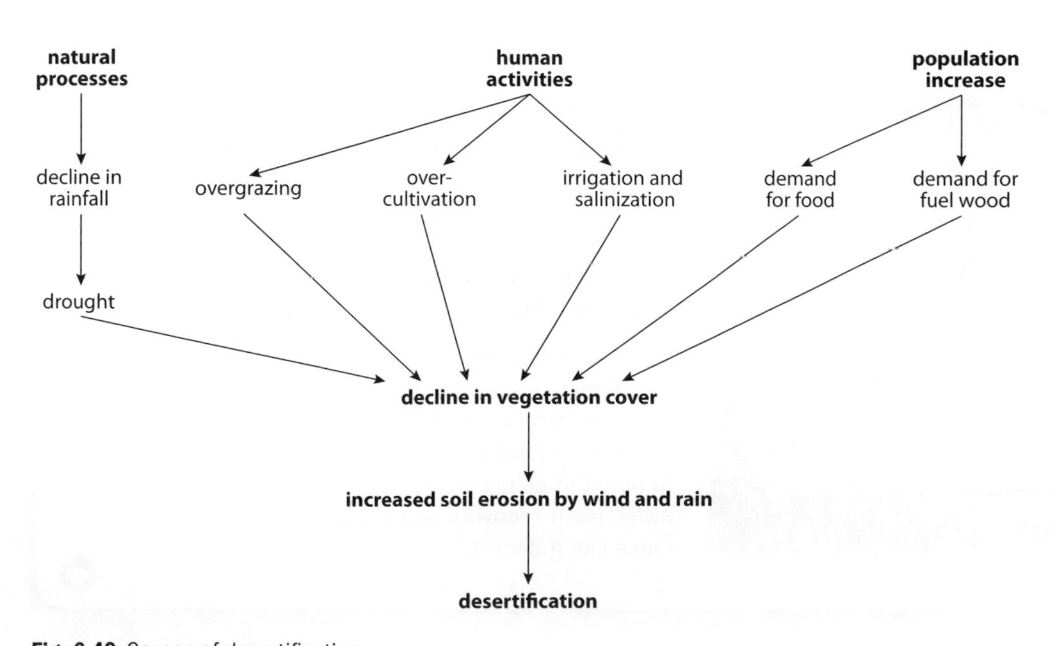

Fig. 6.49 Causes of desertification

CASE STUDY

The environmental risks of economic activities in Chad (an LEDC)

Climatic difficulties

Chad is a country in Central Africa that is basically split into two. The northern part is in the Sahara Desert and has only about 180 mm of rain a year. The southern part is in the Sahel and is semi-desert. About 80% of the population lives in the south and depends on subsistence farming and livestock rearing. Less than 3% of the land is arable.

The mean annual rainfall at Ati is only 393 mm (nearly all of which falls in July and August). Because the rain falls in the summer months, the high temperature evaporates a lot of it before it can sink into the soil or reach the **water table**. In the many years between 1970 and 1993 with below-average rainfall, Lake Chad shrank enormously and the area was thrown into crisis. Crops failed, pastures died and water supplies dried up – leading to insufficient food and water for the people and their animals. There was widespread malnutrition in the country.

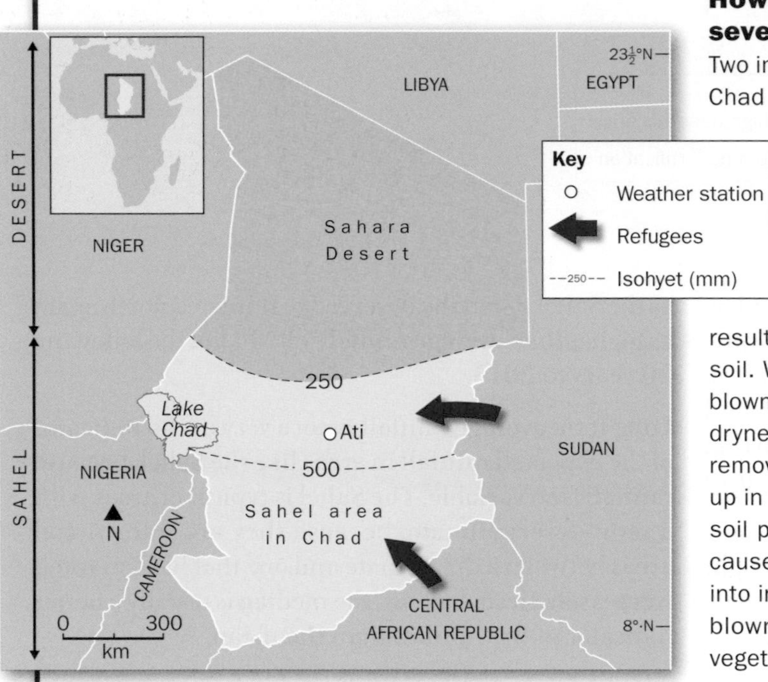

Fig. 6.50 The division of Chad between desert and semi-desert

15 Use the data in Table 6.10 to draw bars to show the variations from the long-term average of rainfall in the Sahel from 1970 to 2013. Arrange your graph with a horizontal line representing 0 (normal rainfall) in the middle with a vertical scale at the left-hand end rising to +3 above it and falling to –3 below it. Label the scale line on the *y*-axis 'Differences in annual rainfall from the long-term average in the Sahel'. Write the years at five-year intervals along the *x*-axis.

Draw vertical bars above the horizontal line for years with higher than average rainfall. Place bars for years with lower than average rainfall vertically below the 0 line.

Lake Chad

Lake Chad is shallow, so a decrease in rainfall quickly reduces its size. It shrank from 25 000 km^2 in size to about 8000 km^2, partly because of the drought years and partly because more water was being removed for crop **irrigation**. Between 2009 and 2011 it increased in size again.

How human activities have contributed to severe desertification

Two important factors that contributed to the problems in Chad were that, during the previous wetter than average years, there was rapid human population growth, plus an increase in livestock numbers of about 35%. Overstocking is a problem in many African countries as cattle are a symbol of wealth.

When droughts occurred, there was no longer enough pasture for the animals. Overgrazing resulted and removed the remaining grass, exposing the soil. With no roots to hold it in place, the soil was easily blown away. The **soil erosion** was worsened by the soil's dryness – dry soil is lighter than wet soil and more easily removed. The overgrazing also led to less **humus** ending up in the soil (humus is decayed vegetation, which helps soil particles to stick together). The reduction in humus caused the soil structure to break down, so it crumbled into individual particles that were lighter and more easily blown away. Once the soil has gone, recovery of the vegetation becomes very difficult.

Chad's human population growth led to an increasing amount of woody vegetation being removed for use as firewood – again leaving the soil more exposed to wind erosion. The water table also fell, because of the lack of rain filtering down to replenish it during droughts – and because the higher population was drawing more water from the boreholes for themselves and their animals. Over-cultivation was another result of population growth. It caused **soil exhaustion**, where the nutrients are removed until all fertility is lost.

Desertification occurred in Chad much more quickly than would have happened had the carrying capacity of the land for both livestock and humans not been exceeded. Overgrazed pastures cannot regrow as there are no roots to hold the soil in place. When the soil has been eroded away, plant recovery is almost impossible and the area looks like a desert.

Year	Difference from long-term average	Year	Difference from long-term average
1970	+0.8	1992	+0.25
1971	+0.25	1993	-0.7
1972	-1.5	1994	+2.3
1973	-1.0	1995	0
1974	+1.2	1996	-0.55
1975	+0.8	1997	-0.8
1976	-0.75	1998	+1.3
1977	+0.25	1999	+2.0
1978	+0.35	2000	-0.6
1979	-0.2	2001	+0.1
1980	+0.1	2002	-0.8
1981	-0.3	2003	+1.0
1982	-1.0	2004	-0.6
1983	-1.5	2005	+0.4
1984	-2.25	2006	+1.0
1985	-0.5	2007	+0.8
1986	-0.1	2008	+0.25
1987	-1.5	2009	-0.1
1988	+1.5	2010	+0.8
1989	0	2011	-0.4
1990	-1.25	2012	+1.95
1991	-0.3	2013	-0.1

Table 6.10 The variations from the long-term average of rainfall in the Sahel from 1970 to 2013

16 **a** Summarise rainfall in the Sahel since 1970 in one word.

b In which period of years was there a prolonged drought?

c Describe the pattern of rainfall between 1988 and 2002.

d Describe the pattern of rainfall since 2000.

The earlier drought years were believed to be caused by a failure of the Inter-Tropical Convergence Zone (ITCZ; see page 191) to reach as far north as the Sahel. The drought was intensified by global warming. However, increased rainfall in recent years may also be caused by global warming (see Chapter 10): the enhanced greenhouse effect has raised air temperatures over the Mediterranean Sea, increasing evaporation from the sea. The resulting moist air moves south as the north-east trade winds, to provide rain in the Sahel.

Widespread drought in the 1970s and 1980s caused mass starvation of humans and cattle. In 2009 the cereal harvests failed again because of insufficient rainfall. This resulted in widespread malnutrition and the loss of over 30% of livestock (780 000 cattle died). Over half of Chad's children suffered from chronic malnutrition. The child mortality rate was very high. Over 2 million people were affected by the drought and crop failure. Even though the 2010 harvest was good (because adequate rains fell), food aid was still needed in 2011.

Sustainable development and management

It is essential that land use in the Sahel must be made sustainable by conserving the precious resources of vegetation, soil, and water. There are ways in which desertification can be stopped – and even reversed. The following examples have all been used successfully in the Sahel:

- Reducing the number of farm animals. This stops overgrazing and allows the protective vegetation to grow back.
- Encouraging mixed farming – growing crops as well as keeping animals. The animal manure is used to fertilise the soil, which helps the crops to grow.
- Planting more trees to protect the soil from wind and rain. The tree roots help to hold the soil together and prevent erosion. Shelter belts of trees reduce wind erosion.
- Building earth dams to collect and store water in the wet season. The stored water is then used to irrigate crops in the dry season.

Chad's population has continued to grow. By 2016 it had reached over 14.5 million, which is seven times larger than it was in 1950. It is still increasing by 3% a year because of very high birth rates and because, since 2003, thousands of refugees have entered Chad (mainly from Sudan). This has put extra strain on Chad's already stretched resources.

Deforestation of the acacia bush in the south for use as firewood has increased by 60% since 1990.

The work of aid agencies

The efforts of aid agencies in Chad have been hampered by a certain amount of lawlessness in the country, and the fact that the problems continue to increase in scale.

The United Nations Refugee Agency has been working to stop desertification by:

→ planting young trees, including woodland trees (like acacia) and fruit trees (like mango)

→ providing the people with firewood, to prevent the existing trees and shrubs from being cut down

→ introducing solar-powered cookers to reduce the need for firewood in the first place.

A sustainable strategy to protect water resources has also been developed. It is planned to replace electric pumps with manual ones, which are cheaper to maintain. Wells have been dug in dry riverbeds, to preserve the water in the **aquifer** beneath the water table deep in the rocks.

The Great Green Wall (GGW)

In the 1980s the head of the state of Burkino Faso suggested the Great Green Wall as a method of stopping the Sahara Desert spreading southwards. In 2005 the Nigerian president promoted the idea to the African Union. It has expanded to involve 20 countries. The aim is to plant a 'wall of trees' across the continent. A mixture of trees, shrubs, and herbaceous plants will be used. Work started in 2012 with the planting of 12 million trees in Senegal – mainly drought-resistant acacia, which will fertilise the soil as their dead leaves decay into it and provide shade to reduce loss of moisture from the soil by evapotranspiration.

Finance for the project comes from many international agencies, including the World Bank, UNFAO, the African Union, the European Union and GEF (Global Environmental Facility). The aims of the project are ambitious. It aims to enable sustainable local economic development by:

→ stopping erosion and desertification to enable the land to be productive again

→ providing vegetables, fruit, and other foods

→ providing firewood to reduce deforestation

→ protecting water sources

→ restoring animal habitats.

It also aims to make social and political improvements by:

→ encouraging cooperation in and between countries

→ reducing poverty

→ stopping migration of young people from the region

→ promoting political stability.

This initiative will reduce threats to the environment and to people caused by the economic developments that have already taken place.

The Bonn Challenge

Many countries are now responding to the Bonn Challenge, which was agreed by world leaders in 2011 to promote sustainable development and management and to encourage restoration of degraded land. Kenya, where 65% of the land area is thought to be restorable, is one country taking part. The brown colour on Fig. 6.51 indicates the large area in which vegetation is not growing well. Areas where vegetation is thriving are green. A comparison of satellite images taken in different years is a useful indication of the spread of desertification.

17 Arrange the following words and phrases into order on a flow diagram to show some of the processes leading to desertification:

- soil erosion and degradation
- desertification
- increased use of marginal land
- overgrazing and over-cultivation
- vegetation damaged and vegetation cover removed
- need for more animals to be kept and more food to be grown
- population increase

LEARNING TIP Rainfall figures from areas in the Sahel and other areas with a semi-desert climate are sometimes quoted as proof of climate change. One article, by an agency working with the government of Chad to address climate change, compared the poor rainfall in Chad in 2009 with the higher rainfall for 1950, as proof that it is getting drier. However, rainfall was higher in 1994 and 2003 as well, which (if used as the comparison) would have led to a different conclusion.

It is only meaningful to make comparisons with the median or mean calculated over a long term. This was emphasised in 2010, when so much rain fell in eastern Chad that flooding was a problem! In four months, nearly 70 000 people lost their homes because of flooding, but there was an above – average harvest to compensate.

Vegetation that was not growing well is shown as brown on the satellite image, but thriving vegetation is green. It is clear that drought was not affecting western Kenya when the image was taken in January 2011.

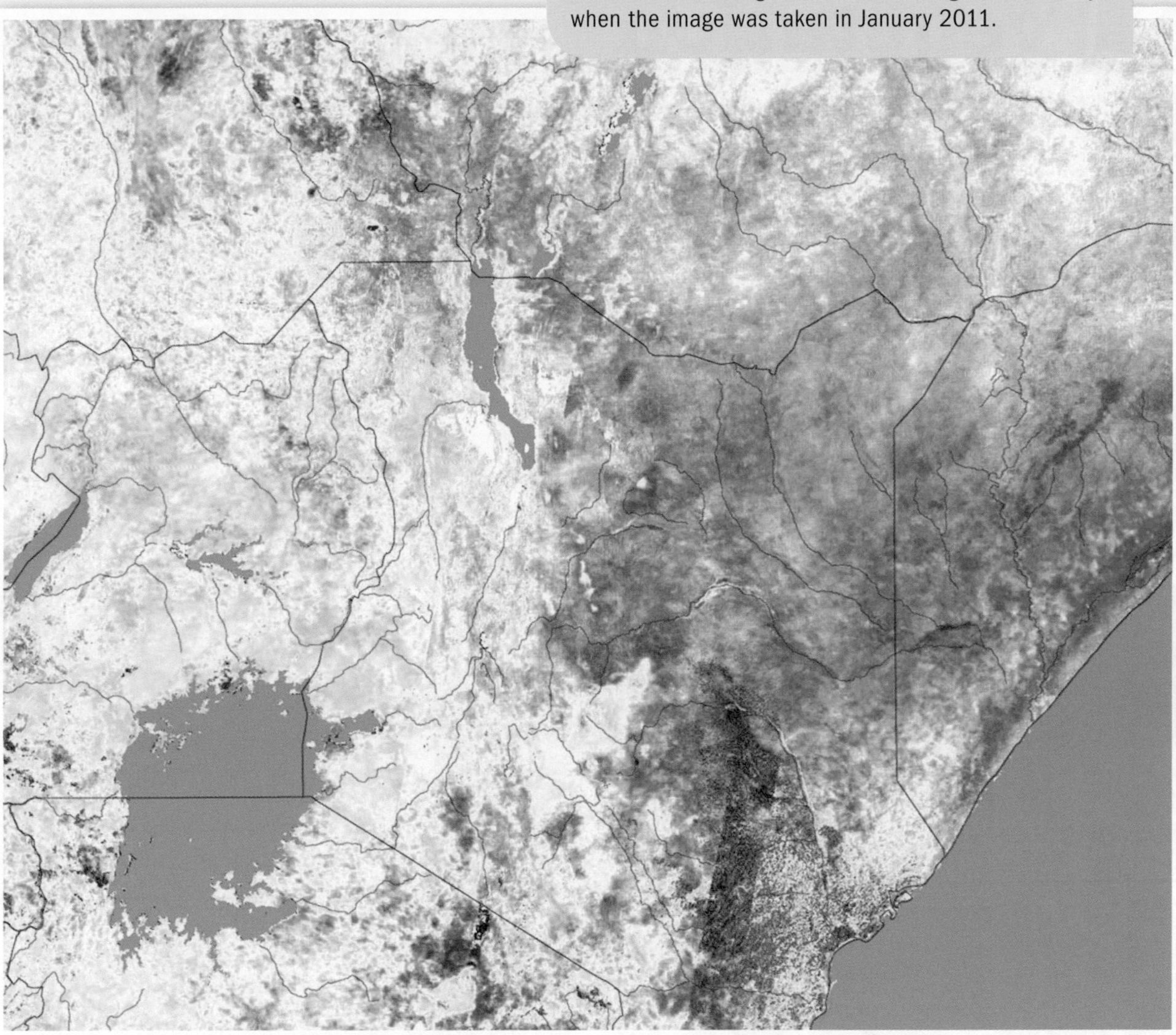

Fig. 6.51 A satellite image of part of East Africa, including Kenya

7 Climate and natural vegetation

This chapter covers the following Cambridge IGCSE® and O Level topic:

- **2.5 Climate and natural vegetation**
- **part of 3.7 Environmental risks of economic development**

- Is it always hot and dry in the Sahara Desert?
- Why does it rain almost every afternoon in equatorial areas? Do you live in one of these areas?
- How do people, plants and wildlife survive in hot deserts without surface water?
- What are the links between the living and non-living world?
- Does our ecosystem stretch as far as the sun?
- The tropical rainforest has the richest vegetation and variety of wildlife, yet it's being cut down at an alarming rate – some countries have very little primary forest left. Should we be concerned about this? What can we do about it?
- Is making money more important than conserving our natural world?
- Will the destruction of the rainforest change the climate?

In this chapter you will learn about:

- the climates in which tropical rainforest and tropical desert vegetation grow, as well as the factors influencing those climates
- climate graphs
- the characteristics and distributions of tropical rainforest and hot desert ecosystems
- the relationships between natural vegetation, wildlife, and climate in the tropical rainforest and hot desert ecosystems
- the need for sustainable development, resource conservation and management in the tropical rainforest and hot desert ecosystems
- tropical rainforest and hot desert areas at risk
- attempts being made to conserve the environments of tropical rainforest and hot desert areas at risk.

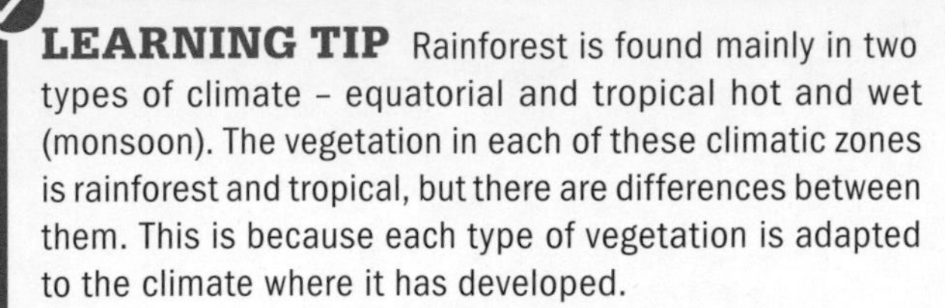

LEARNING TIP Rainforest is found mainly in two types of climate – equatorial and tropical hot and wet (monsoon). The vegetation in each of these climatic zones is rainforest and tropical, but there are differences between them. This is because each type of vegetation is adapted to the climate where it has developed.

Studying climate and vegetation zones in different textbooks and atlases reveals that there are considerable differences in classification – leading to differences in the areas shown on maps.

At a higher level of study, the equatorial and tropical monsoon climates, and their associated vegetations (tropical rainforest and tropical monsoon forest), are dealt with separately. The IGCSE® and O Level syllabuses require knowledge of one rainforest climate. The equatorial climate has been chosen for study in this chapter, because it is the climate normally associated with the term 'tropical rainforest'.

Climate

We have seen that weather can change continuously and that one day's weather cannot be used as a guide to what the weather will be like at any other time. For this reason, records of weather taken over at least 30 years are averaged and used to indicate the climate. **Climate** is therefore an average of the weather. Historical records and other evidence indicate that climate changes over time. For example, it was particularly cold in northern Europe from about 1750 to 1850, but about 5000 years ago it was warmer than it is today. There is also evidence that the Sahara Desert was much wetter in the past than it is today. Fifty years ago, after a period of cooling, it was suggested that another Ice Age could be expected. Now global warming is the main concern.

The CAIE IGCSE® and O Level syllabuses require you to study two types of climate – the climates in which tropical rainforest and hot desert vegetation grow.

The equatorial climate

The global distribution of the equatorial climate

The equatorial climate is hot and wet all year – it has no seasons. It is found in three large areas – the Amazon Basin in South America, the Congo Basin in Central Africa (with an extension westwards along a narrow coastal strip in southern West Africa), and the Malaysian Peninsula and islands of South East Asia, which include Singapore. In Chapter 6 you learned about a day's weather in Balikpapan (on the east coast of Borneo), which is in the same climatic area as Singapore (Fig. 7.1). All of the main areas with an equatorial climate lie within 10° of latitude of the equator.

1 Compare the information in Fig. 7.1 with other climate graphs in your atlas. What characteristics make this climate different?

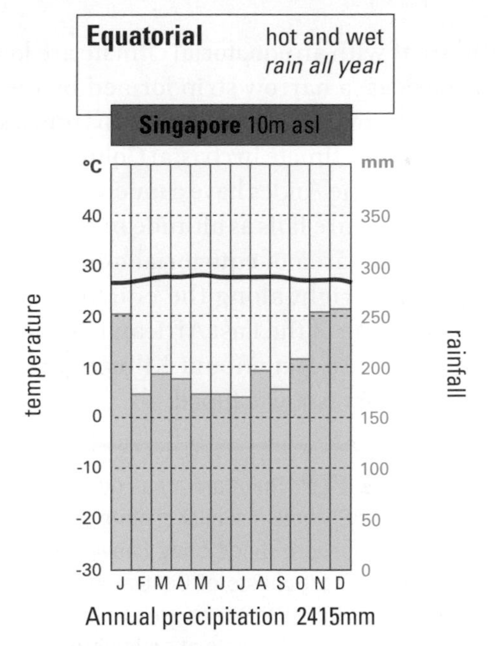

Fig. 7.1 The climate of Singapore

Discussion point

As you can see, Singapore's climate is uniform. Do you think living in an equatorial climate would be boring? Why or why not? Describe the characteristics of your ideal climate to live in.

LEARNING TIP When asked to describe a distribution over time, do not give a month-by-month account. Instead, describe the main patterns or trends and the extremes.

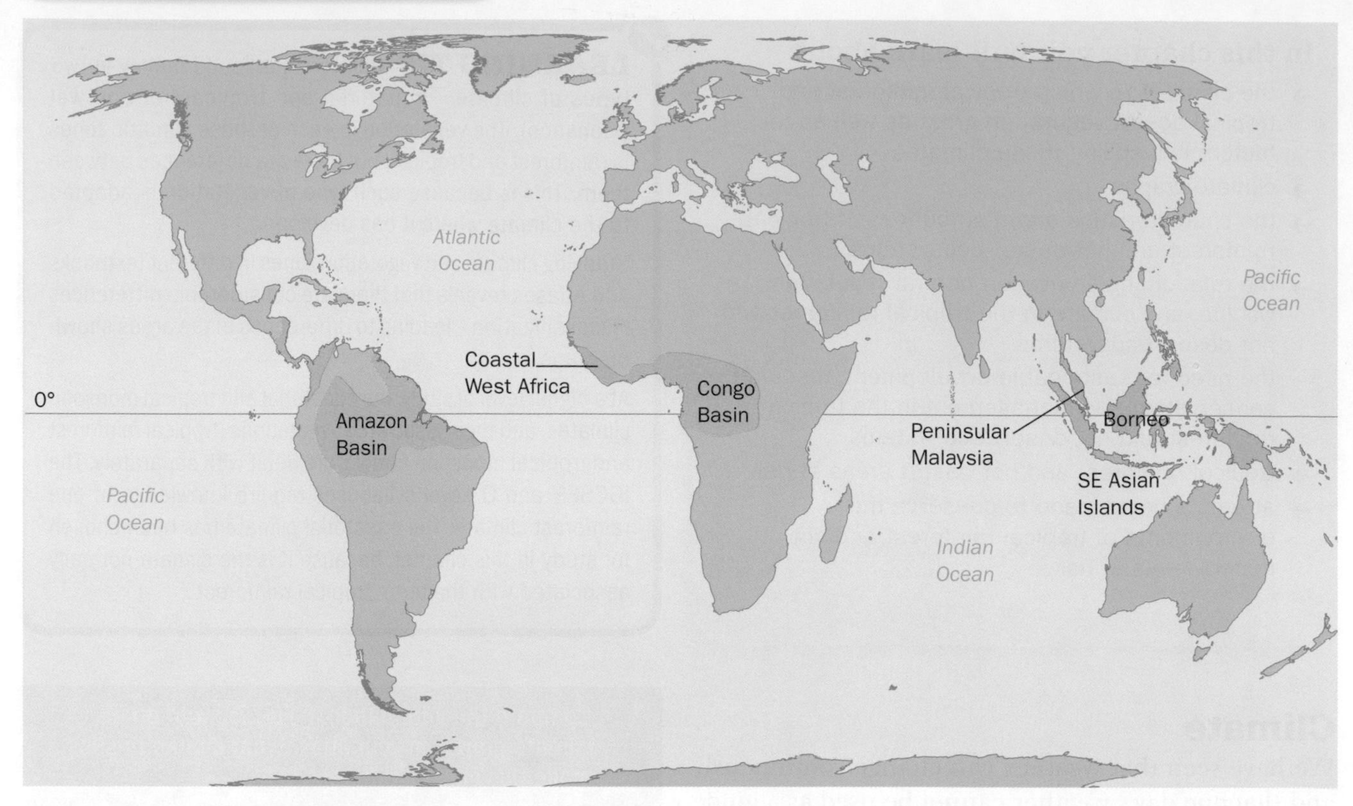

Fig. 7.2 The global distribution of the areas with an equatorial climate

All of the areas with an equatorial climate are lowlands. In the Americas, a narrow strip formed by the Andes Mountains separates the Amazon Basin from a smaller area of equatorial climate in coastal Columbia, Panama and Costa Rica. The Andes have a much colder climate, because temperature falls as altitude increases. For the same reason, the area of equatorial climate in Africa does not extend right along the equator to the east coast - the altitude of the East African Plateau reduces the temperature. In fact, Mount Kilimanjaro (located at only 3°S) is even snow-capped.

LEARNING TIP There are areas outside the zone from 10°N to 10°S where a similar climate exists on east coasts - where onshore winds bring heavy rain all year. This includes the east coasts of Central America, Brazil and Madagascar. However, these areas also experience hurricanes - not a characteristic of an equatorial climate.

Characteristics of the equatorial climate

2 Use the data in Table 7.1 to answer these questions.

a Calculate for Singapore and use a word to describe: (i) the mean annual temperature, (ii) the annual temperature range. (Show your working.)

b In Singapore, the sun rises at 6 am and sets at 6 pm all year round. Look at the data about sunshine hours. Would you regard this as a cloudy climate? Explain your answer.

c Estimate the mean annual relative humidity. What does this tell you about the atmosphere in Singapore?

d Singapore's average annual precipitation is 2415 mm. Describe this total and the distribution of the rainfall throughout the year.

Month	Jan	Feb	Mar	Apr	May	Jun	Jul	Aug	Sept	Oct	Nov	Dec
Mean temp. (°C)	26.5	27	27.5	27.5	28	28	28	28	28	27.5	27	26.5
Mean max. temp. (°C)	30	31	31	31	32	32	32	32	32	31	31	30
Mean min. temp. (°C)	23	23	24	24	24	24	24	24	24	23	23	23
Mean relative humidity (%)	82	79	79	81	81	79	80	80	80	80	82	82
Mean rainfall (mm)	252	169	190	183	175	175	170	197	179	214	253	258
Mean daily sunshine hours	5.1	6.4	6.1	5.9	5.9	6.2	6.2	6.0	5.6	5.3	4.6	4.5

Table 7.1 Climate statistics for Singapore (1° 23' N, 103° 59' E)

Reasons for Singapore's high temperatures

The influence of latitude on temperature

The high temperatures in Singapore result from the sun at midday being directly overhead at the equator on 21 March and 23 September, and very nearly overhead during the rest of the year. This is caused by the orbit of the Earth around the sun, and the fact that the Earth's axis is at an angle of 23½° to the plane of orbit. An orbit takes one year. It results in the sun being furthest from overhead in Singapore on 21 June (when at midday it's overhead at the Tropic of Cancer, 23½°N) and on 22 December (when at midday it's overhead at the Tropic of Capricorn, 23½°S). On these two dates, the sun at the equator is still at a high angle of 66½°.

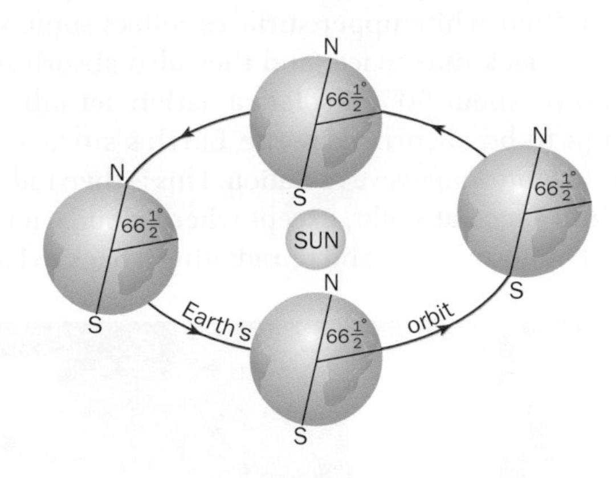

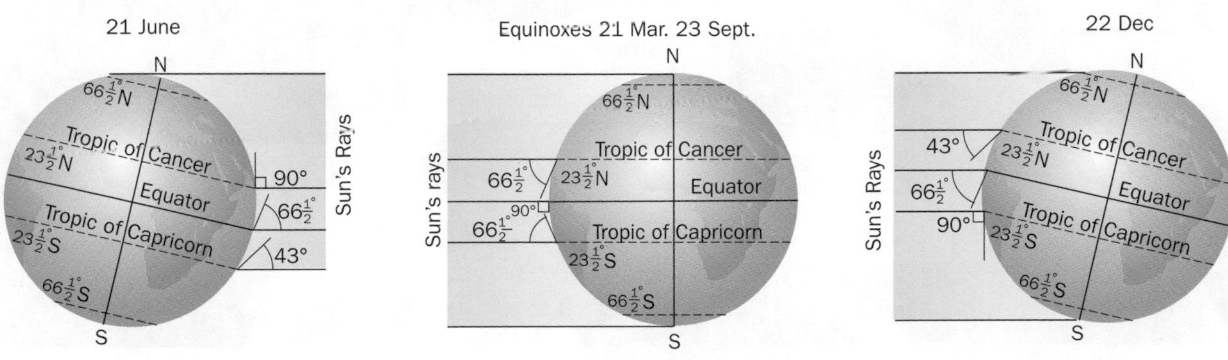

Fig. 7.3 The orbit of the Earth round the sun, causing varying angles of the midday sun on different dates in the year

The higher the angle of the sun's rays, the greater is the heating of the Earth's surface. This is for two reasons:

- As shown in Fig. 7.4, the two bundles of sun's rays have equal amounts of heating power. But, at the equator, the high-angle rays fall on a smaller area of the Earth's surface, so the heat is more concentrated than at higher latitudes.
- Secondly, the rays at the equator have a shorter passage through the Earth's atmosphere. As the sun's rays pass through the atmosphere, they can be reflected and scattered back into space by clouds and dust, or absorbed by atmospheric gases. Therefore a shorter passage through the atmosphere allows more **insolation** (incoming solar radiation) to reach the Earth's surface.

Latitude also influences the length of daylight and darkness, which has an impact on temperature. Imagine a vertical line bisecting the June world in Fig. 7.3 with the right-hand half of it in sunlight and the other half in darkness. The Earth rotates on its axis once every 24 hours. The result is that, although there is always 12 hours of daylight and 12 hours of darkness at the equator, all other latitudes have varying lengths. When the sun is overhead at the Tropic of Cancer, places in the northern hemisphere have longer days and shorter nights - and the reverse occurs when the sun is overhead at the Tropic of Capricorn. At latitude 66½°N, the sun does not set on 21 June and it does not rise above the horizon on 22 December.

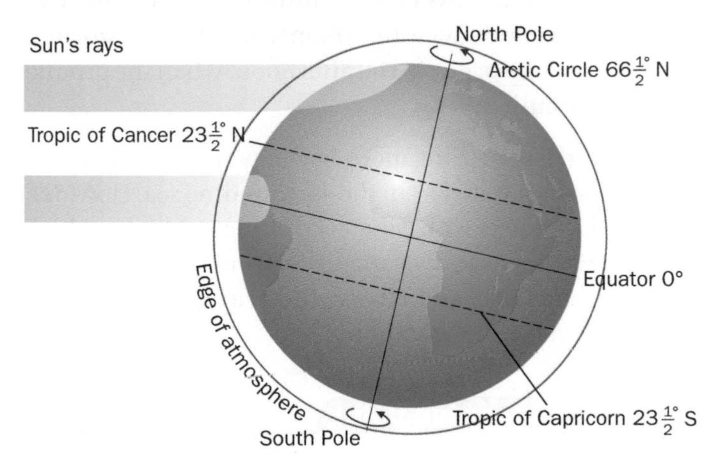

Fig. 7.4 The influence of latitude on temperature

Discussion point

What must the temperature and precipitation be like at the North and South Poles? How long is the period of uninterrupted sunlight at the Poles? For how long does the sun never rise? Look at an atlas map of the relief of northern Canada. Why do people living there find it easier to travel in winter than in summer?

The influence of clouds on temperature

Places near the equator are hot, but - because of the influence of clouds - they are not the hottest places on Earth. During the day, clouds reduce surface temperatures, because their white upper surfaces reflect some solar radiation back into space (and they also absorb some radiation). About 50% of solar radiation actually gets through to be absorbed by the Earth's surface and converted into long-wave radiation. This is then radiated back into space at night, except where clouds act as a blanket by absorbing it and re-radiating it back to Earth.

Water droplets in clouds, water vapour, carbon dioxide, and other gases all contribute to the **greenhouse effect.** The air near the Earth's surface is kept warmer because short-wave solar radiation can pass through the atmosphere relatively easily, but the Earth's long-wave radiation is trapped. This means that, for example, cloudy nights in Singapore, and other places with an equatorial climate, keep it warmer than it would otherwise be. Even so, the diurnal (daily) temperature range of around 7 °C is considerably greater than the annual range of temperature of 1.5 °C.

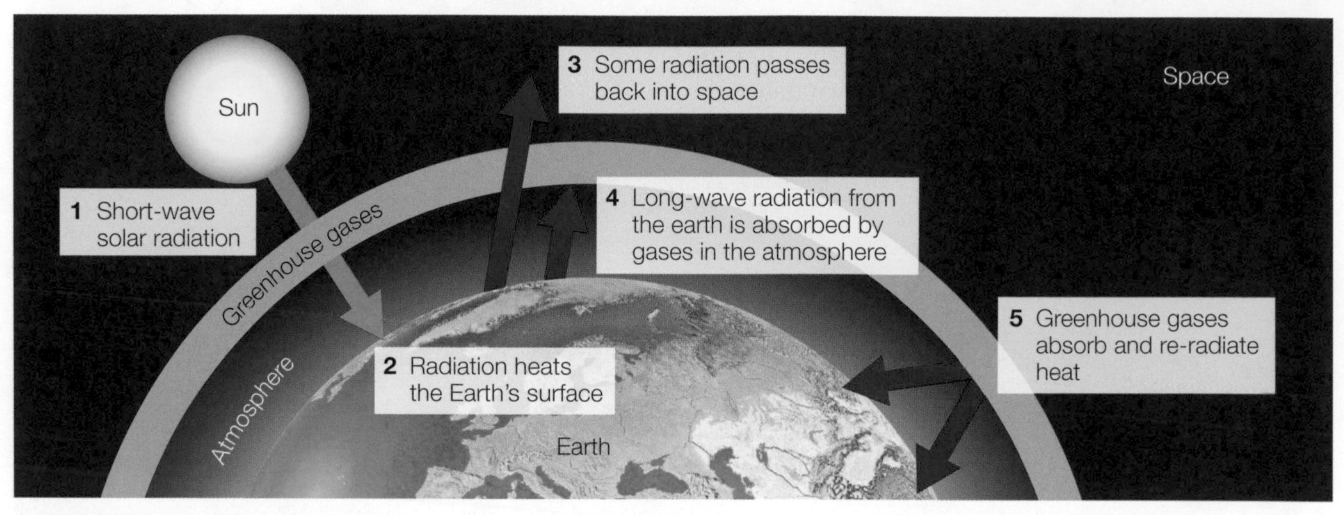

Fig. 7.5 The greenhouse effect

Reasons for Singapore's high rainfall

The combination of high temperatures and air with high moisture content results in **convectional rainfall** (Fig. 7.6). The towering cumulonimbus clouds can produce very heavy rain (often with thunderstorms). These usually occur in the afternoon, when the ground heat has built up.

Being an island, Singapore has very moist air because water is evaporated from the surrounding sea. However, inland locations with equatorial climates also have high humidity because of evaporation from the wet ground, numerous lakes and rivers, and **transpiration** from the dense forest vegetation.

Discussion point

If there were no atmosphere, what would the Earth's day and night time temperatures be like?

LEARNING TIP The factors causing the equatorial climate at Singapore apply to all other areas with an equatorial climate, except its small island location.

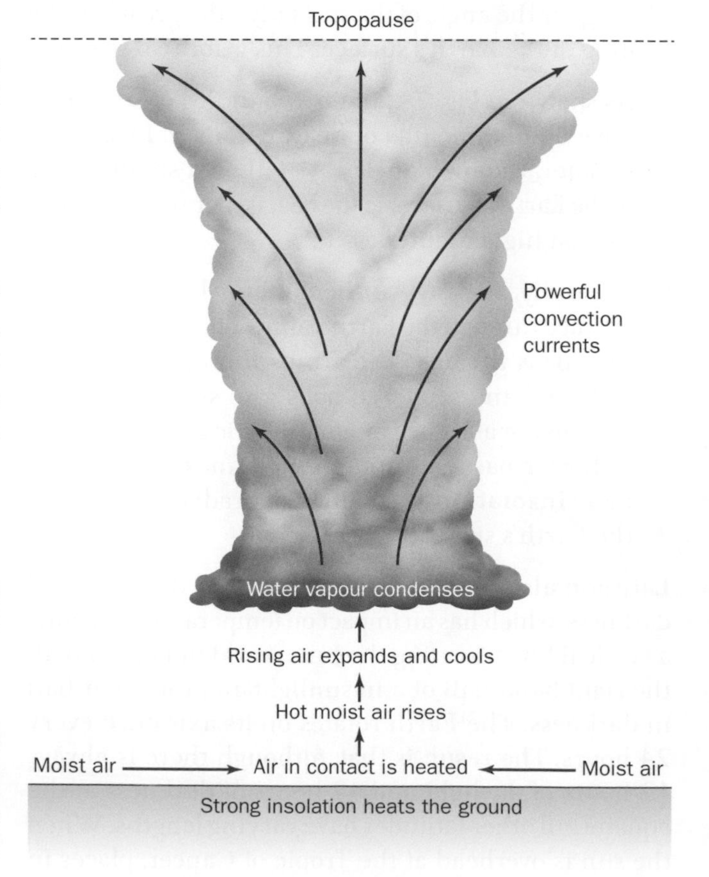

Fig. 7.6 Convectional rainfall formation

Fig. 7.7 An afternoon thunderstorm developing at Panama City. Dense cumulonimbus cloud has formed.

Because air is rising in the equatorial zone, surface winds are light - except when sudden downdrafts from the cumulonimbus clouds produce strong gusts. In the past, when sailing ships reached this zone they would often be stuck for weeks because of the lack of wind. As a result, the equatorial low-pressure areas over the oceans became known as 'the doldrums'.

Because warm air expands, there is less weight in a column of warm air than in colder air, so the pressure at the surface is lower. The equatorial zone is a permanent low-pressure belt. As winds blow into low pressure to replace the rising air, the winds meet and rise in a zone known as the **Inter-Tropical Convergence Zone (ITCZ)**. This zone encircles the world, but its position changes according to the different positions of the overhead sun:

- In March and September the ITCZ is approximately at the equator.
- In December (when the sun is overhead at the Tropic of Capricorn) the ITCZ is further south.
- In June the sun is overhead at the Tropic of Cancer, so the ITCZ moves into the Northern Hemisphere.

As the ITCZ moves over equatorial areas twice a year, many weather stations have double maxima of rain - caused by the increased convectional activity along the ITCZ. The rainfall pattern at Kisangani in the Congo Basin illustrates this (Table 7.2). Kisangani is very near the equator.

3 Using the data in Table 7.2, construct a rainfall graph for Kisangani. On the graph add the labels 'sun overhead at Kisangani' with a vertical arrow pointing to 21 March, and repeat it for 23 September. Put arrows to 21 June and 22 December and label them with 'sun overhead at Tropic of Cancer' and sun overhead at Tropic of 'Capricorn'. Note above your rainfall bars the total mean annual rainfall.

Jan	Feb	Mar	Apr	May	Jun	Jul	Aug	Sep	Oct	Nov	Dec
50	88	175	159	132	104	136	169	186	221	195	83

Table 7.2 The mean monthly rainfall at Kisangani in millimetres (the total mean annual rainfall is 1698 mm)

The hot desert climate

The global distribution of hot deserts

Most hot deserts are located in latitudes from 15° to 30° (astride the tropics) on the western sides of continents. However, the Mojave, Mexican, Iranian and Thar deserts all have hot desert climates but lie north of the Tropic of Cancer - extending to 40°N.

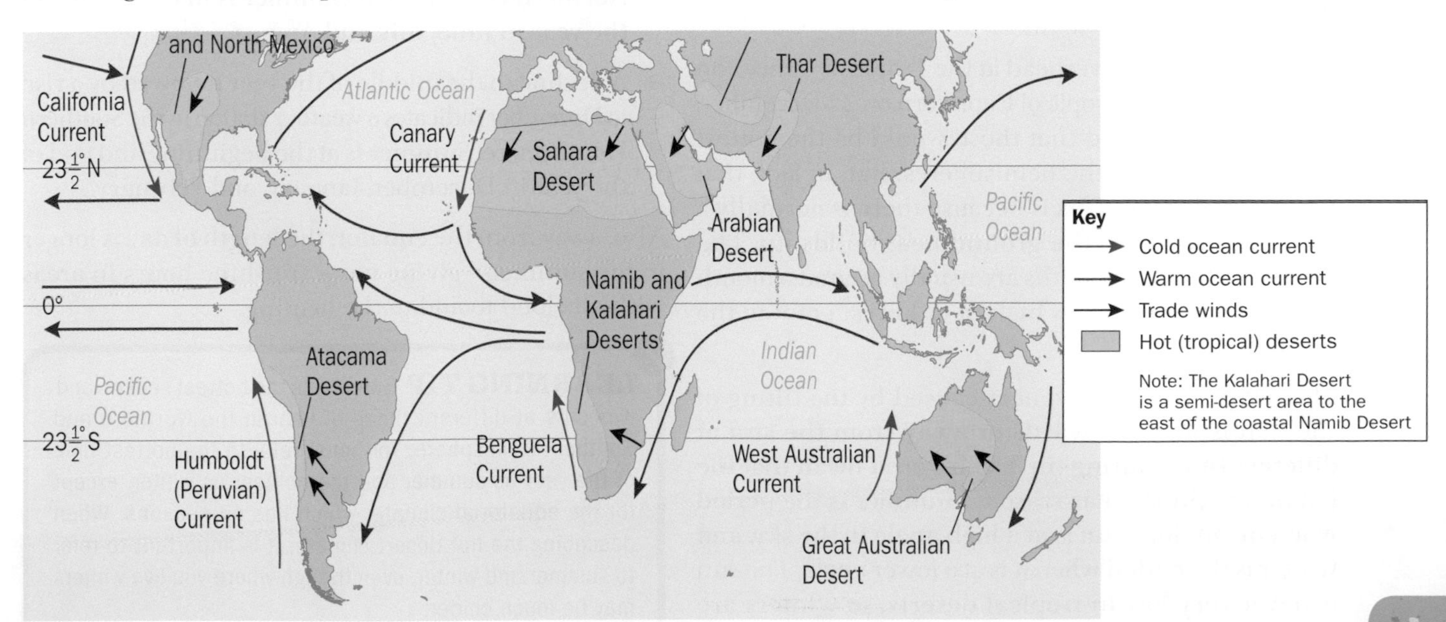

Fig. 7.8 The global distribution of hot (tropical) deserts

Characteristics of the hot desert climate

4 **a** Compare the information in Fig. 7.9 with other climate graphs in your atlas. What characteristics make this climate different?

b Which hemisphere is Tamanrasset in and how does the climate graph indicate this?

5 Using the data in Table 7.3, construct a graph to show the mean monthly temperatures and rainfall for Luderitz. Add the total rainfall of 31.4 mm above your rainfall bars.

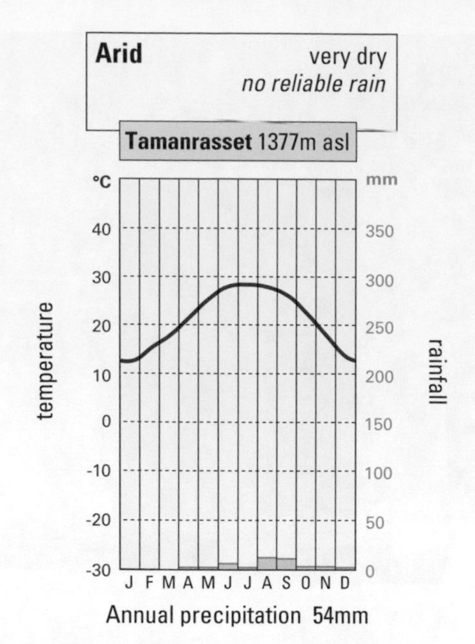

Fig. 7.9 The climate of Tamanrasset in Algeria

Month	Jan	Feb	Mar	Apr	May	Jun	Jul	Aug	Sep	Oct	Nov	Dec
Mean temp. (°C)	19	18.5	18	17	16	15.5	14.5	14	15	16	17	18
Mean max. temp. (°C)	22	22	22	21	20	20	19	18	19	20	21	22
Mean min. temp. (°C)	16	15	14	13	12	11	10	10	11	12	13	14
Mean relative humidity (%)	75	78	75	74	74	66	67	70	74	75	73	72
Mean rainfall (mm)	0.4	1.4	3.6	10.5	4.1	5	2	1.5	1.3	0.7	0.7	2
Mean sunshine hours	10	9	9	9	9.5	9	9	9.5	19.5	10	10.5	11

Table 7.3 The climate statistics for Luderitz (26° 38'S, 15° 10'E) on the coast of Namibia in the Namib Desert

The influence of latitude on temperature

When the Northern Hemisphere is tilted towards the sun, latitudes in that hemisphere receive the sun's rays at a higher angle than the same latitudes in the Southern Hemisphere. As a result, they are hotter. The opposite is true when the Southern Hemisphere is tilted towards the sun.

Because the sun is overhead at the Tropic of Cancer on 21 June, and at the Tropic of Capricorn on 22 December, it might be expected that those would be the hottest months for each of the hemispheres. But, in fact, that is not usually true. This is because there is normally a temperature lag as the ground heat builds up. The hottest and coolest months are usually around a month after the sun reaches its highest or lowest point in the sky at midday.

The temperature differences caused by the tilting of the hemispheres towards or away from the sun at different times during the Earth's orbit mean that the hot desert climate has seasons. Summer is the period when the midday sun is at a high angle in the sky, and winter is the period when it is at a lower angle. The sun is never very low in tropical deserts, so winters are normally hot and summers very hot!

Luderitz is in the Southern Hemisphere and Tamanrasset is in the Northern Hemisphere, so their temperature graphs show different patterns:

- A rise to the middle of the year, followed by a decline to December, indicates a weather station in the Northern Hemisphere. Summer is in the middle of the year in June, July and August.
- A decline to the middle of the year followed by a rise to December indicates a weather station in the Southern Hemisphere. Summer is at the beginning and end of the year in December, January, and February.

Also, away from the equator, the length of day is longer in the summer - giving more sunshine hours in areas without cloud to add to the heating.

LEARNING TIP Summer is the hottest season and it occurs at different times of year in the Northern and Southern Hemisphere. You must refer to the hottest time of the year as summer and the coldest as winter, except for the equatorial climate, which has no seasons. When describing the hot desert climate, it is important to refer to summer and winter, even though where you live winters may be much colder.

Khartoum is located at 15°N, so the sun is overhead twice – in the month before and the month after it is overhead at midday at the Tropic of Capricorn. Khartoum's maximum mean monthly temperature of 34.5 °C occurs in May and the minimum of 23.2 °C in January – giving a moderate annual temperature range of 11.3 °C. However, temperatures vary considerably in different hot desert locations. This is because latitude is not the only influence on temperature.

Month	Jan	Feb	Mar	Apr	May	Jun	Jul	Aug	Sep	Oct	Nov	Dec	Year
Mean max. temp. (°C)	30.8	33.0	36.8	40.1	41.9	41.3	38.4	37.3	39.1	39.3	35.2	31.8	37.1
Mean temp. (°C)	23.2	25.0	28.7	31.9	34.5	34.3	32.1	31.5	32.5	32.4	28.1	24.5	29.9
Mean min. temp. (°C)	15.6	17.0	20.5	23.6	27.1	27.3	25.9	25.3	26.0	25.5	21.0	17.1	22.7
Rainfall (mm)	0	0	0	0.5	4	5	46	75	25	5	1	0	161

Table 7.4 Climate statistics for Khartoum (380 m) in the Sahara Desert

The influence of altitude on temperature

The air temperature decreases as altitude increases. This is because the air becomes thinner and contains less water vapour and other gases to absorb the Earth's long-wave radiation. There is also less dust to scatter it back to Earth.The rate of temperature decrease varies, but averages about 0.6 °C for every 100 m of height gained. (This is not the same rate as the rate of cooling in unsaturated *rising air*, which is about 1 °C per 100 m).

Look again at the temperature graph for Tamanrasset (Fig. 7.9). Its temperatures are considerably lower than those of Khartoum (Table 7.4). This is a result of the fact that Khartoum is at an altitude of 380 m above sea level, whereas Tamanrasset is 1377 m – a difference of almost 1000 m, which is equivalent to 6 °C of temperature.

The influence of distance from the sea on temperature

Because water heats up and cools down more slowly than land, coastal areas have warmer winters and cooler summers than places further inland. This is known as the **maritime** influence, when air blowing in from the sea brings the temperature of the sea to the land.

A comparison of the climates of Luderitz and Keetmanshoop, both in Namibia, demonstrates this effect (Tables 7.3 and 7.5). Find these two locations in an atlas. Keetmanshoop is on the same latitude as Luderitz, but it is about 290 km away from the sea. On average, Keetmanshoop's warmest month is 8 °C warmer than that of Luderitz, and its coldest month is 1 °C colder. This gives it an annual temperature range that is larger by 9 °C than that of Luderitz, because of Keetmanshoop's more **continental** location.

Month	Jan	Feb	Mar	Apr	May	Jun	Jul	Aug	Sep	Oct	Nov	Dec
Mean temp. (°C)	27	26.5	25	22.5	18	14	13	15.5	18.5	21	23.5	26
Mean max. temp. (°C)	35	34	32	30	26	22	20	24	26.5	30	32	35
Mean min. temp. (°C)	19	19	16	15	10	6	6	7	10.5	12	15	17
Mean relative humidity (%)	32	33	35	36	33	34	33	29	24	24	25	26
Mean rainfall (mm)	29	31.5	30	20.5	9	2	2	8	4	4.5	14.5	11.5

Table 7.5 Climate statistics for Keetmanshoop in inland Namibia

LEARNING TIP It may seem strange that it is much colder at the top of high mountains than it is at sea level, because a mountain top is nearer to the sun than the surrounding lowlands. However, that height difference is tiny when compared with the distance of the Earth from the sun. It is important to remember that the air is not heated by the passage of the sun's rays through it, but by contact with the Earth's surface and by absorbing the heat it radiates. This means that the part of the troposphere nearest to the Earth is the warmest.

RESEARCH Find a map in an atlas showing mean January temperatures in Europe. How does the temperature change from the Atlantic Ocean inland along 50°N?

The influence of cold ocean currents on hot desert temperatures

Fig. 7.8 shows that there are cold **ocean currents** off the coasts of hot deserts. These currents are bodies of water that move through the oceans from areas nearer the poles to areas nearer the equator. Winds that blow over the cold Benguela Current (off the coast of Namibia) are chilled by contact with the current. They then carry the cooler air on to the land – lowering the temperature of Namibia's coastal strip. As a result, places like Luderitz have lower summer temperatures than would be expected at that latitude. Another example is Iquique in the Atacama Desert on the coast of Northern Chile.

The influence of a lack of cloud on the temperatures of hot deserts

Desert air has very low relative humidity, so desert skies are often cloudless or have very little cloud. This results in extreme diurnal (daily) temperatures. Without cloud, the maximum amount of solar radiation can reach the Earth's surface, so daytime temperatures are often as high as 38 °C. But, in summer, they can reach over 50 °C. However, at night – without clouds to stop the Earth's long-wave radiation escaping into space – temperatures can fall rapidly to about 15 °C in summer and 5 °C in winter. So, daily temperature ranges in deserts are very large all year round. Low temperatures at night can result in the condensation of water vapour – forming dew (droplets of water on the surface). These are important in assisting plant growth. Fig. 7.10 shows plants growing on boulders in the Mojave Desert.

The influence of aspect on temperature

The **aspect** of a place is the direction in which it faces if it is on a slope. Aspect has little influence on temperature in the tropics, but outside that zone it does.

- In the Northern Hemisphere, north of the Tropic of Cancer, places located on north-facing slopes are cooler than those on south-facing slopes. This is because they are facing away from the sun.
- In the Southern Hemisphere, south of the Tropic of Capricorn, places located on north-facing slopes are warmer than those on south-facing slopes. This is because they are facing the sun.

Discussion point

Why is it only possible to make a simple statement about aspect for places outside the tropical zone?

Precipitation in hot deserts

The average annual precipitation total in hot deserts is less than 250 mm a year. Some places have no recorded precipitation at all. Iquique in the Atacama Desert and Aswan in Egypt are two such places.

Much of the rain that does fall in the desert occurs in torrential convectional downpours – but these are rare and erratic. After a storm, a place might have no more rainfall for years. The main reason why rainfall is so rare in deserts is because these are high-pressure regions with descending, warming air. When it reaches the ground this air moves across the land towards the west coasts as a dry, offshore wind.

Fig. 7.10 High daytime temperatures in hot deserts result from clear skies

The influence of pressure on precipitation in hot deserts

The rising air that leads to so much precipitation in equatorial climates eventually descends to the Earth's surface in the hot deserts. This descending air is a major cause of desert aridity.

- Having risen to the tropopause, the air moves towards the poles and starts to cool - becoming denser as a result.
- It then sinks at about 30°N and 30°S - creating high pressure at the surface.
- The sinking air becomes compressed and that compression causes warming. This results in a decrease in the air's relative humidity.
- After reaching the surface, the dry air moves from the high-pressure area back to the low pressure in equatorial latitudes - as the trade winds. Some air also moves towards the poles.

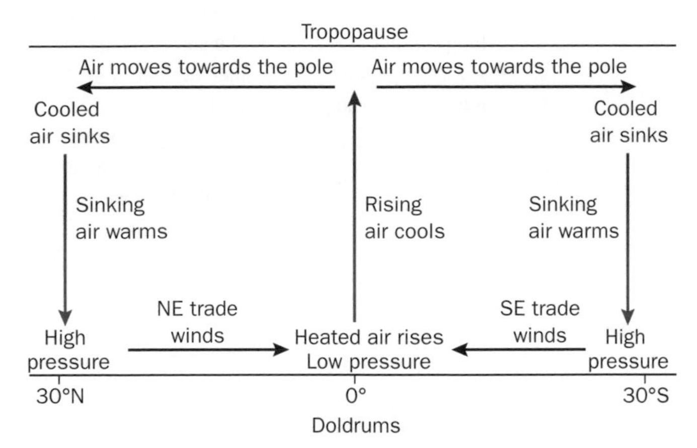

Fig. 7.11 Processes in the Hadley Cells

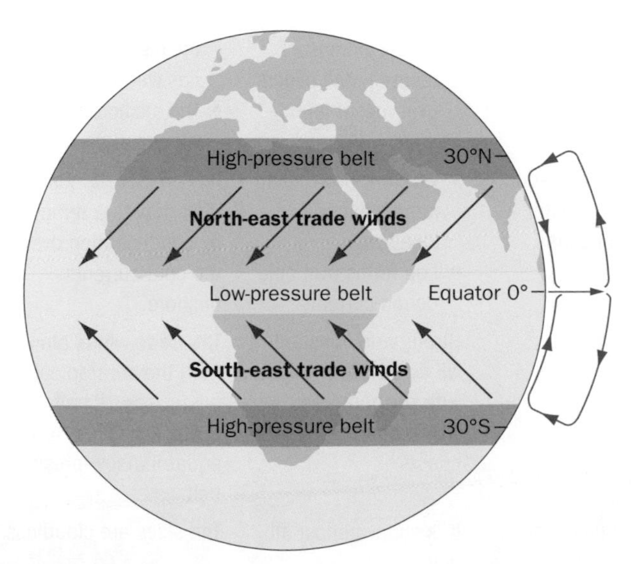

Fig. 7.12 Pressure belts and the air movements which influence tropical climates

The circulations of air between the equator and 30°N and 30°S are known as the Hadley Cells (see Fig. 7.11).

Trade winds are strong and constant and, because they derive from sinking air and blow over land to the deserts, they do not contain much moisture. The combination of sinking air and offshore trade winds leads to very low precipitation in the hot deserts. North-east trade winds blow in the Northern Hemisphere and south-east trade winds blow in the Southern Hemisphere.

The above wind directions are the result of two controlling factors:

- Winds blow out of high-pressure systems into low-pressure systems.
- They are deflected by the Earth's **rotation** as they do so. They deflect to the right in the Northern Hemisphere and to the left in the Southern Hemisphere.

The influence of cold ocean currents on precipitation

When onshore winds blow on to the west coasts of deserts, they can be cooled enough by the cold ocean currents there for condensation to occur. This condensation leads to **fog** (tiny water droplets hanging in the air near the surface), which reduces visibility. Coastal places like Luderitz can have many foggy days and relatively high humidity when the winds are onshore.

The condensation process removes moisture from the air. As the foggy air moves inland, the water droplets quickly evaporate. Daytime warming as the air passes over the land further reduces its relative humidity and the chance of precipitation. That is why Luderitz only has about 30 mm of rain a year.

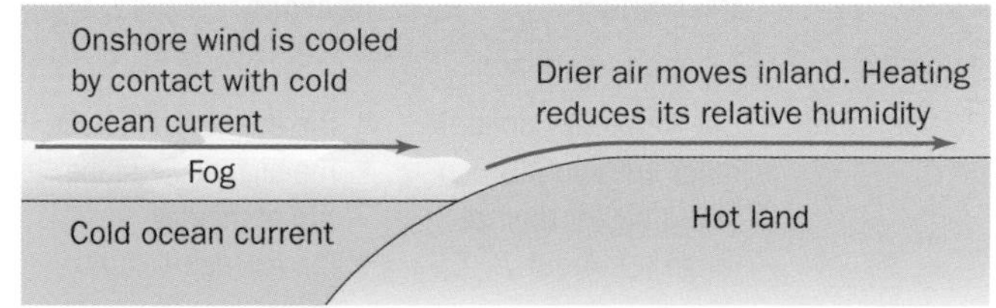

Fig. 7.13 Why cold ocean currents offshore increase coastal aridity

The influence of relief on precipitation

At Keetmanshoop (about 290 km inland from Luderitz) there is more rainfall - about 166 mm a year. This higher rainfall results from the difference in altitude between the two places. Luderitz is at sea level, whereas Keetmanshoop is over 1000 m higher (on the Namibian Plateau). When the air containing moisture from the sea reaches the higher ground, it rises to pass over it. The same processes then occur as for convectional rainfall (Fig. 7.6), but the resulting rain is known as **relief** or **orographic** rainfall (Fig. 7.14).

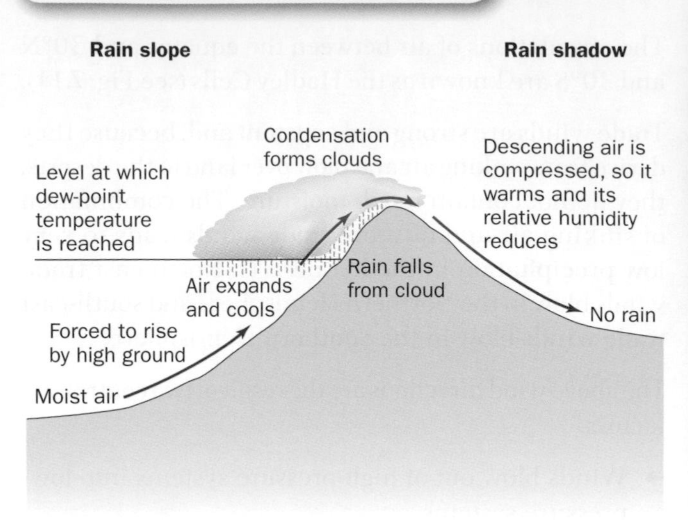

Fig. 7.14 Relief rain and rain shadow

Fig. 7.15 Cloud resulting from the uplift of air on one side of a mountain

The influence of temperature on precipitation

Rain does not fall often in deserts but, when it does, it's usually torrential and often causes flash floods. This is particularly true in the areas nearest to the equator, where occasional convectional storms occur in the summer heat. In these areas, summer is usually the season when most rain falls. Areas on the poleward side of deserts have winter rain.

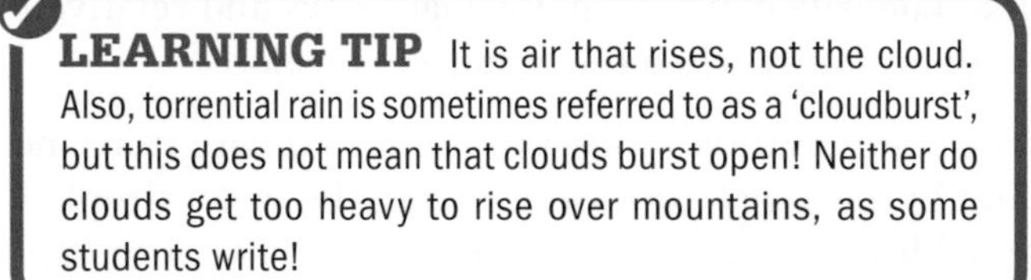

LEARNING TIP It is air that rises, not the cloud. Also, torrential rain is sometimes referred to as a 'cloudburst', but this does not mean that clouds burst open! Neither do clouds get too heavy to rise over mountains, as some students write!

Equatorial climate

	Description	Reasons
Location	Around the equator.	
Temperature	Hot with a low annual range (of about 3 °C) and a higher diurnal range (of about 7 °C).	• The equatorial latitude. • The high angle of the sun at midday. • The extensive cloud cover.
Precipitation	High all year round (over 1500 mm), with a minimum monthly rainfall of 60 mm.	• A high humidity, because of the high evaporation rate. • Convection caused by the air heating up. • The air rising at the ITCZ. • Moist air moving into the low-pressure zone.
Wind	Light and variable.	The rising air.
Sunshine	Usually less than half the daylight hours.	Cloudy for the same reasons as for precipitation.

Hot desert climate

	Description	Reasons
Location	Around the tropics.	
Temperature at inland locations	• Very hot in the summer. • Hot in the winter. • A moderate annual temperature range. • A very high daily temperature range.	• Between latitudes 15° and 40°C. • The midday sun is overhead in the summer. • The sun is still quite high at midday in the winter. • The cloudless skies allow solar radiation to penetrate well during the day but radiate out at night.
Temperature at coastal locations	• Hot in the summer. • Very warm in the winter. • A lower annual range than inland. • A lower daily range than inland.	The maritime influence. The sea is colder than the land in the summer and warmer at night in the winter.
Precipitation at inland locations	• Very low (below 250 mm). • It may increase if relief rainfall occurs over higher land.	• A low relative humidity, because sinking air in the high-pressure belt has warmed. • The offshore trade winds are dry.
Precipitation at coastal locations	• Very low (lower than nearby inland locations). • Fog forms over cold ocean currents.	Any onshore winds have their moisture removed by condensation over the cold currents offshore.
Wind	Strong and constant NE or SE trade winds, with onshore west winds at times near the coasts.	The trade winds blow from the subtropical high-pressure belts at 30°N and 30°S to the equatorial low-pressure belt.
Sunshine	It is sunny almost all day.	The skies are cloudless.

Table 7.6 A summary of the main differences between equatorial and hot desert climates

Fig. 7.16 A satellite image of cloud cover over the Americas

This satellite image shows the equatorial cloud belt covering part of the Amazon Basin and the Panama area. Cloud over the cold Peruvian Current reaches the South American coast, but the Atacama Desert, Mexico and the southwest USA remain under the influence of the subtropical high-pressure belts and are cloudless. A strip of cloud over the northern Andes is caused by the high relief.

6 When this photograph (Fig. 7.17) was taken, a local wind was moving from right to left up the valley, in the middle distance. Explain why and how the cloud formed.

Fig. 7.17 Cloud in a valley

Ecosystems

What is an ecosystem?

A natural **ecosystem** is an area in which plants and animals live in balance with their environment and are interlinked with it. An ecosystem does not change unless external factors influence it. The tropical rainforests and hot deserts are two of the world's major ecosystems or **biomes**. Sand dunes and mangrove swamps are examples of smaller-scale ecosystems. Some of the elements and interrelationships within an ecosystem are shown in Fig. 7.18.

Plants need **nutrients** to survive. These nutrients are minerals released from rocks by weathering (the breakdown of rocks into smaller pieces), which circulate in an ecosystem (Fig. 7.19).

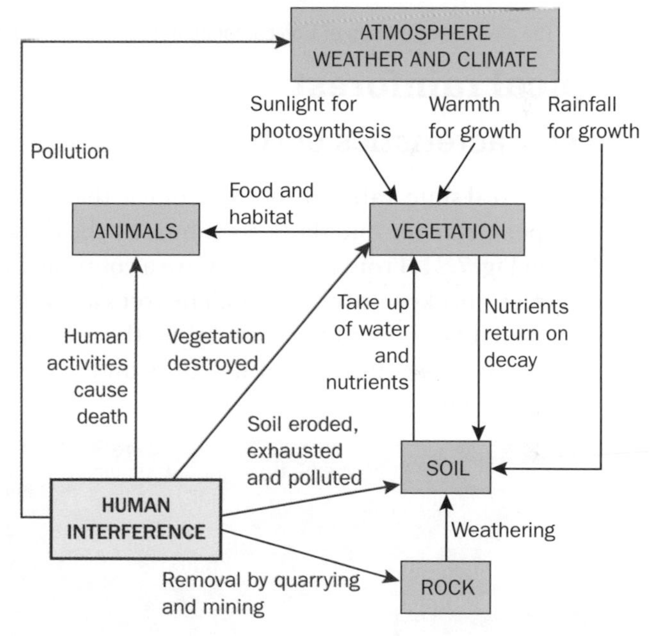

Fig. 7.18 Some links between parts of a natural ecosystem, together with examples of human interference

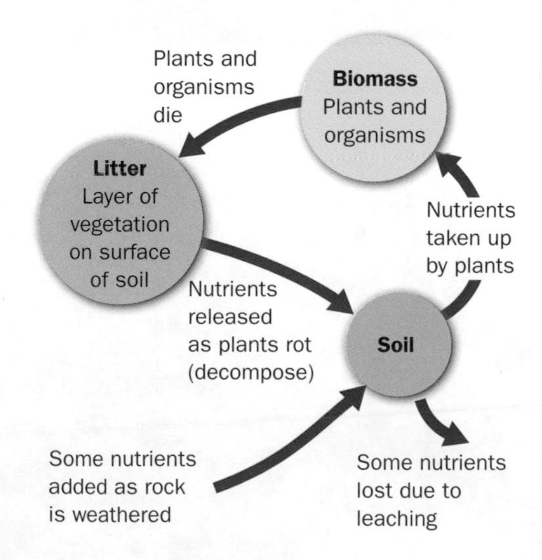

Fig. 7.19 Nutrient recycling

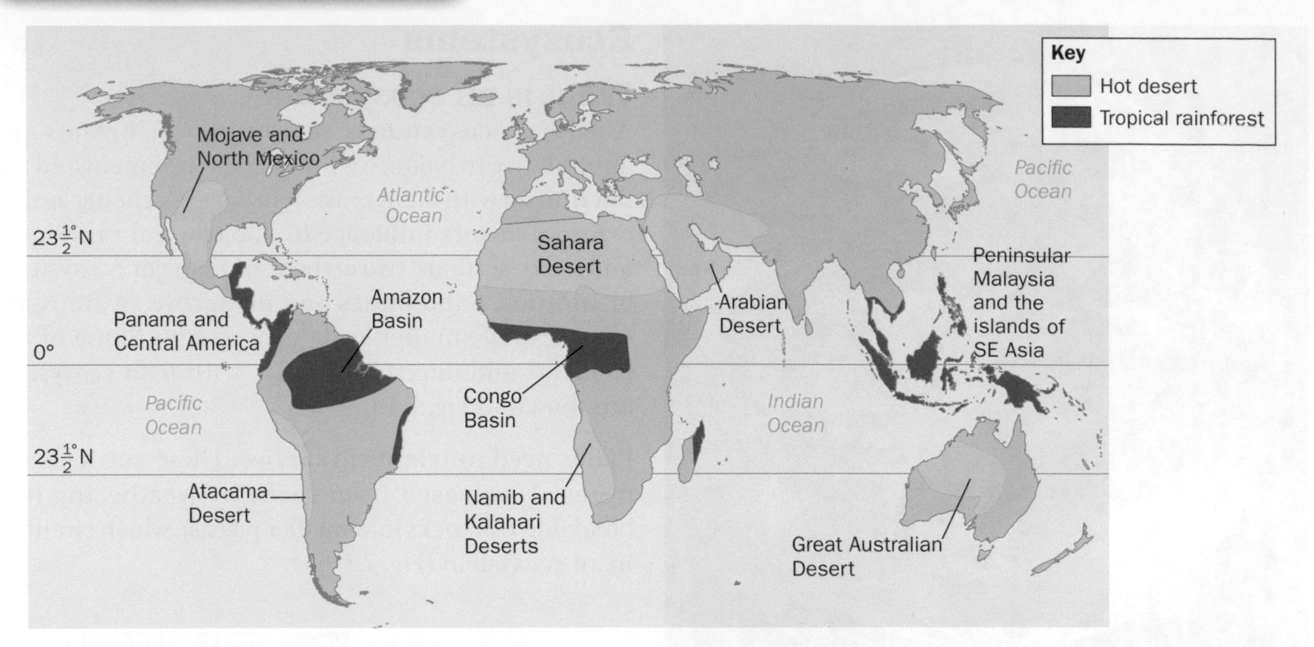

Fig. 7.20 The global distribution of tropical rainforest and hot desert ecosystems (biomes)

Tropical rainforest

The characteristics of the vegetation

In their natural state, rainforests are so dense that light does not penetrate far into them, as shown by the dark patches on Fig. 7.21. From the air they are a continuous mass of trees - broken only by rivers. The trees are very tall with straight trunks and branch only at the top. The forest **structure** is composed of five tiers and is well adapted to the climate in which it grows.

Rainforest tiers

- The emergent layer reaches a height of 30–40 m. Emergent trees are widely spaced.
- The dense main canopy layer is at a height of 20–30 m.
- The under-canopy of sapling trees is about 15 m high. These occur only where light is available.
- There is a very sparse shrub layer in clearings, or where light is available.
- There is a ground layer of tree ferns - but only in clearings and on riverbanks.

7 **a** Use Fig. 7.22 to describe the characteristics of the tropical rainforest vegetation visible in Fig. 7.21.

b Explain why deeper into the forest sapling trees are only found where main canopy trees have died and fallen.

c Describe and explain the difference in shape between the crowns (branching part at the top of the tree) of the emergent trees and those of the trees in the main canopy.

Fig. 7.21 Rainforest in Panama

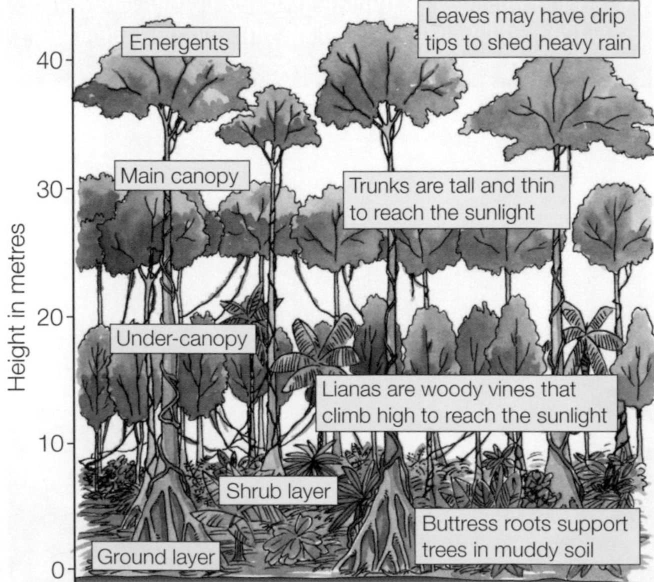

Fig. 7.22 The structure and characteristics of tropical rainforest

There are a large number of species in an area, but they all look alike. Each species is widely spaced apart.

The south-east Asian island of Borneo, for example, has more than 10 000 plant species, including 3000 tree species. They are mainly hardwoods, such as ironwood, mahogany, ebony, and rosewood.

The forest is not seasonal - at the same time, some trees will have flowers and others fruit, while some are losing their leaves. The deciduous trees drop some of their leaves throughout the year, so the forest always has an evergreen appearance. Some trees may have branches with no leaves while others have full foliage. The trees grow continuously all year round until fully grown.

The leaves of rainforest trees are broad and often have a depressed central vein that leads to a drip tip.

Fig. 7.24 Drip tip leaves

The relationship between natural vegetation and climate

8
a Using your knowledge of its climate, state whether X or Y on Fig. 7.23 represents the conditions in which tropical rainforest will grow.
b What is the lowest precipitation required for the growth of tropical rainforest?
c What is the lowest temperature at which tropical rainforest will grow?
d Use the information in Fig. 7.25 to complete the sentences below, which explain how the tropical rainforest is adapted to the climate. You can use a reason more than once.
- The forest is not seasonal because ...
- The forest is dense and has a great variety of species because ...
- Trees grow very tall because ... and ...
- The trunks are straight because ...
- The trees have large leaves because ...
- Leaves have drip tips because ...
- Growth is continuous because ...
- There are few plants below the canopy because ...
- The bark is thin because ...

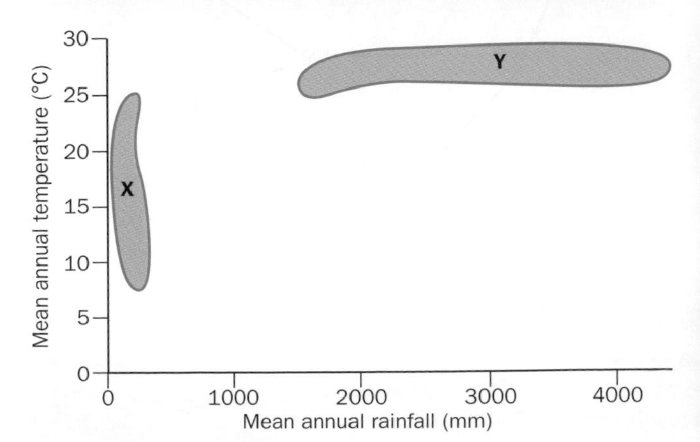

Fig. 7.23 The temperature and rainfall requirements of the tropical rainforest and hot desert biomes

... it is hot and wet all year.
... they need to shed the torrential rain quickly to help transpiration and prevent them breaking off.
... they do not need protection from the cold.
... they have to compete for sunlight.
... there is insufficient sunlight for photosynthesis.
... they need to catch the maximum amount of sunlight in the cloudy climate.

Fig. 7.25 Reasons for the characteristics of tropical rainforests

The relationship between tropical rainforest vegetation and the soil

Rapid chemical weathering processes in the hot and wet climate result in very deep soils, yet the trees of the rainforest usually have shallow buttress roots.

Fig. 7.26 A rainforest tree with buttress roots

Tropical soils are red because they contain a lot of iron near the surface. The trees remove nutrients from the soil – to be returned later when their leaves fall. The leaves decompose rapidly on the forest floor in the hot and wet conditions, releasing nutrients. The deep soil below has few nutrients, because of **leaching**. The nutrients are taken down and removed from the soil by rainwater.

The shallow buttress roots allow the trees to access the nutrients near the surface of the soil. These roots are also essential for the stability of the very tall trees, which need their support in the wet and muddy soils.

Life in the tropical rainforest ecosystem

The tropical rainforest has very rich and diverse animal life because it has constant warmth and water, a variety of **habitats**, and an abundance of vegetation for food. New animal species are discovered every year.

Each layer of the forest has different conditions of sunlight, temperature, and moisture, as well as vegetation characteristics, so the animals that live in each layer are adapted for that environment and live where they can obtain the different types of food they need. Animals that cannot climb live on the forest floor. Monkeys live in any tier above the ground layer as they can easily swing and climb to access the fruits on the trees. Many of the animals are fruit eaters and fruit is abundant all year.

The forest floor

Many insects and fungi live in the decaying leaf litter. Foragers such as anteaters dig their food out of the ground. Pygmy elephants also live here, hiding among the buttress roots.

Fig. 7.27 Rainforest growing on tropical red soils. The two sapling trees on the right of the photograph are beginning to develop buttress roots. There is little leaf litter to be seen on the forest floor.

The under-canopy (understory)

Here there is a little more light, warmth, wetness, and more plants, so there are very many insects, food for the birds, geckos, bats, and tree frogs that live in this layer. Big cats, such as jaguars, and snakes live on the branches where they can spot their prey below.

The canopy

Here dense vines and trees provide shelter from winds, rain, and predators. It has the most abundant wildlife, including many birds. The top two layers of the forest are home to 80% of the forest dwellers. Seeds and fruit provide plentiful food for animals such as monkeys, tree frogs, parrots, and toucans , which in turn provide plenty of food for the snakes living in this layer. Many animals move around by hopping, gliding, flying, or jumping.

The emergent layer

This has the most sunlight and weather variations. Only lightweight creatures that the thin branches can support live here, such as monkeys, birds, and butterflies.

Energy flows, food chains, and food webs in the tropical rainforest

The energy passes in the direction of the arrow. A simple example of a **food chain** in Borneo would be:

sun's energy → plants produce leaves and fruit →
squirrels eat them → raptors eat the squirrels →
on their death, bacteria and fungi decompose the raptors

On the death of any animal or plant, the nutrients they contain will re-enter the soil for new plants to take up. So the two important processes in ecosystems – nutrient cycling and energy flows – are linked.

There are many food chains in the forest. They are interlinked in a complex **food web**. A small part of that web is shown in Fig. 7.28. The links are made because squirrels, frogs and insects are eaten by more than one animal.

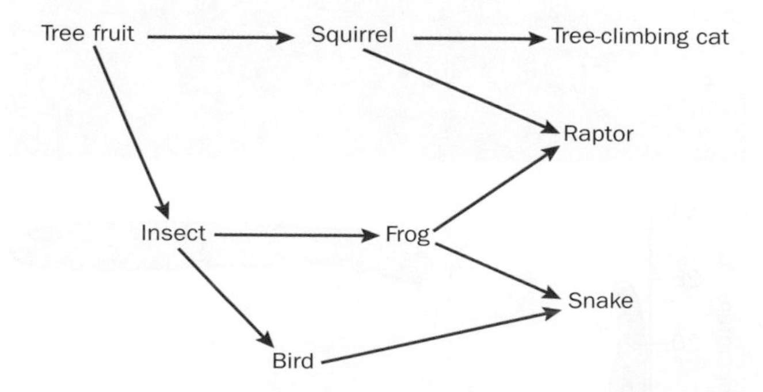

Fig. 7.28 Part of a tropical rainforest food web

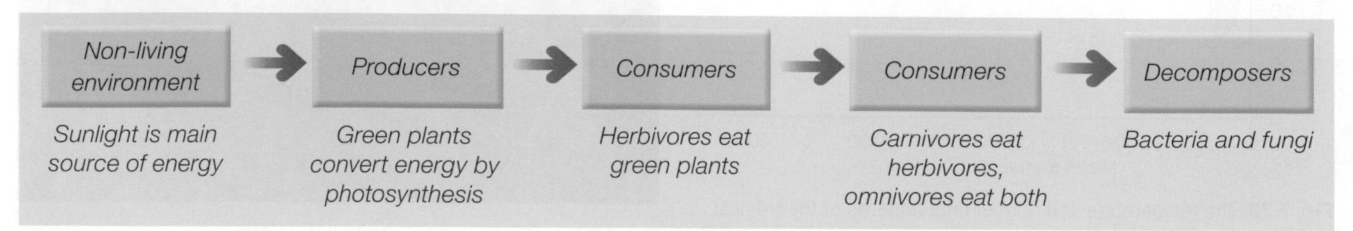

Fig. 7.29 Energy flows through the food chain

Keeping tropical rainforests as a valuable natural environment

Soils

If the forest is left alone, it maintains the little soil fertility it has by returning nutrients to the soil in leaf fall. It also reduces soil erosion in a number of ways:

- Its roots hold the soil in place.
- Its large leaves lessen raindrop impact on the soil by catching them.
- Its roots also take up water from the soil – reducing the chance of a mudflow down a slope. Reducing the amount of water in the soil also reduces the leaching of plant nutrients out of the soil.

Water quality

By keeping the soil in place, forests stop it from being washed into rivers. This keeps them healthy for aquatic life.

Climate

The water taken from the soil by tree roots is passed into the atmosphere through pores in the leaves by the process of transpiration. The water vapour can then be converted as a result of convection into rainfall to provide the necessary water for a healthy ecosystem.

During photosynthesis in the daytime, plants take in carbon dioxide (one of the main **greenhouse gases**) from the atmosphere and release oxygen (a vital gas for human life) into it. The more forests the world has (acting as a **carbon sink**), the greater the chances are of being able to reduce **global warming**.

The dark colour of the forests also absorbs solar radiation. If the forests were replaced with crops, more of the sun's rays would be reflected off those lighter surfaces – causing the surface to be cooler, so reducing convection and rainfall.

The value of tropical rainforests to humans

We have already seen that the tropical rainforest ecosystem contains an enormous diversity of species, many of which may have a value as yet unknown. For example, so far over 7000 medicines to combat illnesses have been derived from tropical rainforest plants (e.g. quinine from the chinchona tree of Africa and Central America is used to cure malaria).

Tropical forests provide timber for construction and furniture, fuelwood for local people and raw material for industry. They also boost the local and national economies by attracting **ecotourists** (tourists interested in nature who do not harm the ecosystem but contribute to the local economy). The rainforests also provide a home for many native peoples.

CASE STUDY

Tropical rainforest in Borneo

The island of Borneo has land belonging to three countries: Brunei in the north, Sarawak, which is part of Malaysia, in the north-west, and the largest area, Kalimantan, in Indonesia.

The diversity of the rainforest in Borneo is the richest in the world, with more than 10 000 plant species (including the world's largest flower), 380 types of bird, and many other animals. This rainforest is believed to have existed for more than 60 million years but it is being destroyed alarmingly quickly by human activity. Many of its species do not exist anywhere else and many may still remain to be found.

The forest layers

Most of the trees are strong hardwoods, much in demand for furniture. Emergents, the tallest trees such as *Shorea*, grow up to an incredible 65 m tall. The middle of the canopy layer is about 30 m above the ground. Here the ironwood tree is common and valuable. It is so hard and resistant to decay that it is in demand for long-lasting piers, docks, jetties and piles. The many species of fig tree provide a plentiful fruit supply throughout the year. The shrub layer includes many shade-tolerant species, including lianas, ferns and orchids. The forest floor has shade-tolerant palms and herbaceous plants.

Animals

Borneo's carnivores include a leopard and bear. Other forest-floor dwellers include herbivores such as the pygmy elephant, rhinoceros, and ant-eating pangolins. Fruit-eating orang-utans move throughout the forest, as do gibbons, macaques, and monkeys. Many birds, reptiles, and amphibians are active during the day, while flying squirrels and flying foxes become active towards night-time.

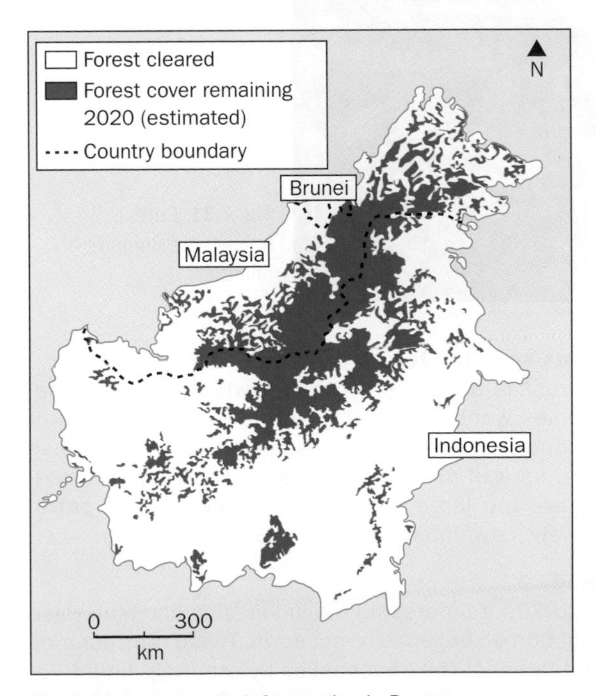

Fig. 7.30 Large-scale deforestation in Borneo

Deforestation in Borneo

Apart from its highest mountains, Borneo was covered in tropical rainforest – until humans embarked on a programme of rapid forest destruction in the 1950s. A hundred years ago, this huge island was inaccessible and unexplored. But, recently, the clearance of vast areas of forest by burning has sometimes covered countries as far away as China in a brown smoke and ash cloud.

Nowhere else in the world has experienced forest clearance as rapid as in Borneo since 1980. Today, only patches of forest remain outside the less accessible forested mountainous interior in Central and Northern Borneo.

9 **a** Construct a line graph to show the loss of Borneo's rainforest over time. The years and percentages of land covered by forest are: 1950 (94%), 1985 (74%), 2000 (57.5%), 2005 (50.5%), 2010 (44.5%).

b Describe the trends the graph shows.

c If the present rate of destruction continues, estimate the likely year when all of Borneo's forest will have been cut down.

LEARNING TIP Although the years are not in regular intervals, it is possible to plot them at the correct point along the *x*-axis, so that the variation in rates over time can be seen.

Fig. 7.31 Cutting trees in the Borneo rainforest

Why has so much forest been cleared?

The forest is a source of wealth, which multinational companies want to exploit to provide raw materials for their industries. Population pressure and poverty are also factors, as well as the vastness of the areas involved, which are too large for protection agencies to patrol effectively. Law enforcement is weak in Indonesia.

Logging

In the 1970s, rainforest in the Indonesian and Malaysian parts of Borneo began to be cut down. Those two countries wanted to boost their economies by exporting timber for uses such as furniture making and pulp and paper manufacture. In the 1980s, the logging industry expanded rapidly as new roads provided access to previously untouched areas of forest. Between 1980 and 2000, more timber was exported from Borneo than from Africa and Latin America combined. The forests removed by logging were usually **clear cut** and not replanted. The use of heavy machinery on wet soils compacts them, so rain cannot sink in and runs off instead, causing soil erosion.

Population pressure

Between 1970 and 2000, Indonesia's **transmigration** programme (see page 23) moved thousands of Indonesians from overcrowded islands like Java to less crowded areas like Kalimantan. This resulted in the clearance of millions of hectares of forest. The new settlers' attempts to live by subsistence farming failed, because the heavy rainfall quickly leached the plant nutrients from the soil once the protective forest cover was removed. This is known as **soil degradation**. Instead, the migrants provided labour for logging companies and the roads made for them opened up the forest to the logging companies.

Fig. 7.32 Logging in Borneo

Plantations

In the 1980s, the **deforestation** was speeded up as vast oil palm plantations were planted. By 2004, these plantations covered one million hectares of Kalimantan. The oil palm is a very productive tree (one hectare yields about 6000 litres of oil), so planting it was very profitable for the landowners. Palm oil is used widely in the manufacture of soap, cosmetics and processed foods.

Fig. 7.33 A young oil palm plantation in Borneo

The replacement of tropical rainforest with oil palm plantations is likely to continue, because palm oil has

now become an economical way of making biodiesel. It is cheaper than conventional oil and the demand for biofuels is soaring in developed countries, where they are subsidised. To meet the growing demand, Malaysia and Indonesia plan to expand the amount of land being used for oil palm cultivation.

Plantations have also been established for rubber, for industrial hardwoods, and for fuelwood and charcoal. Native trees are being replaced by those species more suitable for pulping.

Hydroelectric power (HEP)

In Sarawak (in Malaysian Borneo) a large area of forest was cleared to provide a reservoir for the Bakun HEP scheme. Despite strong local opposition, the Malaysian government plans to build six more HEP plants in Sarawak's forests to make industrial developments possible.

Mining

Some forest inland of Balikpapan has now been turned into an oilfield. Recently, large-scale coal mining (especially in the east and south of Kalimantan) has become a big threat to the forests there. The Indonesian government is promoting economic development by building a railway to link the mines to the ports. Forest conservation is generally less important to an LEDC than increasing the country's wealth, if they face a choice between the two.

Road building

Roads built to access mines, HEP sites and logging areas open up the forest alongside them for further deforestation and industrial and other developments, including settlement.

Consequences of deforestation

Atmospheric pollution

Using burning as a method of forest clearance has become widespread. Every year it endangers wildlife and causes an enormous cloud of ash and smoke to drift across neighbouring countries. Attempts have been made to blame the fires on shifting cultivators, but satellite images prove that clearance for plantations is the main cause. The 1997–8 fires burned nearly 7 million hectares in Kalimantan and were the largest.

Global warming and risks to health

The burning of forests emits a lot of carbon dioxide into the atmosphere. This greenhouse gas absorbs radiation and contributes to global warming. Forests are also thought of as the 'lungs of the world'. When they are removed, less carbon dioxide is taken in and less oxygen is given out. The forests are needed to combat the effects of the extra carbon dioxide in the atmosphere due to the burning of fossil fuels like coal and oil.

Smoke from burning forests is also a health hazard, which – at its worst – was thought to be responsible for one in five of all the deaths in the entire south-east Asian island region.

Loss of biodiversity

Deforestation could result in the loss of plant species which might have unknown uses as medicines, industrial raw materials and foods.

Death of ways of life and unique cultures

Forest clearance in Sarawak for HEP developments, oil palm plantations, and logging may threaten the **hunter-gatherer** way of life of the Penan people. They feed on fruit, nuts, plants and animals from the forest.

In Kalimantan, the Dayak people are also under threat. They are shifting cultivators who need to be able to move to different areas of the forest once their current plot has had its soil leached and exhausted of minerals. When there is plenty of forest, their way of life is **sustainable** – because they only clear small plots and they allow the forest to regenerate and the soil to recover its nutrients before they return to the same plot. It does not harm the environment. However, now deforestation is forcing them to return to each plot before sufficient years of fallow have passed. As a result, crop yields and soil quality quickly deteriorate. This leads to soil erosion and the growth of poorer vegetation than the original forest.

There is also tension between the Dayak people and the migrants who were moved to Kalimantan as part of the transmigration programme, because their cultures are totally different.

The hunter-gatherers and shifting cultivators use the forests in sustainable ways.

Fig. 7.34 Forest clearance by burning

Loss of habitats

The numbers of orang-utan in Borneo have fallen in the last 50 years – partly due to the reduced forest cover. Poaching has become easier. Local people can boost their low incomes greatly by selling young orang-utan as pets, after first killing their protective adults (whom they consider to be pests because they include palm fruit in their food). Many other animals and birds are thought to be near extinction in Borneo.

Fig. 7.35 Orang-utan in Borneo

Fig. 7.36 Ground vegetation by a path

Reduced rainfall in the local area

When forest is removed, moisture is no longer transpired through its leaves to the atmosphere and rainfall is no longer intercepted on its leaves to be evaporated back to the atmosphere. The moisture content of the air is reduced, resulting in less local rainfall and more droughts. Also, ground left without shade under the hot sun becomes hardened, so that rain runs off quickly to the rivers instead of sinking into the soil. This causes a loss of moisture from the local area that could have been evaporated back to the air.

Loss of soil fertility

When soils are degraded as a result of deforestation, any forest that is allowed to regenerate (secondary growth) is always poorer than the original primary forest.

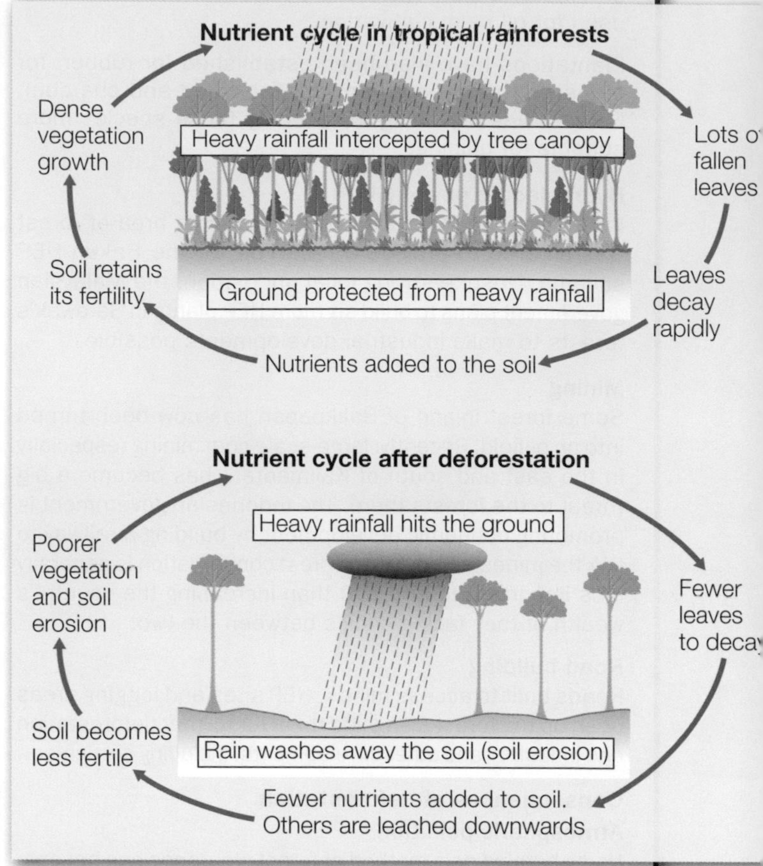

Fig. 7.37 The impact of deforestation on the nutrient cycle

- With no rainforest vegetation to protect the soil, heavy rainfall washes it away. The farmers then need to clear more land.
- When rainforest is cleared and the land is intensively farmed, it loses its fertility within 20 years.
- Deforestation breaks the nutrient cycle, which the soil depends on for its fertility.

10 Look at Fig. 7.36.

a **i** Describe the ground layer vegetation and explain why it is not typical of ground layer vegetation deeper in the forest.

ii Explain why the soil does not have a cover of leaves, even though leaves fall off the plants every day.

b Another consequence of deforestation is soil erosion, leading to the choking of rivers and reservoirs with silt and the consequent loss of aquatic life and flooding. Explain how this occurs.

Attempts to manage Borneo's rainforests

Some areas of Borneo have been designated as **national parks**, which are supposed to have their wildlife and habitat protected. In reality, that is not always the case. For example, when Kutai National Park was formed in Kalimantan, over 300 000 hectares of forest should have been made safe from development. However, timber and oil exploration rights were then granted in the area, and the official size of the Park was reduced by a third. Illegal logging has even degraded that, because it is difficult for park rangers to monitor such a large area.

There is no doubt that the remaining forest should be preserved, but that will be difficult to achieve because a growing population needs a growing economy to sustain it. One third of Indonesians live on less than one US dollar a day.

Large-scale deforestation is bad news. But there are ways of managing rainforests in order to halt or reduce the destruction, and in some cases repair the damage.

Selective logging

This is a technique where individual trees are felled only when they are mature. The idea is that the rainforest canopy is then preserved, which protects the ground below and also helps slower-growing hardwoods, like mahogany. However, the roads left behind by selective loggers allow other people to follow them in and open up the forest further. Also, for every tree that's selectively logged, up to 30 other trees can be damaged or destroyed getting the logged tree out of the forest.

Reducing debt

Conservation swaps, or **debt-for-nature swaps**, are a way of reducing a country's debt and benefiting nature and conservation at the same time. The most common type of debt-for-nature swaps work like this. A country (e.g. the USA) that is owed money by another country (e.g. Peru) cancels part of the debt in exchange for an agreement by the debtor country to pay for conservation activities there.

Promoting responsible management and use

The Forest Stewardship Council is an NGO that promotes the responsible management of the world's forests. Approved companies can use its logo to show that their wood products have been produced responsibly. Consumers can then make a choice between buying approved products, with the logo, or products produced in a less responsible way (hopefully reducing demand for them).

The Forest and Land Restoration Initiative in Kalimantan has a stated aim to restore 900 000 hectares of forest each year by **afforestation** and **reforestation**. It involves local government and villagers. Teak is usually planted because it is valuable and fast growing, so it can be logged at a sustainable rate. However, it is not planted as fast as felling is occurring. Incentives are given to local farmers to replant trees but they are often not big enough to achieve that aim. One disadvantage is that, by selecting to plant teak, biodiversity is reduced and the true tropical rainforest is not restored, because teak is not a tropical rainforest species (it is a monsoon forest species).

Ecotourism

Ecotourism helps the local economy. It is often based in resorts within wildlife reserves, such as Tabin Wildlife Resort in the north of Borneo. Walkways, both at ground and canopy level, are often provided for the visitors – together with information boards and visitor centres.

Organisations like the World Bank are also promoting the development and use of more efficient farming methods or more sustainable ways for farmers to earn a living, such as ecotourism.

Improving the standard of education generally and educating the people about why forest conservation is needed should also help. The use of some forest areas as tribal reserves would benefit both the rainforest and the tribes, because they would use the forest in sustainable ways.

11 **a** Write a speech in which you use two reasons to try to persuade people in a village to save the remaining rainforest in their area. Do not use reasons covered in question 10b.

b Copy and complete Table 7.7 overleaf. You can write 'As for mining' where appropriate.

c Describe and explain the **global** effects of deforestation of the tropical rainforests (i) on the natural environment, (ii) on people.

Fig. 7.38 Ecotourists walking just below the canopy of a tropical rainforest

Cause of deforestation	Economic impacts	Social and political impacts	Environmental impacts
Mining	It boosts the economy with foreign earnings which can be used to pay off debts and support further development.	It provides jobs.	Massive amounts of rainforest are cleared, causing soil erosion, degradation, and flooding. It can also cause drought by reducing transpiration.
Road and railway building	This helps economic development by improving communications within the country.	It reduces the development gap between communities.	It increases access to previously inaccessible rainforest areas, leading to further deforestation.
Shifting agriculture	This is subsistence farming, so it has no economic impact for the country as a whole.	It allows the Dayak people to continue their traditional lifestyle and maintain their unique culture.	Only a small patch of rainforest is cleared at a time, so there is no impact if the forest is allowed time to regenerate and refertilise the soil.
Logging			
Plantation agriculture			
Ranching			
Transmigration (migration into the forest)			
Hydroelectric power schemes			

Table 7.7 The impacts of using deforested areas for other purposes

CASE STUDY

Panama's tropical rainforest

A different management story – but for how long?

A strip of Panamanian rainforest was controlled by the USA until 2000 – the only rainforest controlled by an MEDC. This is because it lies next to the Panama Canal, which runs between Colon on the Caribbean Sea and Panama City on the Pacific Ocean, via Lake Gatun.

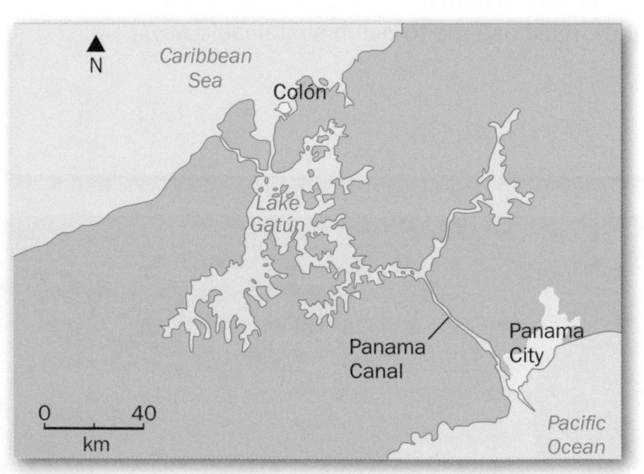

Fig. 7.39 The Panama Canal

The Americans took over the building of the Panama Canal from the French in 1904. The idea was to create a new transport route between the Atlantic and Pacific Oceans, through Central America. The new route would allow ships to avoid taking the stormy passage around Cape Horn (at the tip of South America), saving a lot of valuable time and fuel. The Panama Canal cost the USA $75 million to build, but it saved ships travelling from one US coast to the other over 12 000 km.

A formal treaty was signed between the USA and Panama. It stated that the Panama Canal would belong to Panama, but that the USA would have jurisdiction over the Canal Zone – a strip of tropical rainforest 8 km wide on either side of the canal.

The smoke in Fig. 7.40 indicates shifting cultivation. This is because the USA allowed no other development in the Canal Zone.

Fig. 7.40 Forest in the Canal Zone

- Since 1990, the rate of deforestation in the whole of Panama has been only 0.2% a year.
- Satellite images suggest that deforested patches have been growing in size away from the Canal Zone.
- Most of the deforestation was for cattle ranching, logging, and the growing of cash crops like bananas and coffee.

Panama's rainforest has a wide biodiversity. This is because the **isthmus** connecting North and South America (i.e. Central America) acted as a land bridge, so that species from both continents could mix. However, some of that biodiversity disappeared when a vast area of forest was flooded during the construction of the Panama Canal. The Chagres River was dammed to create the enormous inland Lake Gatun – the largest artificial lake in the world. This lake is above sea level, so two sets of locks are needed at the outlet to each ocean. The lake water is necessary to provide the 230 million gallons of water which spill into the sea each time the locks open. About 50 ships pass through the Panama Canal every day, and there are always vessels waiting for their turn.

What does the future hold for Panama's rainforests?

On 1 January 2000, control over the Canal Zone was transferred back to Panama by the USA. The USA had run it on a non-profit basis, but the canal's first Panamanian director said that in future it would make the most profit possible. With that in mind, a third set of locks has been built to increase the number of ships that can use the canal each day.

World trade is expanding every year, and there is no doubt that a third set of locks would benefit many countries by reducing the queuing time for the canal. If you include the queuing time, the passage through the Panama Canal currently takes a day and a night – but the actual journey through the canal itself takes only 10 hours, so reducing the queuing time would make a big difference.

Panama has made the most of its new opportunity to develop. By 2010 a considerable amount of rainforest near the canal entrance had been destroyed to make way for hotels, apartments, and golf courses. Developments were also taking place on the shores of Lake Gatun (see Figs. 7.43 and 7.44, which show developments in 2007).

Fig. 7.41 Two parallel sets of locks along the Panama Canal lead into Lake Gatun

Fig. 7.43 Forest clearance on the shore of Lake Gatun

Fig. 7.42 The far hillside was excavated for the third set of locks (it has to be terraced to prevent landslides)

Fig. 7.44 Clearance from the lakeshore into the forest

Can Panama's natural environment be saved?

Panama is facing big water pressures, which are made worse by deforestation:

- An enormous amount of water is needed to operate three sets of locks on the canal. There is a real fear that in the drier months Lake Gatun may not be able to supply enough.
- The World Bank has estimated that for every 10 000 hectares of tropical rainforest destroyed, Panama's rainfall is likely to be reduced by 10%. This is because the moisture taken from the soil by the rainforest vegetation is then transpired back into the atmosphere. If the forest is destroyed, this recycling of water can no longer occur and drought will be one consequence. Panama's economy will suffer as a result.
- Panama City has more than one million residents, and it needs 100 times more litres of water a day than that needed to empty a lock.
- Soil erosion is another consequence of deforestation and this could result in the silting up of the canal and its locks.

In 2004 the USA arranged a debt-for-nature swap with Panama. The USA offered to reduce Panama's debt to it in return for Panama's conservation of its tropical rainforest. The government of Panama has now established 14 national parks to protect its biodiversity. It is hoped that the country will follow Costa Rica by creating a successful economy through developing ecotourism.

However, Panama is also under great pressure to build a road south to Colombia to connect the two sections of the Pan-American Highway, so that it will stretch from Alaska to the south of Argentina. A bridge (the Bridge of the Americas) already connects the two continents. If this road is ever built, it is very likely to hasten rainforest destruction in Panama.

Fig. 7.45 Cumulonimbus over the Bridge of the Americas between North and South America

The hot desert ecosystem

The vegetation's characteristics and its adaptations to climate and soils

Although days are very hot in deserts, this does not have a direct impact on the vegetation - except by reducing the rainfall that plants can use by rapidly evaporating it. As desert rainfall is so low anyway, it is not surprising that most plant adaptations are designed for survival with a minimum amount of water.

There are an amazing variety of ways in which desert plants are adapted to the environment. Some plants, like the tall saguaro cactus shown on page 210 and Fig. 7.46, flower only at night. Others open their pores to transpire only at night. Many have pores only on the underside of the leaf where they are in shade. Some plants have light-coloured leaves to reflect the sun.

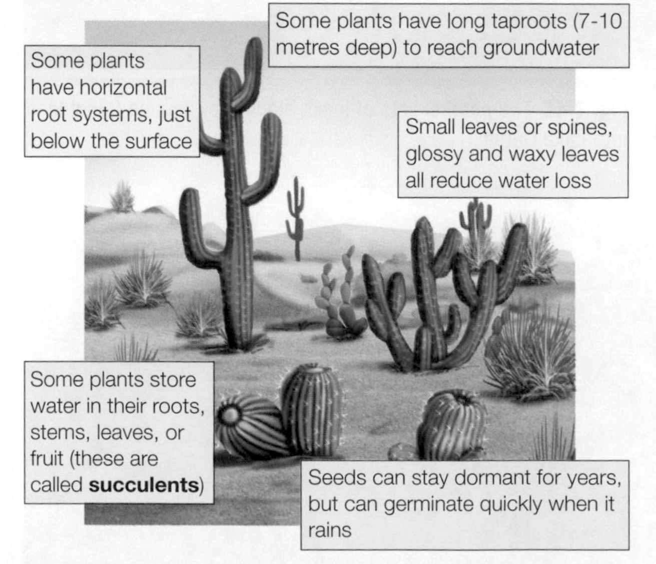

Fig. 7.46 Some characteristics of hot desert vegetation

Fig. 7.47 A prickly pear cactus (Opuntia)

This succulent stores water after rain in its circular stems to support it through long dry periods. It has no leaves but sharp spines to protect it from being eaten by animals. Some cacti have a covering of fine hairs on their stems to minimise transpiration by providing shade.

Fig. 7.48 Desert grasses and scrub

Desert vegetation is sparse. The plants in this photograph are widely spaced because they have to compete for water. Their roots are shallow and wide-spreading to catch water after rain before it evaporates. They are usually low-growing plants, because there is little moisture available for growth. Their leaves are either very small or are thorns to reduce transpiration.

Fig. 7.49 Joshua tree and desert scrub in the Mojave Desert

The Joshua tree, a yucca species, grows only in the Mojave Desert. It has roots up to 10 m deep which also spread. Its leaves are like radiating spikes. It is so well adapted to desert conditions that it can grow 15 m tall over many years.

Fig. 7.50 Water on a valley floor after rain allows denser, taller vegetation to grow

A place where water comes to the surface, usually from a spring, is known as an oasis.

Desert soils also cause difficulties for plants:

- They are either rocky or sandy, and very porous, so water passes very quickly into them after rain.
- Sandy soils are mobile (so plants can easily be covered) and loose (so plants can be uprooted). Nevertheless, some grasses can spring to life after rain, such as in the sands of the Namib Desert (Fig. 7.51).
- Desert soils are also thin and contain very few plant nutrients. This is because very little organic matter is available to decompose into them.
- Many desert soils are grey, because they contain salts that are drawn to the surface in solution after rain and then deposited at the surface when the water evaporates. Only salt-tolerant plants, such as saltbush, can grow in saline soils.
- Because of the harsh conditions, the seeds of some desert plants are forced to lie dormant for years and then flower and fruit very quickly after rain. They have a very short life cycle.

Fig. 7.51 Grass growing in sand in the Namib Desert

12 Describe and explain the nature of the soil and vegetation in the area of the Atacama Desert shown in Fig. 7.52.

Fig. 7.52 Vegetation in the Atacama Desert

The relationship in a hot desert ecosystem between wildlife, soils, vegetation and climate

Adaptions to very dry conditions

- Zebras that migrate in the wet season into the valleys of the Namib Desert are able to detect pools of water below the surface with their nostrils. They use their hooves to dig holes to reach the water.
- Fennec foxes in the Sahara and ostriches in the Namib and Kalahari Deserts can go for long periods without water. The foxes also reduce water loss by being nocturnal.
- Desert tortoises in the Mojave Desert of the US feed on plants in the spring and store the moisture they obtain to last them until the next spring.

Adaptations to the wide range of temperatures

- Fennec foxes have thick fur which keeps their bodies cool in the hot day and warm in the cold night.
- In the Mojave Desert the animals have light-coloured fur or feathers to reflect the sun. This also camouflages the animals against the sand.
- Many desert animals, like fennec foxes, are small and find shelter from the intense daytime sun by staying in cool burrows or hiding under rocks or leaves. Some are nocturnal and hunt in the cool of the night.
- Horned vipers bury themselves in sand, partly to protect themselves from the heat and partly to be camouflaged to ambush prey.
- Desert antelopes and jackals adapt to the seasonal temperature difference by having white summer coats and grey winter coats to stabilise their body temperatures.

Example desert food chains would be:

sun → primary producer → plant eater → predator carnivore

sun → desert grass → springbok → cheetah

sun → desert grass → springbok → lion

Because both lions and cheetah eat springbok, they would be linked on a food web.

Fig. 7.53 A cheetah near Keetmanshoop in the Namib Desert

The camel of the Sahara Desert and other dry areas in North Africa probably has more characteristics to enable it to survive than any other desert animal. Camels have long eyelashes and can close their nostrils for protection in sandstorms. They can go without water for months and can drink a lot of water very quickly when it is present. As food is not readily available, camels store fat in their humps. As this is digested, hydrogen is released and mixes with oxygen to form H_2O - water.

CASE STUDY

The Namib Desert

The Namib Desert extends along the Atlantic coast of south-west Africa from latitude 15° S in Central Angola, through Namibia, to 32° S in South Africa. The hot desert climate in Namibia extends along the entire coast. In the north the area with an annual rainfall of 0 to 250 mm is within 250 km of the coast, but it widens to cover the entire country south of the Tropic of Capricorn. In the north it merges eastwards with the wetter Kalahari Desert, a semi-desert. From the coast the ground rises gently to the Great Escarpment, where it ascends suddenly to over 2000 m.

Climate

The climate of the Namib Desert is hot and dry, and ranges from the very arid coast where temperatures are stable, to wetter conditions inland with greater temperature variations. The main differences are summarised in Table 7.8. If you are not able to give reasons for these climatic characteristics, refer back to pages 192–6.

Climate characteristic	Luderitz (coastal and low)	Keetmanshoop (inland and high)
Mean temperature of the hottest month (December and January, °C)	22 hot	35 very hot
Mean temperature of the coldest month (July, °C)	14 warm	6 cool
Annual range (°C)	8 small	29 large
Annual rainfall amount (mm)	33 very dry all year	166.5 dry all year
Seasonal distribution	No significant difference	

Table 7.8 Climatic differences between coastal and inland locations in Namibia

Morning fogs occur at the coast on half the days of the year and their moisture is very important in helping plants and animals to survive.

Natural vegetation

The coastal strip is sandy desert with sand dunes. Inland, the surface is a gravel plain in the north and rocky desert in the south. All these desert surfaces and the climate are difficult for plants (see pages 208–10). Common vegetation types adapted to the harsh conditions include yellow grasses, short thorny bushes, succulent aloes and occasional acacia trees where there is a water source underground. One aloe, the Quiver Tree, drops its branches in severe droughts to save moisture loss through its leaves.

Wildlife

Predators at the top of the food chain include hyena, lions, and leopards. Many obtain some of the moisture they need from the bodies of their prey.

Meerkat and ground squirrels live in burrows in the sand where the temperature is stable. They emerge well after sunrise and go underground before sunset. On emerging after the cold desert night meerkat warm their bodies by standing upright facing the sun. Both species have fur which helps regulate their body temperature against the extreme air temperature changes during the day and its colour provides a good camouflage in their habitats (Fig. 7.54). Ground squirrels eat grass and other plant stems and leaves and seeds. Their long-haired long tails provide a sunshade as they fan out when raised.

Fig. 7.54 Ground squirrels in the Namib Desert

Other herbivores include the desert elephant and antelopes, which can go for days without water, while Namibian ground squirrels can last the whole dry season.

There are also many species of well-camouflaged and adapted reptiles. Side-winding adders can ascend very steep sand dune faces.

What opportunities do tropical deserts provide for human activities?

CASE STUDY

Management and conservation in the Namib Desert

Namibia is an LEDC and its government is well aware that its desert ecosystem is extremely fragile, and that any development is likely to have serious consequences. Damaged desert vegetation takes a very long time to recover – or might never recover – but this has to be balanced against the need for the country's economic development. Namibia has mineral reserves of diamonds, uranium, copper, lead, zinc, tin, silver, and tungsten, so mining is a major activity there. It provides more than 50% of Namibia's export earnings.

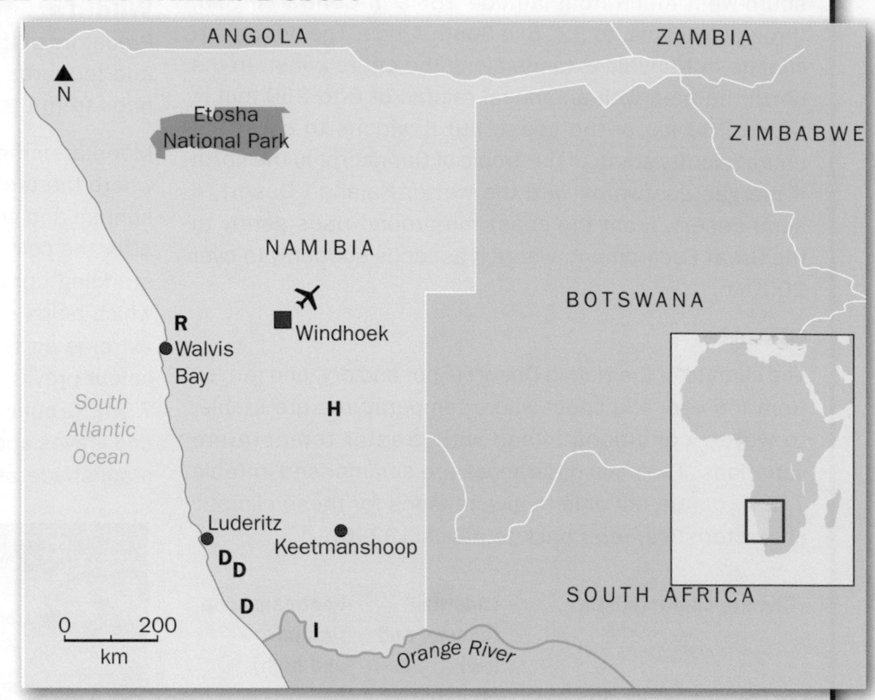

Fig. 7.55 Some of Namibia's economic resources

Mining

There are many working diamond mines near the south-west coast. Mining is not an environmentally friendly activity. Opencast mines are very destructive. This is because vast amounts of sand **overburden** have to be moved in order to access the diamonds underneath. Many animal habitats are destroyed in the process.

Fig. 7.56 A diamond mine in the Namib Desert

All of the uranium mines are in the desert, and the infrastructure they need – particularly water pipelines – has an impact on animals because they cannot cross them.

The giant Rossing opencast uranium mine is located 65 km north-east of Walvis Bay. Only 0.035% of the excavated material is actually uranium.

The scale of Rossing's operation is immense. But Namibia benefits from its taxes, local people gain salaries (97% of the employees are Namibian), and many supplies are bought from Namibian businesses. The company also provides training and education. However, there is a downside too:

- The huge demand for water for the mining industry severely strains Namibia's scarce water resources. Annual rainfall is only 30 mm in the uranium-mining area.
- The mines are so visually polluting that there is a fear that they may deter tourists from visiting the country.
- Also, they will only continue to operate while the mineral price is high enough. When they eventually close, they will leave many enormous ugly holes and a destroyed desert ecosystem behind them.

13 Look at Fig. 7.57.

a Describe the hole left behind by opencast mining. Suggest why it might be dangerous.

b What difficulties are there in doing anything about holes like this after mining has ended?

c The processing of some minerals takes place at the mines. What types of pollution could result from this processing?

Fig. 7.57 An opencast mine in the Atacama Desert

Irrigated farming

Although Namibia has a number of rivers, it has only 80 km^2 of irrigated farmland. Apart from the rivers, it has no permanent natural surface water. On its southern border with South Africa, water is taken from the Orange River for grape production.

Fig. 7.58 Irrigation along the Orange River

Another small irrigation scheme is at Hardap, where a dam across a river has created a reservoir near Mariental. A canal takes water to the fields from the reservoir. Maize, wheat, alfalfa (a crop grown for feeding livestock), grapes, and dates are grown. The reservoir is only full after exceptional rains, but the unpredictability and difficulty of the desert climate were evident in February 2011, when a flood caused extensive damage to the irrigation scheme.

Irrigation has to be managed carefully to prevent soil degradation by salination. This is caused by over-irrigation in which excess water supplied moves down into the rocks and takes in salts, which are drawn up to the surface to contaminate the soil, as explained in Fig. 7.59.

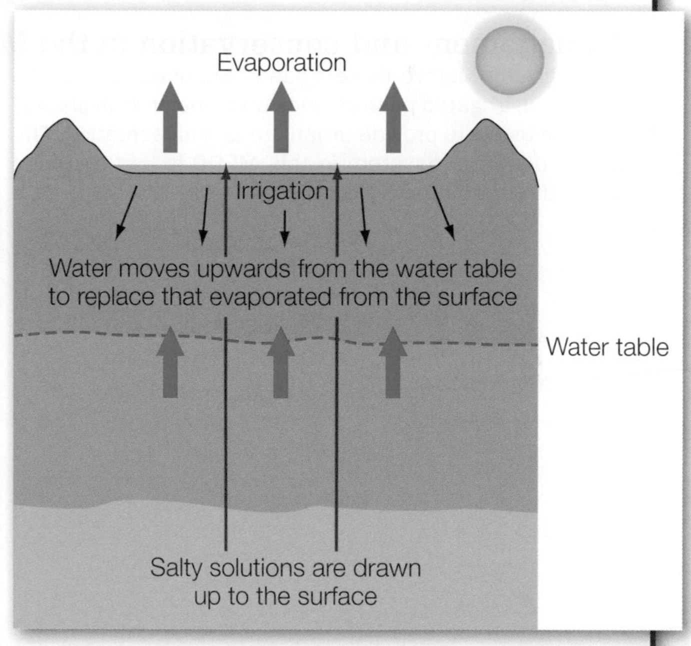

Fig. 7.59 How salinity destroys irrigated land

Tourism

Many of the Namibian population work in tourism. The sunny Namib and Kalahari Deserts, with their interesting wildlife, provide opportunities for tourism. Like mining, tourism can only thrive where adequate water is provided.

Some tourist activities are very damaging to the environment. Of greatest concern is the use of the spectacular dunes near Walvis Bay for tours on quad bikes, motor bikes and off-road vehicles (as well as trekking on foot). The damage done by trampling to the dune vegetation and the animals that depend on it may take hundreds of years to mend. The breeding of beetles, spiders and reptiles on the steep dune faces is affected, and the tyre marks ruin the appearance of the beautiful dunes.

Ecotourism is a very important activity in Namibia. The most visited area is the Etosha National Park, which covers more than 22 000 km^2. There is a vast salt pan there, with a number of waterholes on its southern side where animals come from the surrounding desert grassland and thorn scrub to drink. The pan is a dried-up lake that now rarely contains water. The big attraction of Etosha is that it contains almost every large African animal species, including lions, giraffes, elephants, and rhinoceros. Gravel roads have been laid to the waterholes where most animals can be seen. Rest camps with hotels, chalets, and camping sites have also been provided within the Park.

Fig. 7.60 An off-road vehicle on sand dunes

14 **a** Where will the point of arrival be in Namibia for many European tourists?

b What types of jobs might the people who work in the Namibian tourist industry do?

c In what other ways do tourists benefit the economy?

d How might tourism cause pollution?

CASE STUDY

Management and conservation in the Mojave Desert

Although the Mojave Desert has abandoned gold mines and small, irrigated pasturelands and croplands in places, its main use is to provide urban areas and recreation. The fragile desert ecosystem in this MEDC is just as much under threat as it is in Namibia.

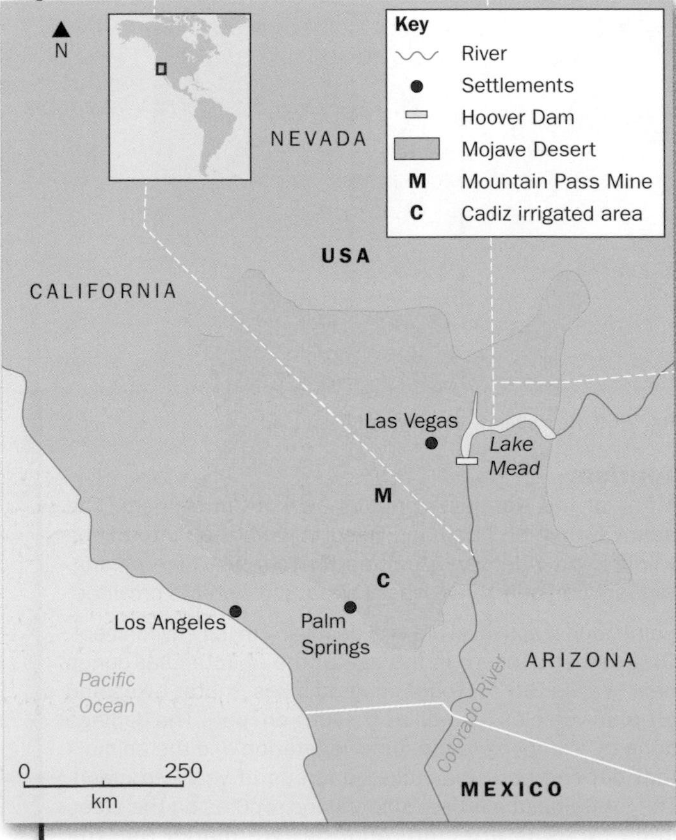

Fig. 7.61 The Mojave Desert in the south-western USA

One irrigated area is at Cadiz, where citrus fruit, grapes, and stone fruit are grown. Other farms are found along the Colorado River. There are also cattle ranches with irrigated pastures.

The first big invasion of the desert occurred when Palm Springs became a popular retreat for Hollywood film stars. The population of Los Angeles soon discovered the delights of spending holidays in Palm Springs in the winter, when the climate there is more pleasant than it is in Los Angeles. Palm Springs now has more than 43 000 inhabitants – mainly tourists and retired people. With its watered golf courses and palm trees, it is like a green oasis surrounded by brown desert and mountains.

Fig. 7.62 Sprinklers irrigating pasture in the Mojave Desert

Fig. 7.63 A hotel in Palm Springs

However, Palm Springs is dwarfed by the massive urban development of Las Vegas. This city, surrounded by desert, has a population of more than two million within its metropolitan area. It developed as a holiday playground for adults.

Fig. 7.64 Las Vegas

Lake Mead, behind the Hoover Dam on the Colorado River, supplies 90% of Las Vegas's water needs. However, after years of drought and high demand, it was only half full by 2009. Some people were forecasting that it would be completely dry by 2020 if its water continued to be used at the same rate. Water used in hotel kitchens and bedrooms in Las Vegas is already being recycled.

The Lake Mead reservoir also serves Nevada, California, Arizona, and northern Mexico. It is estimated that about 80% of the water supplied by the Colorado River is used for agriculture.

A lot of water is also being extracted for agriculture from underground aquifers, using wells. As a result, groundwater levels in the rocks are dropping quickly.

Urban areas are spreading rapidly, as Fig. 7.65 suggests.

The desert is also used for military bases. As in the Namib Desert, vegetation is being damaged by off-road vehicles, and wind and water erosion are occurring.

Discreet desert eco homes planned for the Mojave Desert

A new desert home community has been planned for La Quinta, California, in the Mojave Desert – just outside Palm Springs. The desert landscape, covered in spiny Joshua trees, will be respected in this new discreet eco-community. Homes will be constructed to carefully minimise any environmental disturbance. Each home has been designed for the hot desert climate of the Mojave and includes solar power.

Fig. 7.65 New homes planned in the Mojave Desert

15 Do you consider the building of homes in the desert to be a sustainable activity (one that will not damage the environment)? Explain your answer.

Discussion point

How important is the sun?

RESEARCH www.mongabay.com is a useful website for obtaining information about environmental issues.

8 Development

This chapter covers the following Cambridge IGCSE® and O Level topics:

- **3.1 Development**

- What are More Economically Developed Countries (MEDCs) and Less Economically Developed Countries (LEDCs)?
- Why do people in some parts of the world live in extreme poverty while others are very rich?
- How can economic development be measured?
- How do people's jobs change as a country becomes more developed?
- What is globalisation?
- What are transnational corporations and how do they affect our lives?

In this chapter you will learn about:

- the ways that the lives of people in different countries are not equal and how these inequalities can be measured
- how these inequalities have occurred
- the different types of jobs that people do and how they are classified
- how globalisation has affected people's lives in wealth and employment, culture, communication and migration
- the giant companies known as transnational corporations (TNCs).

LEARNING TIP The terms More Economically Developed Country (MEDC), Less Economically Developed Country (LEDC), and Newly Industrialised Country (NIC) are not in the Cambridge syllabus but they are used in Cambridge examination questions. There are no modern definitions of these terms but they are useful when used in a general way. MEDCs, like those in Western Europe and North America, have high incomes, low birth rates, high living standards and strong infrastructure. LEDCs, like those in Africa, have low incomes, high birth rates, low living standards and weaker infrastructure. NICs include the more recently industrialised countries such as Brazil, Mexico, Thailand and China.

In geography, the word 'development' is generally used to mean the way that a country becomes more advanced in its economy, infrastructure and the economic and social well-being of its citizens. This includes:

- standard of living - to do with money and wealth
- quality of life - to do with the things that affect a person's well-being and happiness.

Measuring development

Gross domestic product (GDP) per capita is calculated by taking the total value of the goods and services produced by a country in any one year and dividing it by the population of the country. It is expressed as $US per person so that countries can be compared. It measures standard of living but not quality of life. It does not take into account goods produced by subsistence farmers and people working in the informal economy and it may underestimate the production of poorer countries.

The Human Development Index (HDI) is an example of a composite index. It takes into account a country's:

- GDP per capita
- adult literacy and educational provision
- life expectancy at birth.

It was developed by Indian and Pakistani economists and is published annually by the United Nations. It is given as a number between zero (very low) and one (very high). It reflects standards of living and quality of life.

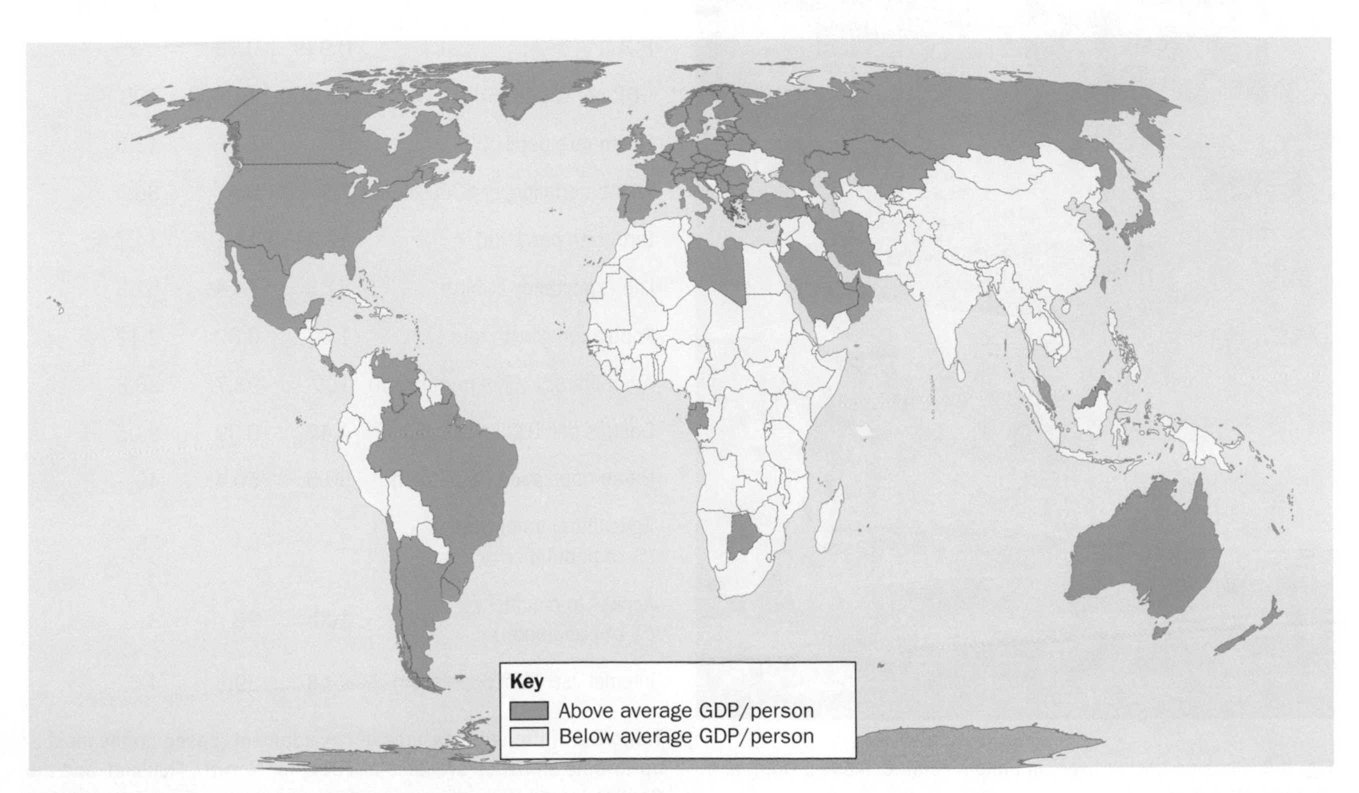

Fig. 8.1 World map of average GDP per person

Reasons for inequalities

Why are some countries much richer than others?

- **Location** - landlocked countries have generally developed more slowly than countries with a coastline.
- **Size** - many small countries have developed more slowly than large countries.
- **Natural hazards** - a country that experiences natural hazards such as earthquakes is less likely to develop rapidly.
- **Climate and soil** - tropical countries have developed more slowly than temperate countries. Tropical soils tend to be infertile and reduce agricultural production. Tropical countries suffer from pests and diseases, which limit population growth and agricultural production.
- **Stable government** - many Europeans and North Americans believe that their democratic political system has stimulated economic growth. Unstable government, poor law and order, and corruption can lead to civil unrest and delay economic and social progress.
- **Economic policies that encourage growth** - investment may come from credit, savings or transnational corporations (TNCs).
- Ability to **trade** - poor countries have traditionally suffered from unfair trading practices such as tariffs and import duties.
- **Population issues** - in Stage 2 of the Demographic Transition Model (DTM; see Chapter 1), if economic development does not keep up with population growth, the increasing population will not have enough food, housing, jobs, or services. Governments can achieve this either by encouraging economic growth or by reducing birth rates to limit the amount of population growth. Once a country reaches Stage 3 of the DTM, the growth in population starts to provide a large and productive workforce and a more wealthy market for goods and services.

Fig. 8.2 Housing beside rivers in Phnom Penh, Cambodia (top), and Chicago, USA (bottom)

Some of these points explain why some countries in Africa are often much poorer than some countries in Europe. However, there are many exceptions. For example, Singapore is a small, island nation on the Equator but it is a wealthy MEDC. Japan suffers from volcanic eruptions, earthquakes, tsunamis, and typhoons but it is one of the wealthiest countries in the world.

Countries at different stages of development

Table 8.1 shows information about three countries. Norway has one of the highest HDIs in the world, Thailand became an NIC in the 1990s, and the Central African Republic has one of the lowest HDIs in the world.

	Norway	Thailand	Central African Republic
HDI	0.949	0.73	0.35
GDP per capita ($US)	69 000	16 800	700
Death rate per 1000	8.1	7.9	13.5
Infant mortality per 1000 births	2.5	9.4	88.4
Birth rate per 1000	12.2	11.1	34.7
Life expectancy at birth	81.8	74.4	52.3
Population growth rate (%)	1.07	0.32	2.12
Adult literacy (% of population)	100	96.7	36.8
Doctors per 1000 population	4.42	0.39	0.05
Urban population (% of total)	80.5	50.4	40
Agricultural employment (% of population)	2	9	58
Access to electricity (% of population)	100	99	3
Internet use (% of population)	96.8	39.3	4.6

Table 8.1 Different measures of development (based on the most up-to-date statistics available in 2017) for Norway, Thailand, and the Central African Republic

> **1** Study Table 8.1.
> **a** Which figures stand out as being different to the general pattern?
> **b** What other measures of development not shown in Table 8.1 would be good indicators of development?

> **RESEARCH** The United Nations Human Development Report can be found at: http://hdr.undp.org/en/2016-report. Look at the countries list at the end of the report. How does your country compare with others?

Regions at different stages of development

The whole of a country does not develop at the same rate. The same differences that are found between different countries are often found within a single country. It is often the central, more accessible areas of a country which develop fastest and its remoter areas which develop more slowly. Fig. 8.3 shows the more and less developed regions of the European Union (EU). Notice that the poorer areas are at the edges of the EU, such as Wales, Portugal, and southern Italy.

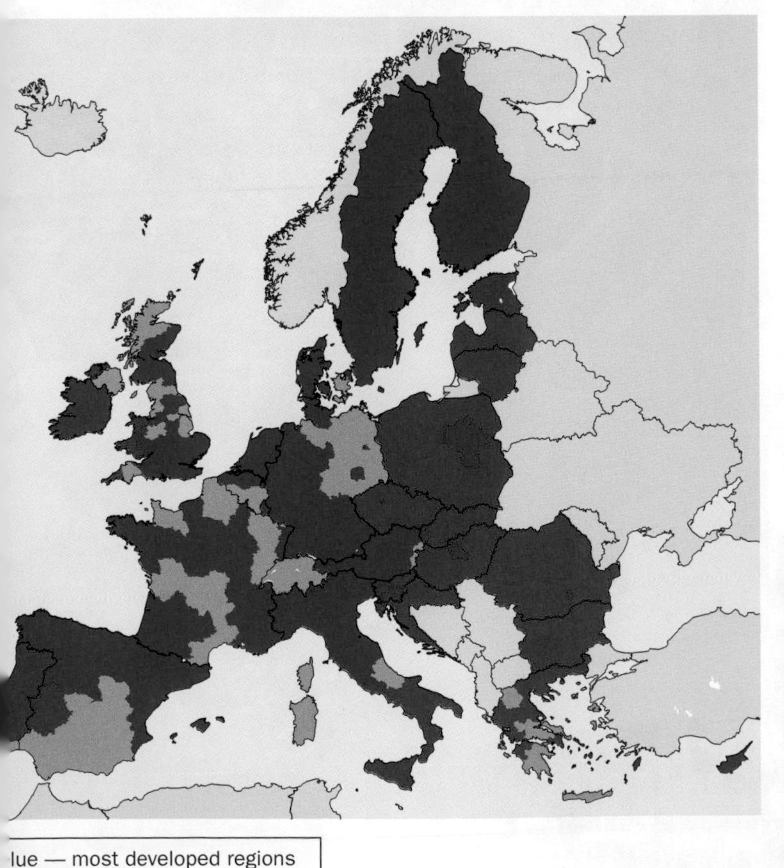

Fig. 8.3 Most and least developed regions of the European Union (source: Cranberry Products at English Wikipedia)

The features of these regions are shown in Table 8.2.

Central ("core") areas	Regional ("peripheral") areas
More urbanised	More rural
More tertiary and quaternary industry (see the next section of this chapter)	More primary industry
Higher incomes and more wealth	Lower wages and higher unemployment
Higher living costs	Lower living costs
Inward population migration	Outward population migration, especially of young educated workers - a "brain drain"
Strong transport systems	Poor accessibility
Home of government and social elite	

Table 8.2 Features of central areas and regions

The rich, core areas tend to get richer and the peripheral areas poorer. The European Union has given financial grants to poorer regions in an attempt to change this. This extra money is to develop transport facilities and support industrial development.

In China, the coastal areas have developed rapidly, for example in the port city of Shanghai. Inland areas lag behind, and there is population migration from inland areas to the coastal areas.

There are also great contrasts in wealth between people in the same region. The greatest contrasts between rich and poor are often in LEDCs.

Fig. 8.4 Luanda, Angola, 2015 - the contrast between rich and poor

Industrial sectors

The jobs that people do can be divided into four groups, or sectors:

Sector	Definition	Examples
Primary	Collection or production of natural resources, food and raw materials directly from the land or sea	Farming, fishing, forestry, mining, quarrying
Secondary	Processing, manufacturing, and assembly of the products we need	Steelmaking, car assembly, paper making, food manufacture such as baking
Tertiary	Providing a service	Health, education, retail, transport, banking, insurance
Quaternary	Modern, hi-tech manufac-turing and service industries	Aerospace, computer science, pharmaceuticals, biotechnology, research and development

Table 8.3 The four sectors of industry

National employment statistics do not always recognise the quaternary sector and quaternary jobs are sometimes included in the secondary or the tertiary sectors.

2 Classify these jobs into the four sectors of employment – primary, secondary, tertiary and quaternary:

- Nurse
- Shop worker
- Worker in a car factory
- Miner
- Teacher
- Accountant

RESEARCH Conduct a survey in your class about the employment sectors of your classmates' families. Convert your results into percentages of the total and then plot them, as either a pie chart or a divided bar graph.

Discussion point

Classifying jobs is not always easy. Imagine a plumber employed in the building of new houses and another plumber repairing faults in existing houses. Are the two plumbers in the same employment sector?

Fig. 8.5 Primary, secondary, tertiary and quaternary industry

Employment structures

The proportion of people working in primary, secondary, tertiary and quaternary activities in any country or region is called the **employment structure**.

As a country becomes more economically developed, the percentage of its population employed in primary industries decreases, while the percentage employed in tertiary industries increases. The percentage employed in secondary industries increases at first, but then decreases (as the tertiary sector continues to grow). The actual percentage figures vary from country to country. For example, today employment in secondary industry is 17% in the Netherlands, 24% in Germany and 28% in Italy – all European MEDCs.

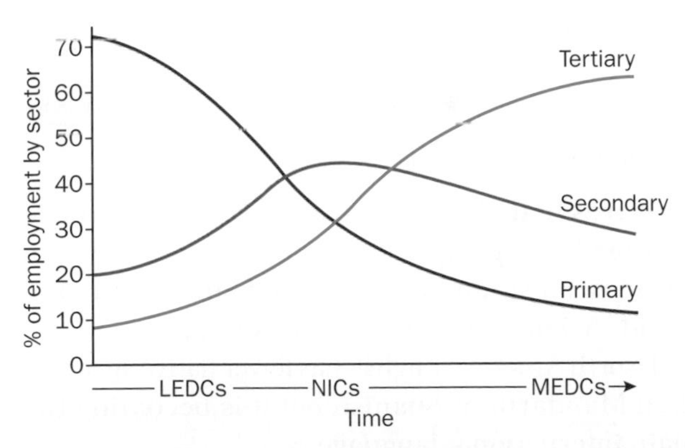

Fig. 8.6 How employment in industrial sectors changes with time as a country becomes more developed

3 Look at the employment statistics in Table 8.4.

a Plot the data as a pie chart (or divided bar graph) for each country.

b Which of the countries is:

i an LEDC?

ii an NIC?

iii an MEDC?

c How might the employment structure of Malaysia change in the future?

Country	Primary %	Secondary %	Tertiary %
Australia	4	21	75
Bangladesh	47	13	40
Malaysia	11	36	53

Table 8.4 The employment statistics for three countries

The quaternary sector grows after industrialisation. This is shown in Fig. 8.7.

4 Describe the changes in industrial structure shown in Fig. 8.7.

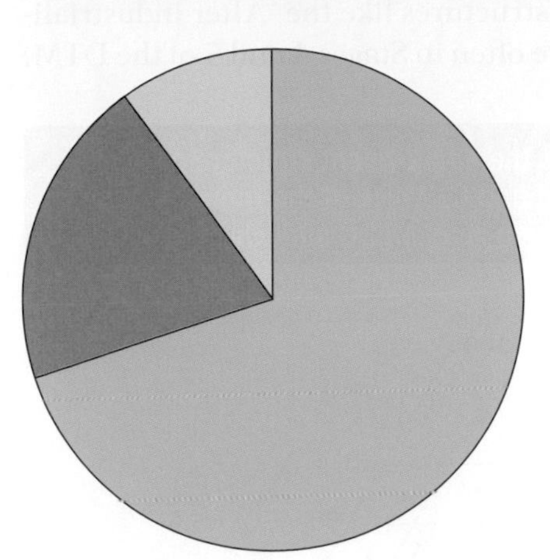

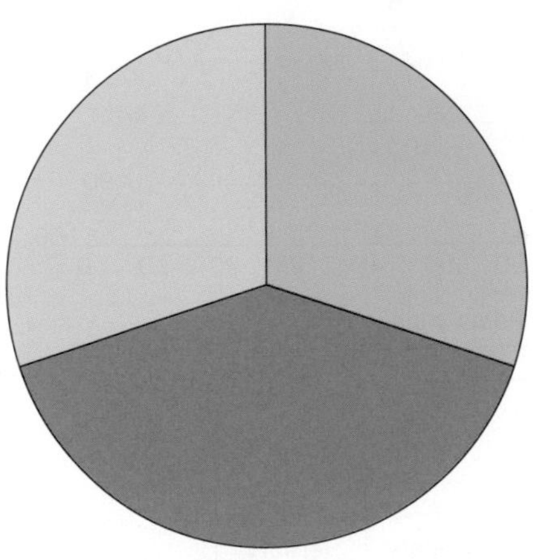

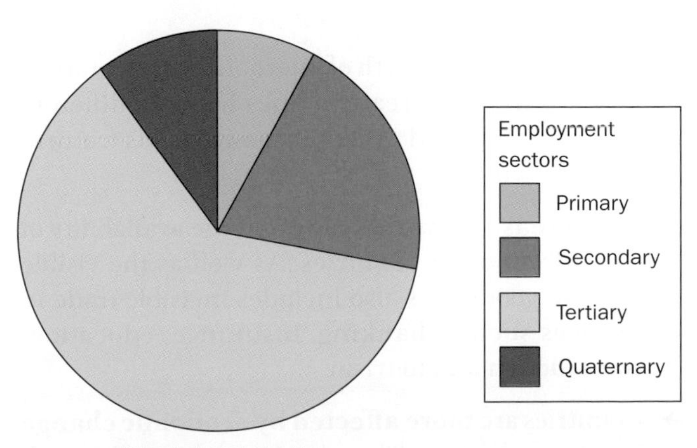

Fig. 8.7 The relative size of employment sectors before, during and after industrialisation

In Chapter 1 the Demographic Transition Model (DTM) was described. Countries with employment structures like the "Before industrialisation" diagram in Fig. 8.7 are generally in Stages 1 and 2 of the DTM. Countries with employment structures like the "Industrialisation" diagram are often NICs in Stage 3 of the DTM. Countries with employment structures like the "After industrialisation" diagram are often in Stages 4 and 5 of the DTM.

5 Fig. 8.8 shows the employment structure of three countries: A, B and C. Which of the three countries is an LEDC, which is an MEDC and which is an NIC?

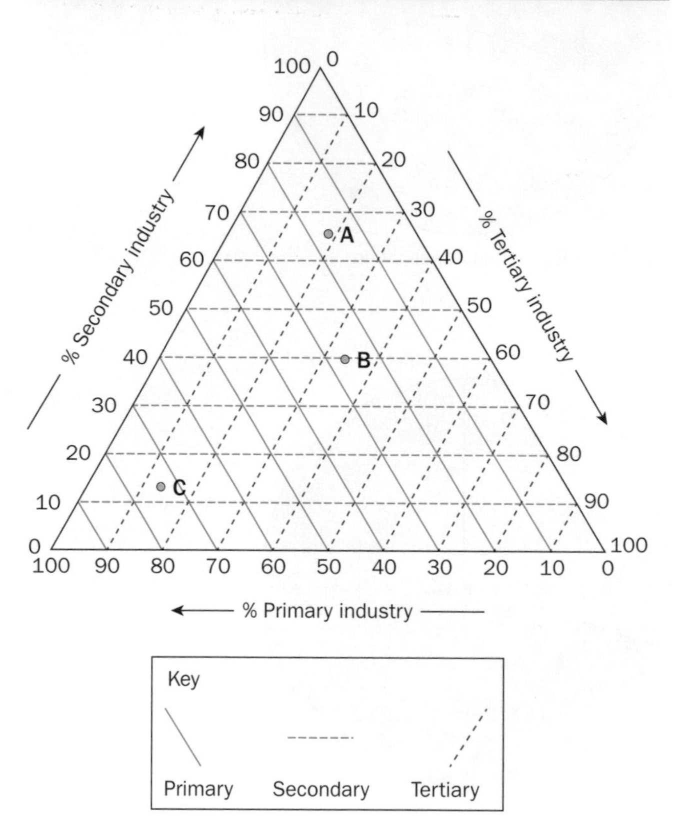

Fig. 8.8 The employment structure of three countries

Globalisation

Globalisation is the growth of international integration, in other words the increase in links between different parts of the world and different countries. Its features are the following:

- An **increase in world trade** and the availability of goods from other countries. As well as the visible trade in goods, this also includes invisible trade in services such as banking, insurance, education, construction and tourism.
- Countries are **more affected by economic change in other countries**. There has been a general growth in trade except for times such as the world financial crisis in 2008–9. This began in the property market in the USA and spread around the world.

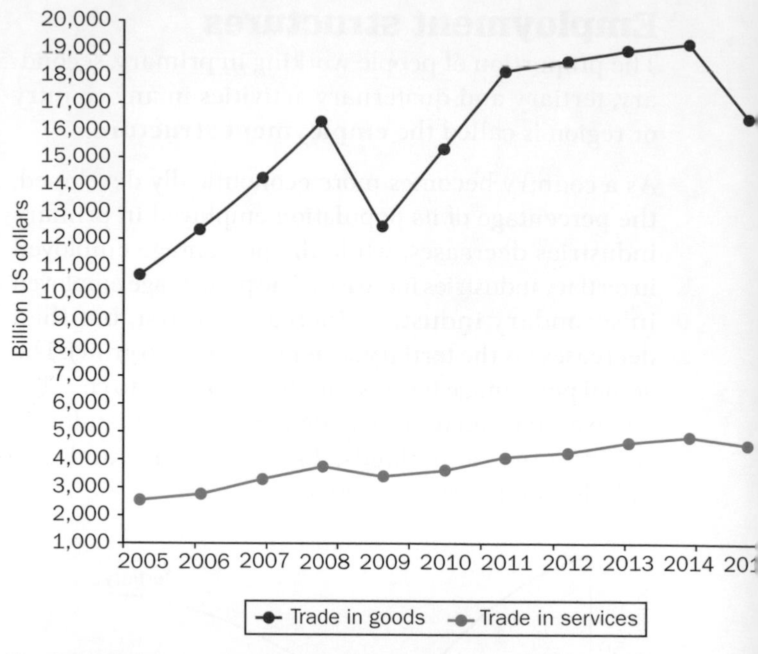

Fig. 8.9 The growth in world trade between 2005 and 2015

- **Cultures in different countries becoming more similar** in languages, food and clothing. Western fashions, music and products are found all over the world. Asian food is now very popular in Europe and North America. English has fewer native speakers than Mandarin or Spanish but it is becoming the main international language.
- There has been a **change in location of some manufacturing industries** from MEDCs such as the UK, USA, and Japan to LEDCs and NICs. This has led to job losses in some countries and new jobs in others.
- **World-wide environmental effects** such as air pollution and global warming. The threat of global warming (see Chapter 10) and atmospheric pollution shows how the actions of one country may affect others. This has led to international action such as the 2016 Paris Agreement, dealing with greenhouse gas emissions. By August 2017, 195 countries had signed the agreement. The 1987 Montreal Protocol has led to international action which has been effective in protecting the ozone layer.
- **International population migration** has increased and people are more likely to travel between countries (see Chapter 1).
- Some of the world's great cities – such as London, New York, Hong Kong, Paris, Singapore, Tokyo, Shanghai, Chicago, Dubai, and Sydney – have become important beyond the boundaries of their own country. They are called **world cities**. Transnational corporations (TNCs, described later in this chapter) have their headquarters in these cities, from where they control their businesses around the world.

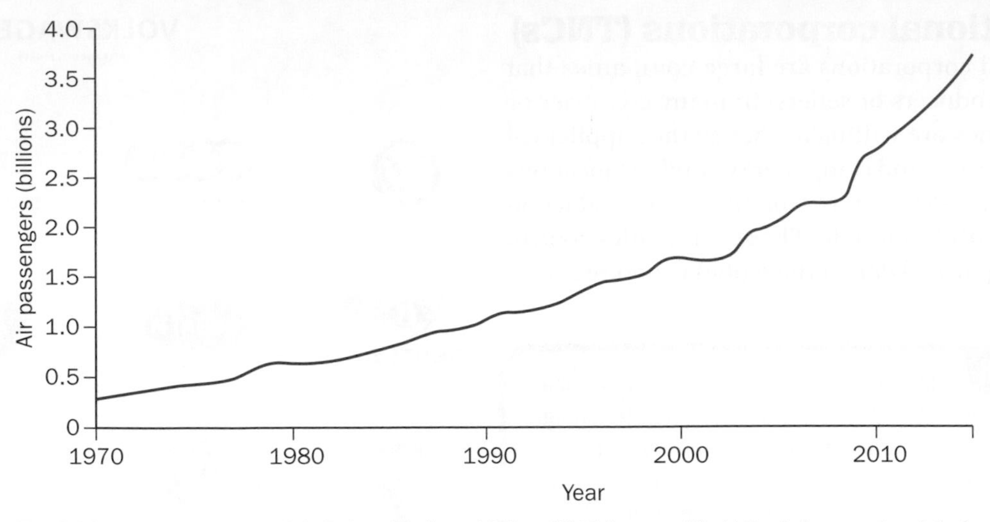

Fig. 8.10 Annual growth in global air traffic from 1970 to 2015 (source: World Bank Group using data from International Civil Aviation Organization, Civil Aviation Statistics of the World and ICAO staff estimates)

Factors which have increased globalisation

- The growth in transnational corporations (TNCs). This is described later in this chapter.
- Advances in transport. This has been particularly so in air travel, as Fig. 8.10 shows. Air travel has become cheaper and accessible to more people.
- Containerisation of freight has allowed large volumes of goods to be moved efficiently.

Fig. 8.11 The world's busiest container port, Shanghai, China

- International organisations, such as the European Union, the United Nations and the Commonwealth of Independent States, involve co-operation between countries in economic and military activities.
- Advances in communications infrastructure, such as the internet and cell phones, allow the rapid movement of knowledge and information.

Fig. 8.12 A woman speaking on a mobile cell phone in rural KwaZulu-Natal, South Africa

Impacts of globalisation

Local level

Discussion point

How does globalisation affect you and your classmates? Discuss this under the following headings: (a) the food you eat and where it is from, (b) where the people you know were born and the languages they can speak, (c) where you go for holidays, (d) the music you listen to and the clothes you wear, (e) the people you communicate with and where they live. Remember that this will be very different from the time when your parents were at school.

National and global levels

6 Using the information in this chapter, describe the impacts of globalisation at the national and global levels using the following headings: (a) the environment, (b) industry and jobs, (c) international organisations.

Transnational corporations (TNCs)

Transnational corporations are large companies that operate (as producers or sellers) in many countries or continents. They are willing to change the suppliers of their raw materials and components – and the locations of their activities – to wherever conditions for production or sales are most favourable. These companies control an increasing proportion of the global economy.

> **RESEARCH** List the TNCs that operate in your local area. Remember that some of them could operate through petrol stations or supermarkets. Also remember that not all are involved in manufacturing. Companies like the travel company Tui (which owns Thomson Holidays and First Choice) and the accountancy and audit firm PriceWaterhouseCoopers are service industry TNCs.

Fig. 8.13 The global brands of one TNC, Volkswagen

The world's top ten companies (as measured by their sales) are shown in Table 8.5. They each have annual sales that are greater than the gross domestic product (GDP) of many entire countries.

TNCs have a strong influence on LEDCs, where they often locate activities like production. There has been some criticism of this, but the presence of TNCs in LEDCs can lead to both advantages and disadvantages for those countries.

Advantages of a TNC for the LEDC

- A TNC provides jobs for local people.
- It provides a guaranteed income for people.
- It improves people's skills.
- It brings in foreign currency, which helps the country to develop.
- The increased employment also increases the demand for consumer goods in the LEDC and helps other industries to develop there.
- It can lead to the development of local raw materials, such as mining minerals or growing crops.
- It often leads to the development of infrastructure projects, such as roads, dams, airports, schools, and hospitals.

Rank	Name	Industry	Sales (million US$)	Number of employees	Location of headquarters
1	Walmart	Retail	485 873	2 300 000	USA
2	State Grid	Utilities	315 199	926 067	China
3	Sinopec Group	Petroleum refining	267 518	713 288	China
4	China National Petroleum	Petroleum refining	262 573	1 512 048	China
5	Toyota	Motor vehicles	254 694	364 445	Japan
6	Volkswagen	Motor vehicles	240 264	626 715	Germany
7	Royal Dutch Shell	Petroleum refining	240 033	89 000	Netherlands UK
8	Berkshire Hathaway	Insurance	223 604	367 700	USA
9	Apple	Computers and office equipment	215 639	116 000	USA
10	Exxon Mobil	Petroleum refining	205 004	72 700	USA

Table 8.5 The world's top ten companies, as measured by their sales, in 2016

Disadvantages of a TNC for the LEDC

- Most of the profits go abroad and are not reinvested in the LEDC.
- The numbers of local people employed can be small.
- The TNC might suddenly decide to leave the LEDC, if conditions inside or outside the country change. This decision is made outside the LEDC.
- Raw materials, such as minerals, are often exported and not processed in the LEDC.
- Levels of pay are lower than elsewhere in the world.
- The operations of the company may cause environmental damage.

Impacts in MEDCs

- Areas involved in manufacturing industries have suffered when TNCs have moved production to places with cheaper labour, often in LEDCs. This has led to unemployment and the economic decline of some regions in an MEDC.
- TNCs have often located their headquarters in "world cities" from where global brands are managed. This has increased skilled employment in management, accountancy, legal services, marketing, and IT. Economic growth has occurred in these cities.

CASE STUDY

Toyota – a leading motor vehicle manufacturer

Toyota worldwide

The Toyota Motor Corporation of Japan has around 40% of the Japanese motor vehicle market, but it manufactures and sells its vehicles in 170 countries. It is the world's biggest car manufacturer (see Table 8.6) and the world's fifth largest company by the value of its sales (see Table 8.5). It conducts its business with 51 overseas manufacturing companies in 26 countries (see Fig. 8.14 and Table 8.7).

The country outside Japan in which most Toyota vehicles were assembled in 2016 was the USA, with a production of more than 1 380 000. China was the second largest overseas producer, with nearly 1 100 000. With more than 600 000, Canada ranked third. By contrast, only 127 000 vehicles were assembled in the whole of the continent of Africa.

Of the ten Toyota plants in China, three assemble vehicles while the other seven make engines and components to supply the assembly plants.

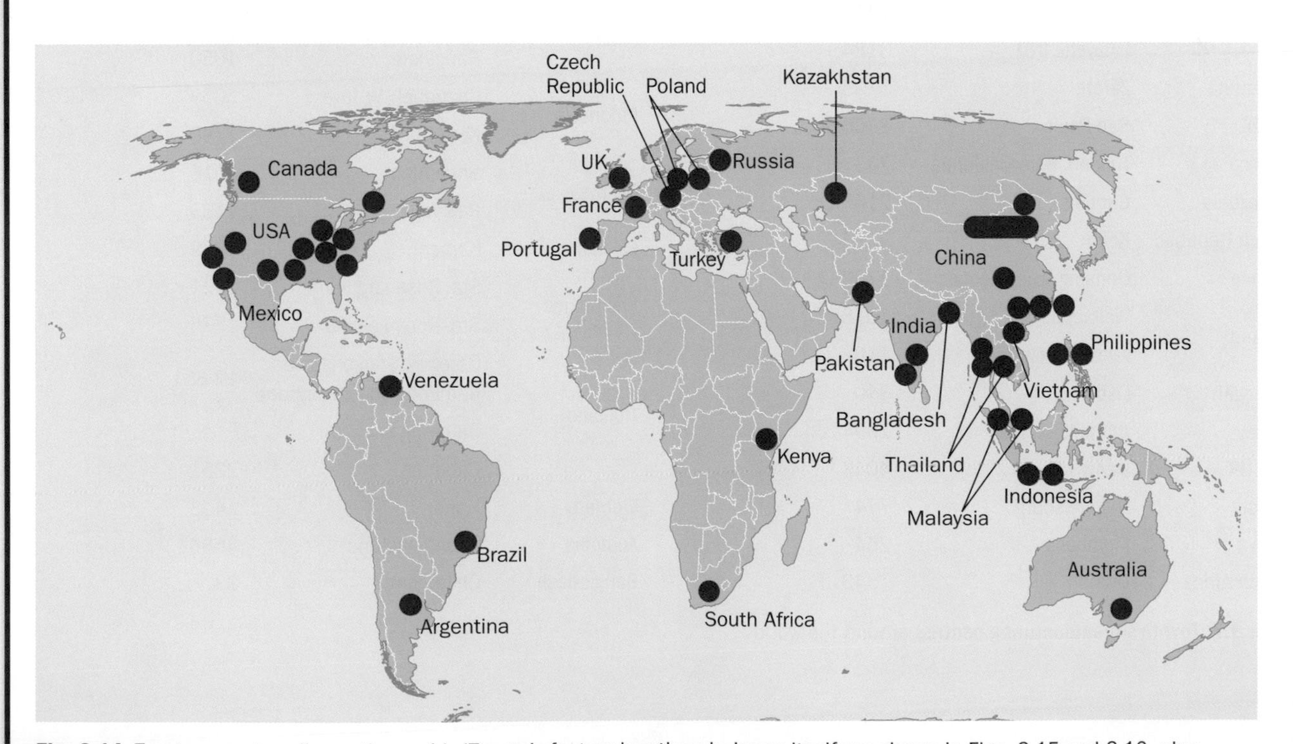

Fig. 8.14 Toyota operates all over the world. (Toyota's factory locations in Japan itself are shown in Figs. 8.15 and 8.16, plus Tables 8.8 and 8.9.)

Rank	Company	Country of headquarters	Number of vehicles produced
1	Toyota	Japan	10 084 000
2	Volkswagen	Germany	9 872 000
3	Hyundai	South Korea	7 988 000
4	General Motors	USA	7 486 000
5	Ford	USA	6 396 000
6	Nissan	Japan	4 544 000
7	Fiat Chrysler	Italy/USA	4 865 000
8	Honda	Japan	4 544 000
9	Suzuki	Japan	3 034 000
10	Renault	France	3 033 000

Table 8.6 The top ten motor vehicle manufacturing companies in 2016

Country		No. of employees
Canada	Delta BC	292
	Woodstock and Cambridge ONT	5919
The USA	Long Beach CAL	533
	Georgetown KEN	7487
	Long Beach CAL	28
	Troy MO, Jackson TN	947
	Buffalo WV	1124
	Princeton IND	4204
	Huntsville AL	796
	San Antonio TEX	2415
	Lafayette IND	3184
Argentina	Zárate	3105
Brazil	Sao Paulo	3306
Mexico	Tijuana, Baja California	743
Venezuela	Caracas	2163
Czech Republic	Kolín	3364
France	Onnaing-Valenciennes	3732
Poland	Walbrzych	2078
	Jelcz-Laskowice	716
Portugal	Lisbon	340
Turkey	Arifiye, Sakarya	2894
The UK	Derby	4043
Russia	St Petersburg	774
Kenya	Mombasa	254
South Africa	Durban	7343

Country		No. of employees
China	Tianjin Jinfeng	385
	Tianjin Fengjin	763
	Tianjin FAW	1898
	Tianjin Forging	235
	Tianjin FAW	12 407
	Changchun	783
	Tianjin FAW	216
	Guangzhou	1300
	Sichuan	2374
	Guangzhou	6321
India	Bangalore	4433
	Bangalore	1050
Indonesia	Cikampek, W Java	5069
	Karawang	7790
Malaysia	Shah Alam	2516
	Rawang	7183
Pakistan	Karachi	1879
Philippines	Sta. Rosa Laguna	1421
	Sta. Rosa Laguna	1375
Thailand	Gateway, Samrong and Ban Pho (Chachoengsao)	12 651
	Samrong	
	Chonburi	2251
Vietnam	Hanoi	1408
Australia	Altona, Victoria	4586
Bangladesh	Chittagong	83

Table 8.7 Toyota's manufacturing centres around the world

Toyota in Japan

Toyota's core production centre is in Toyota City, on the east coast of Japan's main island of Honshu (near Nagoya). Toyota has 12 separate factories in the area, which employ a total of more than 43 000 people. The factories work together to make the components, and then assemble the various models. The company has developed great strength in technological skill and research and development, and has a highly motivated workforce.

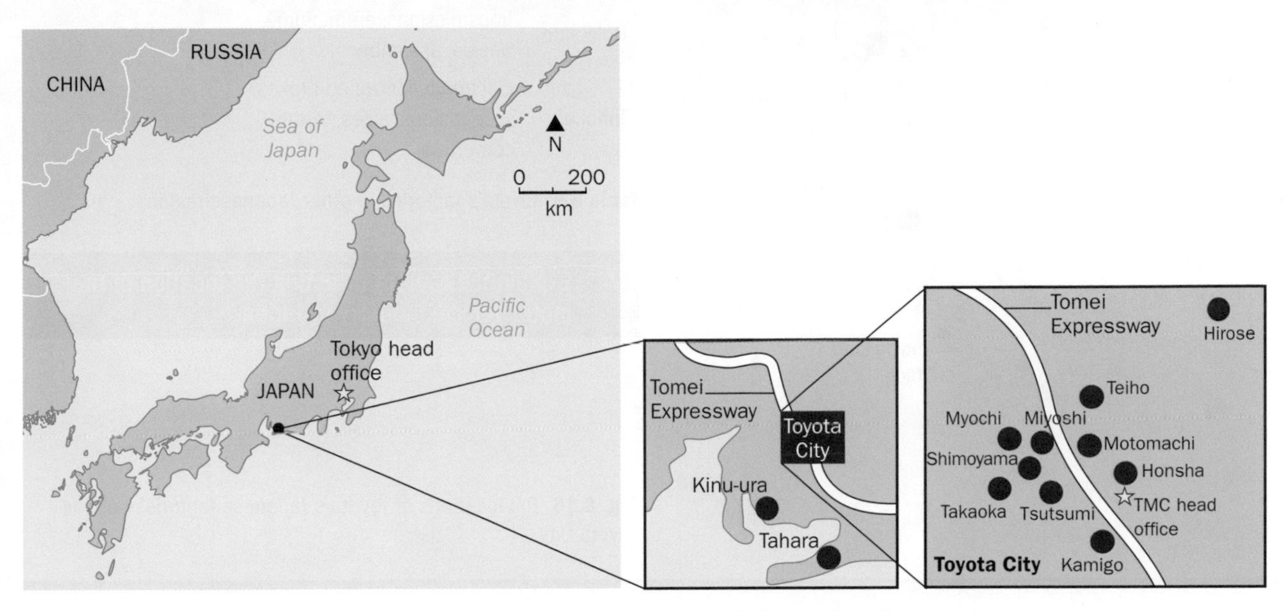

Fig. 8.15 The location of the Toyota City production centre in Japan, plus its individual factories

Factory	Products	Annual vehicle production
Honsha	Forged parts, hybrid system parts	
Motomachi	Assembly	80 000
Kamigo	Engines	
Takaoka	Assembly	267 000
Miyoshi	Transmission-related parts, cold-forged and sintered parts, engine-related parts	
Tsutsumi	Assembly	374 000
Myochi	Powertrain-related suspension cast parts, powertrain-related suspension machined parts	
Shimoyama	Engines, turbochargers, catalytic converters	
Kinu-ura	Transmission-related parts	
Tahara	Assembly and engines	321 000
Teiho	Mechanical equipment, mouldings for resin, and casting and forging	
Hirose	Research and development and production of electronic control devices	

Table 8.8 The Toyota City production centre in Japan

Steel for use in the vehicle manufacturing process is produced nearby at Nagoya, and further away at Kimitsu. The land in this area is flat – an issue in mountainous Japan.

The completed vehicles are shipped to the densely populated Kanto District, which includes Tokyo-Yokohama. Previously the port at Kinu-ura was used, but today the vehicles are transported 29 kilometres to Nagoya Port, from where a fleet of four specialist vessels is used to ship the vessels approximately 350 kilometres to Kanto District.

Toyota also has plants in northern Honshu (Tohuku) and on the islands of Hokkaido and Kyushu.

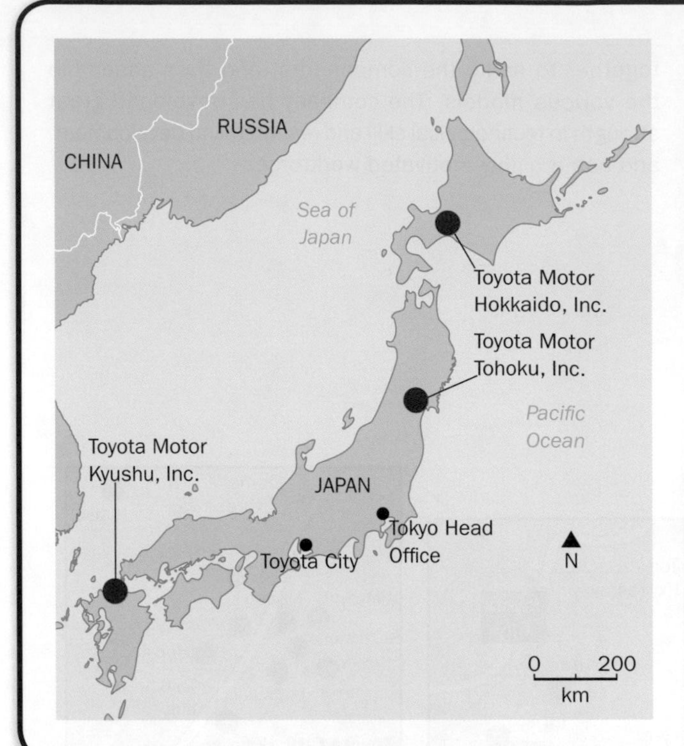

Fig. 8.16 The locations of Toyota's Japanese factories outside Toyota City

Factory	Products	Annual vehicle production
Kyushu	Assembly, engines, hybrid system parts	285 000
Hokkaido	Transmissions, aluminium wheels, assembly	80 000
Tohoku	Electronic controlled brakes, suspensions, axles, torque converters	

Table 8.9 Toyota's factories in other Japanese regions

7 Explain the factors affecting the location of the motor vehicle industry in Japan.

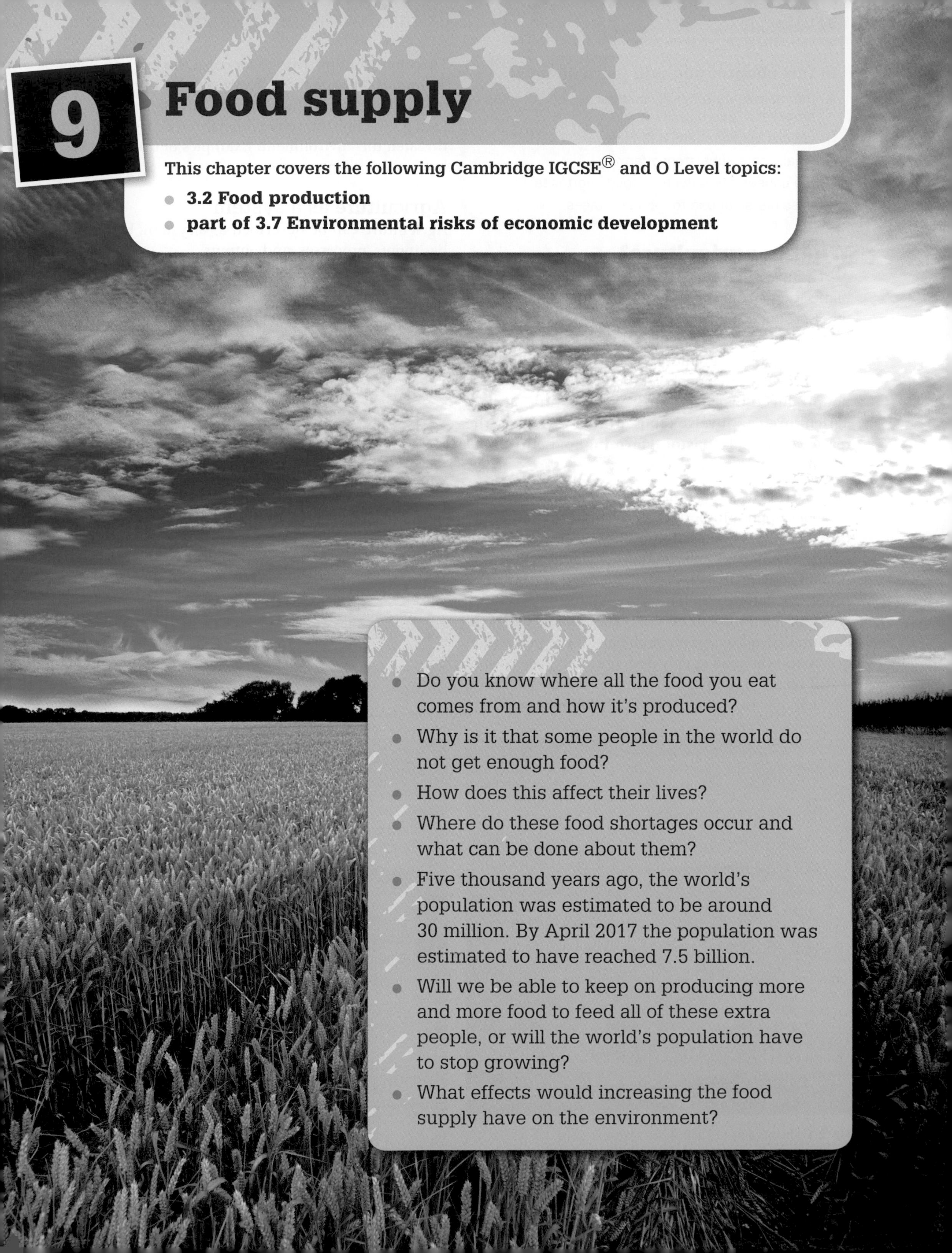

9 Food supply

This chapter covers the following Cambridge IGCSE® and O Level topics:

- **3.2 Food production**
- **part of 3.7 Environmental risks of economic development**

- Do you know where all the food you eat comes from and how it's produced?
- Why is it that some people in the world do not get enough food?
- How does this affect their lives?
- Where do these food shortages occur and what can be done about them?
- Five thousand years ago, the world's population was estimated to be around 30 million. By April 2017 the population was estimated to have reached 7.5 billion.
- Will we be able to keep on producing more and more food to feed all of these extra people, or will the world's population have to stop growing?
- What effects would increasing the food supply have on the environment?

In this chapter you will learn about:

- the main features of agricultural systems: inputs, processes, and outputs
- large-scale commercial farming
- small-scale subsistence farming
- the causes and effects of food shortages
- possible solutions to food shortages.

What is agriculture?

Agriculture is farming. It is the artificial cultivation of plants (crops) and the rearing of animals for food and other products. Our distant ancestors did not practise agriculture. Instead, they lived by scavenging on dead animals, gathering wild plants, and hunting. Today there are very few people left on Earth whose food is not produced by agriculture. As the definition says, agriculture can involve crops (**arable farming**), the rearing of animals (**pastoral farming**), or both (**mixed farming**).

The goal of continuing to develop agriculture to increase the food supply is important for the future of the human race. In 1798, Thomas Malthus (a British scholar) said that, mathematically, the human population will always increase faster than the available food supply. As a result, unchecked population growth in any particular geographical area - or on the planet as a whole - will eventually lead to serious problems. If the population grows and the food supply does not keep pace with it, it will result in increased poverty and even starvation - which may then stop or slow population growth.

Therefore, finding ways to increase the food supply is an important issue today. However, this must be done in a way that is sustainable. Some attempts to increase food production interfere with natural ecosystems and threaten the environment. Examples of this are given later in this chapter.

Agriculture as a system

Agriculture can be described as a system, because it has inputs, processes, and outputs.

Physical inputs

These inputs are provided by nature:

- Climate: temperatures, rainfall, sunshine
- Soil
- Land and its relief

Human inputs

These inputs are provided by people:

- Capital (money)
- Labour
- Machinery and tools
- Seeds
- Social structures
- Government influence
- Market influence
- Fertilisers, pesticides and herbicides
- Irrigation

Fig. 9.1 In this area of South Africa's Western Cape, the natural landscape has been adapted for farming. The hillside in the foreground still has scrub vegetation, but the flat valley floor has now been cleared of its natural vegetation, irrigated, and cultivated.

Fig. 9.2 The Orange River flows through the desert in southern Namibia. Its surrounding area has been irrigated and used to produce grapes.

Processes

These are the methods used by people to produce the outputs:

- Preparation of the land – clearing vegetation, providing terracing, drainage, and irrigation systems
- Ploughing
- Sowing
- Weeding
- Application of fertilisers, pesticides, herbicides, and irrigation
- Harvesting
- Storage and transporting to market

Outputs

These are the products of the system:

- Crops
- Meat
- Milk
- Industrial products, such as cotton, rubber, or leather

Fig. 9.3 Large-scale commercial farming in South Africa's Western Cape. Note the large fields and the large area involved. The farm buildings also cover a large area – storing crops, seeds, fertiliser, and machinery.

Commercial and subsistence farming

Commercial farming

In commercial agriculture, the farmer sells his or her output to make a profit. This is typical of modern, large-scale farming. The crops produced are known as **cash crops** – they are sold for money.

Subsistence farming

Subsistence farming involves growing crops or rearing animals for consumption by the farmer and his or her family. The crops are called subsistence crops. In many subsistence-farming systems, a surplus may be produced from time to time which can then be sold.

Table 9.1 compares the two farming systems in general terms, although individual examples vary (as the following case studies show).

	Commercial farming	**Subsistence farming**
Capital (money)	Large capital input, sometimes from international companies	A complete lack of capital, preventing any increase in output
Land	Large area	Very small farms
Labour	Paid labour (often skilled), much use of research and development	Family labour, relying on traditional methods
Machinery and tools	Much use of mechanisation for all processes	Hand tools, such as hoes, and ploughs sometimes pulled by draught animals
Seeds	Improved varieties and hybrids	Seeds left over from the previous year's crop
Market influence	Production geared to current market demands and prices	No market influence
Fertilisers	Generally used	Not much used, although sometimes animal manure is available
Pesticides and herbicides	Generally used	Not much used
Irrigation in dry areas	Uses complex systems	Either none or very low-technology systems

Table 9.1 Comparing commercial and subsistence farming

Discussion point

Should an LEDC concentrate on subsistence farming, or should it produce cash crops for export? What are the advantages and disadvantages of each policy?

Intensive and extensive farming

These terms can apply to either commercial or subsistence agriculture. Commercial farms may be intensive or extensive, as may subsistence farms - although the latter tend to be intensive.

	An intensive farm	An extensive farm
Area of land	Small	Large
Large machines	Few	Many
Labour input per hectare	High	Low
Fertiliser input per hectare	High	Low
Output per hectare	High	Low

Table 9.2 Comparing intensive and extensive farms

As Table 9.2 shows, on an extensive farm the inputs per hectare are low - as are the outputs per hectare - but this is overcome by using a large area of land. Extensive farming can be highly profitable.

Fig. 9.5 A hillside in Bali, Indonesia, terraced for rice production, which needs an *intensive* input of labour

Fig. 9.6 There is little mechanisation on small-scale, *intensive*, subsistence farms (so this farmer in Rajasthan, India, is using oxen to pull his plough)

Fig. 9.4 Women planting onions by hand in Korea, as part of an *intensive* system

Fig. 9.7 On *extensive* commercial farms, the use of large machines (like these combine harvesters) is common

A large-scale commercial farming system

Natural inputs

The most important natural input in this farming system is land - lots of it! Some large-scale farms cover hundreds of square kilometres. This system might be run on extensive principles (where the large area compensates for a low level of human inputs), but not always.

Human inputs

The most important human input is capital. In some cases, the farm may be backed by a multinational corporation. The capital input pays for the land itself, and also for a labour force that is often highly skilled. A lot of research and development work supports this farming system - leading to the use of the most up-to-date machinery for all processes, improved crop varieties and hybrids, inputs of fertiliser, pesticide, herbicide, and, where necessary, irrigation. Complex systems ensure that production is linked to current market demands, prices, and government policies.

CASE STUDY

Large-scale commercial sugar farming in Swaziland

Sugar can be produced from either sugar beet (a root crop) or sugar cane. In the eastern part of Swaziland (an LEDC in southern Africa), sugar is produced from sugar cane. The sugar cane plant produces sucrose (sugar) in its leaves, which it then stores in its stem.

Sugar cane production is the single biggest industry in Swaziland. Sugar is easily Swaziland's largest export. Large commercial sugar cane estates account for about 77% of Swaziland's production. The rest is in the hands of a great number of smaller growers.

Fig. 9.8 Sugar cane

Natural inputs

- Sugar needs a hot climate. Swaziland's Low Veld area (see Fig. 9.9) has an average monthly **temperature** of 29 °C in summer, and temperatures rarely fall below 15 °C. There are also a lot of **sunshine** hours every day.
- Swaziland has **flat land** for large-scale mechanisation.
- Sugar needs at least 1800 mm of rainfall a year, which Swaziland does not receive (see Fig. 9.10), so **irrigation** is needed from the country's rivers.
- The alluvial **soils** in the river valleys are rich in nutrients and retain moisture.

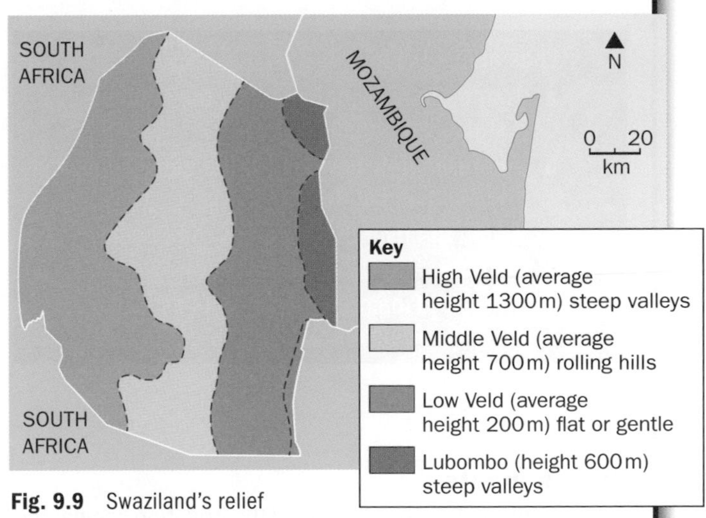

Fig. 9.9 Swaziland's relief

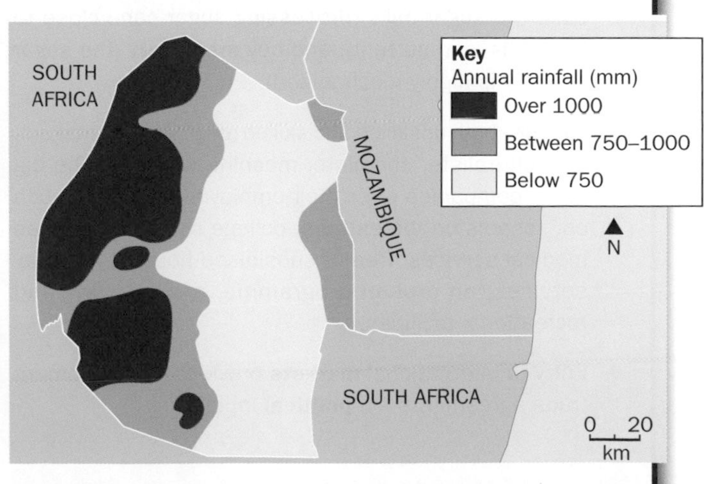

Fig. 9.10 Swaziland's average annual rainfall patterns

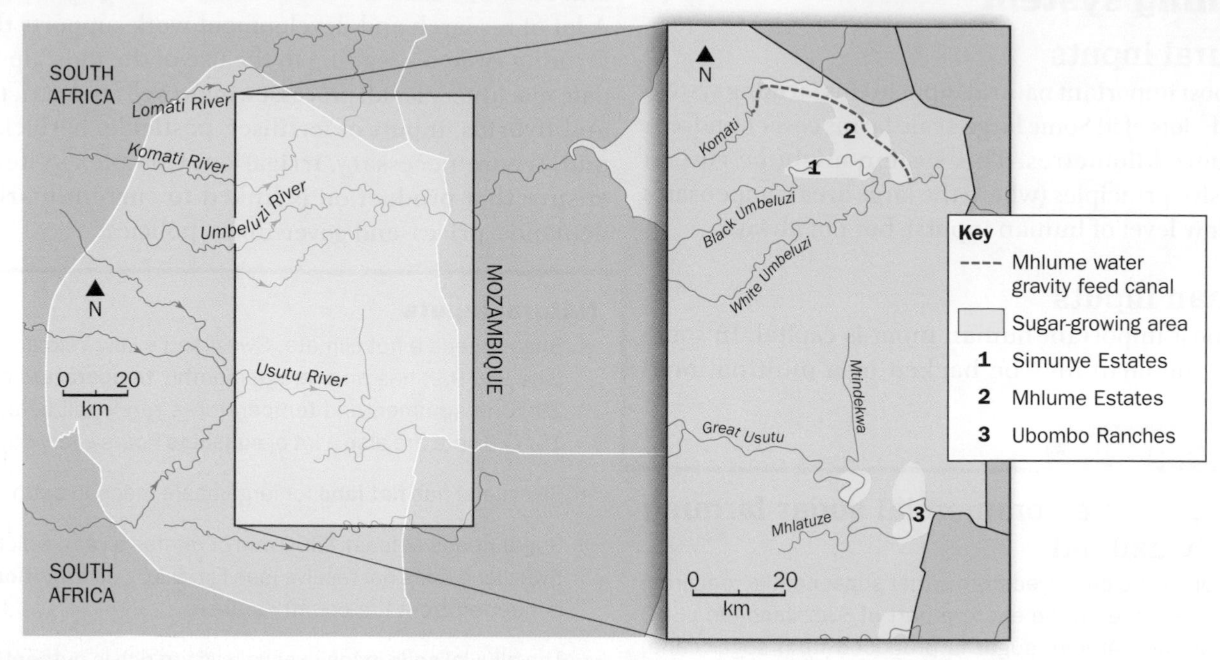

Fig. 9.11 Swaziland's rivers and the location of the sugar-growing areas

1 Look at Fig. 9.11.

a Name the river that supplies the Mhlume Estates with irrigation water.

b Approximately how long is the canal supplying the Mhlume Estates?

c Name the river that supplies the Simunye Estates.

d Name the main river that supplies the Ubombo Ranches.

Human inputs

→ **Capital** – to buy the land, build the irrigation canals (the Mhlume water gravity feed canal was opened in 1958), build the sugar mills (processing sugar cane close to the fields is important), and buy **machinery** (the sugar industry is highly mechanised).

→ **Labour** – both unskilled and skilled (engineers, managers, agriculturalists, chemists, machine workers). The big sugar companies offer their employees benefits, such as: schools on the estates, college scholarships, free medical services, free or subsidised housing, security services, an orphan programme, and sporting and recreational facilities.

→ Entry to international **markets** provided by government trade agreements – a **political** input.

The trade agreements include:

- the SACU Sugar Cooperation Agreement, which allows sugar to enter other countries in southern Africa
- the United States Sugar Program, which allows sugar sales to the USA
- the COMESA agreement, which allows sugar from Swaziland to enter other southern and east African countries
- Currently 39% of Swaziland's sugar is sold to the European Union (EU). Abolition of sugar beet quotas in the EU in 2017 is likely to reduce prices for Swazi sugar, threatening the industry.

2 If you have already learned how to draw pie charts (see Chapter 12), draw one to illustrate the following sugar production figures: Simunye and Mhlume 430 000 tonnes, Ubombo 230 000 tonnes, Tambankulu 62 000 tonnes. First you will need to convert the figures in tonnes to percentages.

Processes

Unlike some large-scale commercial farms, Swaziland's sugar estates are not run on extensive principles. In fact, the inputs of labour, fertiliser and irrigation are quite intensive.

- Irrigation water is taken from rivers by canal. It then reaches the cane by a variety of methods: furrow 39%, sprinkler 54%, drip 4%, and centre pivot 3%.
- The growth of the sugar cane crop in Swaziland takes about 12 months (a relatively short time for sugar cane).
- The ripe sugar cane is first burned in the fields (see Fig. 9.12). This makes harvesting it easier, because it removes all of the leaves – but does not damage the sugar inside.
- The cane is then cut down and taken to the sugar mills for crushing.
- Sugar cane will regenerate for several years before replanting is necessary. New plants are raised in a nursery for replanting in the fields.

Fig. 9.12 Burning sugar cane before harvesting it

Outputs

Swaziland has three sugar mills: Mhlume, Simunye and Ubombo. The first two are part of the Royal Swaziland Sugar Corporation, which operates a 20 000-hectare sugar cane estate with an annual production capacity in excess of 500 000 tonnes.

The outputs are raw sugar, refined sugar, brown sugar, molasses, and bagasse:

- All three mills produce **raw sugar** and **brown sugar**.
- Mhlume and Ubombo also produce **refined sugar**.
- **Molasses** is a sticky substance which does not form sugar grains. It is produced at all three mills. The two main distillers (USA Distillers and RSSC Distillers) use most of the molasses for the production of alcohol. The remainder is sold to small local and foreign customers, who use it as an input for food production and as animal feed.
- The **bagasse** (or fibre) is used as animal feed.

Discussion point

How could MEDCs change their policies to help agriculture in LEDCs?

Sustainable development and resource conservation

Any attempts to increase agricultural production and food supply must be **sustainable**. In other words, they must be achieved in such a way that future generations do not suffer as a result.

One crucial resource that must be **conserved** (protected) is the soil. Soil erosion can be completely natural. However, agricultural practices can cause it to begin or to increase. The case studies on Brazil (pages 238-40), Canada (pages 241-2), and Swaziland (pages 244-8) show that soil erosion is a hazard faced by farmers in three very different climatic zones - using very different farming systems.

Soil erosion

Soil can be eroded by water running down slopes, or by the wind. Both types of erosion are the result of:

- the soil being exposed and not covered by vegetation
- soil which is loose and damaged by poor agricultural practices, so it loses its **structure**.

Soil consists of two parts: minerals and organic matter. The minerals come from the rocks below by the process of weathering. They are the nutrients that plants absorb through their roots and use to survive and grow. Organic matter, including humus, comes from decaying plant matter from the vegetation above. Both of these parts are needed to give the soil structure - in other words, to make it hold together and not become loose.

Soils lose their minerals when plants use them, or when they are washed out by rainwater (in a process called leaching). Farmers can replace the lost minerals by adding fertiliser or animal manure to the soil. They can also replace lost organic matter by adding manure or plant matter.

Soil erosion by the wind

For this to occur:

- rainfall needs to be low, so the soil dries out and is loose
- strong winds need to blow to actually remove the soil.

One example of severe wind erosion is the 'Dust Bowl' in the High Plains and Prairies of the USA and Canada between 1930 and 1936 (in some areas until 1940). The ploughing of the natural grasslands to grow cereal crops allowed the soil to dry out and exposed it to the wind. The same crop was also planted year after year, which removed nutrients and destroyed the soil structure. Then, during a long drought in the 1930s, the exposed soil completely dried out and turned to dust. It was then blown away for thousands of kilometres towards the south-east, and ended up in the Atlantic Ocean. In the frequent dust storms, people could not see more than a few metres. About 400 000 km^2 of farmland were destroyed, and hundreds of thousands of people were forced to leave their homes.

The following case study on Canada (pages 241-2) shows how farmers today are dealing with the threat of soil erosion by the wind.

Soil erosion by running water

For this to occur:

- slopes must be steep enough for water to run down them due to gravity
- rainfall must be too heavy for all of it to soak into the ground, so that surface run-off occurs down the slopes - either in sheets of water (on gentler slopes) or concentrated into channels (on steeper slopes).

The following case studies on Brazil and Swaziland give examples of soil erosion by running water.

Discussion point

Which types of climate are likely to be affected by soil erosion, and why?

LEARNING TIP Soil erosion is part of Section 3.7 of the IGCSE® and O Level Geography syllabuses. Make sure that you know about a case study of soil erosion. Questions often ask about the causes of soil erosion and what can be done to prevent it. To get full marks you need not just to name an example but to be able to explain it thoroughly, giving some details about your chosen example.

Soil conservation

Table 9.3 describes some of the methods used to conserve soil. Examples of many of them (but not all) can be found in the case studies.

Method	Erosion prevented		Description
	Wind	Water	
Terracing		✓	A series of retaining walls is built on a slope, with the soil piled up and flattened behind each one. This prevents water from running down the slope and carrying soil away. The water is trapped and soaks into the ground. This method is shown on Figs. 9.5 and 9.26.
Contour ploughing		✓	Ploughing takes place across a slope, rather than up and down it. This means that water does not run down the furrows and wash soil away. The water is trapped and soaks into the ground.
Crop rotation	✓	✓	A different crop is grown on a plot of land each year for three or four years, before the first crop is grown again. The different crops take different nutrients from the soil. This means that the soil does not become exhausted, lose its structure, and become loose and easily eroded.
Fallow periods	✓	✓	A piece of land is 'rested' every few years. This allows it to regain lost nutrients. This means that the soil does not become exhausted, lose its structure, and become loose and easily eroded.
Strip cultivation and inter-cropping	✓	✓	Different crops are grown in narrow bands in a single field (often at right angles to the prevailing wind). The crops are harvested at different times, so the field is never left completely bare. Any soil blown from a bare strip by the wind is trapped by the crop in the next strip.
Cover cropping	✓	✓	This works on the same principle as strip cultivation. Usually a fast-growing crop is planted after the main crop has been harvested. Sometimes this is a 'green manure' crop, which is then ploughed back into the soil to add nutrients. In this way, the soil is left bare for the minimum time.
Reducing stock density	✓	✓	By having fewer livestock, a piece of land does not become overgrazed. There is always a cover of vegetation to protect the soil. Although there are fewer livestock, they may well be of better quality and value. It is often achieved by having fenced fields or paddocks and using rotational grazing.
Check dams		✓	Where gully erosion has occurred (see Fig. 9.27), small walls are built across the gullies. This reduces the speed of run-off down each gully and prevents it from becoming larger.
Filling gullies		✓	Gullies are filled with rocks and other materials as soon as they form – to prevent them from becoming larger.
Afforestation	✓	✓	Planting large areas of trees reduces soil erosion in various ways. The trees stop strong winds and heavy rainfall from reaching the soil, and the roots hold the soil together.
Shelter belts (windbreaks)	✓		Rows of trees are grown at right angles to the prevailing wind on the side of the field that the wind is blowing from. The trees reduce the speed of the wind, so that it is not strong enough to pick up the soil.
Dry farming	✓		This is a series of methods rather than a single one. They are described in the case study about the Canadian Prairies on pages 241–2.
Irrigation	✓	✓	Careful irrigation keeps the soil moist and prevents it from being picked up, especially by the wind.

Table 9.3 Some of the methods used to conserve soil

CASE STUDY

Large-scale commercial beef farming in Brazil

Brazil (an NIC in South America) is the world's second-largest beef producer – and the largest beef exporter. In 2010 Brazil's share of the world's beef export market was 25% and this is likely to increase.

Physical inputs

The most important physical input is large areas of **land**. The density of the beef cattle is often as low as a single animal per hectare.

Brazil's beef industry is not restricted to a single climatic zone:

- Much of the North Region (part of the Amazon Basin) has a tropical rainforest **climate** and **vegetation**. In 2000 the region contained just 13% of Brazil's total beef herd, but this has increased dramatically and it is now the largest producing region. Cheap (and sometimes illegally occupied) land has been made available for farming in this region.
- The Northeast Region has a dry tropical climate and wooded grassland – known as *cerrado*.
- The South Region has a temperate climate.

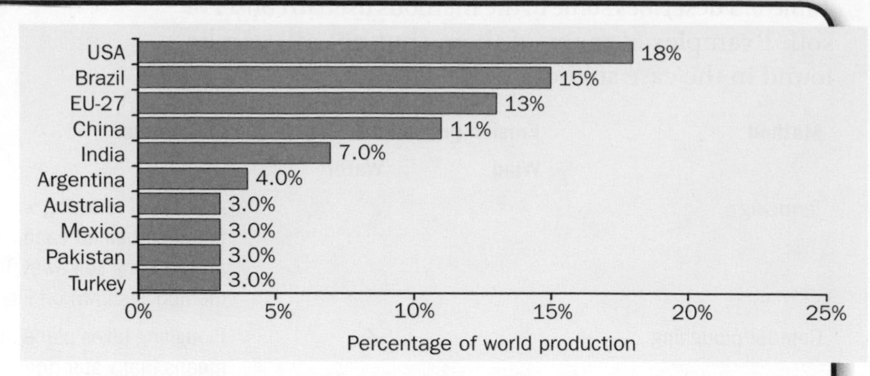

Fig. 9.13 The top ten beef-producing countries in 2016

3 In 2016, an inspection scandal affected Brazil's beef exports.

a Plot the data in Table 9.4 as a line graph.

b Describe the general trend in the graph.

	July 2016	Aug 2016	Sept 2016	Oct 2016	Nov 2016	Dec 2016	Jan 2017	Feb 2017	March 2017	April 2017
Beef exports change in value (%)	-14	+8	+11	-8	-6	+8	-4	-8	+24	-27

Table 9.4 Changes in the value of Brazil's beef exports

LEARNING TIP Describe a trend in words, but illustrate your answer with figures (unless the question asks otherwise).

Fig. 9.14 Brazil: regions

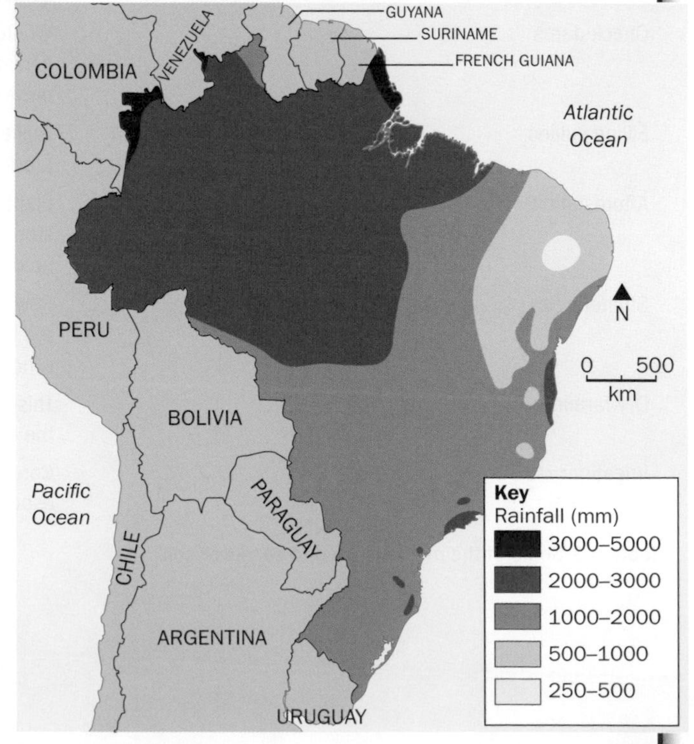

Fig. 9.15 Brazil: rainfall

Human inputs

Market is a critical factor. Beef consumption increases as countries become more affluent. Brazil is the world's fifth most populous country and, as an NIC, has a growing domestic market. Some of the major international retail chains operate there, including Carrefour (France), Walmart (USA), Royal Ahold (The Netherlands), and Sonae (Portugal). Traditional butchers and small retailers tend to be replaced by large supermarket chains in large urban areas. This has led to modernisation and an increase in the scale of operations for meat processing and distribution.

Fig. 9.16 Cattle grazing on land in the Southeast Region

JBS-Friboi is the largest Brazilian multinational in the food industry. It has established itself as the world's largest company in the beef sector, and is a global meat exporter. Its production and exports both increased greatly from 2000 onwards, but concerns about foot and mouth disease (a highly infectious cattle disease) have led to problems exporting to the European Union.

The growth in the industry has attracted multinational companies to invest **capital**, because large profits can be made.

In the south of the country, where beef rearing has gone on for many years, there is a supply of skilled **labour**. In the North and Northeast Regions, where there are high levels of poverty, labour has been acquired very cheaply.

Research and development have been important inputs to many of the processes listed below.

Processes

To sell its beef on the global market, Brazil has to place an emphasis on meat quality, traceability and standardisation of the product. To improve the quality of the meat there are crossbreeding programmes, usually involving crossing native and European breeds.

There are differences in the methods of production between the regions:

- In the North Region, the felling of rainforest is followed by seeding and grazing. Little fencing or improvement of pastures is carried out. Producers specialise in breeding (the first six to eight months of the life of the cattle).
- In the *cerrado* area of the Northeast Region, natural tropical grasslands are used as pastures.
- The Southeast and Central-West Regions are closer to the markets in the densely populated areas around Rio de Janeiro and São Paulo, so they concentrate on the later stages of growth and fattening of the cattle. Production here is well organised and large breeders have invested in improved pastures. This has resulted in a drop in slaughter age from four to three years, and improved the quality of the meat reaching the market. The Southeast Region has some production based on fattening cattle in pens (feedlots), a system more common in the USA. The slaughterhouses and meat-processing industry are concentrated in these regions.

Outputs

The Brazilian system is very extensive, with very low productivity. The global export of beef has been made possible by refrigeration and canning. Brazilian cattle products now end up in a wide range of consumer goods.

Fresh beef is converted into burgers sold in fast-food restaurants and grocery stores across Brazil, Russia, Venezuela, and a number of other countries. **Processed meat** finds its way into canned and ready-meal products in Europe and America. **Leather** goes to China, Italy, and Vietnam, where it is used to make shoes and clothing sold all over the world. The result is a boom in global demand for Brazilian beef products.

The impact of cattle on the Amazon Basin

Cattle ranching is the biggest cause of deforestation in the Amazon Basin (in the North Region). Beef ranches account for about 70% of the cleared forest. Results from a 2006 agricultural census showed that, over the previous 20 years, cattle ranches had caused the destruction of more than 20 million hectares of Amazonian forest. However, only 6% of Brazilian beef comes from ranches created by deforestation, and it must be remembered that a lot of beef comes from the country's other climatic zones.

The global effects of deforestation are described in Chapter 7, but the immediate local effects include changing the traditional way of life of local people and causing soil erosion (mainly by running water, because of the heavy rainfall of the tropical rainforest climate falling on the bare soil).

Fig. 9.17 Soil erosion caused by deforestation and cattle ranching in the Amazon Basin

Making beef production sustainable

Brazil intends to increase its beef production and gain over 44% of the international export market by 2020. However, it needs to do so without any further negative impacts on the environment. This could be done by making better use of land that has already been deforested – increasing production in those areas and taking better care of the soil and pastures. However, to achieve this, greater government control over future deforestation will be vital.

Where it remains cheap – or even free – to occupy land, it will always be more economical to cut down trees than to invest in increased productivity in areas that have already been cleared. The system must become more intensive, with more animals raised in smaller areas.

Brazil's biggest domestic beef buyers – the supermarket chains Walmart, Carrefour, and Pão de Açúcar – have announced that they will suspend contracts with suppliers found to be involved in Amazon deforestation. The shoemakers Adidas, Clarks, Nike, Geox, and Timberland now have sourcing policies in place to ensure that their products do not use leather produced in areas of Amazon deforestation.

4 Explain why beef production in Brazil is described as (a) commercial and (b) extensive.

Discussion point

From what you have learned from this chapter and Chapter 7, why has clearing areas of the tropical rainforests raised strong feelings in people all around the world?

CASE STUDY

Large-scale commercial wheat farming in Canada

Canada is an MEDC in North America. Chapter 2 explained how the country's three Prairie Provinces (Alberta, Saskatchewan and Manitoba) were important for the production of wheat. The Prairies are the plains of central North America, which – before farming began – were natural temperate grasslands. The three Prairie Provinces make up about 75% of Canada's farmland and are one of the largest cereal production regions in the world. Canada's neighbour, the USA, is the world's leading wheat exporter.

Physical inputs

- The **climate** is critical for the growth of wheat. The very warm, sunny summers allow optimum growth in the short growing season. Rain also comes in the summer growing season. During the cold winter, frost breaks up the soil and kills pests that might attack the crop.
- The **land is flat**, or gently undulating, which allows the use of large machines like combine harvesters.
- This area used to be a natural grassland, so beneath the grass a black, humus-rich **soil** (called a chernozem) developed. This provides natural fertility with the minimum input of fertilizer.
- There is a **large amount of land** available. The Prairies stretch 1500 km from east to west – and a typical farm is about 300 hectares.
- The annual **rainfall** with its summer maximum is enough, although there are drier areas in the north and south where there can be problems in drier years.

Human inputs

- The **labour force** is small and highly skilled. There may be as few as two or three people operating a farm.

Fig. 9.18 Wheat

- Farming is highly mechanised and uses a large proportion of **capital** on the purchase and repair of equipment. A large combine harvester, operated by one person, can harvest 30 tonnes of wheat a day.
- Although extensive farming systems usually involve few inputs of **chemicals** like fertiliser, herbicides (weed-killer), and pesticides, the use of these has increased with time.
- The **research and development** of new strains of wheat has been important. One aspect of this is disease resistance and another is the time needed for the wheat to reach maturity. The new varieties are able to grow in regions in the north and south where the growing season is shorter and there is less rain.

Fig. 9.19 The wide open Prairies of Manitoba with a grain elevator in the background

Processes

The crop is spring wheat. In other words, it is sown in the spring, grows through the summer and is harvested in the autumn. Most European wheat grows through the winter.

After harvesting, the grain is taken to grain elevators (storage silos) next to railways. There are 2000 railway shipment points in the Prairies. From there, the grain is taken by rail to the ports shown on Fig. 9.21.

Outputs

The main output is unprocessed wheat grain. Some other cereals, such as barley, are also produced in smaller quantities.

5 Describe how grain is exported from the Canadian Prairies.

Fig. 9.20 A grain elevator

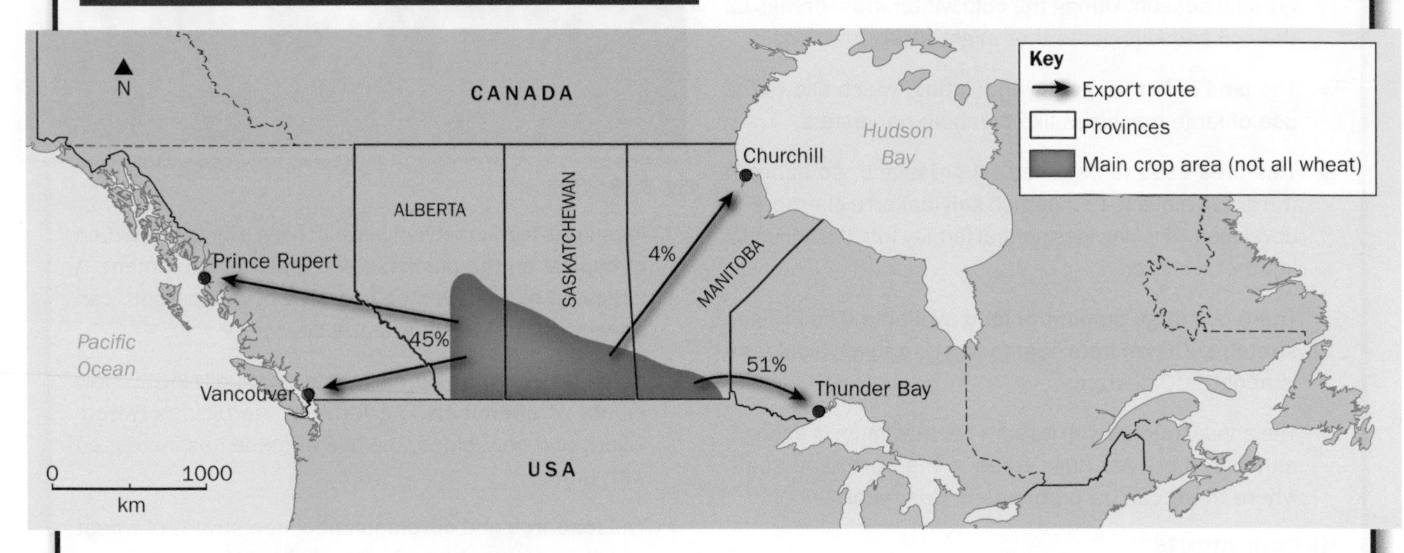

Fig. 9.21 The wheat-growing areas and export routes

Problems

Like all exporting countries, fluctuations in world prices and demand are issues facing Canada. However, the main problem experienced is soil erosion by the wind, especially in the drier, western areas. Any drought when the soil is bare leads to the erosion of loose particles by the strong winds blowing across the treeless plains.

Although there is some irrigation, the main solutions come under the general heading of **dry farming methods**. These include:

- **Stopping ploughing.** Cultivators cut weeds below the surface and seeds are sown into the stubble of the previous year's crop.
- **Fallowing.** The land is cultivated only every other year, to conserve the soil moisture.
- **Ripping.** A machine rips up the frozen ground in winter into large chunks, which block the wind close to the surface.
- **Strip fallowing.** Growing wheat in strips at right angles to the prevailing wind with fallow strips in-between. The fallow strips trap any soil that is blown by the wind.
- **Growing drought-resistant varieties.**

Fig. 9.22 Average annual rainfall in the Prairies

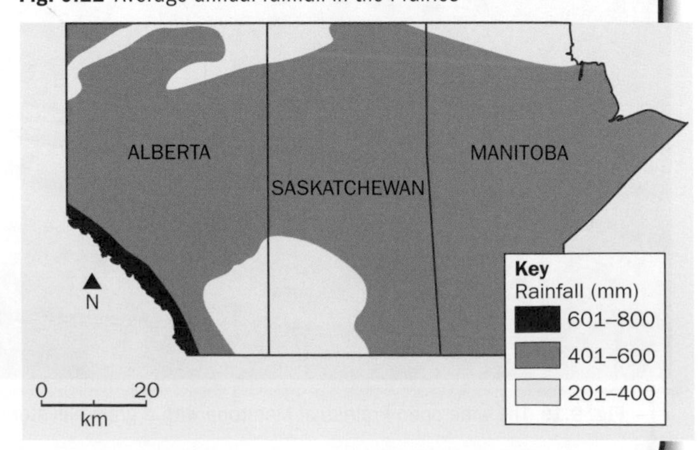

> **6** Study the data in Table 9.5.
>
> **a** How has world wheat production changed in the last 10 years? What are the differences in the changes between countries?
>
> **b** Suggest reasons for the changes shown.

Country	2008	2009	2010	2011	2012	2013	2014	2015	2016	2017
China	112.5	115.0	115.2	117.4	125.6	121.7	126.2	120.5	130.2	131.0
India	78.6	80.7	80.7	86.9	94.9	93.5	94.5	94.8	86.5	96.0
Russia	63.7	61.7	41.5	56.2	37.7	52.1	59.7	37.7	61.0	69.0
USA	68.0	60.3	60.1	54.4	61.8	60.0	55.4	61.7	56.1	49.6
France	39.0	38.3	38.2	38.0	40.3	38.6	39.0	40.3	38.5	36.8
Canada	28.6	26.5	23.2	25.3	27.0	37.5	29.3	26.0	27.6	28.3
Pakistan	21.0	24.0	23.3	25.2	23.5	24.2	26.0	25.8	25.1	26.2
Australia	21.4	21.7	22.1	27.4	29.9	22.9	25.3	29.9	24.5	25.0
Ukraine	25.9	20.9	16.9	22.3	15.8	22.8	24.1	25.0	27.2	25.0
Germany	26.0	25.2	24.1	22.8	22.4	25.0	27.8	26.4	25.3	24.6
Turkey	17.8	20.6	19.7	21.8	20.1	22.1	19.0	19.0	19.5	18.0
Iran			15.0	13.5	13.8	14.0	13.5	13.5	14.2	15.0
Kazakhstan	12.7	12.7	9.6	22.7	13.3	13.9	13.0	15.0	14.0	13.0

Table 9.5 World wheat production (million metric tons)

> **LEARNING TIP** Question 6b is a 'suggest question', so you are allowed to include any reasonable answers. Some answers are not definitely right or wrong and you will gain marks for any sensible suggestions.

Small-scale subsistence farming

Natural inputs

In most examples of subsistence arable farming in the world today, the area of each unit is small - perhaps as small as 1-3 hectares. This land is cultivated intensively. In the case of subsistence pastoral farming, areas of land can be much larger, especially in the case of pastoral nomadism - where people move from place to place with their animals.

There are still some areas where **shifting cultivation** is practised. In these areas, a plot of land is cleared and the ash from burning the vegetation is used as fertiliser. The land is cultivated for a few years in the traditional manner, until it is exhausted (the plant nutrients - minerals in the soil - have been used up) and crop yields decline. The people then move to another area, often building a new settlement, and repeat the process - not returning to the original plot for perhaps 20 years. This system is still used in some tropical areas like rainforests, where the soil fertility is low and minerals are leached by heavy rainfall.

The Chitimene system practised in the Miombo woodland in Zambia, central Africa, was an example. Because of population growth, the breakdown of tradition and economic factors, the Chitimene system can no longer be practised. Finger millet, a grain relatively rich in protein, has been replaced by maize - leading to a poorer diet and resulting in the need for measures to restore soil fertility periodically.

> **7** **a** Why does population growth sometimes mean that people change from shifting cultivation to permanently settled agriculture?
>
> **b** Suggest social and economic reasons for this change.

Human inputs

A lack of **capital** input is an issue that prevents many subsistence farmers from increasing their output. The ways of doing this (described later) are often impossible because of poverty. Family **labour** is generally used (relying on traditional methods). **Tradition** also fixes the roles of men and women in different ways in different societies. There are few **machines** - hand **tools** (like hoes) are used, with draught animals (like oxen) being used to pull ploughs. **Seeds** left over from the previous year's crop are used for the next year, which prevents the use of improved varieties. The only **fertiliser** used might be animal **manure**, although in many areas this is used as a cooking fuel instead (which prevents soil improvement). Where **irrigation** is used, very low technology systems are in place - usually draining water in channels from a nearby stream.

Discussion point

What effects have population growth had on agricultural systems? How has this affected the natural environment?

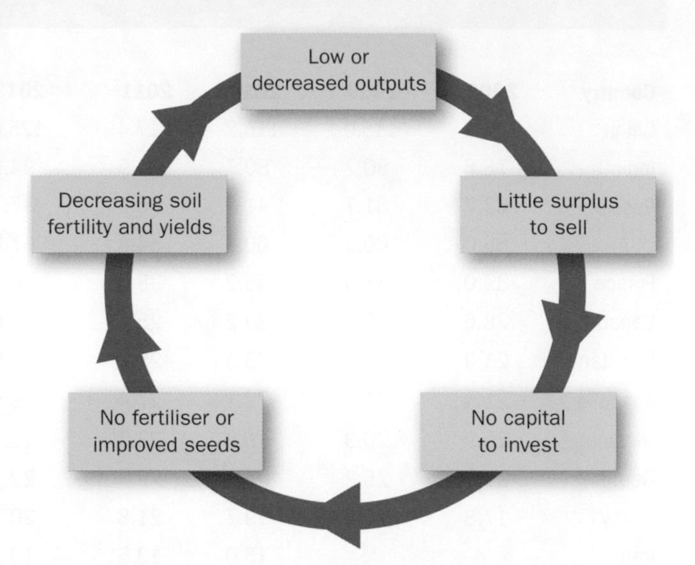

Fig. 9.23 The vicious circle of low outputs and a lack of capital

Improving subsistence agriculture

Many people who practise subsistence agriculture are stuck in a vicious circle of poverty (see Fig. 9.23). A lot of attention is therefore given to improving the lives of subsistence farmers and breaking the circle. This is discussed in the following case study and in the section about improving food supplies.

CASE STUDY

Small-scale subsistence farming in Swaziland

Most of the people in rural Swaziland are subsistence farmers. Many of them live on Swazi National Land, which makes up over 60% of the country's total land area. It is held in trust by the King for the Swazi people, and supports about 70% of the population. However, the Swazi National Land system sometimes leads to the fragmentation of already small farms – with individual farmer's fields separated from each other and from the household. Some farmers have to walk for up to 3 km just to reach their fields.

Under the traditional land system, farmers cultivate small plots – averaging less than 3 hectares in size – and have no right to sell the land. Most of the subsistence farming is based on the production of maize, without irrigation. Raising cattle is also important.

For many communities around the world, farming forms an important part of their culture and Swaziland is no different. Cattle are an important form of wealth and the farming tasks performed in the different seasons are an important part of people's cultural beliefs.

Fig. 9.24 A girl cultivating a food plot in Swaziland

Physical inputs

As you have already seen in Fig. 9.9, Swaziland has four zones. As well as the natural inputs of soil and rainfall, the input of land is usually small units of up to 3 hectares.

Zone	Average annual rainfall (mm)	Soil	Relief
High Veld	1270	Soils are often thin, **leached** or eroded	Areas of steep slopes
Middle Veld	940	Some areas of rich soils, especially in the river valleys	Gentle slopes
Low Veld	660	Some rich alluvial soils	Flat
Lubombo Uplands	787	Some good red clay soils but some are thin	Areas of steep slopes

Table 9.6 Swaziland's four zones and their features

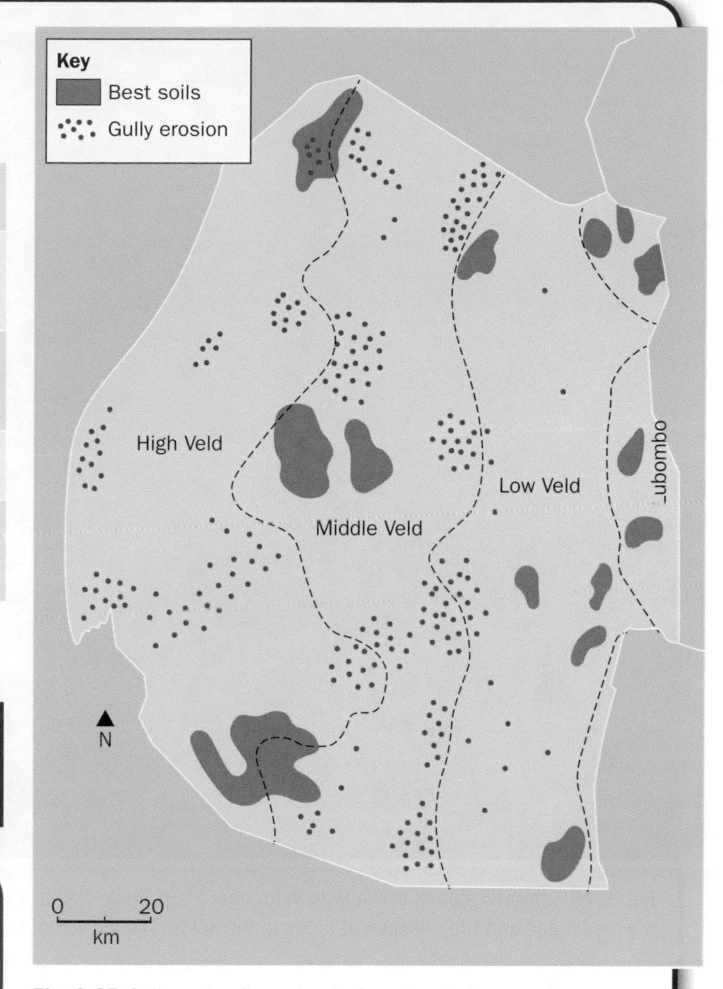

Fig. 9.25 Soils and soil erosion in Swaziland's four zones

8 Use the data in Table 9.6, plus information from Fig. 9.25, to explain the advantages and disadvantages of each zone for farming.

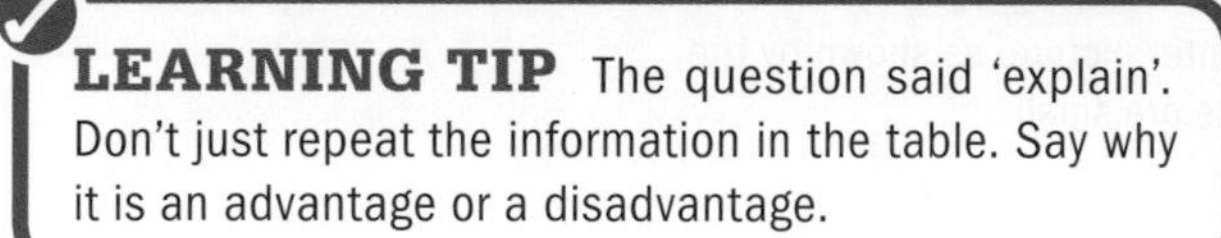

LEARNING TIP The question said 'explain'. Don't just repeat the information in the table. Say why it is an advantage or a disadvantage.

Human inputs

- **Labour** is generally from family members. In recent years, women have played a particularly important role, because many men have left the country to work in the mines in neighbouring South Africa.
- Many farmers also try to find casual work in the towns – to provide extra income to support their families and also a small amount of **capital** for their farms.
- **Traditional knowledge** is important and many people are still guided by religion and custom in how they live their lives.
- Mostly hand **tools** like hoes are used, although the government has operated a tractor hire scheme.
- There is little use of **fertiliser**, although there may be some animal manure from stubble grazing by cattle in the winter.
- The **social/political system** of the Swazi National Land (described earlier) is an important input.
- There is simple **irrigation** in the Middle and Low Velds, which involves diverting water from nearby streams.

Processes

Many Swazi farms are self-sufficient areas of 1–3 hectares. Their main crop is maize, but the farmers also graze a few cattle. The individual subsistence farmers are encouraged to join up and form farmers' associations, which help to spread information about new processes (under the direction of field officers working for the Department of Agriculture).

- A lot of cattle are grazed on the Swazi National Land. The stock is guarded and controlled on the unfenced land by herd boys.
- In the dry winters, pastures are burned to get rid of the coarse dry grass and allow new nutritious shoots to emerge.
- Swaziland's rains come in the summer (starting roughly in October), so this is when the maize seeds are sown.
- The lands are ploughed in June, before the rains are expected.
- Following the harvest in April and May, cattle may graze the fields.
- The steeper slopes in the High Veld are terraced.

Fig. 9.26 Terraced slopes in the High Veld, near Piggs Peak. This is a winter picture, as shown by the dry landscape and the absence of crops in the fields. The individual plots are small.

Outputs

The harvested maize is milled at home to produce flour, which is then cooked and eaten by the family. There is also some production of millet and vegetables (such as cabbages, tomatoes, and pumpkins) to vary the family's diet. The cattle produce meat and milk.

Problems

Erratic rainfall is a problem. There was a **drought** in the growing season of 2015–16, which led to a national emergency. The **irrigation water** that could enable farmers to increase their productivity is not available to many. Water rights from the Usutu River have already been allocated.

Farmers lack **capital** and do not have money to buy improved seeds and fertiliser. Those who do not have teams of oxen for ploughing find it difficult to rent tractors.

As in many countries in Africa, the **younger people** are often not interested in agriculture. Some of the vulnerable people who live in Swaziland's rural areas have been discouraged by the prolonged drought and the burden of **HIV/AIDS**, which reduces and weakens the workforce.

Roads connecting farmers to supplies and markets are not good in remote areas of the north and south.

Overstocking of the pastureland is a major issue – often caused by the shortage of land. It means that the cattle raised are of poor quality, and they also use up valuable land that could be used for cultivation. In addition, overgrazing leaves the land bare and leads to soil erosion.

Animal diseases have also weakened the herds. The regions of Manzini, Lubombo and Hhohho had been free of foot and mouth disease in cattle for more than 20 years, but there was an outbreak in November and December 2000 – just after Swaziland had been designated by the World Organisation for Animal Health as a 'FMD Free Country Without Vaccination'.

Soil erosion is a problem that leads to the formation of gullies (or dongas as they are known in southern Africa), especially in the High Veld. This leaves the land unusable for agriculture.

The soil erosion is caused by a variety of factors, including:

- torrential rainfall, which cannot soak into the ground quickly enough and runs off the surface. This is a particular problem at the end of the dry season, when the soil is exposed
- pasture burning, which again leaves the soil exposed
- steep slopes leading to rapid surface run-off of the rainfall
- loose sandy soils
- the overgrazing of cattle as a result of overstocking and uncontrolled grazing – leaving the soil exposed. This particularly occurs around waterholes. Under the open-access grazing system, cattle are allowed to graze on Swazi National Land at no direct cost to the cattle-owning household
- leaving the soil exposed after the harvesting of crops
- destroying the soil structure by overcropping, which leaves it loose and easily removed
- ploughing up and down slopes, so that rainwater runs down the furrows.
- adopting a **monoculture** of maize, which leads to soil exhaustion and destroys the soil structure.

Solutions to the problems

More investment in dams and canals would help to extend irrigation in the Middle and Low Velds. Some micro-irrigation schemes for holdings of 1–5 hectares have been planned (often for co-operatives of women farmers). Where irrigation is not possible – and the average rainfall is less than 800 millimetres a year – alternative crops to maize are being investigated. They include sorghum, sweet potatoes, cassava and groundnuts.

Subsidised farming inputs like seed and fertiliser, especially for farmers of small areas, would help to overcome the lack of capital. One scheme requires farmers to pay for one-third of the cost of the inputs.

Measures are also needed to improve livestock production. A fence has been proposed along the border with Mozambique – to help control foot and mouth disease. All cattle within a 15-km-wide buffer zone would then be vaccinated. Reducing stock density would also help, but many farmers resist this. Rotational grazing under fenced conditions would lead to higher grass production and healthier cattle.

Programmes of education and training have been set up for farmers. For example, there are residential training courses at the Veterinary and Farmer Training Centre at Mpisi. This organisation has established units to teach farmers about the fattening of cattle, dairy production, and also broiler, egg, and pig production.

Fig. 9.27 Gully erosion

Farm mechanisation is a major issue when farmers lack capital. A programme to hire out tractors could succeed only if the service was subsidised. A service to provide small capital loans would require farmers to make a contribution and prove their creditworthiness. The formation of co-operatives is important for getting bank credit.

Improved markets for crops would stimulate production. The National Maize Board, whose main storage facility is located in the centre of Swaziland, is to be decentralised – with storage facilities being built in all four provinces. Schemes to generate interest in agriculture and stimulate younger people to become farmers may help what is a problem in Swaziland and in many other parts of the world.

There are a variety of measures to control soil erosion, some of which have already been mentioned. They include:

- inter-cropping (growing other crops which mature at different times between the rows of maize)
- terracing (see Fig. 9.26)
- contour ploughing across the slope, rather than up and down it
- crop rotation (changing the crop on a plot every year for three or four years before the first crop is grown again)
- reducing stock densities
- the careful use of fertiliser and manure
- stopping the burning of grass in the High Veld.

9 **a** The causes of soil erosion in Swaziland have already been explained. For each of the solutions listed above, explain how it will help to reduce the problem.

b Using the three case studies, make a table to compare the causes of soil erosion in the Canadian Prairies, the Amazon Basin, and Swaziland.

Discussion point

Many countries report that young people are not interested in being farmers but would rather have an office job. Why do think that this is so? Is it a good thing?

Food shortages

At the beginning of the chapter, the importance of increasing world food supply to keep pace with the growing population was mentioned.

Poor nutrition is particularly noticeable in statistics about children's health. Well-nourished children perform better in school, grow into healthy adults and give their own children a better start in life. Well-nourished women face fewer risks during pregnancy and childbirth, and their children get a better start in life.

UNICEF, the United Nations Children's Fund, believes that undernourished children have lowered resistance to infection and are more likely to die from common childhood ailments like diarrhoea and respiratory infections. Frequent illness saps the nutritional status of those who survive - locking them into a vicious cycle of recurring sickness and faltering growth.

Poverty, low levels of education, and poor access to health services are major contributors to childhood under-nutrition. Not surprisingly, the highest levels are found in LEDCs.

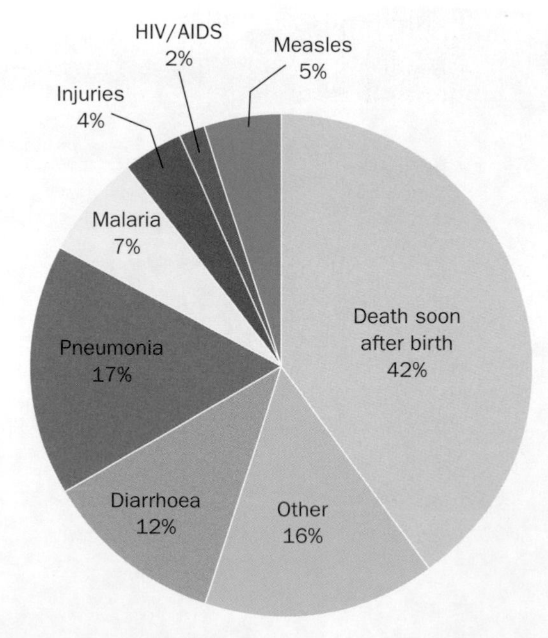

Fig. 9.28 The causes of death of children under the age of five in 2015 (globally, more than one-third of child deaths are attributable to under-nutrition)

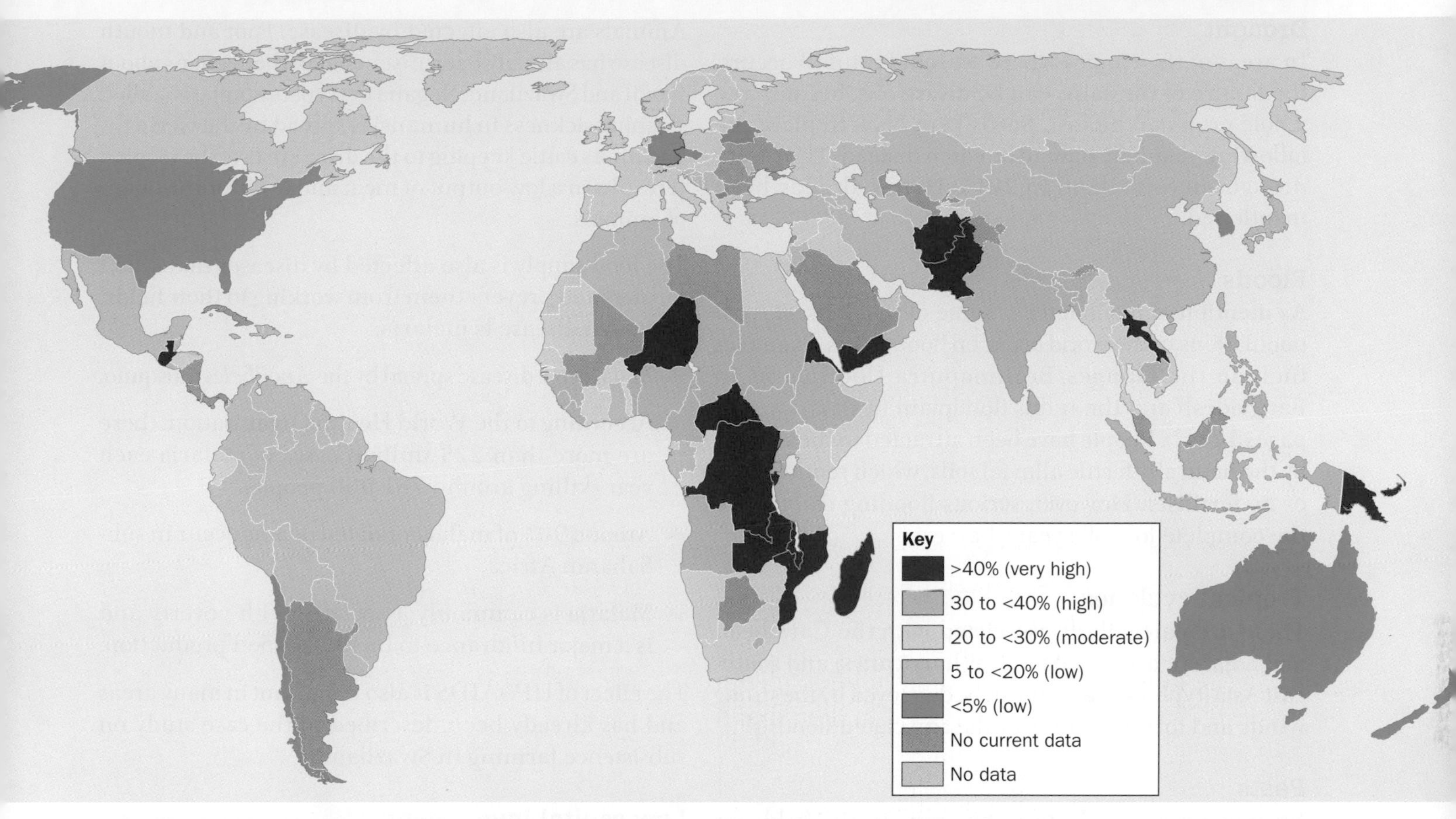

Fig. 9.29 The global distribution of child under-nutrition. This information, compiled by UNICEF, shows the percentage of children in each country who were moderately or severely underweight (based on the best statistics available up to 2017)

10 Describe the global distribution of child under-nutrition.

As well as the common childhood illnesses made worse by under-nutrition, there are also specific diseases caused by a poor diet. There are two particular diseases linked to protein deficiency. **Marasmus** is most common in very young children and results in babies being very thin. **Kwashiorkor** results in children having swollen bellies, round faces and changes in hair colour.

Effects of food shortages

Food shortages lead to many problems:

- Increased death rates, especially infant mortality
- Diseases linked to protein deficiency
- Common illnesses such as diarrhoea having a greater effect
- Increased expenditure on health services
- Fewer children being able to complete schooling
- A weaker, less productive workforce, in both non-manual and manual work such as agriculture
- Slower economic growth and output because the weaker workforce results in a drain on the economy
- Increased dependence on international aid
- Rural-urban migration to escape rural poverty, which may have a negative effect on the rural economy.

Causes of food shortages

Soil exhaustion

Soil exhaustion leads to low crop yields. It is a result of overcropping (growing too many crops on a plot) and monoculture (growing the same crop every year, using up the same minerals). Insufficient fertiliser and manure are added to replace the lost minerals. Heavy rainfall also results in the leaching of minerals.

Drought

In areas of the tropics where seasonal rainfall occurs, the failure of the rains can be disastrous, because the whole crop can be lost. Seeds kept back to plant the following year may have to be eaten instead. The severe drought in Swaziland in 2015-16 has already been mentioned.

Floods

As mentioned in Chapter 1, some of the densest rural populations in the world occur on floodplains. Examples include the Ganges/Brahmaputra floodplains in Bangladesh and the Indus floodplain in Pakistan (see pages 121-2). People have been attracted to these areas by the naturally fertile alluvial soils, which require little or no fertiliser. However, serious flooding can lead to the complete loss of a year's harvest.

Tropical cyclones

These are particularly associated with the Caribbean and adjacent areas of America (hurricanes) and south-east Asia (typhoons). Crops can be destroyed by the strong winds and torrential rain, or the associated floods.

Pests

Various pests can destroy the crops in the fields (or harvested crops during storage). Among the most serious pests are the locust swarms that occur in many parts of the world, but are most destructive in the subsistence-farming regions of Africa.

Desert locust swarms:

- could threaten the economic livelihoods of 10% of the world's population.
- can be 1200 km^2 in size, and contain between 40 and 80 million insects in less than one km^2.
- of this size eat about 200 million kg of plants a day.

Less spectacular are the more everyday pests like birds, which eat crops in the fields and require scaring away, or rats that eat stored crops.

Disease

A variety of diseases can destroy crops in the fields or during storage. For example, in West Africa cereals and legumes are frequently contaminated with mycotoxins (fungal growths caused by storage in humid conditions). Many rural West African children have excessive levels of mycotoxins in their blood, which can cause serious health effects. Bananas (often a subsistence crop in humid tropical areas) suffer from 'banana bunchy top virus', which stunts and kills plants. Bacterial wilt causes bananas to ripen prematurely, which can wipe out up to 90% of a crop.

Animals are also affected by disease. Foot and mouth disease has already been discussed in the sections about Brazil and Swaziland. Nagana or trypanosomiasis (called sleeping sickness in humans) is spread by the tsetse fly, and limits cattle keeping to the drier areas of the tropics. It results in a low output of meat and milk, or the death of animals.

The food supply is also affected by diseases that afflict farmers and prevent them from working in their fields. One such disease is malaria:

- Malaria is a disease spread by the *Anopheles* mosquito.
- According to the World Health Organisation, there are more than 225 million cases of malaria each year (killing around 781 000 people).
- Around 90% of malaria-related deaths occur in sub-Saharan Africa.
- Malaria is commonly associated with poverty and is a major hindrance to increased food production.

The effect of HIV/AIDS is also important in many areas and has already been described in the case study on subsistence farming in Swaziland.

Low capital investment

The vicious circle of low output and poverty is described in Fig. 9.23.

Poor transport

This reduces productivity in two ways:

- Farmers in remote areas find it difficult to receive supplies and information about possible improvements to farming practices and processes.
- They cannot sell surpluses easily and are therefore less likely to raise capital and more likely to remain poor.

Wars

Many of the countries with the highest rates of poorly nourished children (see Fig 9.29) have been badly affected by conflicts. The prolonged conflict in the eastern part of the Democratic Republic of the Congo is an example. If people are forced to leave their homes and become refugees with an uncertain future, it will obviously have an affect on their ability to make long-term investments in increased food production.

Increased use of biofuels

Between 2000 and 2016 world production of biofuels increased by a factor of 10 and land previously used for food production was changed to produce crops for biofuel production instead. This led to increases in world food prices and decreases in the food supply.

CASE STUDY

Food shortages in Swaziland

Some of the problems in Swaziland that have led to food shortages there have already been discussed. In addition, unemployment in neighbouring South Africa has now risen, and many migrant workers from Swaziland are returning to their homes. This has increased unemployment in the country and caused even greater poverty and food shortages. Chronic under-nutrition is a particularly important problem for children – leading to slower growth and increased vulnerability to disease.

- In 2016 Swaziland's GDP per capita (a measure of wealth) was just $9800.
- Between 1970 and 1990, life expectancy at birth rose from 48 to 61 years. But, by 2016, it had dropped back to 51.
- The Swazi population has been badly weakened by HIV/AIDS. In 2016 27% of the population were infected (the highest prevalence rate in the world).

The extent of the food problem

In recent years Swaziland has been affected by a series of droughts. The last of these was in 2016 and was caused by an **El Niño** event.

- The government of Swaziland declared a national emergency.
- The harvest (in April) produced 10% of the required amount of food, mostly the staple food crop maize.
- 40 000 cattle died.
- 25% of the population were short of food and water.
- The UN World Food Programme (WFP) fed more than 200 000 people.
- Much of Swaziland's income comes from sugar production, but irrigation stopped on 30% of the cane fields.

Food aid or not?

There are two types of food aid: emergency aid to deal with natural disasters such as drought and longer-term aid to deal with more general poverty and poor nutrition. For the last few years, the UN's WFP and other organisations have been moving away from distributing food aid towards programmes that encourage self-sufficient food production instead. Previously, they supplied food aid directly during emergency situations, and via governments during non-emergency times. They also supplied seeds for sorghum, beans or maize.

This international approach led some farmers to think that they would always be supplied, so they became dependent on the aid. They sometimes even sold what they had been given. It is important that farmers contribute and have a sense of ownership, rather than just receiving something for nothing. In Swaziland, some children born since the 1990s – when drought crippled Swaziland's agriculture – do not know anything other than food aid, because their parents have given up on farming. They never get to acquire much-needed agricultural skills at home.

Between 2006 and 2010 world food prices rose dramatically. In many countries there were protests in the streets about the price of food (e.g. Indonesia, the Philippines, and Egypt).

Food crops or inedible cash crops?

Small-scale farmers often have to decide whether to grow food crops to feed their families directly, or cash crops that they can then sell for money to buy food and other goods. In Swaziland, many small-scale farmers have decided to produce cotton (in the drier areas) and sugar cane, rather than staple foods like maize.

Sugar is Swaziland's biggest industry, employing 93 000 people (see the case study on pages 233–5). Over the last 30 years, many small-scale farmers have been persuaded to abandon food crops such as maize and join co-operatives growing sugar cane instead. Up to now, these co-operatives have relied on Swaziland's guaranteed access to markets in other countries to repay the bank loans they took out to irrigate their land and buy equipment. However, the problem for these small farmers is that if world sugar prices fall, or the trade agreements change and cut off their key markets, they may be in severe difficulties.

11 a Why is a high rate of HIV/AIDS a problem for food supply and agriculture in Swaziland and elsewhere?

b Why is food aid not necessarily the best way of improving supply, and what is the alternative?

c List the causes of food shortages in Swaziland.

Discussion point

How would you attempt to solve world food shortages?

10 Industry and tourism

This chapter covers the following Cambridge IGCSE® and O Level topics:

- **3.3 Industry**
- **3.4 Tourism**
- **part of 3.7 Environmental risks of economic development**

- Why is it that some countries can build flourishing economies while others remain poor?
- Which countries are developing their industries very quickly?
- Why can some industries only be located in certain areas?
- How does industry produce problems for the environment?
- What are the advantages and disadvantages of different types of industry?
- How can we clean up after industry?
- Should we be using up our finite mineral resources, or should we save some for future generations?
- Are tourists an intrusion into the lives of local people?
- Do they spread bad habits to other cultures?
- Do they damage the environment?

In this chapter you will learn about:

- the way the industrial system works
- how different industries have different requirements
- how leisure and tourism have grown to become very important over the last 50 years
- the advantages and disadvantages of the tourist industry
- why resources need to be conserved and why development must be sustainable
- the effect of industries on ecosystems.

Types of industry

Manufacturing

This is the production of new products by industrial processes, usually in large quantities in a factory using machinery. Domestic industry is small-scale production in the home, mainly using hand tools.

Processing

Processing is the preparation of a raw material into a different state for three purposes:

- for direct sale to consumers, such as the making of cheese from milk
- to change a perishable natural product into a longer-lasting form. The tanning process converts skin into leather, which will not decay.
- to turn it into a form which can be further changed in a manufacturing industry, such as leather to shoes.

Processing of raw materials may also be done to reduce transport costs. For example, some waste is removed from mineral ores near mines to reduce bulk so that it is cheaper to transport for manufacturing.

Assembly

This is the putting together of component parts to make a product. A vehicle is made from parts made elsewhere and put together on an assembly line.

High-technology industry

This is explained later in the chapter.

RESEARCH Research a processing industry in your country. Describe and explain the processes that the natural product goes through.

Industrial systems

Inputs

The inputs of an industry are the things that are required before it can take place. They have varying importance for different industries.

Raw materials

These are the commodities that will be changed into the finished product. Heavy industries, such as iron and steel, require raw materials that are bulky and relatively expensive to transport. Light industries, such as electronics, use materials that are small in volume but of very high value, so transport costs are less important in these industries.

Water supply

This may be important to certain specialist manufacturing industries, such as paper or chemicals. The water may be required in high quantities in a very pure form.

Site

This is an important factor for large-scale manufacturing, such as motor vehicles, oil refining, and chemical manufacture. The large size of the factory or plant means that very large areas of flat land are needed.

Labour

The overall size of the labour force is important to many industries - where there can be literally thousands of employees. However, particular types of skill and flexibility may be more important to some industries.

Capital

The start-up costs of an industry can come from other businesses, banks, or governments.

Energy

For many industries, a link to the electricity grid is sufficient. For a few manufacturing industries (e.g. aluminium smelting), there are greater requirements.

Processes

These are all the activities that go on in different types of industry, whether it's a farm, a quarry, a factory or an office. Agricultural processes have already been discussed in Chapter 9. The motor vehicle industry involves a wide range of complex processes - first to produce the components and then to assemble the final product.

Outputs

Products

These are the most obvious industrial outputs. They are the commodities that are sold on the market to make a profit and raise capital for reinvestment.

Waste

Waste materials are produced by many manufacturing processes. In some cases, they may be sources of air or water pollution. Disposal of the waste can be costly or even dangerous, e.g. in the nuclear industry (see Chapter 11).

1 Copy and complete Table 10.1, using the information on the first few pages in this chapter.

Inputs	Processes	Outputs

Table 10.1 Inputs, processes, and outputs of industrial systems

Factors affecting industrial location

A number of different factors affect industrial location. The influence of each one varies, depending on the industry concerned. This has been illustrated in Chapter 8, when an example of heavy manufacturing industry is discussed (the motor vehicle industry); high-technology industry will be discussed later in this chapter.

Physical factors

Raw materials

The availability of raw materials used to be the single most important factor in the location of heavy industries (such as iron and steel and chemicals). These industries were located close to iron ore and coal mines, or (if imported raw materials were used) close to the importing ports. However, transport systems are now much more efficient. It is economically possible to transport bulky commodities, like iron ore, across the world in large ships.

Market and government influences have become much more important factors in choice of industrial location than the presence of the raw materials. For industries with low-bulk and high-value goods, the supply of raw materials is not a factor affecting their location.

Site

Large factories need flat, well-drained land on solid bedrock. It is also helpful if the land is cheap. Because factories are unattractive, they tend to be located away from higher-class residential districts.

Energy

Nearby energy supplies were an important factor in the location of industries in the past. Sites next to fast-flowing rivers or coal mines were favoured. However, today's electricity grid systems have largely overcome this need. But, for some industries, like aluminium smelting, that require very large amounts of power, access to cheap energy is important. Iceland has plentiful and cheap hydroelectric and geothermal power, so aluminium ore is brought by ship all the way from Australia to be smelted in Iceland.

Water supply

The manufacture of paper, certain chemicals, and metals requires more water than can be provided by a normal mains supply. These industries might need to be located where they have their own water supplies from rivers or boreholes.

Natural harbours and route centres

Ports are favoured locations for many industries because raw materials can be imported – and finished products exported – more easily and cheaply. Major roads and railways often follow natural routes such as valleys.

Human and economic factors

Capital

The finance to establish an industry might be more freely available in one country or area than another. This is often connected to the political factors described later.

Labour

The quality of the labour force is just as important as its size. As well as skill levels, the labour force needs to be adaptable to changes in circumstances. It might need to relocate to another area, or adopt new technology or working practices. The reputation of the workforce in an area is very important today.

Transport

Although less important today, transport is still an important factor in the location of industries with bulky goods. It may become more important again, as attempts are made to reduce the use of fossil fuels. Access to air transport is important for the location of high-technology industries and some tertiary activities.

Markets

The case study of the motor vehicle industry back in Chapter 8 illustrates how important access to people willing to buy the products can be.

Political influence

Governments can directly influence the location of industries by providing financial incentives to companies to locate in particular areas. The tax systems of countries are also an important factor in the decisions taken by multinational companies.

Quality of life

A highly skilled professional workforce will favour areas with good housing and leisure facilities.

> **2** List the ways in which governments and markets can affect the location of industries.

Fig. 10.1 The informal sector – a person selling chestnuts in Hunan Province, China

The informal sector

In LEDCs in particular, many people rely on the informal sector for their livelihoods. This sector is not taxed or monitored by government and includes street sellers, shoe cleaners and small workshops such as hairdressers, shoe repairers, and dressmakers. The people are self-employed and the activities are usually on a small scale. There are advantages and disadvantages for the people and the country from these activities.

Advantages of the informal sector for the people or the country

- The sector employs and supports a large number of people with no alternative employment.
- Little capital is needed, so it's fairly easy to start a business.
- Often, cheap or recycled raw materials are used.
- Informal jobs develop skills that could lead to better employment in the future.
- Any profits will be used locally and will stimulate other local activities.

Disadvantages of the informal sector for the people or the country

- Often the activity is illegal.
- No taxes will be raised for the country.
- The standard of the technology and goods is low.
- The income generated is uncertain and irregular.

High-technology industry

Processes

High-technology (or high-tech) industry uses the most advanced technology to make products that may, or may not, be high technology themselves. Pharmaceutical products, for example, are low tech but produced using high-tech methods.

A high degree of research and development is involved, because companies are always trying to keep ahead of competitors by developing new products and designing new machines to make them. The manufacturing process is highly automated and computerised.

Outputs

Products include pharmaceuticals, medical, optical and other precision instruments, computers, televisions, mobile phones, and aircraft. Biotechnology companies develop new kinds of food, drink, and vaccines.

Inputs and their effects on location

Capital

Large amounts of capital are required in this industry, because the best brains are needed to do the research and development work - and top-level, university-educated employees like this demand high salaries.

The headquarters and research and development units tend to be concentrated in MEDCs, where they can afford the expense of the highly educated top-level engineers and scientists. Most of these companies are based in the USA, Japan and Western Europe. However, as well as the creative work being carried out in MEDCs, high-tech companies may also have branches in LEDCs to assemble or manufacture their products. This is because land in LEDCs costs less than land in MEDCs, and labour there is also easily available and cheaper.

There is usually no shortage of financial backing, because investors can make a good profit if the company creates a successful new product. Many governments also provide financial support. The governments of some NICs have encouraged high-tech companies to develop there, e.g. Samsung in South Korea.

Raw materials

The silicon or silicon chips and electricity inputs needed are not expensive to transport, so the industry is not tied to a location near the silicon. It is described as 'footloose'.

Labour

Labour is the most important locational factor for high-tech industry. Because the industry needs large numbers of highly skilled and well-qualified workers, high-tech companies tend to locate in or near cities and towns with universities, engineering colleges and technological research institutes. They are frequently located in the suburbs, because this is where their workers prefer to live. Some high-tech companies even locate in pleasant rural areas on 'greenfield' sites.

To attract the necessary highly qualified workforce, industries need to locate where the climate is pleasant and the scenery attractive, and where there are good-quality housing and shopping opportunities, as well as cultural facilities and entertainment.

Land

High-tech companies usually locate themselves in science parks or technology development parks that have been developed by governments and local authorities to promote the growth of the industry. These parks need large areas of flat land.

Market

Nearness to market is not an important influence for high-tech industries, because many of their products are exported worldwide. However, in NICs a very large population encourages governments to finance new local high-tech firms.

Transport

Rapid transport is important, both for the management and the product, which needs to beat its competitors to the markets. Most companies locate near international airports, expressways, or railways.

Political influences

Government influence is enormous in funding the science parks, providing the necessary infrastructure for economic growth, and attracting investors by tax incentives. Since the Indian government started to develop the electronics industry in Bangalore in 1971, its efforts have been very successful - the electronics industry in India has grown faster than any other.

It is in a government's interest to encourage the development of high-tech industries, particularly those connected with aerospace and weapons.

Links to universities

Universities provide an input into high-tech industries in a variety of ways. Science parks were first set up in the USA so that the knowledge of university scientists could be used to develop high-tech industries. For example, in Baltimore the infrastructure for the science park was provided by the local authority, but the university decided which staff to employ and what laboratories and equipment were needed. Research papers are also available in university libraries, and companies are able to keep ahead of their competitors by being close to the latest research.

Pre-existing high-tech industries

High-tech companies tend to locate in clusters. Being close to other high-tech companies is important for promoting the competition that leads to the development of new ideas. If an area has a reputation as a location for successful high-tech industry, new companies are likely to be attracted to it.

3 Make a table with two columns to summarise the influences on the location of high-technology industry. Head the left-hand column 'Important factors' and the right-hand column 'Unimportant factors'. Make sure your table has a title.

Discussion point

How do the inputs of high-tech industries differ from those of most manufacturing industries?

The global distribution of high-tech industries

Fig. 10.2 shows that the headquarters of high-technology companies tend to be located in countries with high standards of education, plenty of money to invest, and excellent transport and telecommunications systems – as well as populations with strong purchasing power. Their branch factories (where the assembling of the finished products is carried out), by contrast, are often located in LEDCs.

This is changing as people trained in universities and high-technology firms in MEDCs return home and start up similar companies.

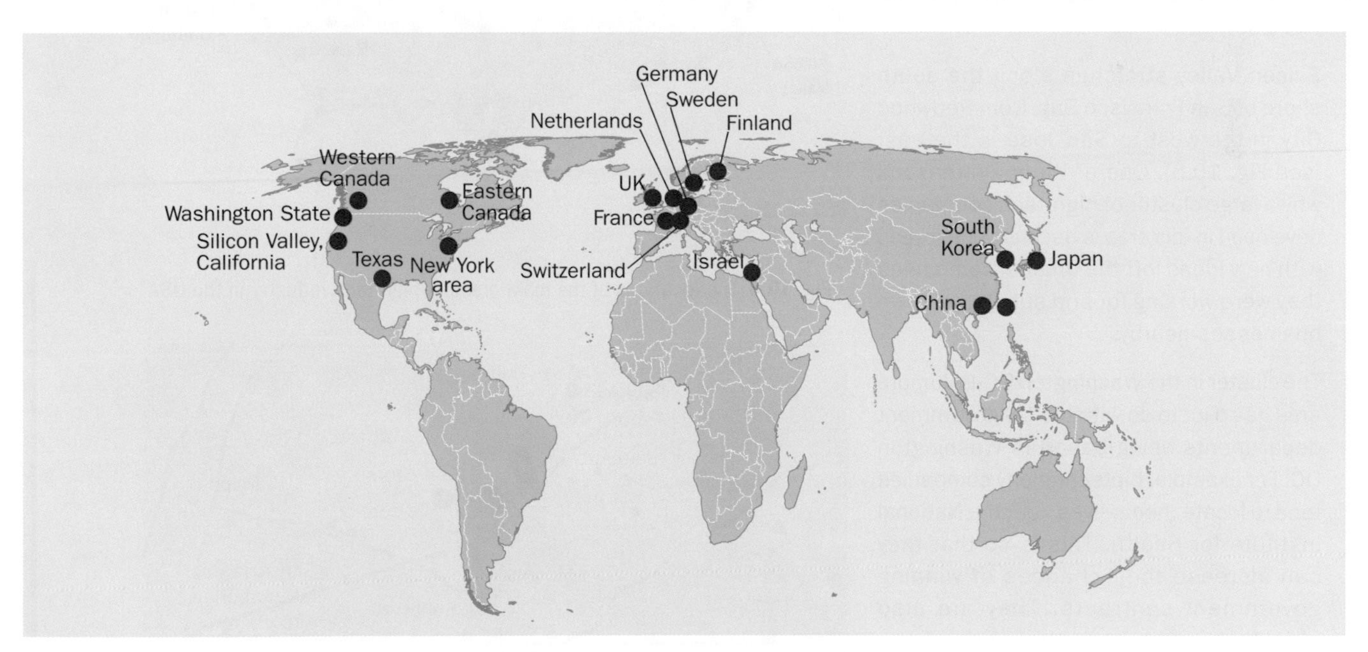

Fig. 10.2 The locations of the headquarters of major global companies making computer software, hardware, and electronics. See Fig. 10.4 for detail of the locations in the USA.

4 **a** Describe the distribution pattern shown on Figs. 10.2 and 10.4. Use the information on this page to describe the distribution in the USA.

b Suggest why there are no headquarters of major high-technology companies in Africa and South America. Use economic information from your atlas or another source to support your ideas.

Fig. 10.3 Assembling electronics at a branch factory in an LEDC

CASE STUDY

High-technology industries in the USA

Although there is a fairly widespread distribution of high-tech companies in the USA, there are several areas where they are particularly numerous. Silicon Valley in California is the largest, followed by the Washington-Baltimore area and places along Route 128 in Massachusetts (see Fig. 10.4).

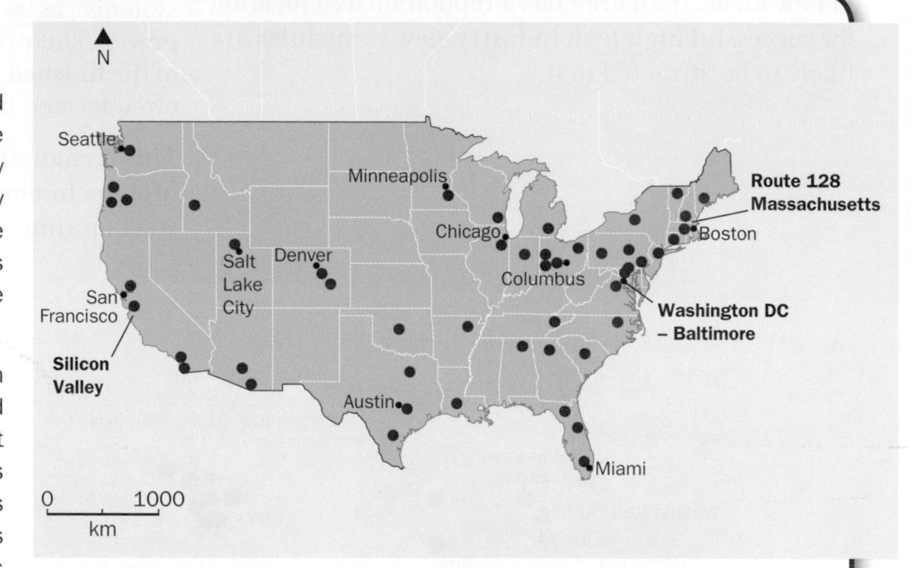

Fig. 10.4 The locations of the main areas of high-tech industry in the USA

Silicon Valley stretches along the south shore of San Francisco Bay, from Redwood City in the west to San Jose in the east (see Fig. 10.5). One of the main reasons why a large cluster of high-tech companies developed in this area is because employees with new ideas left the original companies they were working for and set up their own businesses nearby.

The cluster in the Washington DC-Baltimore area has a lot to do with the US government departments being based in Washington DC. For example, biotechnology companies tend to locate themselves near the National Institute for Health. This is so that they can increase their chances of winning government contracts. They are also able to benefit from the work of the government's research institutes. There are more than 70 high-tech companies in Washington DC.

The cluster of high-tech companies in Massachusetts is close to Boston and its world-famous universities, in particular the Massachusetts Institute of Technology (MIT).

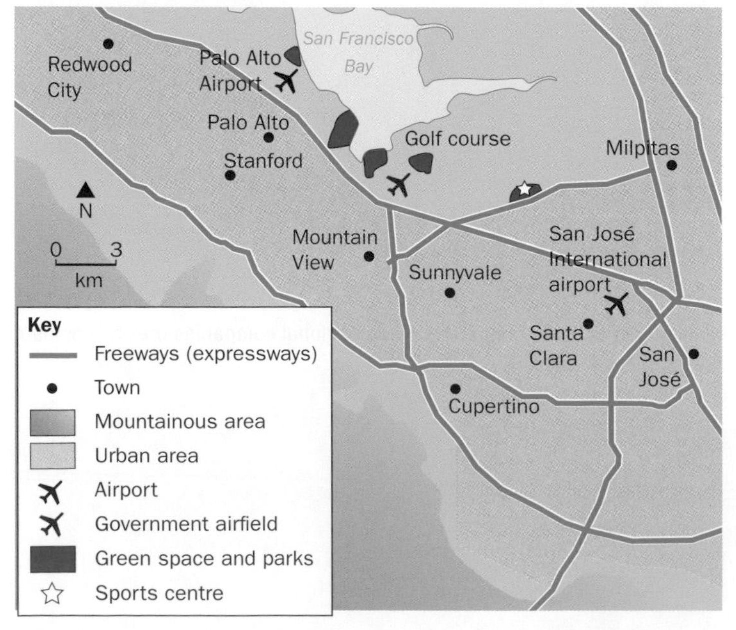

Fig. 10.5 Silicon Valley, California

Silicon Valley

Table 10.2 shows that clusters of companies making software, hardware and consumer electronics can be found in the towns of Silicon Valley. (Consumer electronics are electronic products that the general public buy, such as PCs, mobile phones, and televisions.)

Supply links exist between the different companies. For example, computer manufacturers like Hewlett-Packard and Apple Inc. need microprocessors and silicon chips. AMD and Intel Corporation make microprocessors. Cypress Semiconductor and National Semiconductor make silicon chips. Symantec, McAfee and Mozilla all develop security software for computers. Google and Yahoo are both search engines. Other high-tech companies in the area make computer games, storage devices, and other computer necessities.

These companies all have access to highly qualified workers, because there are many universities and technical institutes in the area. Stanford University at Palo Alto is one of the world's leading research institutions. There are also university buildings at Cupertino and Santa Clara.

Town in Silicon Valley	High-tech company making software, hardware, and consumer electronics
Cupertino	Actel, Apple Inc.
Milpitas	Adaptec, JDSU, LSI Corporation, Maxtor, SanDisk
Mountain View	Antibody Solutions, Google, Intuit, Mozilla Foundation, Symantec
Palo Alto	Aricent, Facebook, Hewlett-Packard
Redwood City	Electronic Arts, Oracle Corporation
San José	Adobe, Business Objects, Cisco Systems, Cypress Semiconductor, eBay, Hitachi Global Storage Technologies, Xilinx
Santa Clara	Aeria Games and Entertainment, Agilent Technologies, Applied Materials, Foundry Networks, Intel Corporation, McAfee, National Semiconductor, Nvidia, Sun Microsystems
Sunnyvale	AMD, Amdahl Corporation, Atari Inc., Juniper Networks, Maxim Integrated Products, NetApp, Yahoo

Table 10.2 The locations of firms making software, hardware, and consumer electronics in Silicon Valley

Fig. 10.6 The Pacific Shores Business Park at Redwood City

RESEARCH Research the climate of California in your atlas. Explain why the 'Sunshine State' has a desirable climate in which to live.

Fig. 10.7 The headquarters of Hewlett-Packard at Palo Alto

5 **a** Use all of the information provided so far to list the reasons why Silicon Valley is the home of so many high-tech companies. Start with the natural environment and then include the other factors.

b Use Figs. 10.5 and 10.6 to describe the advantages of the location of the Pacific Shores Business Park at Redwood City for the workers there.

c Look at Figs. 10.7 and 10.8. Describe the features of the headquarters buildings of these two high-tech companies and their surroundings. Suggest why they have the appearance and surroundings you have described.

Fig. 10.8 The headquarters of Apple Inc. at Cupertino

CASE STUDY

High-technology industries in China – an NIC

Towards the end of the 20th century, the Chinese government decided that it needed to encourage the development of high-technology industries in China. The country has some clear incentives for achieving this goal. It has an enormous population of nearly 1.4 billion, so there is a vast potential domestic market for consumer electronics. China's urban population is growing quickly. With better salaries, urban workers have more purchasing power.

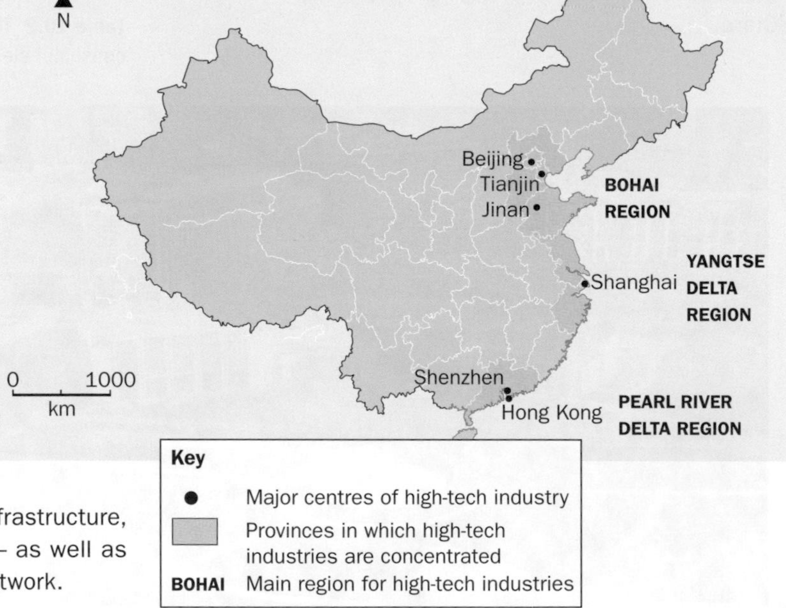

Fig. 10.9 The areas with the highest concentrations of high-technology industry in China

What has the Chinese government done?

- The government has made enormous improvements to the country's transport infrastructure, by building new airports and fast roads – as well as developing a good telecommunications network.
- It has also encouraged education, especially in science and technology. In the 1990s it sent thousands of researchers to the West to gain research experience.
- It has established major technological development zones in Beijing, Shanghai, and Shenzhen. Shenzhen is near Hong Kong (a leading financial centre). In Shanghai and Shenzhen, multinational companies from Europe, the USA, and Japan played a large part in the developments.
- It has invested massively in research and development. As of 2017, China has 17 national high-tech innovation parks and almost 150 regional ones.
- It has recruited foreign investors to encourage the growth of silicon chip manufacturing. They are often given exemption from income tax for the first few years after the business makes a profit, with reduced taxes after that.
- Foreign scientists have also been encouraged to work in China.
- Bank loans are available for new businesses.

There are over 100 million new subscribers to telecommunications every year in China. China now produces more computers, mobile phones, antibiotics, and vaccines than any other country.

The Shanghai area concentrates on pharmaceuticals. China is particularly keen to develop biotechnology. As in the USA, its computer manufacturers cluster with software and hardware manufacturers.

Jinan – a high-tech industrial development zone

Jinan is the capital of Shandong Province. The high-tech industrial zone there was started in 1991. It covers 83 km². Its publicity describes it as 'an exquisite environment' with green mountains to the south of the flat site and many trees. It is near an expressway and 15 minutes away from the international airport. It is also close to the Beijing–Shanghai super-highway, and a three-hour drive along an expressway from the port of Quingdao (one of world's largest seaports for handling containers). The main railway station is only 20 km away. A modern school, nursery and housing have been provided. Jinan has 37 universities and colleges, 11 of which are within the zone. There are over 500 software factories on the Qilu Software Park, which specialises in information technology, biopharmaceuticals and precision machinery. More than 10 multinational companies have set up joint ventures there.

Fig. 10.10 Jinan

6 **a** Use your atlas and Fig. 10.9 to describe the locations of China's high-technology industries. Suggest reasons for the distribution you have described.

b Use Fig. 10.10 to suggest why Jinan is a suitable location for a high-tech industrial development zone.

c Use all of the information provided to draw a spider diagram with 'High-technology industry in Jinan' in the centre and notes stating the main reasons for its development on the end of lines radiating out from the centre.

Fig. 10.11 An advertisement for a new science and technology park in China

Threats posed by industry to the natural environment and to people

The influence of inputs, processes, and outputs

Some industries use substances or materials - the inputs - that require a great deal of processing and can lead to a lot of pollution. The outputs include waste, as well as marketable products. Some examples of the varying effects of inputs, processes, and outputs for various industries are summarised in Table 10.3.

Industry	Inputs	Processes	Outputs	Possible adverse results
Quarrying and opencast mining	Dynamite	Blasting	Blocks of rock	Noise pollution, visual pollution, dust particulates
Iron and steel manufacturing	Iron ore, coke, limestone to separate the iron from impurities in the ore, water, recycled scrap iron for special steel: alloys, e.g. chromium, cobalt	Heating of ore to separate the iron by burning coke, rolling into sheets, cutting into lengths	Cast iron and pig iron, waste: slag, gases (sulfur dioxide, carbon dioxide, nitrous and nitric oxides, hydrogen sulfide)	Noise, large ugly buildings, slag heaps, dust, air pollution, water pollution (contaminated cooling water and **scrubber** effluent), risk of fires and explosions
Oil refining	Petroleum (hydrocarbon), fuel to heat, water to cool	Refining by heating the oil until it vaporises, then collecting the separate hydrocarbons as it cools	Various oils, bitumen, gas (sulfur dioxide, nitrous and nitric oxides, carbon dioxide, particulates)	Noise, visual pollution, odours, air pollution, water pollution (cooling and processing water, **scrubber** effluent), risk of fires and explosions
Leather	Hides, water, chemicals for tanning and dyeing	Treating, tanning, dyeing	Leather, chromium compounds	Odour, pollution of processing water with sulfates and chromium

Table 10.3 The relationships between inputs, processes and outputs in certain industries

Some of the outputs are waste products that are harmful to life. For example, the chromium compounds produced in leather manufacture can cause lung cancer and ulcers. Some of the air pollutants that leak accidentally from chemical industries are carcinogenic (e.g. benzene) or toxic.

Even high-tech industries have the potential to pollute. Accidental spills and leaks of solvents and acids can cause toxic substances to pollute both air and water.

Sulfur dioxide is produced when fossil fuels are used in smelting, refining, and electricity generation. After mixing with water droplets in the atmosphere, it eventually falls back to earth as acid rain, often far from the source of pollution. During the 1980s sulfur dioxide emissions from the UK acidified soils, rivers and lakes in Norway. Fish were poisoned. Now much of the sulfur is removed by scrubbers from the gases emitted from chimneys.

In MEDCs, pollution is controlled by strict laws to ensure that dangerous waste does not normally enter the air, seas, or rivers. Many modern industries use scrubbers on chimneys to remove harmful gases.

In countries such as the UK, government agencies monitor air and water pollution. They have powers to fine polluting industries and to shut them down if they do not conform to the required standards.

7 Imagine that there are proposals to build an oil refinery near your home. What advantages and disadvantages would result from it?

Discussion point

Look at the case study about Karachi. Should MEDCs pay to replace outdated technology in LEDCs? Could people in MEDCs benefit from this?

Fig. 10.12 Air pollution is also a form of visual pollution

CASE STUDY

Pollution in the industrial zone of Karachi, Pakistan

In Pakistan there are almost 600 leather tanneries, mostly in the Karachi industrial area. Tanning is one of the most toxic industries because it uses large volumes of polluting chemicals. Leather is an important export, but the tanning process (the conversion of skin into leather) is affecting the health of the people in the area and harming the environment. Many workers have respiratory illnesses and skin infections.

Tanning causes three types of pollution:

- The skins are washed with large quantities of water containing chromium compound and treated with arsenic to preserve them. The untreated effluent is then discharged into open channels, which empty it into the sea – endangering the health of people living in coastal villages.
- The foul-smelling waste water also contains large amounts of fat, hair and other solid waste from trimming and shaving the skins.
- Finally, ammonia is released during deliming and hydrogen sulfide is released into the air during tanning at the dehairing, colouring, and drying stages.

It is believed that the water and air pollution is directly affecting the health of more than a million people in Korangi Township (a coastal area within Karachi). The tanning factory owners have plans to reduce these pollutants, but financial difficulties have caused the measures to be delayed. Small businesses cannot afford more environmentally friendly technology.

Three other major polluting industries, which discharge large quantities of effluent, are located near the port (to make exporting easier). They are a steel mill, a power plant and oil refineries (together with many smaller industries).

High levels of heavy metals – such as arsenic, lead, and chrome compounds – are found in sea animals in the area. If humans eat sea life containing lead, brain damage, anaemia, and kidney failure can occur. The lead may be from pesticides used to preserve the hides during transport to the tannery.

The marine ecosystem is also being damaged because of pollution caused by shipping, especially importing oil and exporting oil products, moving in and out of the port. A lot of oily discharges are pumped out within the port area. All of these marine pollutants are threatening the mangrove ecosystem of the Indus delta.

The need for sustainable development and management of economic activity

An activity is sustainable if it does not spoil the chances of future generations being able to meet their needs. Certain resources are finite and, when used up or destroyed, are gone for ever. Some activities have harmful effects that could make life more difficult for future generations. Table 10.4 summarises case studies about these impacts.

Activity	Problem	Conservation methods
Opencast mining (Chapter 7, Namib Desert)	Finite resources Destruction of the landscape	Restoration of the landscape for future use
Rock and mineral use for industry (Chapter 10, Pollution in Karachi)	Finite resources Air and water pollution	Recycle into new products Legislation to enforce recycling Use renewable substitutes
Fossil fuel use for energy (Chapter 11, Energy in Germany)	Finite resources Air and water pollution	Use renewable substitutes Educate the public to save energy
Deforestation (Chapter 7, Deforestation of Borneo; Chapter 9, Cattle in the Amazon Basin)	Loss of habitats and food Extinction of species Atmospheric pollution Global warming Soil degradation	National Parks Selective logging, afforestation and reforestation Substitutes for wood-burning stoves
Depletion of water sources (Chapter 7, The Mojave Desert)	Groundwater supplies and reservoir water used faster than can be replenished by rain	Prioritise and ration water use Transport water from wetter areas
Soil erosion, degradation and desertification (Chapter 9, Wheat farming in Canada, Subsistence agriculture in Swaziland; Chapter 7, Desertification in Chad)	Soil is a finite resource and takes many years to form Decreased yields and food stocks leading to malnutrition and starvation if severe	Numerous measures, including controlled grazing, reducing stock densities, contour ploughing, crop rotation, terracing, inter-cropping, dry farming techniques
Mass tourism (Chapter 10, Tourism in Jamaica)	Replacement of natural environment with tourist accommodation Air pollution	Limit development Ecotourism

Table 10.4 Case studies of economic activities that can be made more sustainable

How economic activities cause global warming

Average global temperatures are rising but there is disagreement about what is causing it. The consensus of scientific opinion is that the main culprit is carbon dioxide.

Look back at Fig. 7.5 on page 190 showing the *natural* greenhouse effect that keeps our planet warm because greenhouse gases naturally present in the atmosphere absorb the Earth's long-wave radiation. Human activities are increasing this process by adding greenhouse gases through atmospheric pollution, causing the **enhanced greenhouse effect**.

LEARNING TIP Sun's rays that are *reflected* off surfaces are long-wave and escape more easily to space. Only rays *radiated* by the Earth are trapped.

The CO_2 that humans and animals add to the atmosphere and the CO_2 taken in by trees used to be in balance, but continued industrialisation, increased transport and forest clearance now add more CO_2 to the atmosphere. Fig. 10.14 gives information about the addition of greenhouse gases by human activity.

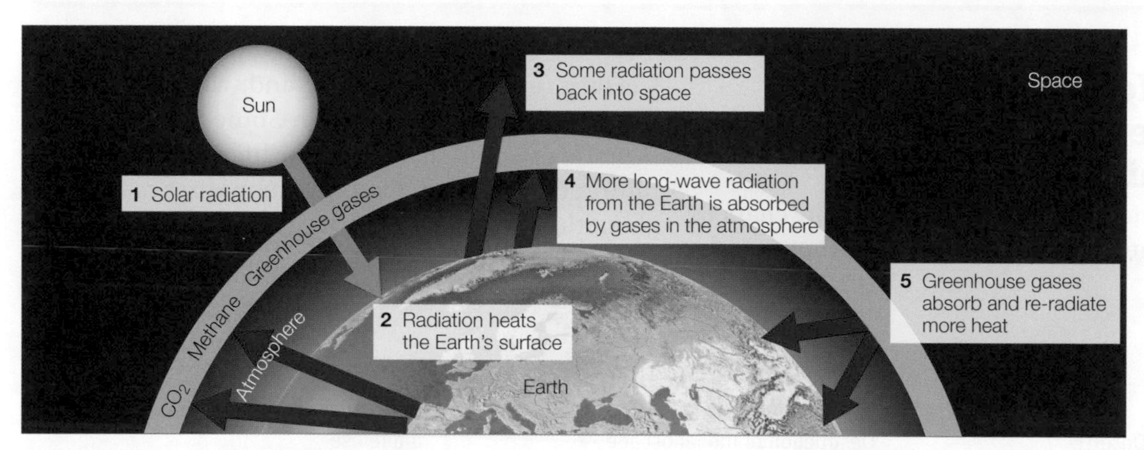

Fig. 10.13 The enhanced greenhouse effect

Greenhouse gas	Atmospheric concentration	Lifetime in atmosphere (years)	Contribution to the enhanced greenhouse effect	Sources from human activity
Carbon dioxide	403 ppm* 2017 Increased from 400 ppm 2014	50–200	The main greenhouse gas	Burning wood and fossil fuels Deforestation
Methane	1800 ppb** Increasing by up to 2% per year	12	The most potent gas. Its heat-trapping effects are 25 times stronger those of CO_2	Cattle and sheep guts Decomposition of organic matter in landfill, wet rice fields and bogs
Nitrous oxides	332 ppb	114	In small quantities but its impact is 300 times that of CO_2	Nitrate fertiliser Burning fossil fuels, especially in diesel engines and burning vegetation
CFCs	1863 ppt***	up to 50 000 (variable)	Very efficient absorbers of long-wave radiation	Old aerosols and refrigerants (not used in new products)

* ppm = parts per million, **ppb = parts per billion, *** parts per trillion

Table 10.5 The greenhouse gases

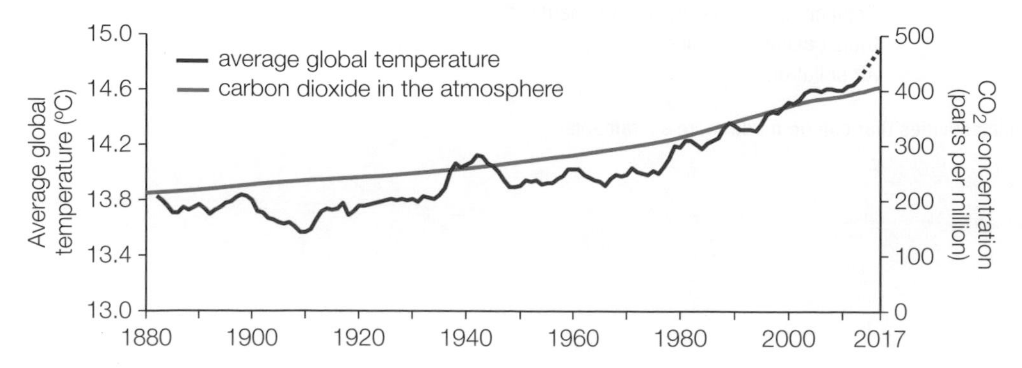

Fig. 10.14 Changes in average world temperature and carbon dioxide concentrations in the atmosphere 1880–2017

8 Use the data from Fig. 10.14 to answer the following:

a By how much has the world average temperature increased since 1860 and 1950?

b Describe the main changes in global average temperatures since 1860.

c How do the changes in carbon dioxide concentrations since 1860 differ from the changes in the global average temperatures?

d Comment on the conclusion that carbon dioxide is the main pollutant causing global warming.

Why continued increase in global average temperature is not sustainable

An increase in average global temperature is not sustainable and has numerous consequences, which are summarised in Fig. 10.15.

Facts about the threats of global warming to the natural environment and people

- Arctic sea ice has decreased by about 13% every ten years.
- Sea levels have risen by 3.4 mm a year since 1993. Low-lying coasts, such as in Bangladesh, experience more frequent flooding.
- Malarial mosquitos have moved north into Southern Europe.
- Melting permafrost in Siberia and Canada is defrosting bogs, releasing methane.

As different parts of the world may become cooler, warmer, wetter or drier, global warming is often referred to as climate change.

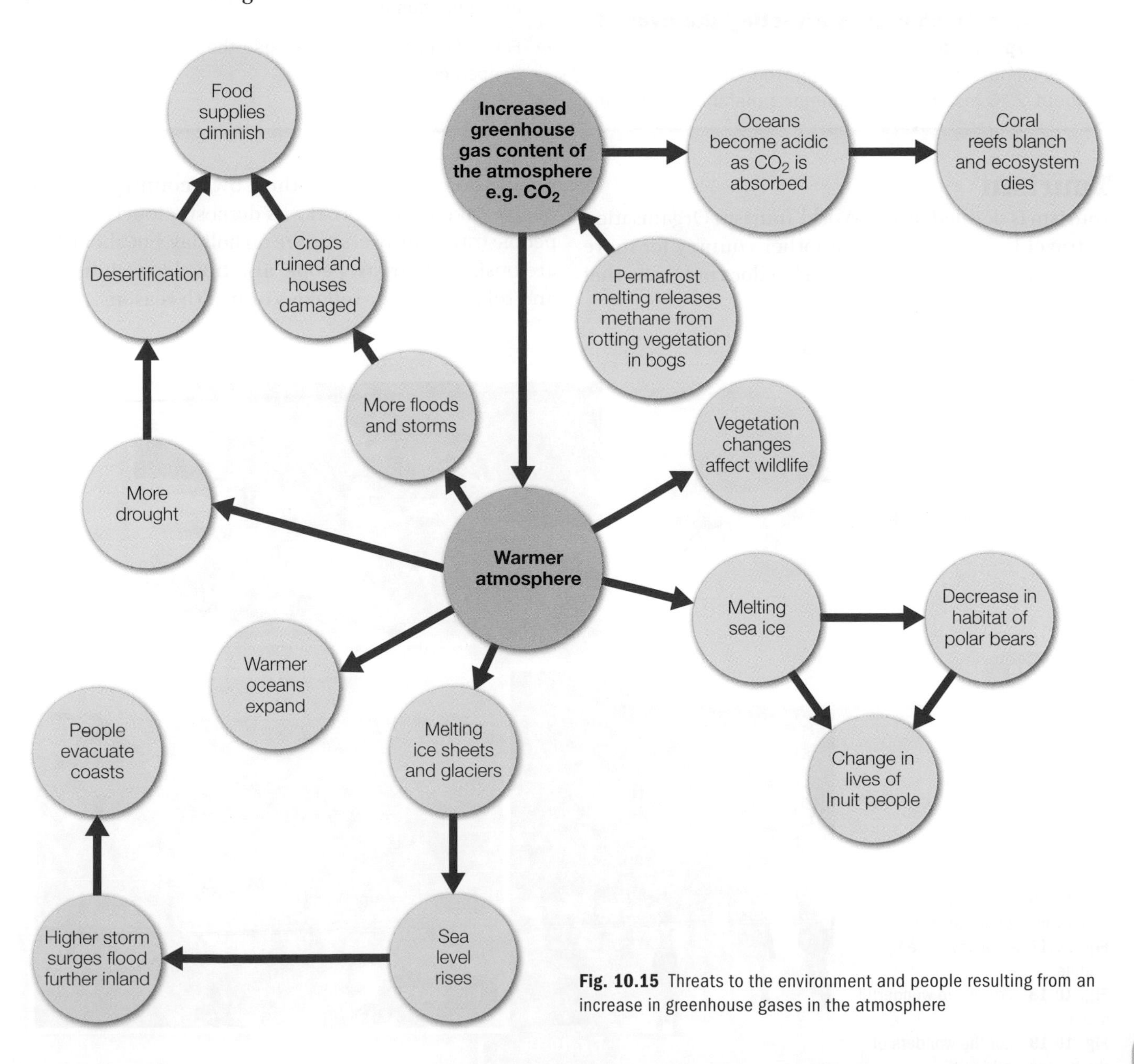

Fig. 10.15 Threats to the environment and people resulting from an increase in greenhouse gases in the atmosphere

CASE STUDY

The effect of climate change on the Inuit people of Northern Canada and Greenland

Inuits live on the coasts of Southern Greenland and Northern and North-eastern Canada, where winters were snowy, with temperatures below freezing, and sea ice extended south to reach the coast.

Traditional way of life

The Inuit's subsistence living has been based on hunting for meat and skins. In the winter they travel on sea ice on snowmobiles to trade, hunt polar bears and seals and to catch fish through holes cut in the ice. Their traditional way of life depends on the ice being strong and intact. Travel is much more difficult in the summer when the surface layer melts and the ground becomes boggy with many pools.

How climate change is affecting the lives of the population

Their climate has been changing. The Arctic has warmed about 2 °C in 20 years. Summer sunshine is rarer and fog, cloud and rain more frequent. Winters are shorter and warmer with less snow, so the ice is thinning and breaking up into floes. In some years it does not reach the coast.

This climate change is affecting the Inuits' traditional way of life:

- There are fewer months when it is safe to go on the sea ice. Deaths are increasing due to ice breaking.
- Polar bears are forced to remain on land in winters when the edge of the ice is too far north.
- The polar bear's habitat is disappearing. They need to eat enough food in winter to keep themselves alive for the rest of the year but seals are harder to find. There are increasing reports of polar bears attacking humans.
- Fewer fish are caught because they are moving into deeper, cooler waters.

Tourism

Tourism is defined by the World Tourism Organisation as travel for any reason to another country for more than a day but less than a year. That does not mean that people do not holiday in their own country – many people take leisure breaks as domestic tourists. Most people travelling overseas are on holiday, but about 15% are business tourists. People also travel to visit friends and relatives, or for religious or health reasons.

Different types of holiday

Fig. 10.16

Fig. 10.17

Fig. 10.19

Fig. 10.18

Fig. 10.16 A beach holiday or a holiday on a cruise ship
Fig. 10.17 A holiday to see wildlife...
Fig. 10.18 ...or to see wonderful scenery
Fig. 10.19 ...or the wonders of the human world

Fig. 10.20 Take a city break
Fig. 10.21 Visit historic buildings
Fig. 10.22 Have a theme park holiday
Fig. 10.23 Learn about other cultures
Fig. 10.24 Have an adventure holiday in natural surroundings

The photographs shown in Figs. 10.16–10.24 give a flavour of the different sorts of holiday that people can take. Other types of holiday are also possible. Sports tourism has grown rapidly, because almost all sports have gone global by holding international competitions and world championships. For example, many people travel to watch football matches in other countries.

9
a i Explain why people of different age groups are likely to go on the two types of holiday shown in Fig. 10.16.
ii Why would both types of holiday shown on Fig. 10.16 appeal to workers in a city?
b What type of people would enjoy the activities shown in Figs. 10.22 and 10.24?
c Why might the woman in Fig. 10.23 have mixed feelings about tourism?
d Describe the growth of world tourism, and of earnings from tourism, using the information in Table 10.6.

What has caused the tourism boom?

Tourism has increased fast in recent decades. People have been able to travel more because they have:

- a choice of air services to an increasing number of destinations
- more affordable flights, because of the development of larger, more economical aircraft and the growth of budget airlines
- more leisure time, with longer weekends and longer holidays
- higher wages and more money available to spend on things like holidays. There has been a large increase in Chinese tourists since 2000. In 2009, their expenditure on tourism increased by 21% on the 2008 figure. Only Germany, the USA, and the UK spent more on tourism than China
- better health, especially in later life, so more older adults and pensioners now have adventurous holidays (they're often referred to as the 'grey market', spending 'grey pounds')
- the Internet, which has helped people to book cheaper travel deals themselves, rather than using travel agencies.

Another reason for the growth in tourism is that there are now more retired people who are free to take holidays. Also there has been an increase in marketing. Countries have formed tourist boards to promote their destination. They advertise in magazines, newspapers, and brochures, as well as on TV and the Internet.

Discussion point

What type of holiday would you most like to have and where would you choose to go? What would you like about your chosen type of holiday?

Year	World total tourist arrivals	Earnings from tourism (US$ millions)
1950	25 280 000	2 100
1960	70 000 000	6 900
1970	166 000 000	18 000
1980	286 250 000	105 200
1990	459 200 000	264 700
2000	681 000 000	478 000
2010	1 006 000 000	870 000
2016	1 235 000 000	122 000

Table 10.6 World tourist arrivals and earnings from tourism

There has also been an increase in 'built attractions', such as theme parks and water parks. Singapore - a small country with few natural attractions - frequently has to develop new attractions so that visitors will return. That was why its Night Safari Zoo came into being. Singapore is a frequent stopover destination for people taking the long-haul route from Europe to Australia and New Zealand. But it faces competition as a stopover from neighbouring Malaysia, as well as cities like Hong Kong and Dubai.

Fig. 10.25 The elderly are increasingly adventurous

Which are the most popular countries for tourists?

Table 10.7 lists the top ten tourist destination countries in 2016. For some people these countries will be short-haul destinations (reached by a flight of less than three hours).

10 Which groups of people would probably prefer to go to short-haul destinations? Give reasons for your answers.

Discussion point

Members of the class could describe the attractions of a place they have been to on holiday. Find the places on a map and discuss where the most popular destinations are and why that is so.

Rank	Country	Tourist visitors (million)
1	France	82
2	The USA	75
3	Spain	75
4	China	59
5	Italy	52
6	The UK	35
7	Germany	35
8	Mexico	35
9	Thailand	32
10	Turkey	31

Table 10.7 The top ten countries for tourist arrivals in 2016

Where tourism really matters

Tourism matters a lot to many of the world's poorer countries. It can make a big difference to their economies, by bringing jobs and income to places where they are badly needed. Money from tourism can also pay for big **infrastructure** projects, like new roads and bridges and improved water supplies and energy systems. All of these things can help a country to develop.

The percentage of a country's GDP provided by tourism shows how important tourists are to its economy (GDP is a measure of a country's wealth).

CASE STUDY

Lanzarote – turning a desert island into a tourism-based economy

The small Spanish island of Lanzarote lies off the west coast of North Africa – near the Sahara Desert. It's a barren and volcanic desert island, without any natural surface water. Before tourists began to arrive, it had a small population and a very poor economy – mainly based on fishing. Apart from sunshine and a spectacular coast, the island had little to attract tourists.

How did it all change?

In the 1960s, it was decided to turn the island into a tourist resort.

- Good roads were constructed.
- The airport was improved.
- Desalinisation plants were built to produce fresh drinking water out of seawater.
- Artificial beaches were constructed, using imported sand from the Sahara Desert.
- Hotels and holiday apartments were built.
- All buildings had to be painted white (to provide a contrast with the black volcanic lava), and only green doors were allowed. Buildings on the island were also limited to two floors, so that they would not intrude on the landscape.

These changes and improvements produced benefits for the local population, as well as for tourists. Not only was the infrastructure put in place to support tourism, but the natural landscape was used as much as possible, as the photographs in this case study show.

Fig. 10.26 White buildings brighten the barren volcanic landscape (this was the home of the architect who designed the island's artificial attractions)

Fig. 10.27 A viewing platform and a restaurant with glass picture windows on its seaward side was built at Mirador, so that tourists could enjoy the view while having refreshments

Fig. 10.28 A decorative pool was added to a tunnel in the lava at Cuerva de los Verdes (Green Cave)

Fig. 10.29 This extinct volcanic crater was converted into a cactus garden

Tourists are taken in coaches on a volcano tour of the Timanfaya National Park. They end up at the top of Montanas del Fuego, where a restaurant serves food cooked on a metal grill placed over a hole – using heat from the volcano! Outside, tour guides demonstrate that the volcano (which last erupted in 1730–31) is still hot:

- They put dry wood into a small hole and it catches fire immediately.
- They also pour water into a tube going down into the earth and a few seconds later a mini geyser shoots up.

Fig. 10.30 On top of the Montañas del Fuego (Fire Mountains), at one of Lanzarote's volcanoes in the Timanfaya National Park

Fig. 10.31 The view from the Montañas del Fuego of one of the roads specially built across the lava field to enable tourists to reach the top of the volcano

Fig. 10.32 Some local people earn a living by keeping camels and taking tourists for rides on them

11 Look at Table 10.8.

a Calculate the annual rainfall of Lanzarote and compare it with rainfall in the deserts you have studied. Explain why it is so dry.

b Lanzarote is a very popular all-year-round destination. Use the table to explain why this is.

c Many people do not want the island to be developed any further. Suggest reasons why.

d Make a list of the jobs involved with tourism on Lanzarote. The photographs may suggest some and you may be able to think of others. Write them in two lists – one for jobs directly involved with tourism and the other for jobs indirectly involved with tourism.

e In what other ways would the islanders benefit directly or indirectly from the growth of tourism?

f Local people benefit from infrastructure and facilities provided for tourists. Suggest three different types of facility that develop because of tourism, but which local people can also use.

Month	Average temperature (°C)	Average rainfall (mm)	Average sunshine hours
January	17.0	24	7
February	17.5	14	8
March	18.5	15	8
April	19.0	6	8
May	20.2	2	9
June	21.9	0	11
July	23.8	0	10
August	24.7	0	11
September	24.4	2	8
October	22.5	7	7
November	20.3	12	6
December	18.1	27	7

Table 10.8 The climate of Lanzarote

Disadvantages of tourism for Lanzarote

Tourists:

- create noise pollution and litter
- can upset the locals by dressing inappropriately when off the beach
- cause pavement and road congestion – and add to air pollution by using vehicles
- use a lot of water. The tourist resorts have priority over the available water supplies in times of shortage. This means that the locals may not have access to irrigation water for their vines and vegetables at these times
- occasionally cause the island's sewerage system to become overloaded. Discharges from sewers straight into the sea can cause water pollution
- lead to an increase in the price of goods and services for local people. This raises their cost of living.

As well as potential conflict between tourists and locals, there can also be conflict between different groups of tourists – especially if they have different social norms and cultures.

Although Lanzarote is an all-year-round destination, it does not have as many visitors in the winter months as in the other months – so there is some seasonal unemployment. Some facilities, like the water parks, are underused then.

Fig. 10.33 An animal show at the Rancho Texas Park, a big water park where visitors can enjoy water slides and other water features designed for their leisure

Fortunately, the developers of Lanzarote were determined that it would not become spoilt by tourism, as Spain's Costa del Sol has been. Unplanned development there has created a long – almost unbroken – strip of concrete buildings, many of them high-rise, which have obliterated animal habitats, natural vegetation, and covered fertile soil. The natural beauty of the coast has been destroyed and many tourists no longer find it attractive. The Spanish authorities are now trying to improve the area by pulling down some buildings.

Lanzarote is an example of how careful management is needed if problems like those of the Costa del Sol are to be avoided. Most islanders think that enough of the physical landscape has now been covered with buildings and infrastructure, and that no further damage should be allowed.

CASE STUDY

Jamaica – totally tropical tourism

Jamaica is one of the Caribbean's top tourist destinations. One reason for that is its hot, tropical climate. At sea level it's hot all year, and there's plenty of sunshine too (for at least seven hours a day). But Jamaica does have rain, as Fig. 10.34 shows.

Most tourists stay on the north coast – in resorts such as Ocho Rios, Montego Bay and Negril. Here the beautiful sandy beaches have been developed into tourist resorts.

Fig. 10.35 A beach in Jamaica

Fig. 10.34 A climate graph for Montego Bay, Jamaica

Over 2 million tourists visited Jamaica in 2016, compared with 0.6 million in 1982.

Because of the large numbers of tourists involved, mass tourism can have major effects (both good and bad) on tourist destinations and the people who live in them.

How does mass tourism affect Jamaica's economy?

Tourism brings in a lot of money for Jamaica – about 15% of Jamaica's GDP in 2016, when tourists spent 2.6 billion US dollars there. Tourism accounts for nearly half of the money Jamaica earns from abroad.

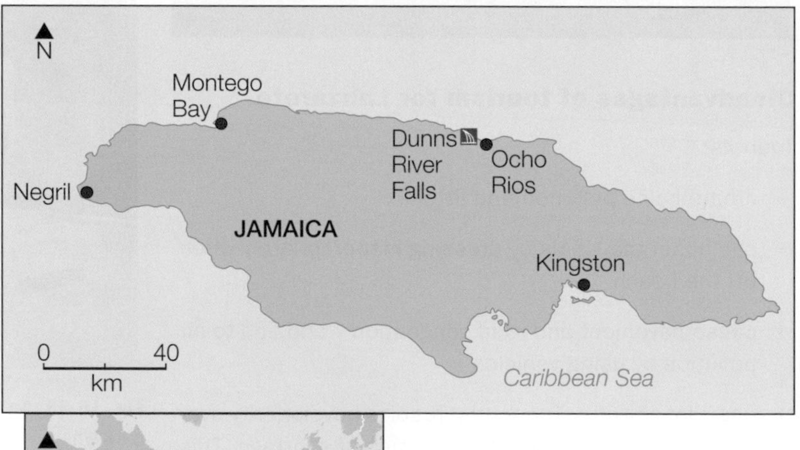

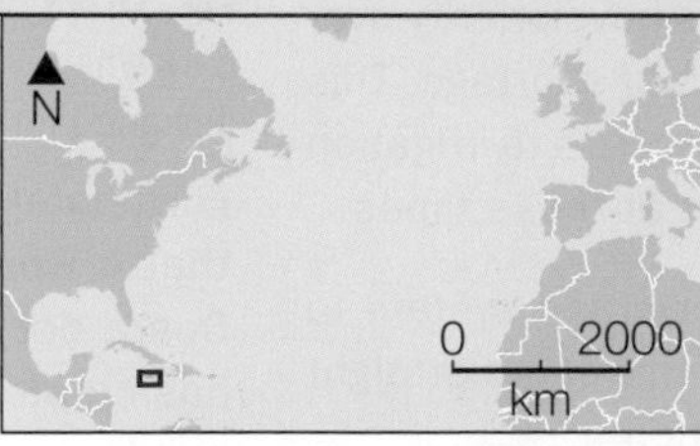

Fig. 10.36 The location of Jamaica (a long way from Europe, from where many of the tourists come)

Positive economic effects	Negative economic effects
• The money spent by tourists makes tourism businesses, like hotels, profitable. • Those tourism businesses employ many local Jamaican staff. • The Jamaican tourism workers spend their wages in other Jamaican businesses, which in turn become more profitable and employ more local staff. • The taxes paid to the Jamaican government by businesses, workers, and tourists provide money which helps Jamaica to develop. • Jamaicans learn skills in the tourism industry that can be used in other parts of the economy. • Many tourism jobs pay well by Jamaican standards. • Tourist resorts and the people who live there become richer.	• Many tourism businesses are owned by foreign companies, so most of the profits end up abroad. This is called economic leakage. • Some tourism staff are foreigners. They also send their wages home. This is economic leakage too. • Economic leakage also means less tax revenue for the government to develop Jamaica. • Jobs in tourism are often seasonal. • Some skilled Jamaicans leave to work abroad for more money. • Tourist destinations attract Jamaicans from poor inland areas, where businesses lose out. • Tourist jobs and money are concentrated in the resorts, so inequalities with other parts of the country increase.

Table 10.9 The effects of tourism on Jamaica

12 a i Use Fig. 10.34 to describe Jamaica's climate.

ii Explain why Jamaica's climate attracts holidaymakers.

b Work as a group of four. Each adopt one of the following roles:

Myron – a Jamaican hotel receptionist

Larry – a Jamaican hotel manager

Leon – a Jamaican shop owner in an inland town

Tom – a tour guide from Canada

You have been asked to contribute to an online discussion forum about whether more tourism would be good for Jamaica's economy. What will you say? Write a two-minute talk about your views.

How does mass tourism affect the environment?

Mass tourism has mostly affected the areas around Jamaica's north coast resorts, east of Negril and as far as Ocho Rios. They've become built-up, congested and polluted. Most tourists arrive at Montego Bay's international airport. Their environmental impact on Jamaica begins here, although their journey this far has already made a difference to each person's **carbon footprint**. Fig. 10.37 illustrates the environmental impacts of mass tourism on a taxi minibus journey from Montego Bay international airport to Negril.

Carbon footprint

A person's carbon footprint is a measure of the amount of carbon that their lifestyle adds to the atmosphere, and travel is part of that. A tourist visiting Jamaica will have travelled to the airport in their own country and then flown to Jamaica – possibly right across the Atlantic or Pacific Oceans. Both parts of the journey will have emitted carbon dioxide and other greenhouse gases into the air.

Fig. 10.37 Jamaican tourism's environmental trail

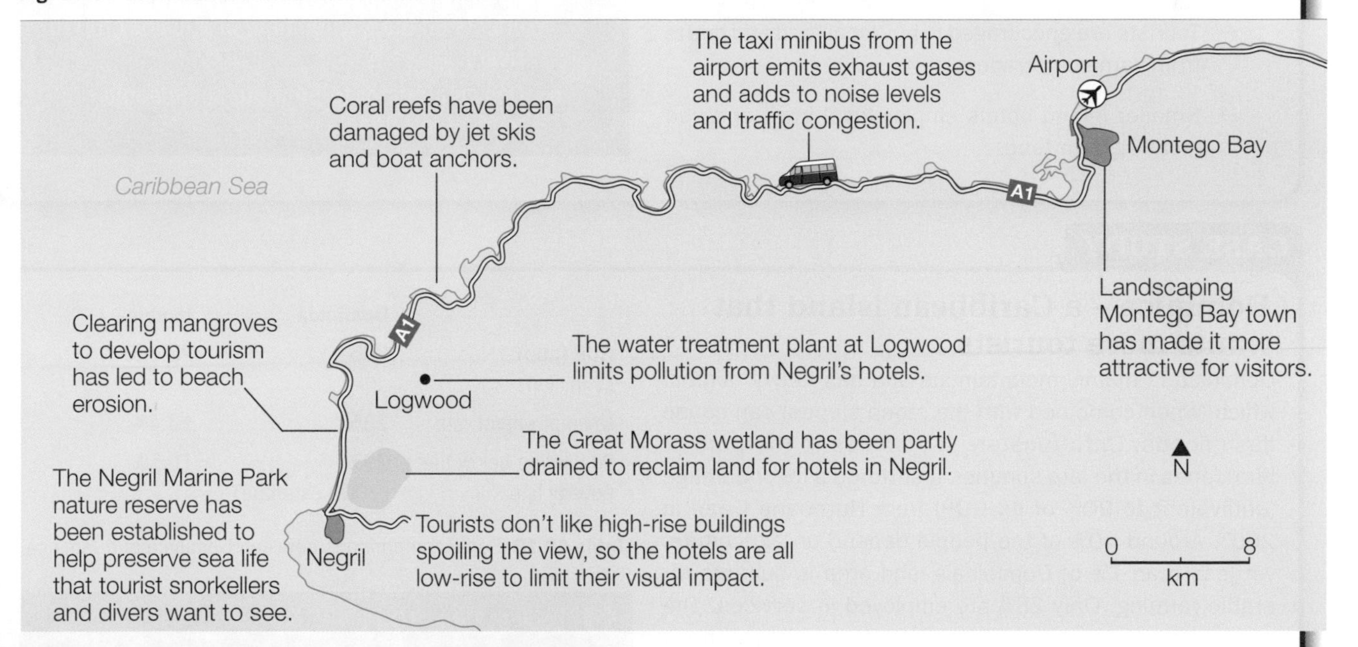

Fig. 10.38 Honeypots are a consequence of mass tourism. People swarm like bees to beautiful attractions like Dunn's River Falls in Jamaica.

How can tourism grow and be sustainable?

Jamaica is a relatively poor country – its GDP is only about US $9000 per person per year, compared with the UK's US $42 500. Increasing tourism could help to raise the standard of living of Jamaica's people.

However, mass tourism also has some negative effects, as you've already seen. Many Jamaicans have come to dislike it. They don't see how it helps them or the country. There is therefore a puzzle for the government to solve – how can they develop tourism but avoid its negative effects?

They need to find sustainable solutions – ways for tourists to visit Jamaica without damaging its future and the future of its people.

What's been tried?

The Jamaican government has been following a Master Plan, to try to develop sustainable tourism. Its three main ideas have been to:

- limit the development of mass tourism to existing resorts, like Ocho Rios
- spread small-scale tourism to other parts of the island
- involve local people more.

As part of its Master Plan, the government has encouraged:

- community tourism – local people running small-scale guesthouses. This helps to bring tourists to less developed towns, such as Port Antonio, without mass tourism's negative effects.
- responsible tourism, which involves local people and aims to do as little harm as possible. For example:
  - Local guides take visitors to off-the-beaten-track attractions, such as the Rio Grande River.
  - Tourists are encouraged to buy local food and crafts from Jamaican traders.
  - Smaller inland hotels employ local staff and use locally grown food.

Fig. 10.39 A sustainable tourist activity – rafting on the Rio Grande River

- tourists and local people to get in touch with each other through the Jamaican Tourist Board's 'Meet-the-People' website initiative. This helps both visitors and locals to understand each other, which is an important benefit of tourism.
- educating tourists and locals about how to avoid negative environmental effects.

13 **a** Use the information in this case study to draw up a table showing the positive and negative effects on the environment of a holiday trip to Negril.

b Design a poster to promote sustainable tourism in Jamaica.

CASE STUDY

Dominica – a Caribbean island that wants more tourists

Dominica is mainly mountainous and has heavy rainfall, which (when combined with the steep slopes) can cause flash floods. Unfortunately, it also lies in the path of hurricanes in the late summer. It suffered a lot of damage (equivalent to 20% of its GDP) from Hurricane Dean in 2007. Around 40% of the people depend on agriculture, yet less than 7% of Dominica's land area is suitable for arable farming. Only 28% are employed in services. The government is now seeking to increase tourist revenues by promoting the island as an ecotourism destination.

Fig. 10.40 A street in the capital, Roseau, where some live in poverty

	Dominica	Jamaica
GDP (ppp) ($ million)	805	25 390
Unemployment rate	23%	13.4%
Population below the poverty line	29% (2009 estimate)	16.5%

Table 10.10 Some comparisons between Dominica and Jamaica

14 Look at Table 10.11.

a Describe the pattern of tourist arrivals since 1990. (Remember to generalise and to state highest and lowest, together with any anomalies.)

b Is the government achieving its aim to increase income from tourism?

c Consider the effect on visitor numbers that the following appear to have had (if any): the economic depression in Asia in 2000; the SARS outbreak in Asia in 2003; Hurricane Dean in 2007; the worldwide economic recession that started in 2008 (this led to a worldwide decline in tourism of 4.2% in 2009); and the A(H1N1) influenza outbreak in 2009. Remember that people often book a holiday some time before they take it.

Year	Visitor arrivals	The contribution of tourism to Dominica's GDP (%)	Year	Visitor arrivals	The contribution of tourism to Dominica's GDP (%)
1990	152 200	4.7	2001	276 000	7.4
1991	155 600	5.2	2002	208 000	7.7
1992	159 000	5.4	2003	254 000	8.3
1993	175 900	5.7	2004	466 000	9.1
1994	192 800	6.2	2005	381 000	8.1
1995	203 000	8.0	2006	465 000	9.7
1996	261 000	7.7	2007	437 000	9.3
1997	299 000	8.3	2008	545 100	8.1
1998	312 000	7.9	2009	614 000	7.6
1999	280 000	8.2	2010	623 900	7.5
2000	312 000	7.5	2016	720 000	11.4

Table 10.11 Visitor arrivals and the contribution of tourism to Dominica's GDP, 1990-2016

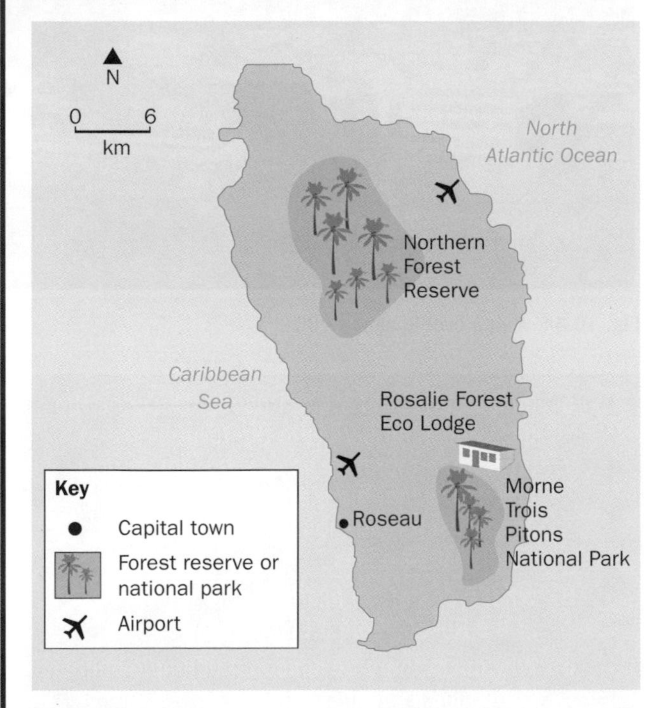

Fig. 10.41 Dominica

What does Dominica have to attract tourists?

Dominica is known as the 'Nature island of the Caribbean', because of its lush tropical rainforest vegetation and varied plant and animal life. Its forests are still fairly intact and teem with bird and animal species – some of which are found nowhere else. For example, there are many lizards and 55 species of butterfly. It has no poisonous snakes or spiders (but it does have boa constrictors). The rainforest is protected in two large areas – a forest reserve and a national park.

Dominica has also become known for whale watching. It also offers diving and snorkelling activities. Tourists are urged to 'look but don't take anything but memories', to preserve shells and other natural features.

Fig. 10.42 Dominica is beautiful and unspoilt, even close to the capital

Dominica has a hurricane-free dry season from January to May. It has sheer cliffs and rivers with large amounts of water and many waterfalls. The scenery is stunning. It also has geothermal areas with hot springs, and a feature called Boiling Lake.

Ecotourism in Dominica

Dominica was the first country to be officially made a standard for ecotourism by Green Globe 21. **Ecotourism** aims to contribute to sustainable development by conserving the natural environment and improving the standard of living of local people. It is small-scale tourism where people visit the rainforest in small groups, causing as little disturbance and harm as possible to the environment and local people. Small groups are not likely to do much harm when they trample vegetation and paths.

When ecotourism is important for a local economy, the trees are not cut down for short-term gain. In Dominica the forest is conserved for the future, because it's an important attraction and economic asset.

Dominica has a number of eco lodges. Rosalie Forest Eco Lodge is in the middle of the rainforest on the edge of the Morne Trois Pitons National Park. Its environmental and social sustainability policy is:

- to protect the environment
- to have a sustainable lifestyle by: using no chemicals, minimising the use of resources, using renewable energy (hydro, solar, and wind), composting kitchen waste, shopping locally, not buying packaged goods if possible, recycling, using biodegradable products, and keeping waste to a minimum
- to employ only local people.

In addition, the owner of the Rosalie Forest Eco Lodge boosts the income of local people by arranging one-night stays for guests in village family homes. Unfortunately, in 2017, Dominica was devastated by Hurricane Maria, which is likely to have a serious effect on tourism earnings for several years.

Fig. 10.43 Taxis and coaches waiting for tourists at the quay at Roseau

Ecotourism provides a larger market for local farmers. Sales of handicrafts also increase, because tourists buy them as souvenirs. The local economy improves as a result, and more people can afford to buy consumer goods like televisions. A better local economy also means that more education and health services can be provided. And ecotourism helps to keep the age groups balanced, because young adults are less likely to move away for work.

One negative impact of ecotourism is that any exposure to outside influences can change traditional ways of life, especially among young people.

Cruise ship tourists

Cruise ships carrying thousands of passengers frequently tie up at the quayside in the heart of Roseau. Lines of coaches and taxis wait for those passengers who have booked an excursion to the island's attractions.

Cruise ship passengers do not contribute to the local economy as much as other tourists do, because they may have their meals on board ship – but they do buy souvenirs and pay entrance fees. The cruise ship company also pays harbour fees.

By encouraging the development of tourism, the government of Dominica will strengthen the country's economy and enable it to purchase more goods from abroad. It will increase its foreign exchange. Tourists have to change their money into the East Caribbean dollar (the local currency) in order to be able to spend – and they not only pay a fee to do this, but the exchange rate (how much they get in exchange for their money) is always favourable to the country. The country can then benefit from this by saving the foreign money until the value of the currency they have purchased goes up, so they are able to import more for it.

Another benefit of tourism is that old buildings are preserved, because tourists like to visit them. The old colonial market in Roseau has been restored.

Fig. 10.44 A view over Roseau's CBD

15 **a** Imagine that you live in Roseau. Describe the differences for people in the town between days when there are no cruise ships in port and days when there are. What problems might result from thousands of tourists visiting all at once?

b What types of job might people who work at the Rosalie Forest Eco Lodge do?

c Why might there be seasonal unemployment in Dominica's tourist industry?

Discussion point

Should ecotourism be the only type of tourism allowed?

Overall, does tourism have more beneficial impacts than adverse effects? Apart from all the economic benefits that it brings, and the developments that take place as a result of it, any activity that improves people's understanding of different cultures and preserves attractive old buildings and the natural environment must be beneficial.

Fig. 10.45 This sign on the pathway back to the cruise ship terminal in a small Mexican town says it all!

Energy and water

This chapter covers the following Cambridge IGCSE® and O Level topics:

- **3.5 Energy**
- **3.6 Water**
- **part of 3.7 Environmental risks of economic development**

- Energy and water supplies are controversial subjects. Globally, we are using more and more energy, but to generate it we are also using up the Earth's resources – like oil and coal. Power stations, like the one in this picture, also cause a lot of pollution – especially of the atmosphere.
- Should we just reduce our global energy use?
- The poorer countries need to use more energy to help them develop economically – would it be fair for the richer countries to stop them developing?
- Can the renewable 'green' sources of energy replace coal and oil? Should we develop more nuclear power, or is that too risky?
- Some areas have high rainfall and surplus water, so water supply is not a problem. However, areas like California in the USA have low rainfall which is not enough to supply the needs of its large population. Also, in some less economically developed areas, people do not have access to clean, piped water for their most basic needs.

In this chapter you will learn about:

- world energy consumption – how it is changing and the benefits it brings
- non-renewable fossil fuels
- the growth in the use of renewable energy sources
- the different types of electricity generation, and where the power stations are located
- water supply, water shortages, and the management of water resources.

World energy consumption

Energy consumption depends on the economy - both of individual countries and of the whole world. Between the years 2000 and 2007, world energy consumption increased by about 5% a year, as the global economy grew. The global economy then slowed down in 2008 and the growth in energy consumption also slowed to 1.2%. Then energy consumption actually decreased by 2.2% in 2009. However, the US Energy Information Administration has predicted that world energy consumption will increase by about 28% between now and 2040.

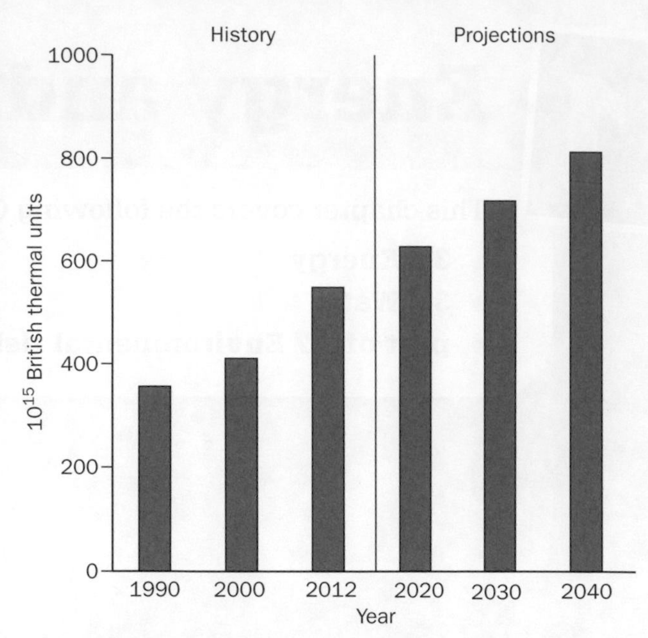

Fig. 11.1 Total world energy consumption, 1990-2040 (estimated), (source: US Energy Information Administration (2017))

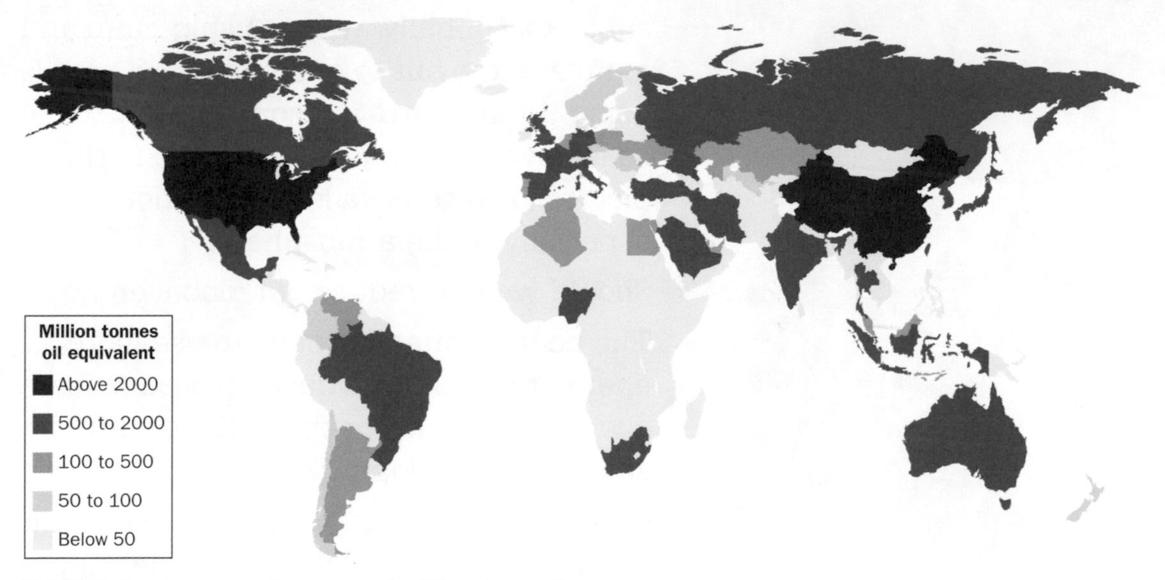

Fig. 11.2 Energy consumption per year in different countries

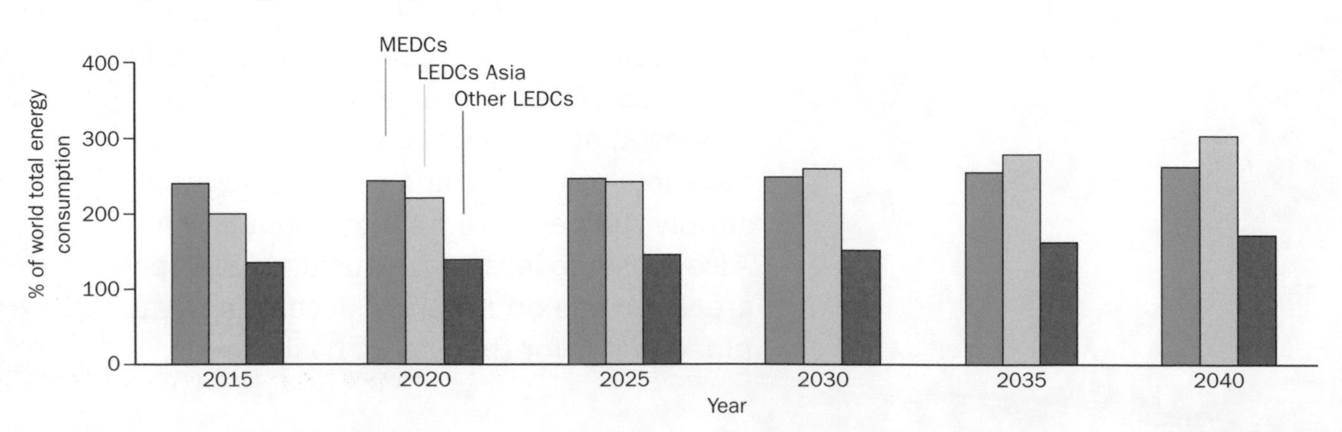

Fig. 11.3 World energy consumption for different groups of countries between 2015 and 2040 (estimated), (source: US Energy Information Administration (2017))

Until recently, MEDCs had a bigger share of world energy consumption than LEDCs. Most of the world's energy growth will occur in LEDCs where strong economic growth and growing populations increase the demand for energy. LEDCs in Asia (including China and India) are expected to account for more than half of the world's total increase in energy consumption over the 2015 to 2040 period. By 2040, energy use in LEDCs in Asia will exceed that of all MEDCs.

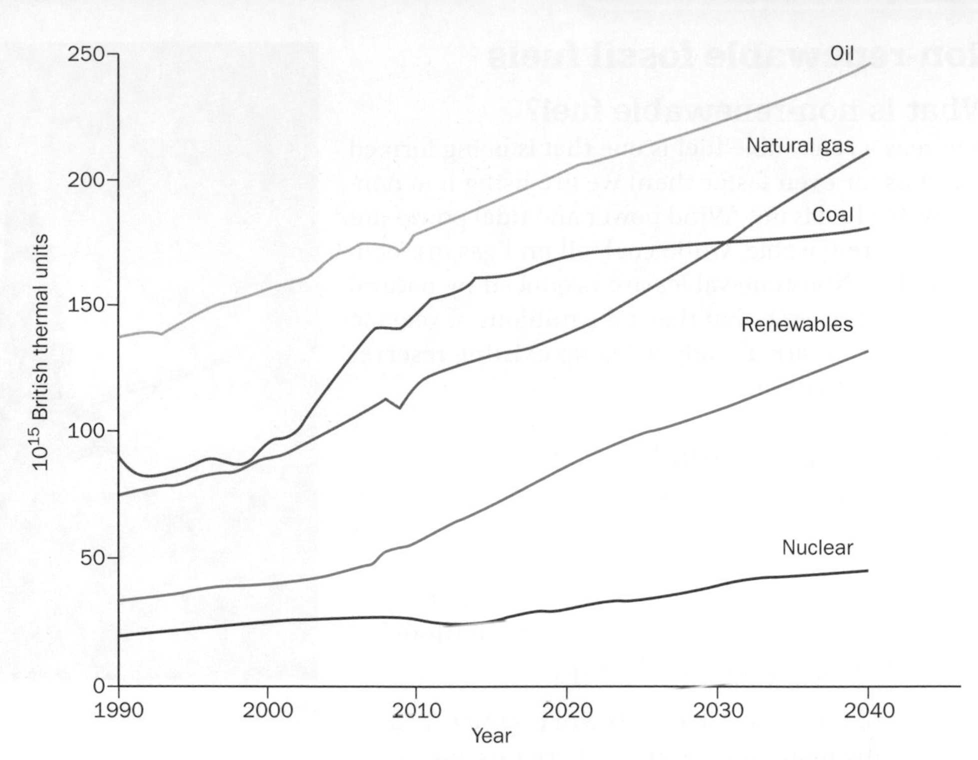

Fig. 11.4 World energy consumption by fuel type, 1990–2040 (estimated), (source: US Energy Information Administration (2017))

China and India lead the world's economic growth and increase in energy consumption. Together, they accounted for about 10% of the world's total energy consumption in 1990. This had become 32% by 2016.

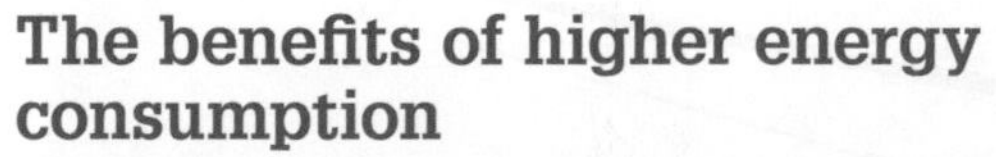

The benefits of higher energy consumption

The growth in energy consumption brings great benefits that improve people's lives. Some of them are listed below, but you may be able to think of others.

- Electricity makes daily household tasks easier - it provides heat, light in the evenings, television, and computers. People do not have to collect fuel or use candles or lamps at night.
- Modern transport systems are based mainly on oil (petroleum) use. They allow people to travel widely for work or pleasure, and also allow goods to be moved over great distances easily and quickly.
- Industry requires energy to make it work. Without it the economy cannot grow, wealth cannot be increased and people's lives will not be improved.

All of these points are particularly relevant to improving the lives of people in LEDCs. Without increasing their energy consumption, these countries will always lag behind economically.

The problems of higher energy consumption

- There are concerns that we are using non-renewable energy sources too quickly, and that we will soon run out of supplies.
- Many scientists worry that the use of fossil fuels is resulting in air pollution (see Chapter 2) and an increased rate of global warming (see Chapter 10).
- The inter-dependence of countries on each other for energy supplies, e.g. of oil and gas, can lead to conflicts.
- There are concerns about the safety of nuclear power (see later in this chapter).

Discussion point

Do you think that there should be limits on energy consumption? How could such limits be enforced?

1 Using Fig. 11.4, describe the global consumption patterns of different types of fuel, both (a) today and (b) if they change as predicted in the future.

LEARNING TIP Sometimes examination questions will ask you to quote figures in your answer and sometimes they will tell you not to. When in doubt, do both - in other words, describe the features in words and support your answer with figures.

Non-renewable fossil fuels

What is non-renewable fuel?

Whereas a renewable fuel is one that is being formed as fast as (or even faster than) we are using it, a non-renewable fuel is not. Wind power and tidal power are obviously renewable, while coal, oil and gas are non-renewable. Non-renewables are produced by natural geological processes but they take millions of years to form - and we are already using up existing reserves much faster than that.

What is a fossil fuel?

Coal, oil and gas are produced from organic material (plants and animals) that was growing millions of years ago. To grow, this organic matter got its energy from the sun. When we use these fuels, we are actually using the sun's energy from millions of years ago (stored in the fossil fuels). It is fossilised energy.

Reserves of non-renewable fuels will eventually run out. At some point in the future, alternative energy sources will need to take over.

Oil

Crude oil, or petroleum, is a mixture of different **hydrocarbons**. It was formed from **plankton** that previously floated in the oceans. When the plankton died, it fell to the seabed and was buried in mud. Geological processes then converted the plankton into crude oil, which is now found soaked into porous rocks. To extract it, a well (a **borehole**) is drilled into the ground and the oil either comes out under its own pressure, or needs pumping out. **Oil rigs** to do this may be located on land or at sea.

Fig. 11.7 An oil refinery in Singapore

Fig. 11.5 An offshore oil rig

Fig. 11.6 A 'nodding donkey' oil pump in the United Arab Emirates

After extraction, the crude oil needs refining to produce petrol for vehicles ('gasoline' or 'gas' in the USA), diesel fuel, aviation fuel, and heating oil. (Although it is not relevant to energy supplies, the **petrochemical industry** gets many of its raw materials from oil refineries. Most plastics are produced from oil.)

It is predicted that oil consumption will increase more slowly than the overall increase in world energy consumption. However, oil is still expected to remain the largest source of energy. When oil prices rise, electricity-generating companies have the option of switching to alternative fuels. However, the use of oil in transport continues to rise, and significant technological advances - such as the development of electric cars - will be needed to change this.

Advantages of oil

- It is easy to transport by pipeline or bulk tanker.
- It is the only fuel in mass use for motor vehicles.
- It is less polluting than coal when burned.
- It is also a raw material in the chemical industry.

Fig. 11.8 An oil pipeline

Disadvantages of oil

- Burning oil produces greenhouse gases and can lead to increased global warming (see Chapter 10).
- Oil spills from leaking tankers and pipelines can cause pollution, which kills wildlife.
- World oil production is concentrated in a small number of countries, which control both the supply and the price.
- Work on board oil rigs, especially those offshore, can be dangerous.

Fig. 11.9 Seabirds killed by an oil spill

2

- **a** Why should world energy production be increased?
- **b** What are the problems in increasing world energy production?
- **c** What is a fossil fuel?
- **d** Why are fossil fuels classed as being non-renewable?
- **e** What advantages does oil have over coal?

Discussion point

What are the sources of fuel and energy used by people in your class? This could be for transport, heating, cooking, or lighting. Where do your electricity supplies come from, and what fuel is used at the power station?

CASE STUDY

The 2010 Gulf of Mexico oil spill

On 22 April 2010, an explosion occurred on the Deepwater Horizon oil rig in the Gulf of Mexico. The rig sank and 11 workers were killed. Crude oil (up to 5000 barrels a day) began leaking from the damaged oil well on the seabed (1524 m down). The delicate ecosystem of the Gulf coastline is rich in wildlife, including pelicans, many species of duck, turtles, and whales. This environment was damaged by the oil spill and brought back memories of the *Exxon Valdez* spill off Alaska in 1989. The Gulf of Mexico spill was from an oil well that exploded, but the *Exxon Valdez* was an oil tanker that ran aground.

As well as the value of the oil that was lost, the clean-up operation in the Gulf of Mexico cost many millions of dollars. The Gulf's important leisure and tourism industry was affected, because tourist beaches were covered in oil – leading to fears of lost income and job losses. The fishing industry was also badly hit, because the oil contaminated stocks of fish and shellfish. The oil that rose to the ocean's surface created a slick that covered many km^2.

Emergency teams used a number of methods to try to contain the oil spill:

- → Four robotic submersibles tried to install a blow-out preventer (a set of huge valves designed to seal the well).
- → A relief well was drilled to tap into the leaking well and take the pressure off the broken wellhead.
- → Skimmers were used to skate over the water brushing up the oil. More than 90 000 barrels of oil-water mix were removed in this way.
- → Around 300 km of floating boom were used as part of the efforts to stop the oil from reaching the fragile coast. A US charity made booms out of nylon tights, animal fur and human hair.
- → Dispersant chemicals, rather like soap, were sprayed from ships and aircraft in an effort to break down the oil, despite worries about their toxicity.
- → Controlled burning was also used to tackle the spill, although it was difficult to carry out and produced toxic smoke.

3 Look at Fig. 11.10.

a Estimate the area of the oil slick in the Gulf of Mexico, in square kilometres, by 15 May.

b What distance did the oil travel from the damaged oil rig to the north by 15 May?

c How quickly, in kilometres per day, did the oil travel to the north?

d List the problems caused by the explosion.

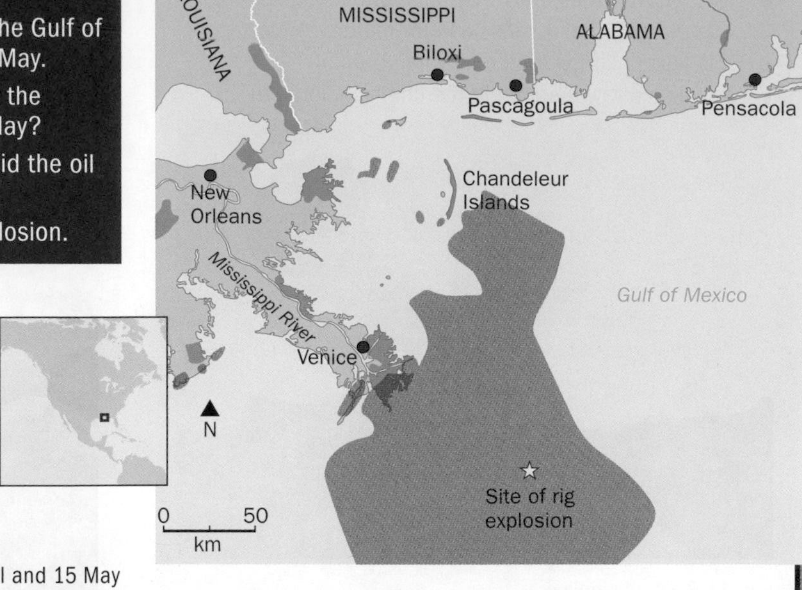

Fig. 11.10 The extent of the oil spill between 22 April and 15 May

4 Study Tables 11.1 and 11.2. Work out which countries were (a) the biggest oil exporters, and (b) the biggest oil importers, in 2016.

Hint: You will need to calculate the difference between production and consumption for each country – to see whether they needed to import oil in 2016, or whether they had spare oil that they could export. (Some countries just appear in one table.) In reality, the situation is more complicated than this, because there are different types of oil. A country producing lots of one type of oil may still need to import some oil of a different type.

Rank	Country	Production (1000 barrels/day)
1	Russia	10551
2	Saudi Arabia	10460
3	United States	8875
4	Iraq	4451
5	Iran	3990
6	China	3980
7	Canada	3662
8	United Arab Emirates	3106
9	Kuwait	2923
10	Brazil	2515

Table 11.1 The top ten oil producers in 2016

Rank	Country	Consumption (1000 barrels/day)
1	USA	18961
2	China	10480
3	Japan	4557
4	India	3660
5	Russia	3493
6	Brazil	3003
7	Saudi Arabia	2961
8	Germany	2435
9	Canada	2374
10	South Korea	2328

Table 11.2 The top ten oil consumers in 2016

Fig. 11.11 Clean-up workers on the beach after the Deepwater Horizon spill

Coal

Coal is a **sedimentary rock** that formed from trees growing in tropical swamp forests. The layers of coal (seams) are 1–4 m thick and are found between other sedimentary rocks, such as sandstones and shales. The two main methods of extraction are deep (underground) mining and opencast mining (quarrying).

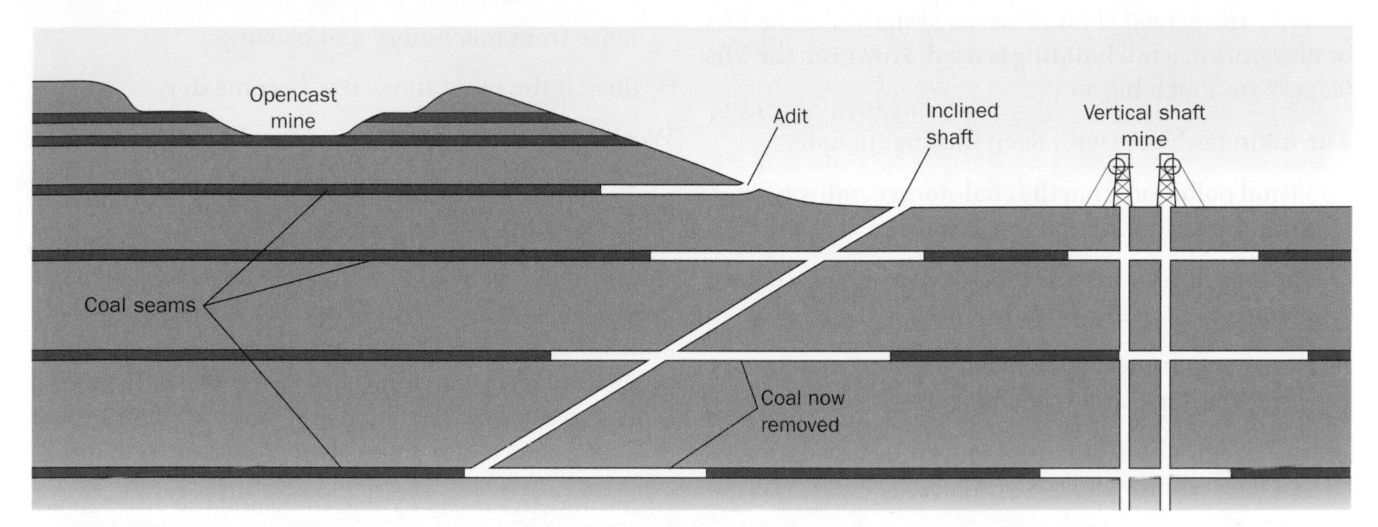

Fig. 11.12a Types of coal mine

Fig. 11.12b Opencast mining or quarrying

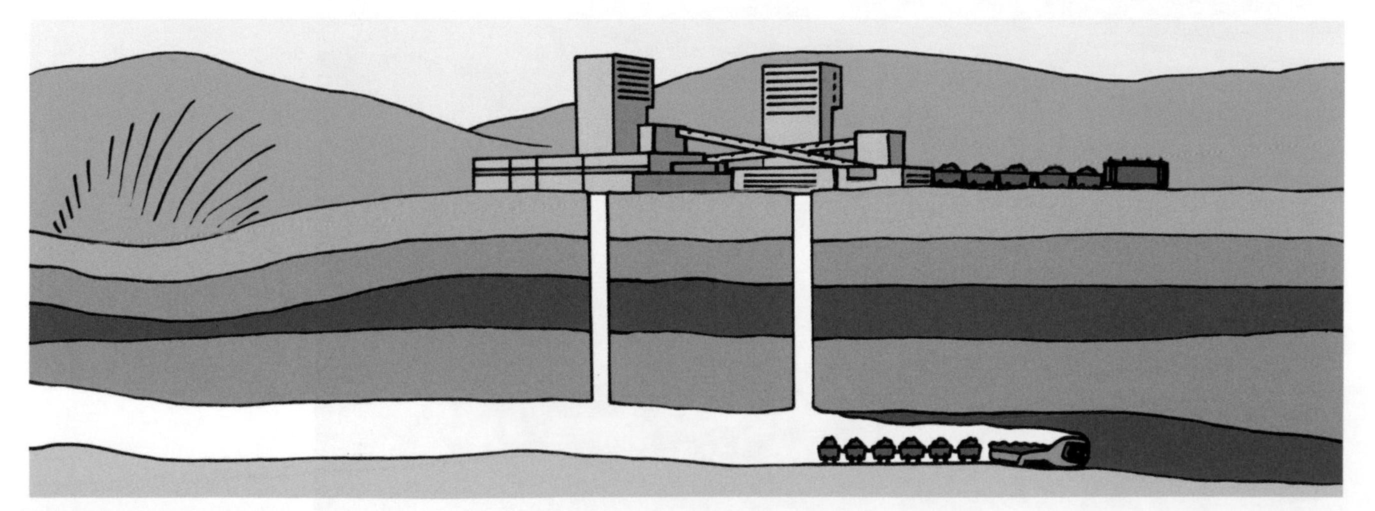

Fig. 11.12c Shaft mining

Deep mining methods are used to extract coal from fairly thick, but deeply buried, coal seams - often up to 1000 m below the surface. The shafts are either inclined (drift mines) or vertical. In inclined shaft mining, there are conveyors in the shafts to bring the coal up to the surface. In vertical shaft mines, a system like the lifts or elevators in a tall building is used. However, the lifts (cages) are much bigger.

The main problems with deep mining include:

- visual pollution from the coal storage, railway lines, and mine buildings on the surface
- the possibility of subsidence, when the surface collapses into old workings
- dangers to miners from accidents with machinery, gas explosions, and roof collapses
- the greater initial capital input compared with opencast mining.

Opencast mining is used for coal seams that are closer to the surface and possibly thinner. The soil and overlying rock are removed and stored. Once the seam is exposed, the coal can be extracted. Huge earthmoving equipment is used in the operation. After the coal is extracted, the waste rock and soil are put back and the land is returned to other uses.

Potential problems from opencast mining include:

- visual pollution from the enormous pit that is excavated
- the temporary loss of the land for other uses while the mining takes place
- noise from machinery and blasting
- dust, if the pit is allowed to become dry.

World coal consumption increased by 35% between 2002 and 2007 - largely because of a big growth in demand from China. There was a slight decline between 2007 and 2009 because of the world **recession**. Consumption peaked in 2013 and since then there has been a slow decline. This is because of concerns about atmospheric pollution and global warming. As well as being used as the fuel in thermal power stations, coal is also used in the steel industry, in domestic fires and as a raw material for some chemical processes.

China and, to a lesser extent, India and the other Asian LEDCs consume coal instead of more expensive fuels like oil. Coal was used to generate 42% of the world's electricity in 2007. But, as well as the problems listed on the left, electricity generation from coal is a major source of air pollution. This is discussed in more detail in the section on thermal power on pages 294-5.

5 **a** Why is coal sometimes mined by opencast methods and sometimes by deep mining?

b What problems are caused by opencast mining and by deep mining?

c What are the main uses of coal?

Discussion point

What sources of fuel are available in the country where you live? Does your country import fuel or energy? Are there any alternatives available?

Fig. 11.13 The Maxim Gorky opencast mine in the Czech Republic

Natural gas

Natural gas (methane) can form from plankton in the same way as oil. However, a lot of existing natural gas was formed from plants that grew in tropical swamps. Like oil, it accumulates in porous rocks (e.g. sandstone and limestone) and is extracted in the same way.

Gas is an important fuel for electricity generation, so its consumption is expected to continue growing - especially if oil prices remain high. However, it is likely to become increasingly expensive to obtain supplies of natural gas in the future (as easily extracted reserves are used up), so the growth of gas consumption will begin to slow down (see Fig. 11.4).

Advantages and disadvantages of natural gas

The advantages and disadvantages of natural gas are the same as for oil. Natural gas also has these advantages:

- Electricity generation using natural gas is less expensive than using oil.
- Gas-fired generating plants are less expensive to build than plants that use coal, nuclear, or most renewable energy sources.

Fuelwood in LEDCs

In many LEDCs, fuelwood accounts for about 70% of energy supplies. The more rural a country is, the greater its dependence on wood. Fuelwood has the advantage that it's often free for the user and does not require high-technology equipment. It provides an accessible source of fuel for heating and cooking. Also, if there is enough land, wood can be a renewable, sustainable energy source - but this is not always the case. For example, in the highlands of Lesotho there is little wood left. For rural people who live near towns, surplus wood is often collected and sold to townspeople, so it also becomes a cash crop rather than just a matter of subsistence. In many societies, the collection of wood for heating and cooking is a regular task for family members (particularly women and children).

Using fuelwood can cause problems:

- In some areas, natural woodland is being cut more quickly than it can grow back. So longer and longer distances have to be walked to collect wood - meaning a lot of hard work and more time taken up with this task. Often, children miss out on their education as a result.
- Deforestation may lead to the exhaustion of soils - and soil erosion - so that the forest cannot grow back. It could even lead to desertification.
- Burning wood in confined spaces on inefficient stoves leads to respiratory illnesses.

Schemes have been developed to improve the system of using fuelwood. They usually involve:

- planting more trees, often on a 'woodlot' system where there is a constant cycle of replanting
- managing the woodland and using systems such as careful pruning and thinning to encourage more growth
- the introduction of new fast-growing species
- the introduction of new fuel-efficient stoves, which cause less smoke.

Fig. 11.14 Women carrying fuelwood, Lesotho

Renewable energy supplies

These include hydroelectricity, geothermal power, wind power, solar power, and biofuels. The classification of nuclear power is discussed later.

Fig. 11.4 shows that renewables are expected to be one of the fastest-growing sources of future world energy, with consumption increasing by 2.6% per year. The reasons for this are:

- anticipated increases in oil prices
- concern about the environmental effects of fossil fuel use
- concerns about the sustainability of fossil fuels - they will eventually run out
- government incentives for increasing the use of renewable energy in many countries.

Hydroelectricity and wind are expected to provide the largest shares of the projected increase in total renewable generation. However, this will vary from country to country. In many MEDCs, the majority of exploitable hydroelectric sources have already been developed.

Geothermal power

Geothermal energy is energy extracted from hot rocks or water beneath the Earth's surface. As you descend down into the Earth, temperatures rise on average by about 25 °C for every kilometre you go down. This is noticeable in underground mines, which are often warm places! However, in volcanic areas, the increase in temperature may be as much as 70 °C for every kilometre you descend. These hot areas tend to be near plate margins, and it is in these areas that the prospects for geothermal energy are greatest.

Volcanic sources

- In volcanic areas, the groundwater (water stored in the pore spaces in the rock) is heated by magma (molten rock) beneath the Earth's surface, at temperatures of up to 1000 °C. The groundwater is usually under pressure, so it doesn't boil.
- When a borehole is sunk into the rocks, the heated groundwater rushes up and turns to steam, because of the reduction in pressure.
- This steam can then be used to drive turbines to produce electricity – although it's normally used to heat water to drive turbines, because the groundwater steam is corrosive.
- The used water is then pumped back into the ground. Countries where this is done include Japan, the Philippines and Iceland.

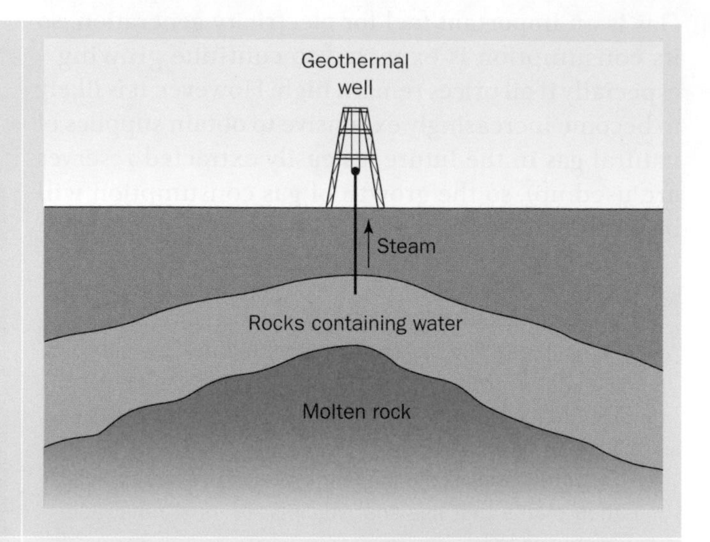

Geothermal aquifers

These are layers of rock in non-volcanic areas that contain hot water at depth. The hot water is pumped out – passed through a heat exchanger to extract the heat – and the cold water is then pumped back into the ground. Schemes like this are operating in the Paris area of France.

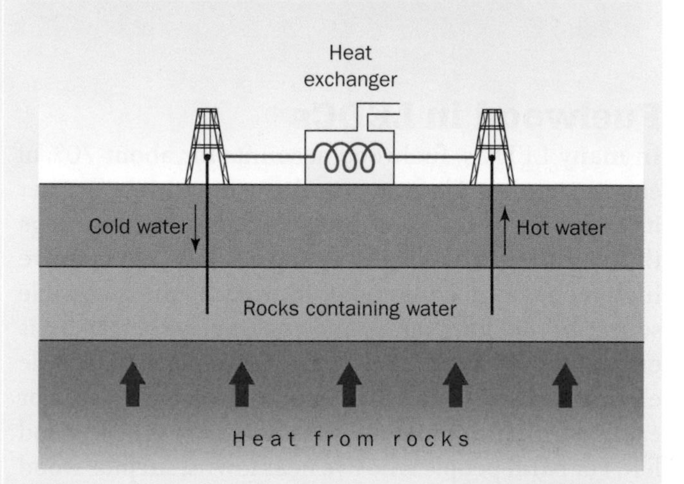

Hot dry rocks

This source is not being exploited yet, but it may be in the future.

- It involves hot rocks at great depths, like granites, which do not have lots of pores full of water.
- Boreholes would be sunk into the ground and the hot rocks would be fractured using explosives.
- Cold water would then be pumped down one borehole and the heated water pumped back up another.

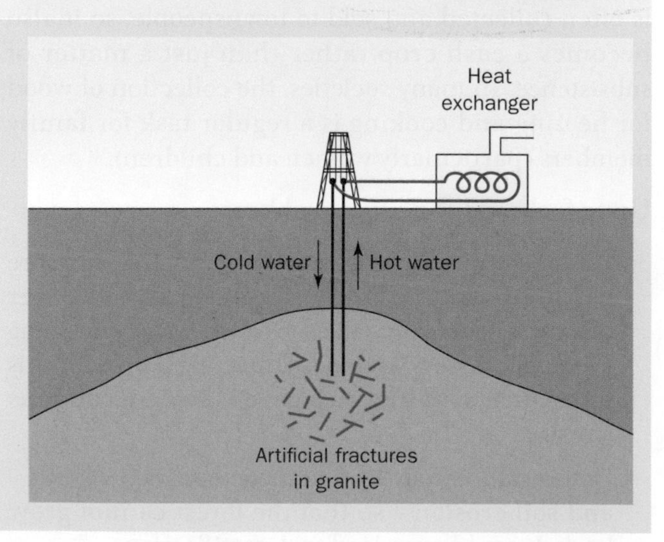

Fig. 11.15 Methods of geothermal energy extraction

Advantages of geothermal power

- It is extremely cheap and reduces the dependence on fossil fuels.
- It does not produce greenhouse gases.
- The water is pumped back into the ground and re-used.
- Unlike other types of renewable energy, it can operate at any time of the day and year – and is not affected by the weather.

Disadvantages of geothermal power

- It is restricted to areas with suitable geology.
- Areas with suitable geology are sometimes affected by earthquakes and volcanoes.
- Although it is usually classified as renewable, each well may be used for only about 25 years.
- The groundwater is saline and often poisonous.

CASE STUDY

Geothermal energy in Iceland

Iceland lies on the divergent plate margin of the Mid-Atlantic Ridge, an area with abundant hot rocks (see page 286). It is the world's leading producer of geothermal energy, with 87% of all Icelandic homes getting their heating and hot water from this source. The methods used to extract the energy are described opposite (in the section on volcanic sources).

Five major geothermal power stations exist in Iceland; 24% of the country's electricity is produced from geothermal energy, 75% from hydroelectricity and 1% from fossil fuels. Plans are under way to turn Iceland into a 100% fossil fuel-free nation in the near future.

The **Krafla geothermal power station** is located near the Krafla Volcano. Its development was threatened by seismic and volcanic hazards in the 1970s but, since it opened in 1978, these hazards have decreased.

Fig. 11.16 The Krafla geothermal power station in Iceland

The **Svartsengi geothermal power station** is located near the Keflavík International Airport at the Reykjanes Peninsula. It extracts hot water at 90 °C, and surplus mineral-rich water from the plant is used to fill up the Blue Lagoon (a nearby tourist bathing resort).

The **Reykjanes Power Station** is a geothermal power station located at the southwestern tip of Iceland.

The **Nesjavellir geothermal power station** opened in 1990 and is the second largest geothermal power station in Iceland. The facility is located near Thingvellir and the Hengill Volcano. The station produces approximately 120 MW of electrical power and around 1800 litres of hot water per second. It serves the hot-water needs of the area around the capital, Reykjavík.

Eleven kilometres away is the **Hellisheiði geothermal power station**, which is the largest in Iceland. Once fully operational, it will probably be the largest geothermal power station in the world.

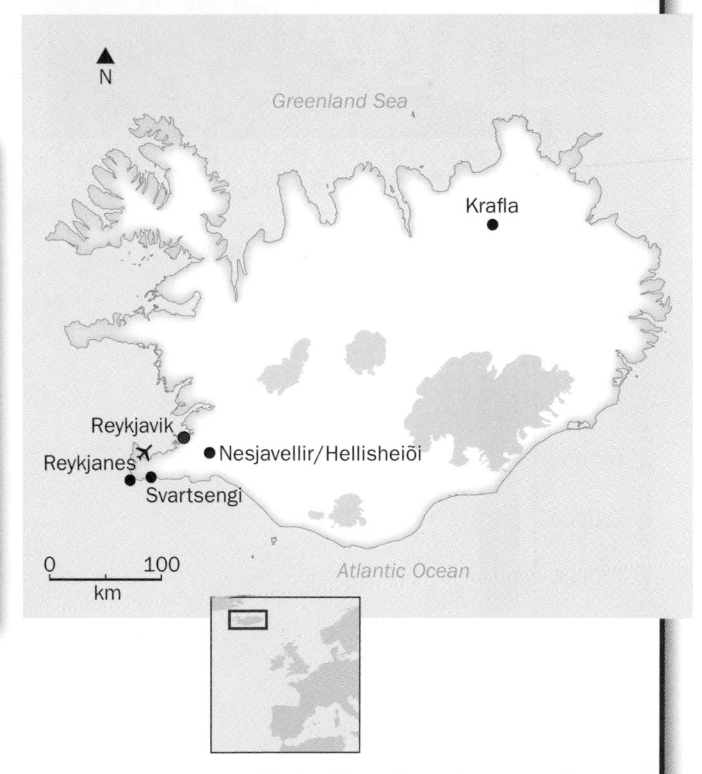

Fig. 11.17 The location of Iceland's geothermal power stations

Wind power

The use of wind power has increased dramatically since 2001 (see Fig. 11.18). Its development has been particularly rapid in the USA, China, India, and in European MEDCs (see Fig. 11.19).

Modern commercial wind farms work by having large numbers of windmills driving huge turbines to produce electricity. At first, wind farms were located in wide-open spaces on land, where the wind was likely to be strongest (e.g. near the coast or on hills). But, nowadays, wind farms are also being built offshore around the coasts of the Netherlands, Denmark, and the UK.

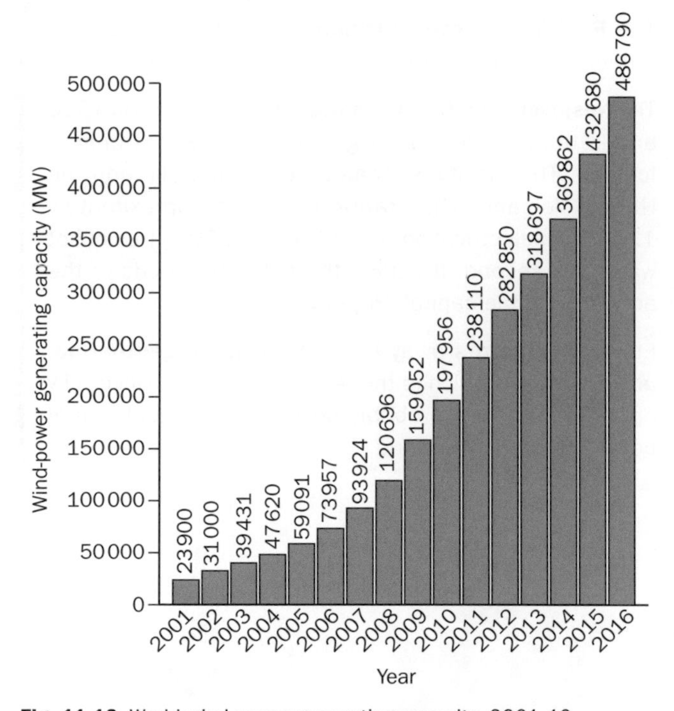

Fig. 11.18 World wind-power generating capacity, 2001–16

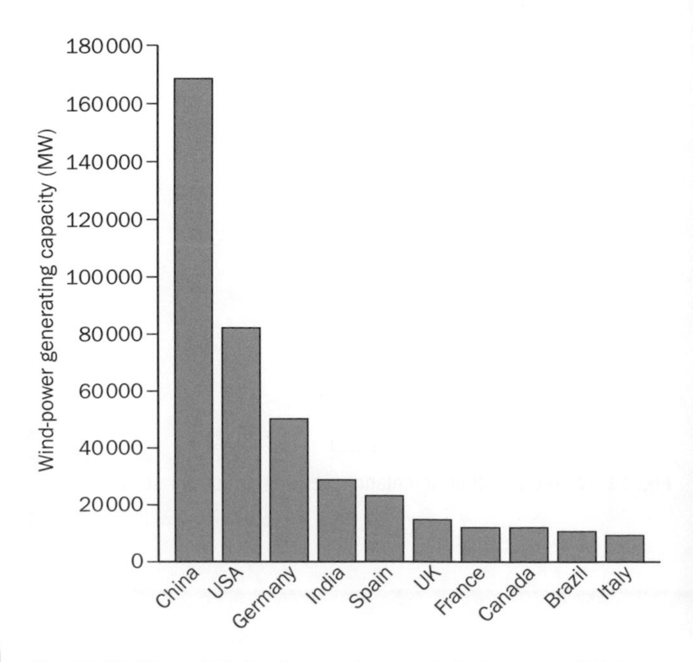

Fig. 11.19 The world's leading producers of wind power in 2016

Advantages of wind power

- It does not cause air pollution, global warming, or acid rain.
- It has very little effect on the local ecosystem, except very occasionally killing birds that get caught in the blades.
- In Europe, the wind is strongest in the winter, when demand for electricity peaks.
- After the initial capital input, production is cheap because the fuel is free.
- Wind farms may provide a small source of income for farmers.

Disadvantages of wind power

- It cannot be used during calm periods or storms.
- Many people consider wind farms to be a form of visual pollution – especially in areas of natural beauty.
- The technology is relatively new and at present very large numbers of turbines are needed to generate fairly modest amounts of electricity.

Discussion point

If someone wanted to build a wind farm near where you live, would you be in favour? What are your reasons?

Fig. 11.20 A wind farm in the south of Spain

Solar power

The sun gives out incredibly large amounts of energy, but trapping it directly is not easy. Light needs to be converted into electricity. The normal method is to use **solar panels (photovoltaic cells)**. When more solar energy is generated than is being used, it can be stored in a battery or exported to the national utility grid.

In many countries, the greatest use of solar power has been by private individuals or companies, rather than as a contribution to national electricity grids.

Fig. 11.21 Houses in California with rooftop solar panels

Advantages of solar power

- It is safe and pollution-free.
- After the initial capital input, production is cheap because the fuel is free.
- It can be used effectively for low-power uses, such as heating swimming pools or central heating.
- Its greatest potential is in warm and sunny countries, or in LEDCs where people live in locations that are isolated from the national electricity grids.

Disadvantages of solar power

- The initial capital input is high.
- It is not as effective in cloudy countries.
- It is less effective in high-latitude countries, where more power is needed in the winter but the days are shorter and the sun is lower in the sky – giving less light.
- It is less effective for high-output uses, such as powering colour TVs.

Wave power

Energy can be captured from waves in the sea for electricity generation using a machine called a wave-energy converter. Unlike tidal power, wave energy is not constant and depends particularly on wind strength. Wave power is not used commercially today. The world's first marine-energy test facility was established in Orkney, UK in 2003 to help the development of the industry in the UK. This site faces the Atlantic Ocean and waves up to 19 m high have been recorded. The first experimental wave farm was opened in Portugal in 2008, at the Aguçadoura Wave Park.

Tidal power

Although not yet widely used, tidal energy has great potential for future electricity generation. A tidal generator converts the energy of tidal flows into electricity. The potential of a site is increased by greater tidal height variation and higher current velocities.

The first tidal power station was opened at La Rance, France in 1966. The largest is the Sihwa Lake Tidal Power Plant in South Korea, which opened in 2011.

Types of tidal power systems

→ **Tidal stream generators**
These are turbines that work in the same way as wind turbines. They can be built into bridges or submerged, avoiding visual pollution. High current velocities through inlets are an advantage.

→ **Tidal barrages**
These are dams across a tidal estuary. They make use of the difference in height of water on either side of the dam.

→ **Dynamic tidal power**
These are dams built from coasts, out into the sea without enclosing an area. Where currents are parallel to the coast, there are height differences either side of the dam.

→ **Tidal lagoons**
These are like tidal barrages, but are circular and not necessarily across an inlet. The proposed Tidal Lagoon for Swansea Bay in Wales, UK is an example.

Advantages of tidal power

- Tides are more predictable than the wind and the sun.
- Tidal power is practically inexhaustible (a renewable resource).

Disadvantages of tidal power

- The cost of developing new technology is relatively high.
- There is limited availability of sites with sufficiently high tidal ranges or flow velocities.
- Some types involve the creation of lakes or dams, which disrupt existing ecosystems.

Biofuels

This term includes any fuel which comes from biomass (plant material). It includes liquid fuels (bioethanol and biodiesel), biogas, and solid biofuels, including fuelwood (see page 285).

Bioethanol is an alcohol made by fermenting the sugar in plants such as maize. Technology is also being developed to allow material such as trees and grasses for bioethanol production. Bioethanol can be used as a fuel for vehicles but it is more frequently added to petrol to increase octane values and improve vehicle emissions. Bioethanol is widely used in the USA and in Brazil.

Biodiesels are made from vegetable oils (like rapeseed oil) and also from recycled used cooking oils from restaurants and kitchens. Like bioethanol, it is often used as an additive to other fuels. It reduces levels of air pollution emitted by diesel-powered vehicles.

Biogas is methane produced by the breakdown of organic materials by bacteria. It can be produced either from waste materials or by the use of energy crops. The solid by-products can be used as a biofuel or a fertiliser. Biogas is also produced naturally (in a less clean form) in landfill sites. However, if it escapes into the atmosphere it is a potential greenhouse gas.

Solid biofuels (often referred to as biomass) can be used in power stations and in the heating systems of houses and other buildings. Special fuels and boilers are needed to make use of this energy source. Russia has 22% of the world's forests and is a big biomass producer. A large factory at Vyborg near the border with Finland produces wood pellets for export that can be used in heat and electricity generation.

Biofuels provided 4% of the world's transport fuel in 2016. Research and development into biofuels is increasing and they are likely to have greater significance in the future. Currently the world leaders in biofuel development and use are Brazil, the USA, France, Sweden, and Germany.

Advantages of biofuels

- Prices could be more stable than world oil prices.
- Supplies can be more secure and reduce reliance on imported fuels.
- Fewer pollutants are produced than by fossil fuels (see below).
- They are 'carbon-neutral', because the growing source crops absorb carbon dioxide from the air, which balances the emissions from the burning fuel.

Fig. 11.22 Oil seed rape growing on farmland in the United Kingdom. This crop is used for biodiesel production.

Disadvantages of biofuels

- In the period from 2008 to 2011, some land previously used for the production of food was changed to produce crops for biofuel production instead. This led to increases in world food prices and decreases in the food supply (see Chapter 9).

6
- **a** What are biofuels made from?
- **b** Why did the production of biofuels increase after 2000?
- **c** What is the main problem in increasing the world production of biofuels?

7
- **a** What are the advantages of using wood as a fuel in LEDCs?
- **b** What problems does the use of wood as a fuel create?
- **c** List the types of renewable fuel.
- **d** Although the use of renewable fuels is growing, why is it difficult for renewables to replace non-renewables quickly?
- **e** What was the wind-power generating capacity in 2016 of the USA, India, and France?

Discussion point

Do you think that the increased use of renewable energy sources is a good thing? Which types of renewable energy would work best in the area where you live, and why?

Power stations

Power stations are the places where electricity is generated. For several decades now, electricity has been the world's fastest growing form of energy use - and this growth is likely to continue.

- The strongest growth in electricity generation is likely to be in LEDCs. As their standards of living begin to rise, there will be an increasing demand for home appliances (like televisions and computers) as well as services (like hospitals, office buildings, and shopping malls) - all of which consume electricity.
- Most MEDCs have their electricity distributed by a national grid system, which allows the power from all of the power-producing plants to be treated as one, and for power to be sent to where it is required.

In the section on renewable energy, power stations fuelled by the sun, wind, biomass, and geothermal energy were all discussed. However, most electricity is still generated from running water (hydroelectric power), coal, oil, gas, and nuclear power (thermal power).

Hydroelectric power stations

Ideally, the site for a hydroelectric power station should have:

- a large river
- a large falling distance (head) of water
- a constant flow of water thoughout the year
- a narrow valley to provide a good dam site
- impermeable rocks, so that the reservoir does not leak
- stable geological conditions, so that the dam and sides of the reservoir do not collapse
- sparsely populated land, so that large numbers of people do not have to be moved to create the reservoir.

A dam is first built across a river. This holds back the river's water, which is then stored in a reservoir behind the dam. Near the bottom of the dam wall is a water intake, which controls the amount of water allowed to leave the reservoir. Gravity causes the water to fall through the **penstock** (the pipe inside the dam). At the end of the penstock is a turbine propeller, which is turned by the moving water. The shaft from the turbine goes to a generator, which produces the power. Power lines are connected to the generator to carry the electricity to where it is needed. The water then continues past the propeller into the river downstream from the dam.

Because the demand for electricity is not constant - going up and down during the day and night - some hydroelectric power stations use a system called pumped storage. During the night, when there is less demand for electricity, some of it is used to pump water that has already flowed through the turbines back up to a storage pool above the power station. When demand is high, the water is then allowed to flow back through the turbine generators to produce electricity again. In this way, the reservoir acts much like a battery - storing power in the form of water when demands are low, and producing maximum power during peak periods. Pumped storage reservoirs are usually relatively small.

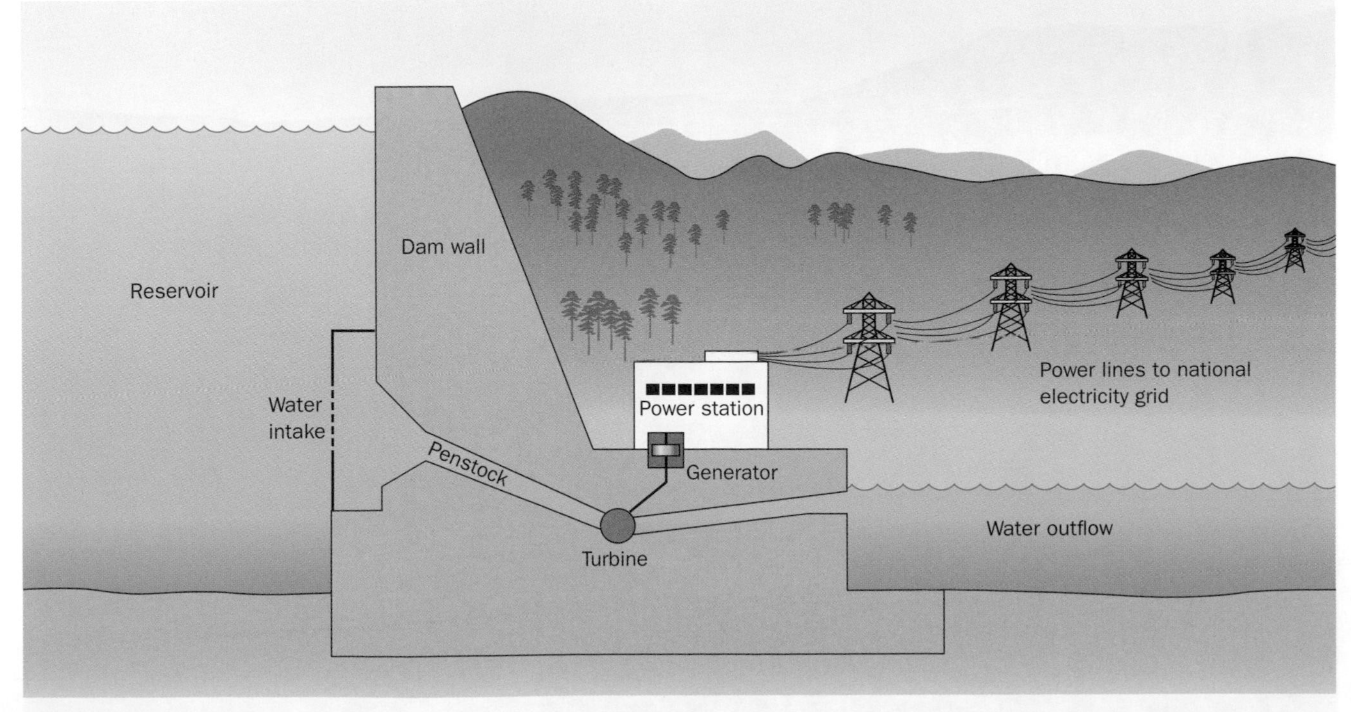

Fig. 11.23 How a hydroelectric power station works

Advantages of hydroelectricity

- Once a dam is constructed, electricity can be produced at a constant rate.
- The power stations can respond quickly to changing demand, as explained above.
- There are no fuel costs.
- The reservoir that forms behind the dam can be used for water sports and leisure activities.
- The stored water can also be used for irrigation and other purposes.
- There is no atmospheric pollution.

Disadvantages of hydroelectricity

- Dams are extremely expensive to build, and they must operate for many decades to make a profit.
- The flooding of large areas of land means that the environment is destroyed, along with natural habitats and historical or archaeological features.
- People living in the villages and towns of the valley to be flooded must move. In some cases, people are forcibly removed so that hydroelectric power schemes can go ahead.
- The building of large dams can cause serious geological damage. For example, the building of the Hoover Dam in the USA triggered a number of earthquakes.
- Although modern planning and design of dams is good, in the past old dams have been known to collapse. This has led to deaths and flooding.
- When a river flows through two or more countries, a dam in one of the upstream countries will affect the flow of the river in the downstream countries. This can lead to serious disputes between the countries concerned.
- Building a large dam alters the level of water in the ground. The building of the Aswan Dam in Egypt caused damage to some ancient monuments, through damp and the effects of salts in the water.
- Dams catch sediment and prevent it from flowing down the river and increasing the fertility of soils on floodplains downstream. The trapped sediment also reduces the reservoir capacity.

Fig. 11.24 Aerial view of the Three Gorges Dam in China, nearing completion

Fig. 11.25 The Katse Dam in Lesotho, southern Africa – the subject of the case study at the end of this chapter

Station	Country	River	Capacity (MW)
Three Gorges Dam	China	Yangtse	22 500
Itaipu Dam	Brazil/ Paraguay	Parana	14 000
Xiluodu Dam	China	Jinsha (tributary of the Yangtse)	13 860
Belo Monte Dam (fully open 2019)	Brazil	Xingu (tributary of the Amazon)	11 181
Siang Upper HE Project (fully open 2024)	India	Siang (tributary of the Brahmaputra)	11 000
Guri Dam (Simón Bolívar)	Venezuela	Caroni (tributary of the Amazon)	10 200

Table 11.3 The hydroelectric power schemes (existing and planned) with outputs of over 10 000 MW

Although the Three Gorges Dam in China has the world's largest generating capacity, during the course of a year the Itaipu Dam generates more electricity. This is because the Three Gorges Dam experiences six months a year when there is less water available to generate power. The Parana River which feeds the Itaipu Dam has a much lower seasonal variation in flow. The Itaipu power station on the Brazil-Paraguay border currently produces the most hydroelectric power in the world from a single dam.

RESEARCH The Internet contains information about many of the major dam schemes. Make your own case study of a hydroelectric power scheme. Concentrate on (a) the dam site, (b) the advantages and (c) the disadvantages of the project. Table 11.3 will give you some ideas about which scheme to choose.

In LEDCs, hydroelectric power is the main source of renewable energy growth. Many hydroelectric power schemes also supply water. The case study on the Lesotho Highlands Water Project is an example (see pages 302–3).

8 **a** Are the main hydroelectric power schemes limited to a particular climate?

b Are the main hydroelectric power schemes limited to LEDCs or MEDCs?

Fig. 11.26 The Itaipu Hydroelectric Dam on the Brazil-Paraguay border

Thermal power stations

The thermal power stations discussed in this section are those powered by fossil fuels - coal, oil, and gas. Nuclear power stations are also thermal, but they are often described separately. The fuel in thermal power stations is burned to produce heat, which turns water into steam. The steam is then used to turn turbines to produce the electricity. However, this process is not very efficient in its use of energy, and the steam has to be turned back into water at a temperature low enough for it to be disposed of without causing environmental damage.

Factors affecting the location of thermal power stations

- Coal is bulky and is normally transported by rail. For this reason, coal-fired power stations are often located close to the coal mines.
- Oil and gas can be transported relatively easily by pipeline, so their power stations do not need to be located close to the oil and gas wells. Nevertheless, oil-importing countries often locate oil-fired power stations at oil refineries close to the port where the oil arrives in the country.
- A supply of cooling water is needed, which is why so many thermal power stations have large cooling towers. Many power stations are located along rivers to get cooling water, but they must be spaced out so that too much warm water is not returned to the river to damage aquatic life. Sea water is not a good coolant, because the salt content attacks pipes.
- A large flat site is needed for the plant, cooling towers, fuel storage, and railway lines.

Making thermal power stations cleaner

Coal burning is estimated to be responsible for 40% of the 30 billion tonnes of CO_2 emitted by human activity every year. Attempts are now being made to develop cleaner coal-fired power stations to cut greenhouse gas emissions. Many developments are being researched, but few of them have been connected yet to a full-sized power station. For example:

- Carbon capture involves trapping the CO_2 before it escapes into the atmosphere, and then disposing of it elsewhere. It could reduce CO_2 emissions to the atmosphere by approximately 80–90%.
- However, capturing and compressing CO_2 requires a lot of energy, so it would increase the fuel needs of a coal-fired power station and raise the cost of energy.

Advantages of thermal power

- Many countries still have large reserves of fuel, e.g. coal in South Africa and Germany.
- Coal is also used to make coke for the steel industry, and oil is the basis of the petrochemical industry.
- Oil and gas can be transported efficiently by pipeline.
- Oil and gas are cleaner than coal because they produce less air pollution. (However, they do still produce carbon dioxide.)

Disadvantages of thermal power

- Power stations are major sources of air pollution. The problems caused by the emission of carbon dioxide, nitrogen oxides and sulfur dioxide are described in Chapter 2 (page 67).
- The pollution caused by one country's power stations is often 'exported' to another by prevailing winds. For example, gases released by power stations in the UK are blown to Norway.
- World coal reserves may last for only another 300 years, and oil and gas for an even shorter time.
- Deep mining is dangerous and careful health and safety measures are required.
- Over-reliance on imported fuels can cause problems. It makes countries vulnerable to sudden increases in price and political and even military threats from the exporting countries.

- Also, storage of the CO_2 is envisaged either deep in the Earth or in the deep oceans. In the case of deep ocean storage, there is a risk of making the oceans more acidic. Long-term predictions about submarine or underground storage are very difficult. CO_2 might leak from the storage into the atmosphere.

Nuclear power stations

Nuclear and fossil-fuelled power stations have many features in common. Both require heat to produce steam to drive turbines and generators. However, in a nuclear power station, uranium replaces the coal, oil, or gas. Uranium is a radioactive metal that can be used as an abundant source of concentrated energy. Uranium atoms split (nuclear fission) and release energy in the form of heat. In a nuclear reactor, the uranium fuel is assembled in such a way that a controlled fission chain reaction can be achieved.

Uranium ore usually occurs in the ground at relatively low concentrations, so most is mined by open-pit mining. Only a small number of countries mine uranium. Kazakhstan, Canada, and Australia are the top three producers – together accounting for 63% of world uranium production. Other important uranium-producing countries are Namibia, Russia, Niger, Uzbekistan, and the USA.

Factors affecting the location of nuclear power stations

- Like other power stations, large flat sites are needed for the plant and for cooling towers.
- The volume of raw material is so small that this is not a factor.
- Pure water is needed for cooling. Sea water will not do, unless it is desalinated. However, sea water has been used in emergencies.
- Some nuclear power stations are built on the coast to dispose of very low-level liquid, radioactive waste.
- In countries like the UK, concerns about the safety of early nuclear power stations meant that they were located in places far away from areas of dense population, e.g. Dounreay in the north of Scotland and Calder Hall in Cumbria. This is not the case today.

Discussion point

Nuclear power is controversial in many countries. Are you in favour of it or against it? Give your reasons.

Fig. 11.27 A nuclear power station in Iran

Fig. 11.4 shows that electricity generation from nuclear power is expected to increase. This is due to high fossil fuel prices, concerns about dependence on imported coal, oil, and gas, and also greenhouse gas emissions from thermal power stations. Production is expected to increase in China, India, and Russia.

However, the disadvantages listed below mean that there is still a lot of uncertainty about the future of nuclear power. In many countries, public concerns may prevent plans for new nuclear power stations. Despite this, the governments of several countries have announced changes in their positions since 2009, including Belgium and Italy, which have both become slightly more in favour of nuclear power.

Advantages of nuclear power

- Only very small amounts of uranium are needed to produce large amounts of energy.
- Uranium ore will not run out in the foreseeable future (some people even classify nuclear power as renewable).
- It does not produce greenhouse gases and acid rain.
- The safety record of nuclear power stations has improved and the industry is highly regulated.

Disadvantages of nuclear power

- There have been serious incidents at nuclear sites which have led to leaks of radioactivity, e.g. at Chernobyl in Ukraine in 1986, Three Mile Island in the USA in 1979 and Windscale in the UK in 1957. Radioactivity is a known cause of diseases such as cancer and leukaemia.
- The earthquake in Japan in March 2011 caused an explosion and leakage of radioactive material at the Fukushima nuclear power station. This raised questions about the safety of nuclear power stations in earthquake zones.
- The cost of shutting down old nuclear plants (decommissioning) is very high.
- The radioactive waste from power stations remains a health hazard for hundreds of thousands of years. It requires careful storage and is difficult to dispose of safely.
- Nuclear power stations produce material that is also the raw material for nuclear weapons, so there can be serious security concerns.
- The capital costs of building nuclear power stations are extremely high.

Discussion point

In this chapter there are various photographs of power stations. How visually polluting do you think they are? Do you think that some look better than others?

CASE STUDY

Electricity in Germany – a European MEDC

Germany is the largest consumer of electricity in Europe. The main fuels for German power stations are coal, nuclear, and gas.

Thermal power stations – coal

Germany has reserves of two types of coal – lignite and bituminous coal:

- Lignite (brown coal) contains less carbon, gives out less heat when burned, produces more ash, and causes more air pollution. It is extracted in Nordrhein-Westfalen and Sachsen (in western Germany) and Brandenburg (in eastern Germany).
- Bituminous coal is more carbon-rich and generally of better quality. It is mined in Nordrhein-Westfalen (the Ruhr Coalfield) and Saarland.

Fig. 11.30 shows that the lignite power stations are located close to the coal-mining areas. The Schwarze Pumpe power station is one example. Transporting lignite over long distances is not economic, because of its low purity.

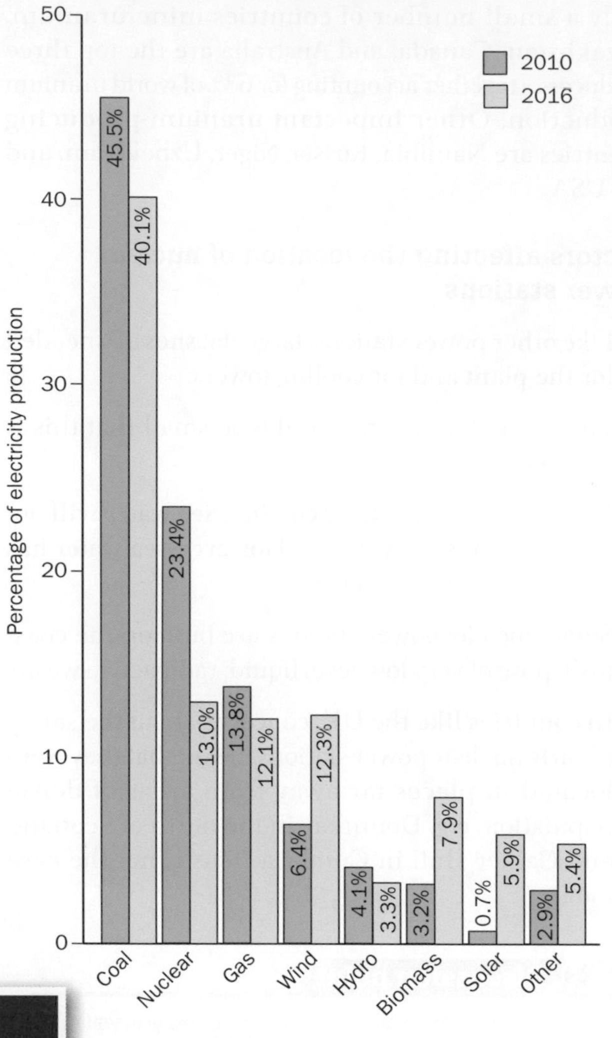

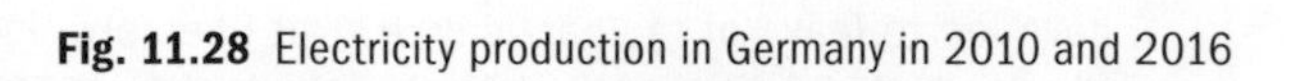

Fig. 11.28 Electricity production in Germany in 2010 and 2016

The bituminous coal power stations are partly located on the coalfields, to save transport costs (e.g. the Scholven power station), but they are also more widely located throughout the country. The power stations on the coast and on navigable rivers (e.g. the Rostock Power Station) use imported coal.

Fig. 11.29 also shows the importance of large rivers for providing cooling water.

Fig. 11.29 The Heilbronn coal-fired power station on the River Neckar. Notice the river to provide cooling water (and sometimes transport for coal). Also notice the tall chimneys to disperse oxides of nitrogen and sulfur dioxide, together with the cooling tower in the background.

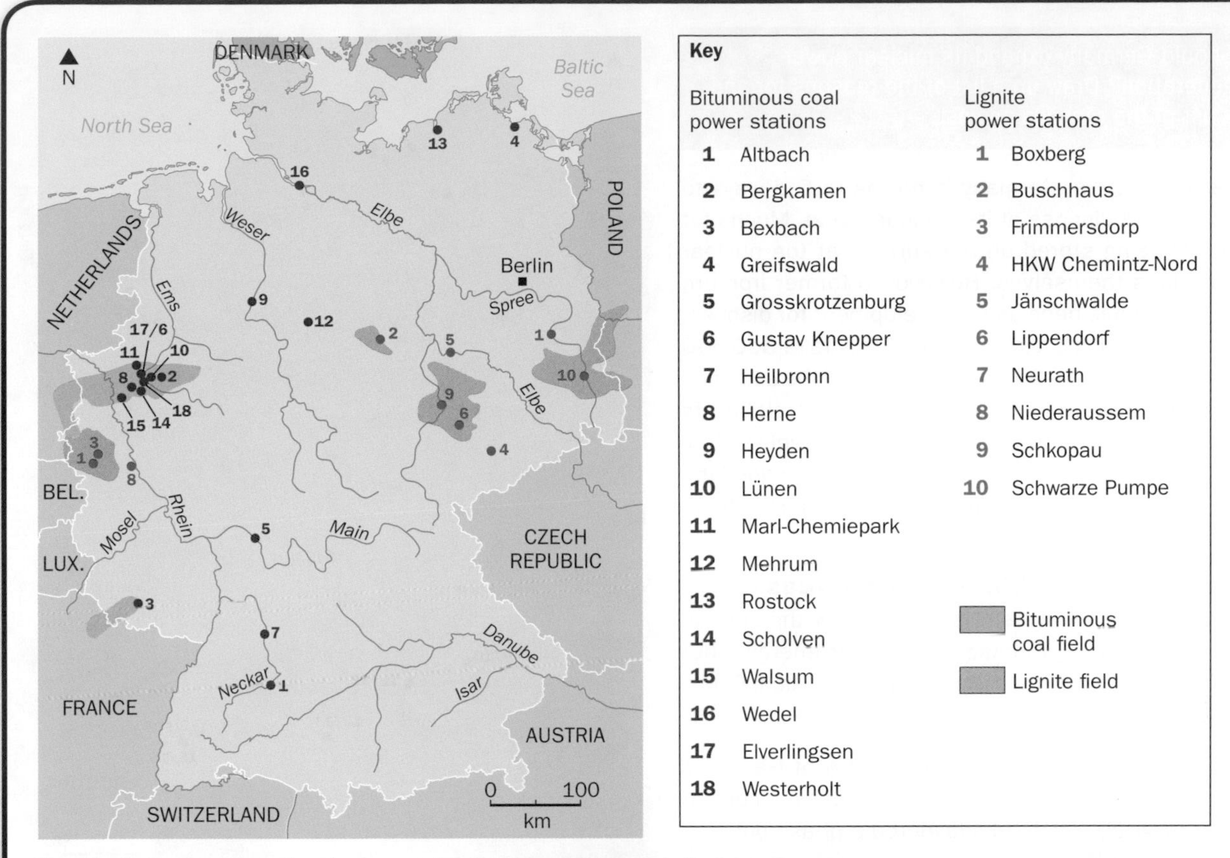

Fig. 11.30 The locations of coal mines and thermal power stations in Germany

Pollution from coal-fired power stations

Germany is committed to a reduction in its greenhouse gas emissions. One example of how this might be done is shown by the Schwarze Pumpe power station. The lignite fuel is burned in the presence of pure oxygen – releasing water vapour and carbon dioxide (CO_2). By condensing the water in a pipe the company that owns the power plant captures and isolates nearly 95% of the CO_2. That CO_2 is then compressed into a liquid and sold (e.g. to companies making fizzy drinks).

Nuclear power stations

Fig. 11.32 shows that Germany's 17 nuclear power stations are distributed throughout the country. Many are located near large rivers to supply cooling water (the factors affecting the location of nuclear power stations have already been discussed).

9 Look at Fig. 11.28.

a Calculate the percentages of electricity production from renewable sources in 2010 and 2016.

b Describe the changes in electricity production between 2010 and 2016.

Fig. 11.31 Niederaussem power station, a source of pollution

10 Should Germany expand its nuclear power generation? Draw up a list of the reasons for and a list of the reasons against.

Like many countries, Germany is having to find ways to safely store and dispose of its nuclear waste. Much of it is currently being stored on the surface at the nuclear power stations themselves. However, a former iron ore mine at Konrad has been under development for disposal of nuclear waste since 1975. It will initially take 300 000 cubic metres of waste – 95% of the volume of the country's waste. An underground disposal site at Morsleben in eastern Germany (for low- and intermediate-level wastes) was closed in 1998. It is said to be in poor condition, and is being stabilised with concrete at a cost reported to be US$3 billion.

The move to renewable energy sources

Energy policy is a controversial issue. In 1998, the German government decided to phase out nuclear energy – but a new government in 2009 cancelled this decision. Public opinion is often divided:

- A poll, early in 2007, found that 61% of Germans opposed the government's plans to phase out nuclear power by 2020, while 34% favoured a phase-out.
- Another poll, in mid-2008, showed that 46% of Germans want the country to continue using nuclear energy; another 46% said they supported the nuclear phase-out policy, and 8% were undecided.

If Germany were to phase out nuclear power production and continue to reduce its carbon emissions, it would need to import large amounts of electricity. This would put Germany in much the same position as Italy today – being dependent on neighbours such as France, the Netherlands, Poland, and the Czech Republic for electricity (much of which would be produced by nuclear power stations).

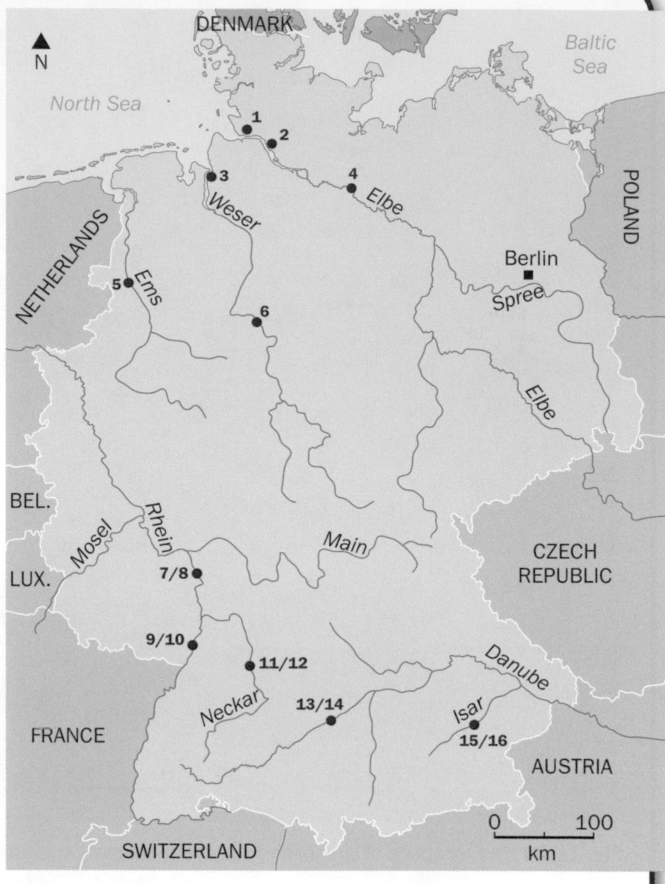

Key
Nuclear power stations

1	Brunsbüttel	**5**	Emsland	**11/12**	Neckar
2	Brokdorf	**6**	Grohnde	**31/14**	Gundremmingen
3	Unterweser	**7/8**	Biblis	**15/16**	Isar
4	Krümmel	**9/10**	Philippsburg	**17**	Grafrheinfeld

Fig. 11.32 Nuclear power stations in Germany

The tsunami damage to the Fukushima nuclear plant in Japan led to public protests in Germany. On 31 May 2011, the German government announced that it would close all nuclear power stations in the country by 2022. Overall electricity consumption would be cut by 10% and the use of coal and renewables would be increased. Germany may also want to reduce its dependence on gas imported from Russia.

Fig. 11.33 The Brokdorf nuclear power station on the River Elbe. Notice the lack of any tall chimneys to disperse pollutants.

The water supply

The water supply comes from two different types of source, both of which come from rainfall (or snow melt).

Surface water is water from rivers and lakes. Because river flow is sometimes variable, rivers are often dammed to create **reservoirs** (artificial lakes, also known as dams in some parts of the world), which store water for dry periods.

Groundwater is water held within the spaces of porous, permeable rocks (in the same way that oil and gas are held). Layers of rock that contain water are called **aquifers**. The water is extracted either by digging **wells** or by the more modern method of drilling **boreholes**. An electric pump then brings the water to the surface. This may use wind or solar power.

Water is in demand for:

- **agriculture**. Chapter 9 identifies areas of the world where the rainfall is so low that it's difficult, or even impossible, to grow crops without irrigation
- **domestic use**. In MEDCs, people use large volumes of water each day for washing, flushing toilets, watering gardens and even washing cars. In many LEDCs, this luxury is not available
- **industrial use**. Many industries use large volumes of water in processing (e.g. paper manufacture) and cooling (e.g. power stations).

The balance between these uses varies greatly from region to region.

Although water use per person is much greater in MEDCs, water shortages occur in both MEDCs and LEDCs and have an impact on the local people and the potential for development. This leads to competition for the use of the available water resources, which requires careful management.

As areas become more economically developed, the demand for water increases. Some areas have a **water deficit** and others have a **water surplus**, so there is a need for water to be transferred, sometimes over large distances. Major dam/reservoir construction projects raise economic, social, and environmental issues. Water supplies should be sustainable - they should not run out and their extraction should not endanger biodiversity in an area by lowering river levels so that plants and animals cannot survive. As you have already seen in the hydroelectricity section, when rivers flow across international boundaries, the use of water by one country can affect other countries downstream.

It is important that water supplies should be clean. Where people consume infected water they are prone to diseases such as cholera and diarrhoea. This is a major cause of death to children in some LEDCs. Lack of clean water also hinders economic development.

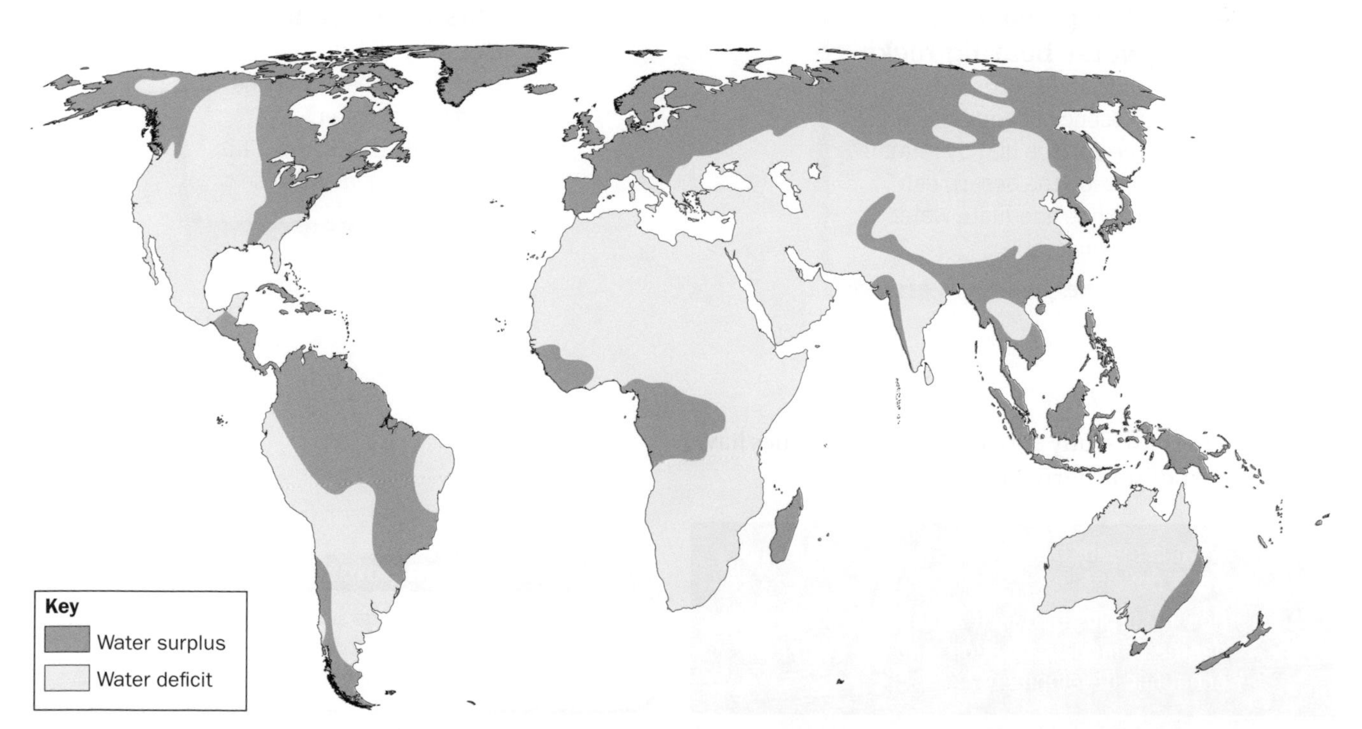

Fig. 11.34 Regions of water surplus and water deficit

The following factors affect whether there will be a water surplus or deficit:

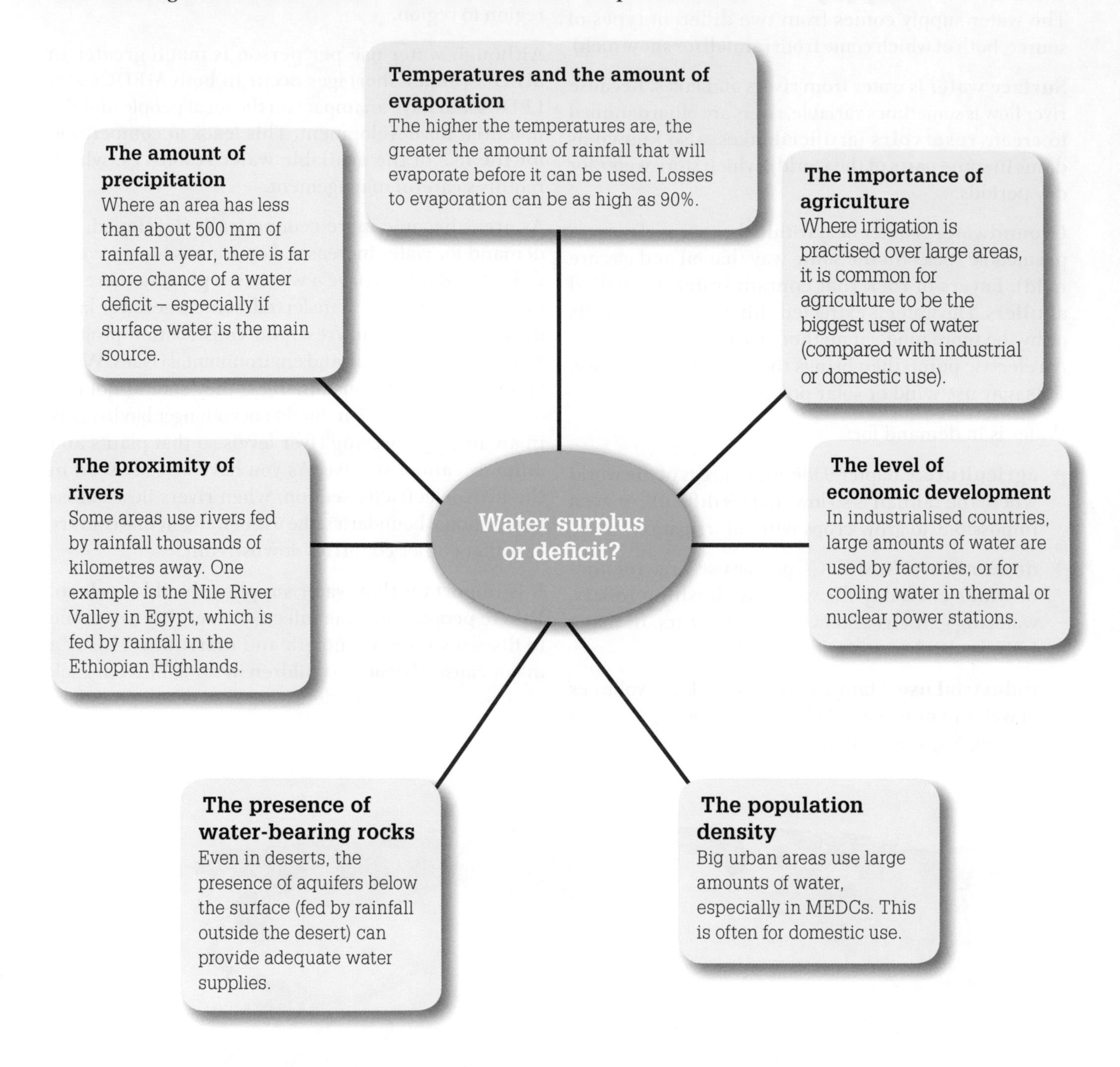

The following case study explains how these issues have affected part of southern Africa.

11 **a** What is meant by the terms water surplus and water deficit?

b How is surplus or deficit affected by climate, the degree of industrialisation, and the degree of economic development?

Discussion point

Do you live in an area of water surplus or water deficit? What are the geographical reasons for this? Where does your water supply come from and how does it reach you?

Desalination

This is the removal of salts from saline (salty) water to produce water suitable for human consumption. This is generally sea water but some groundwater can be saline. Desalination is used on ships and submarines.

Desalinating sea water uses a lot of energy and therefore costs more than fresh water from rivers, groundwater, or water recycling. For this reason, it is more likely to be used for domestic purposes than the larger quantities used in irrigation.

Currently, few people rely on desalination for their water supplies. However, desalination is being regarded as an ideal solution for rich countries with dry climates such as Australia, Saudi Arabia, and the United Arab Emirates. Kuwait produces all its drinking water by desalination.

The single largest desalination project is Ras Al-Khair in Saudi Arabia, which produced 1 025 000 m^3 per day in 2014.

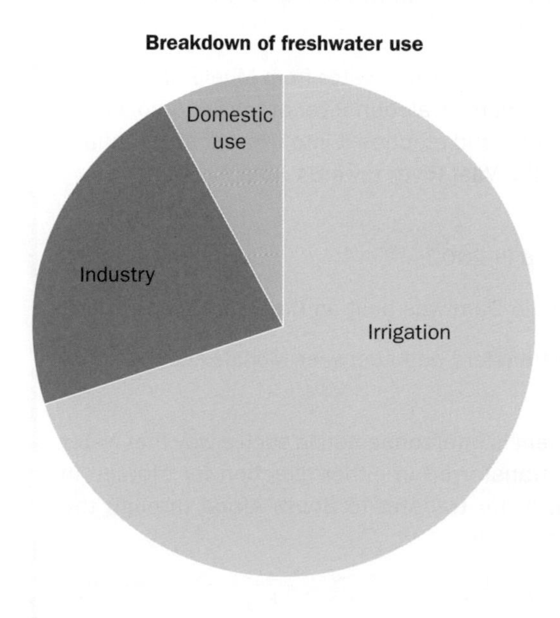

Fig. 11.35 World water consumption

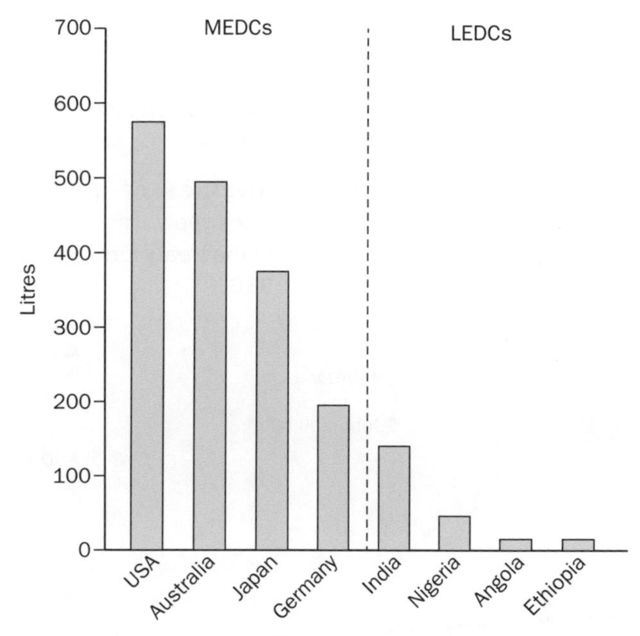

Fig. 11.36 Water consumption per person per day in selected countries

12 a Using Fig. 11.35, what percentage of water resources are used in agriculture?

b Describe the difference in water consumption between MEDCs and LEDCs.

c List the things that you use water for each day.

CASE STUDY

The Highlands Water Project in Lesotho – water supply and hydroelectricity

Lesotho is an LEDC in southern Africa with an abundance of water. It is completely surrounded by South Africa, which is a much richer country and also short of water. This circumstance has led to the ambitious **Lesotho Highlands Water Project** – one of Africa's largest engineering schemes.

The project involves damming some of Lesotho's major rivers – most of which flow south into the Senqu (known to South Africans as the Orange River). This damming process has created large artificial lakes. The water from the lakes is then diverted north (through tunnels under the mountains) to South African rivers. These rivers lead to the Vaal Dam, which supplies the densely populated urban and industrial region around Johannesburg and Pretoria.

South Africa pays Lesotho royalties of around US$1.5 million a month for this water supply. The project was first thought of in the 1950s, but the treaty formalising the project was not signed until 1986.

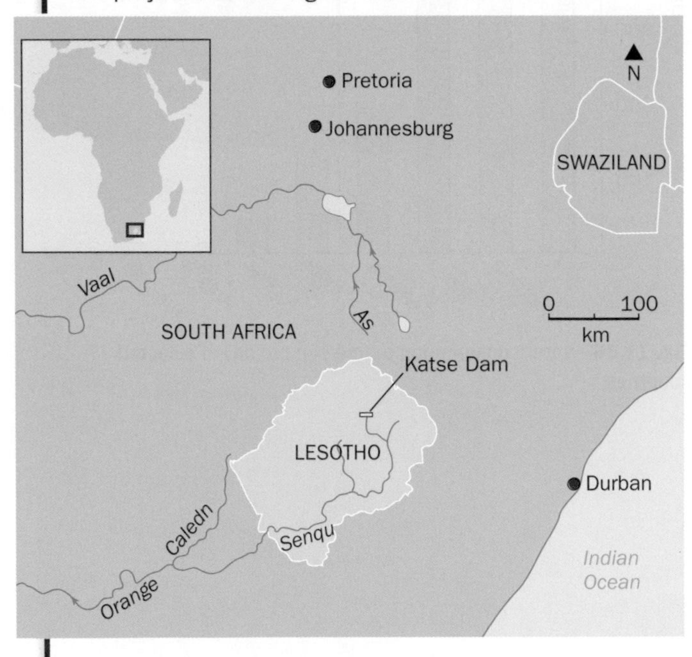

Fig. 11.37 Lesotho is entirely surrounded by South Africa

The project's objectives

- To provide funds for Lesotho by transferring water from the catchment of the Senqu/Orange river in Lesotho to meet the growing demand for water in South Africa's major industrial and population centres.
- To generate hydroelectric power for Lesotho.
- To promote the general development of the remote and underdeveloped mountain regions of Lesotho.
- To provide water for irrigation and drinking water supply.

The project's progress

Phase IA

- Completed in 1998.
- The Katse Dam was built on the Malibamatso River (see Fig. 11.25). It is the tallest dam in Africa at 186 m high – the size of a 52-storey building.
- The Matsoku weir and tunnel collect water from an input on the Matsoku River to the east.
- A 48.2 km-long tunnel transfers water from the Katse reservoir to the Muela hydroelectric power station in northern Lesotho.
- A delivery tunnel draws water from Muela and carries it underneath the Caledon River at the Lesotho/South Africa border and empties it into rivers that eventually flow into the Vaal River towards Johannesburg.

Phase IB

- Completed in 2002.
- The Mohale Dam was built on the Senqunyane River.
- A tunnel transfers water between Mohale and the Katse reservoir.
- The system is interconnected in such a way that water may be transferred in either direction for storage in Mohale, or for transfer to South Africa through the Katse reservoir.

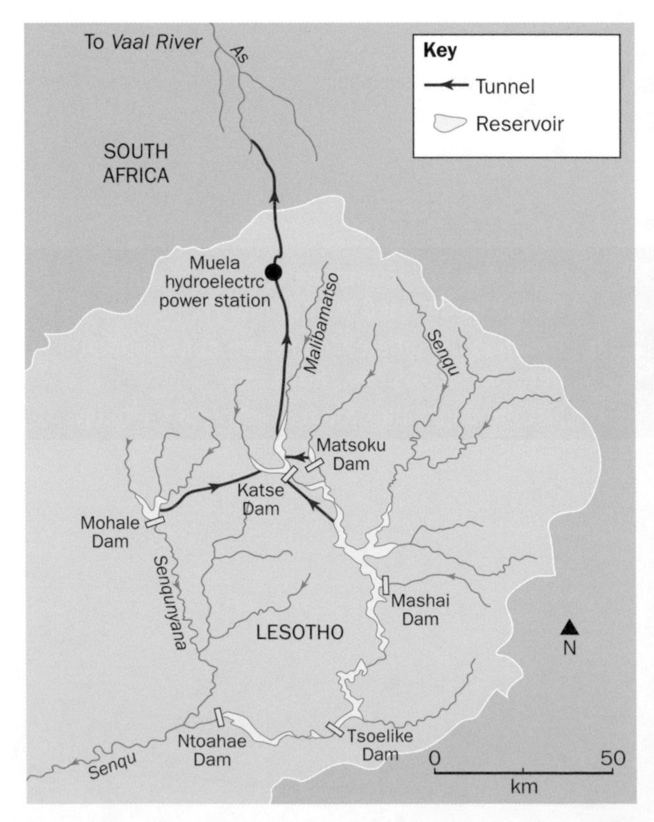

Fig. 11.38 A map of the Lesotho Highlands Water Project

Later phases

- Initially, three more dams were proposed further downstream after the Malimbamatso joins the Senqu River – at Mashai, Tsoelike and Ntoahae.
- In 2007, further studies resulted in changes to the plan and proposed instead a dam on the Senqu, upstream from the confluence with the Malibamatso.

Why was Lesotho chosen for the project?

- The mountains of Lesotho have an average annual rainfall of over 1000 mm. This means that about 50% of the water flowing down the Senqu/Orange River has fallen as rain in Lesotho. Lesotho's water resources far exceed its possible future requirements, even allowing for possible future irrigation projects and improving living standards.
- The water originating in the mountains has good chemical quality and low sediment content.
- The project's dams have strong foundations on either basalt (Katse, Mohale, Matsoku Dams) or the underlying hard sandstone.
- Deep, steep-sided valleys provide excellent dam sites.
- Local dolerite and basalt rocks can be crushed to produce good concrete for dam construction.
- The project is located in an area of low earthquake risk. One earthquake occurred in 1996, during the filling up of the Katse Dam at Mapeleng.

The impacts of the project

Conflicts in land use

Flooding such large areas of the country has led to conflicts with the local people. The rural mountain people farm and herd animals and are proud of their ability to survive in the harsh conditions. Phase 1A took away the houses, land and resources of 20 000 people, while Phase 1B affected another 7000. It severely strained the families and society of nearby villages. Various compensation arrangements have been put in place for villagers whose homes, fields, or grazing areas have been flooded, although not unexpectedly there are grumbles that these promises have not been met, or that resettlement plans are unsatisfactory.

Positive impacts for people

The project provides money for the Lesotho government to spend on other development projects. In recent years, water from the scheme has been discharged into the Mohokare (Caledon) River to provide water for the capital, Maseru, in times of shortage. The Muela power station provides electricity for Lesotho. Hundreds of kilometres of tarred roads were built in order to improve access to the different construction sites. Together with gravel 'feeder' roads around the dams, these tarred roads continue to provide much improved communication for many villages in the mountains. Around 4000 people got temporary jobs at the Katse Dam. Some were employed in construction and hundreds more flocked to the area to work in services for the workers, as food vendors and shopkeepers. A thousand local people also worked at the Mohale Dam.

Vegetation

The new dams threatened the important alpine fauna of the Lesotho Highlands. The 17-hectare Katse Botanical Gardens was established in 1995, to rescue 149 plant species from the flooding. The gardens try to promote enjoyment and a knowledge of the alpine flora of Lesotho through conservation, cultivation, and propagation of the native plants. The gardens serve as an educational centre for local communities, students, and scientists.

Animals

Flooding has a potential impact on land animals, and also on animals that live in the rivers whose water flow was affected by the dam construction. Any reduction in the flow of the river will endanger the existing species. An example is the Maluti Minnow, a small fish that is less than 5 cm in length. It needs very high-quality water and is therefore a good indicator of the river's water quality. The population of this fish is being monitored as an indicator of the effects of the dam construction.

Fig. 11.39 A herdsman above the Katse Dam Lake in the Lesotho Highlands

Fig. 11.40 A village above the Mohale Dam Lake

13 Write an account of the Lesotho Highland Water Project using the headings: Location, Advantages for Lesotho, Disadvantages for Lesotho.

12 Geographical skills

This chapter covers the geographical skills needed for Cambridge IGCSE® and O Level Paper 2.

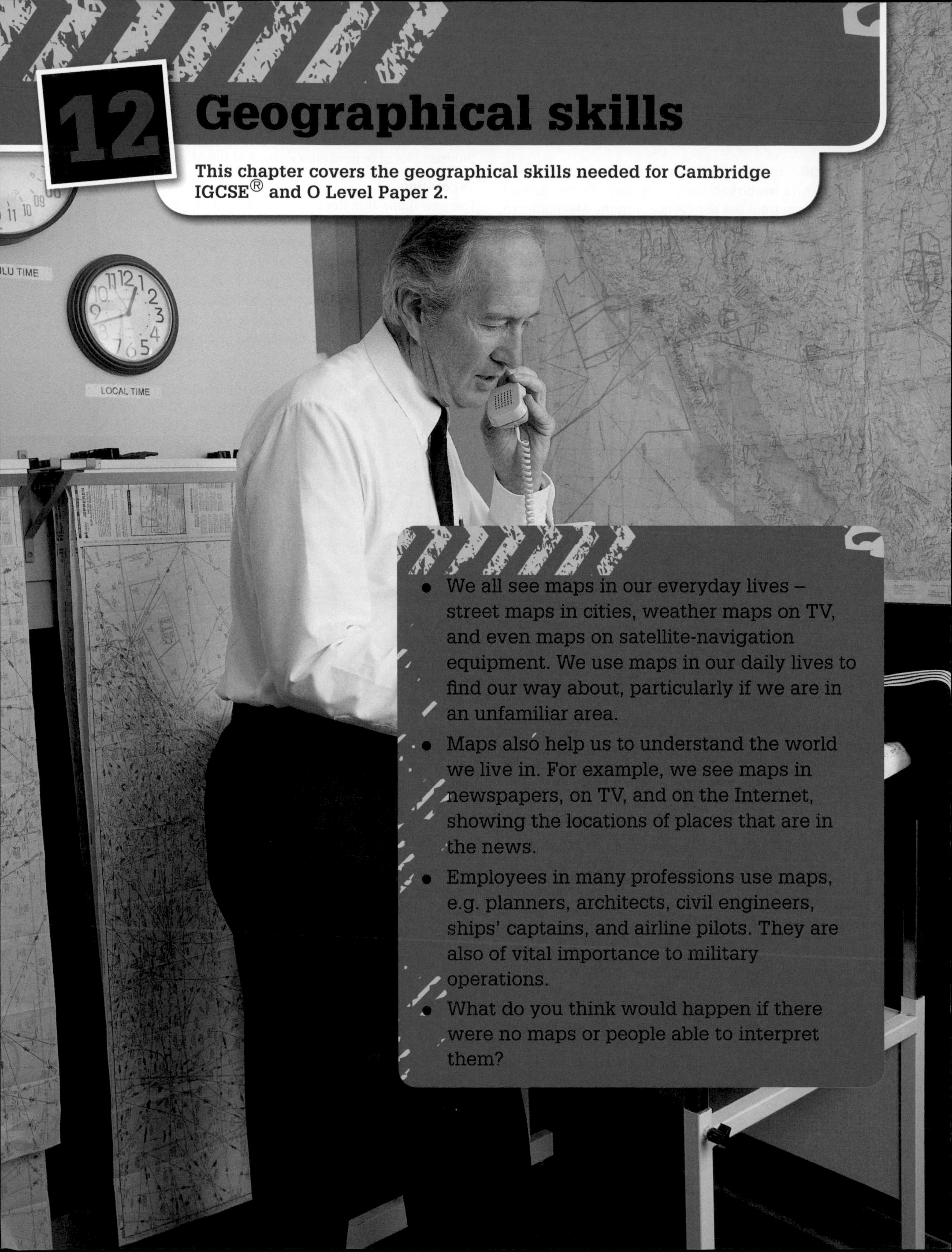

- We all see maps in our everyday lives – street maps in cities, weather maps on TV, and even maps on satellite-navigation equipment. We use maps in our daily lives to find our way about, particularly if we are in an unfamiliar area.
- Maps also help us to understand the world we live in. For example, we see maps in newspapers, on TV, and on the Internet, showing the locations of places that are in the news.
- Employees in many professions use maps, e.g. planners, architects, civil engineers, ships' captains, and airline pilots. They are also of vital importance to military operations.
- What do you think would happen if there were no maps or people able to interpret them?

> **In this chapter you will learn about:**
> - how to answer examination questions about survey map extracts
> - how to answer examination questions about photographs
> - the types of diagram that you can find in examination questions.

Survey maps (topographic maps)

These are examples of large-scale maps. This means that they show a relatively small area of land in great detail. They show the surface features of an area, including relief, drainage, land use, settlement, and roads. This is a compulsory element of the IGCSE® and O Level Paper 2.

Using the key and symbols

The positions of different features on a map are shown by symbols. Different countries use different symbols on their maps, so it is always best to check the meaning of a symbol using the map's key (a list, usually at the side or bottom of the map, which explains the meaning of each symbol). Where a label is written next to a symbol, the feature is located at the position of the symbol (e.g. look at Fig. 12.1).

> **LEARNING TIP** Care is needed when two symbols are shown on the same line. For example, the motorway symbol on this map key (Fig. 12.2) also shows the symbol for a bridge, and the Main B road symbol also shows the symbol for an embankment.

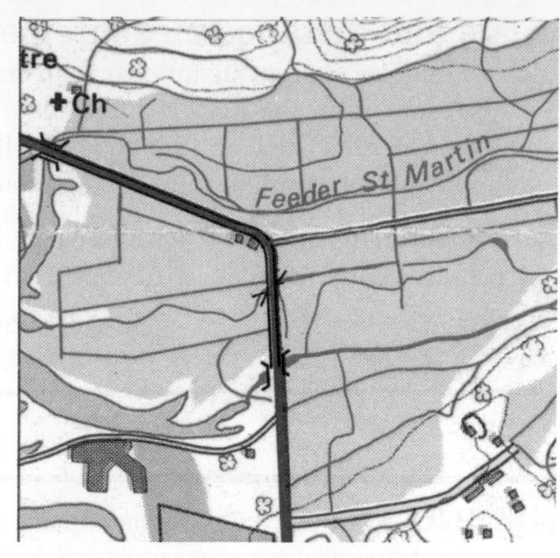

Fig. 12.1 Symbols on a map of part of Mauritius (the church in the north-west is located where the cross-shaped symbol is, not where the letters Ch are)

> **1** Look at Fig. 12.2.
>
> **a** What is the symbol for:
> a main A road?
> a bridge?
> a ruin?
>
> **b** What is the difference between a cutting and an embankment?
>
> **c** What are the symbols for a police station and a post office?
>
> **d** Draw a square 10 cm by 10 cm. In your square, draw a map of an area showing a motorway crossing over a Main B road at a bridge. Also show a town at another crossroads. The town has a church, a town hall, and a hotel.

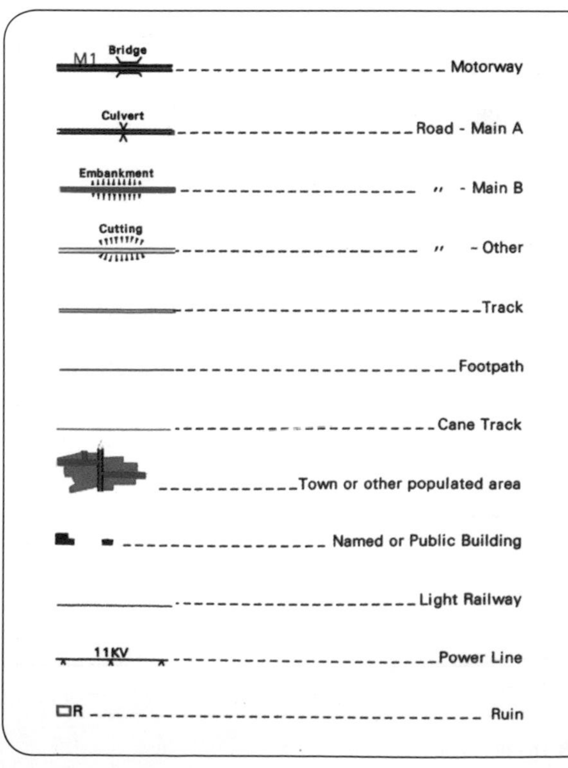

Fig. 12.2 An example of part of a map key (for the map extract shown in Fig. 12.1)

Map scale and distance measurements

The scale of a map shows how distance on the ground has been represented on the map. A large-scale map might show a small area (such as a school or a village), whereas a small-scale map might show a whole country.

The scale is shown on the map in two ways (see Fig. 12.3). One is the representative fraction, in this case 1:25 000. This means that 1 cm on the map equals 25 000 cm on the ground. This works out at 4 cm on the map for 1 km on the ground. The other common map scale is 1:50 000, where 2 cm on the map equals 1 km on the ground. The other way of showing the scale is the scale line (also shown in Fig. 12.3).

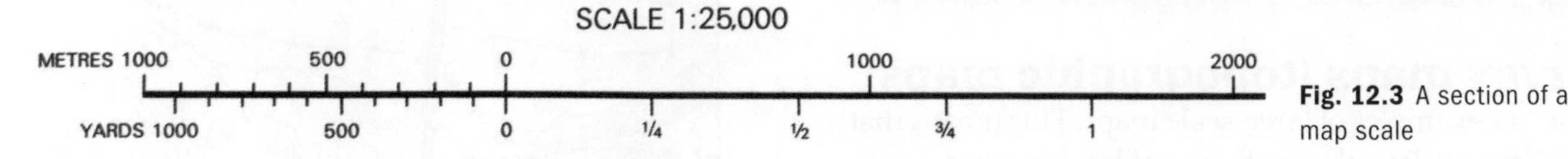

Fig. 12.3 A section of a map scale

LEARNING TIP Use the scale line to measure distances accurately without doing any calculations. Simply put the straight edge of a piece of paper along the line to be measured and mark the two end points on the edge of the paper (see Fig. 12.4). Then place the piece of paper along the scale line and read off the distance in whatever units you require.

If you need to measure a distance that is not a straight line (e.g. along a winding road), divide the curve into straight sections and measure the length of each straight section by rotating the edge of the paper along the curve using the point of a pencil.

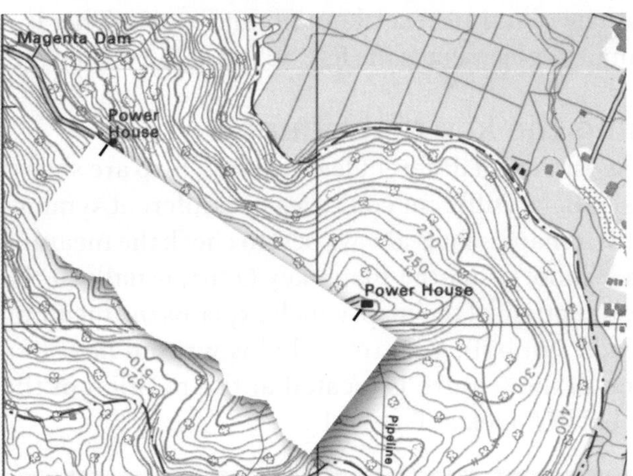

Fig. 12.4 Measuring distance using the edge of a sheet of paper

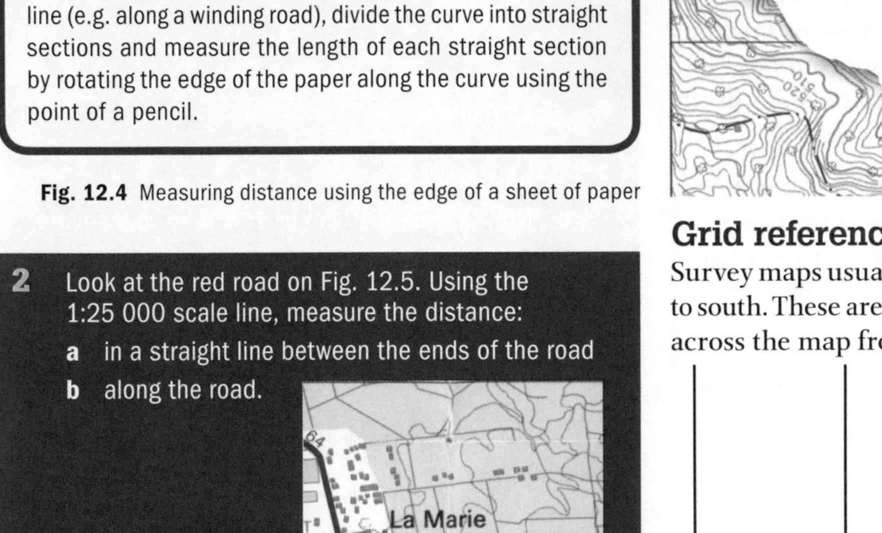

2 Look at the red road on Fig. 12.5. Using the 1:25 000 scale line, measure the distance:

a in a straight line between the ends of the road

b along the road.

Fig. 12.5 A road in Mauritius

3 Now look at the red road on Fig. 12.6. Using the 1:25 000 scale line, measure the distance:

a in a straight line between the ends of the road

b along the road.

Fig. 12.6 Another road in Mauritius

Grid references

Survey maps usually have a grid of lines running north to south. These are called the eastings and show distance across the map from west to east.

Fig. 12.7 The eastings on a map

Another set of lines run from west to east. These are called northings and show distance from south to north.

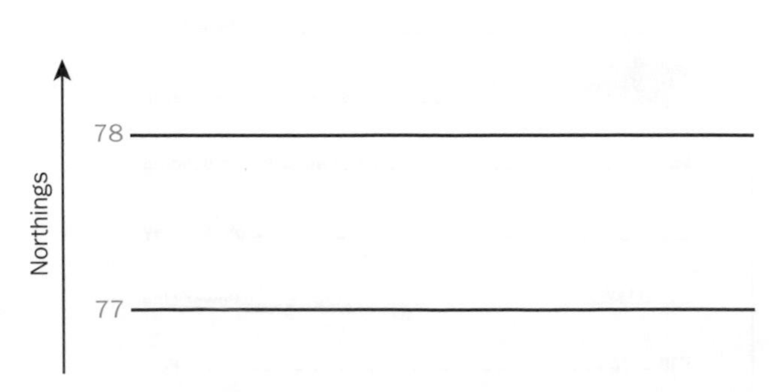

Fig. 12.8 The northings on a map

The eastings and northings form grid squares, which represent 1 km by 1 km. On a 1:25 000 map, the squares measure 4 cm by 4 cm. On a 1:50 000 map, the grid squares measure 2 cm by 2 cm.

The grid is slightly different from latitude and longitude (true north or geographical north), and also slightly different from magnetic north (the direction in which a compass needle points). The differences between the three are often shown on the map using a small diagram (Fig. 12.10). The differences are not constant from place to place, and magnetic north slowly changes its position. Survey map questions will always use grid north.

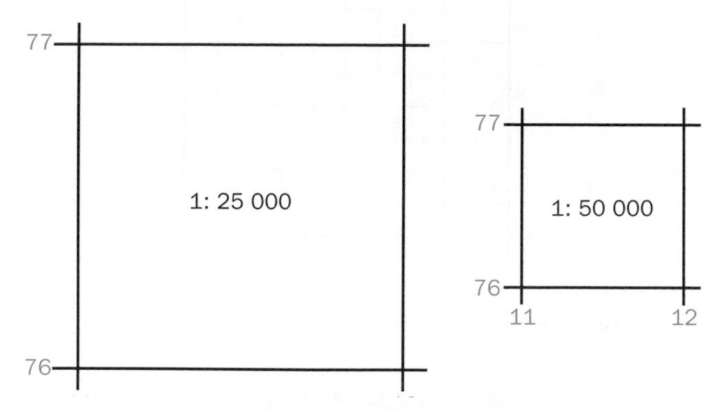

Fig. 12.9 Grid squares on 1:25 000 and 1:50 000 maps

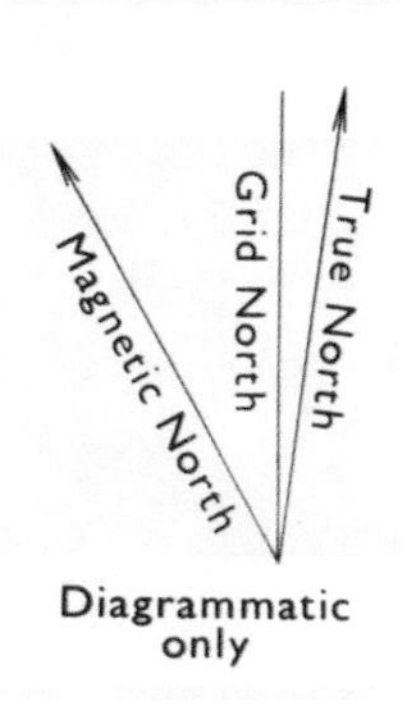

Fig. 12.10 The different types of north

Four-figure grid references

The grid allows locations on a map to be described precisely. A four-figure grid reference fixes a point within a 1-km grid square. It is the point at the south-west (bottom left) corner of the square. The easting is always given before the northing. One way of remembering this is the phrase 'Always go along the corridor before going up the stairs'.

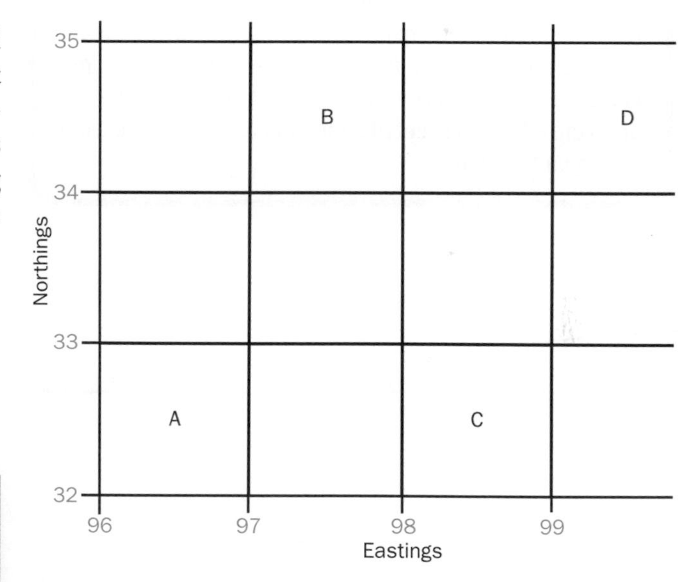

Fig. 12.11 Examples of four-figure grid references

4 Look at Fig. 12.11. The four-figure grid reference for A is 9632 and for B is 9734. What are the four-figure grid references for C and D?

5 Look at Fig. 12.12. What are the four-figure grid references for:

a the Red Knight mine?

b the ruins?

c the school?

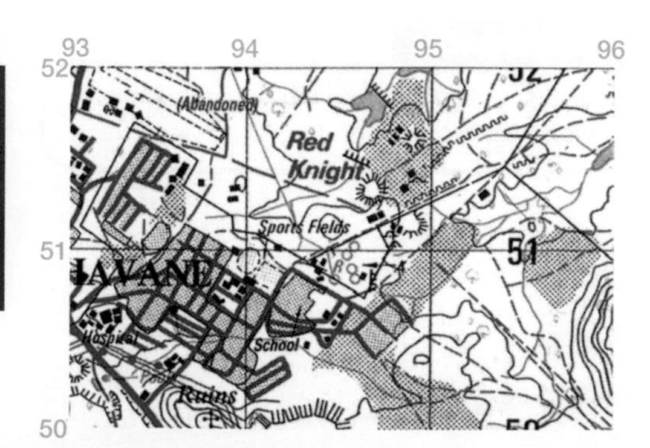

Fig. 12.12 Four-figure grid references in part of Zimbabwe

Six-figure grid references

Six-figure grid references allow locations to be fixed within an area of 100 m by 100 m. The first, second, fourth and fifth figures show the grid square and the third and sixth figures show the precise location within that square. Make sure that you know exactly how the third and sixth figures are measured.

6 Look at Fig. 12.13. The six-figure grid reference for E is 960321 and for F is 963323. What are the six-figure grid references for G and H?

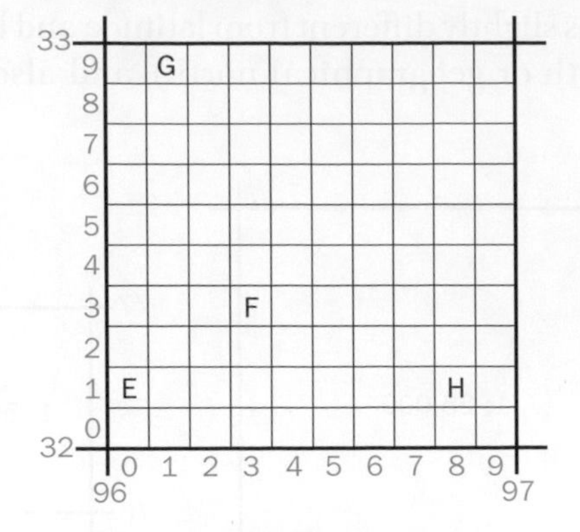

Fig. 12.13 Examples of six-figure grid references

7 Look at Fig. 12.14. What are the six-figure grid references for the following buildings:

a St Clement Chapel?

b the Central Electricity Board?

c the Mosque (Mos)?

Fig. 12.14 Six-figure grid references in part of Mauritius

LEARNING TIP Candidates often get the third and sixth figures of grid references wrong, because they do not use the correct method of measuring the tenths. Note that the first tenth of the square is 0 (zero), the second tenth is 1, the third tenth is 2 and so on. The last tenth is 9. You can copy the subdivisions shown for 1 km on the scale line on to the edge of a sheet of paper and use that to measure tenths on the map.

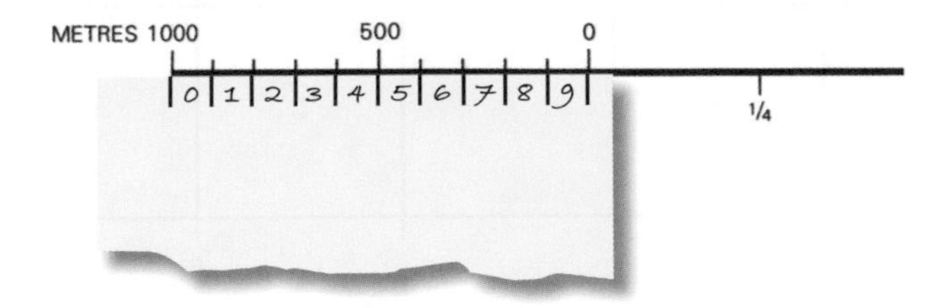

Fig. 12.15 Making a scale on the edge of a piece of paper to measure the tenths

Discussion point

How does the time of day change as you move from east to west around the world, and why?

Latitude and longitude

Don't confuse the grid lines on a survey map with the lines of latitude and longitude on an atlas map. Lines of latitude and longitude are curved lines on the globe. Longitude measures the distance in degrees east and west of the Greenwich Meridian (0^0) that runs through London. Latitude measures the distance in degrees north and south of the equator:

- Equator 0°
- Tropic of Cancer 23½°N
- Tropic of Capricorn 23½°S
- Arctic Circle 66½°N
- Antarctic Circle 66½°S
- North Pole 90°N
- South Pole 90°S

Compass directions

Directions can be given using the points of the compass (Fig. 12.16).

360⁰ bearings

A second way of showing direction is by a 360⁰ bearing from grid north. To do this, you need to use a protractor - making sure that the centre point of the protractor is exactly over the point you wish to measure from, and that your protractor is aligned north-south along the grid lines (as shown in Fig. 12.18). This is measured clockwise from north. In Fig. 12.18, the bearing of the trigonometrical station from the market is 143⁰.

For bearings of more than 180⁰, i.e. to south-west, west and north-west, measure the angle from south and add on 180⁰ (as shown in Fig. 12.19). The bearing of the second trigonometrical station from the market is $62^0 + 180^0 = 242^0$.

LEARNING TIP When answering questions about directions, make sure that you know whether the question is asking for the direction ***to*** a place or the direction ***from*** a place.

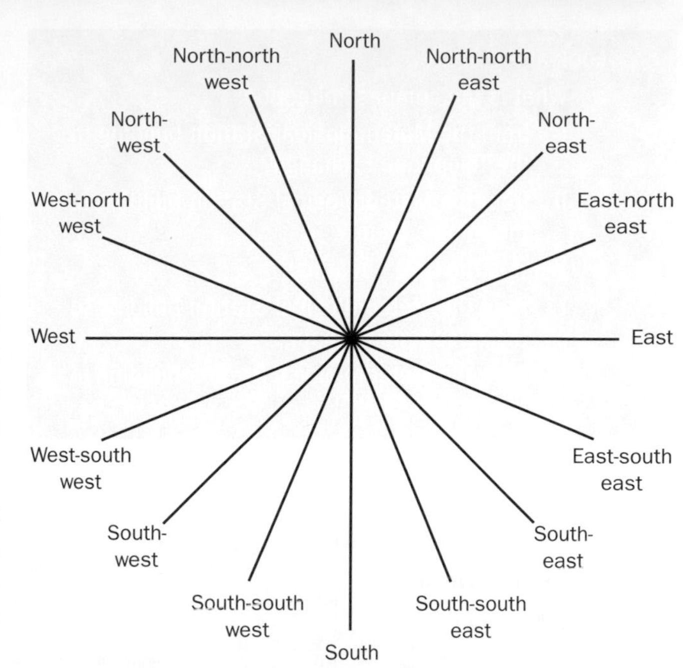

Fig. 12.16 The points of the compass

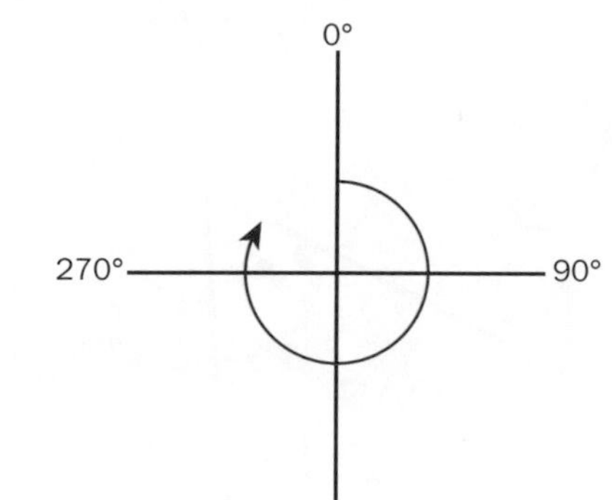

Fig. 12.17 360° bearings

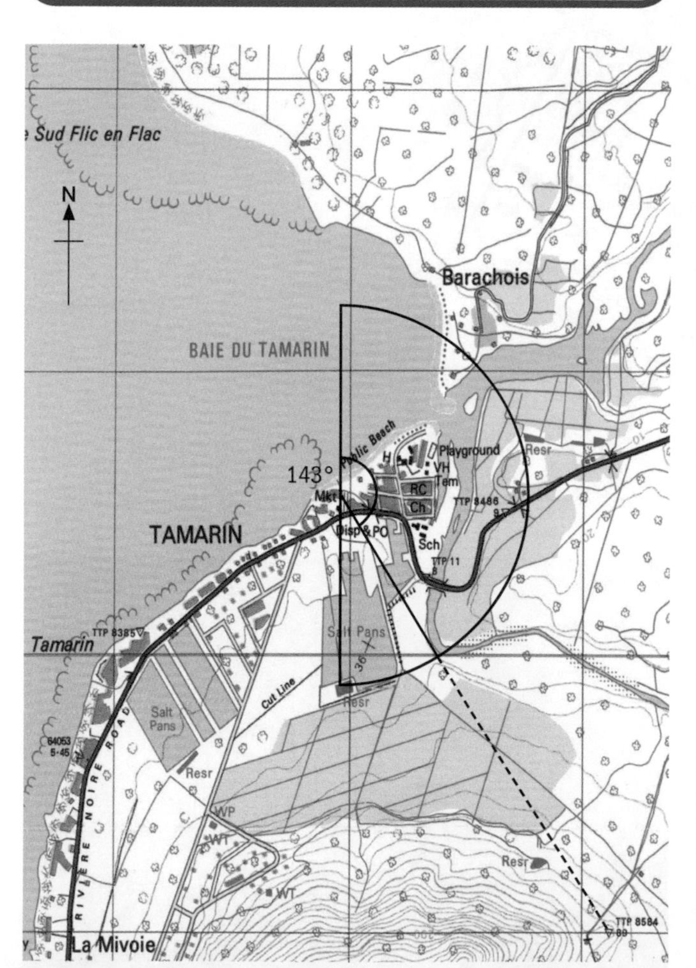

Fig. 12.18 How to use a protractor to measure bearings between 0°and 180°

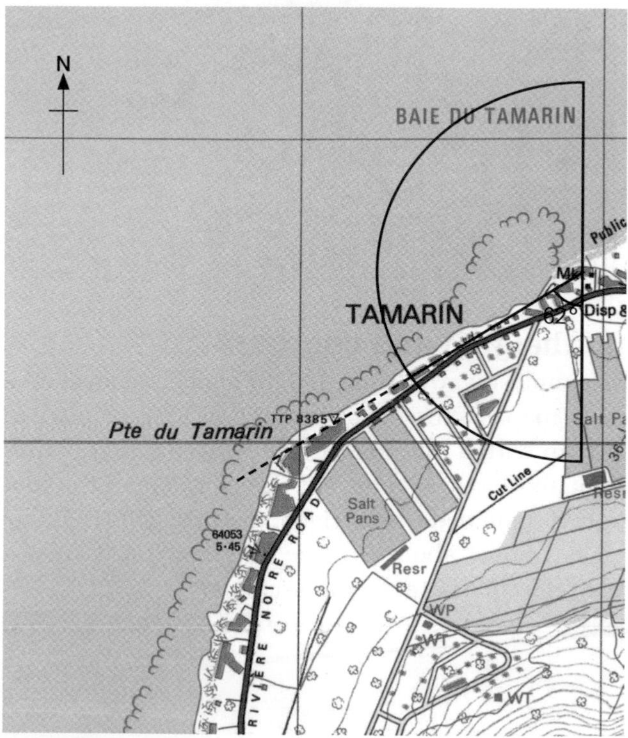

Fig. 12.19 How to use a protractor to measure bearings between 181° and 360°

8 Look at Fig.12.20.

a What is the compass direction:

i from the Meteorological Station building to the Telephone Exchange?

ii from the Meteorological Station building to the temple (Tem)?

b What is the 360° bearing:

i from the Meteorological Station building to the Telephone Exchange?

ii from the Meteorological Station building to the temple (Tem)?

Fig. 12.20 Part of Mauritius

Spot heights and contours

The simplest way to show height above sea level on a survey map is to use a spot height. A dot is printed on the map with a number beside it, giving that point's height above sea level (in metres). Sometimes the spot height is combined with a trigonometrical point (station). This is a pillar, about a metre tall, which is used as a fixed reference point by mapmakers.

Fig. 12.21 A trigonometrical point (station)

9 Look at Fig. 12.22. What is the height above sea level of the trigonometrical station at the top of the hill? Remember to state the correct units in your answer.

Fig. 12.22 The location of a trigonometrical station in Mauritius

A contour on a map is a line (often brown in colour), which joins places of equal height above sea level. The difference in height between the contours (sometimes called the contour interval) varies, but it is often 10 or 20 m. Important contours (such 100 m, 200 m, 300 m, etc.), are often shown by a bold line. These principles apply to any *isoline map* - maps with lines joining places of the same value, e.g. temperature (isotherms), atmospheric pressure (isobars), rainfall (isohyets), or earthquake intensity (isoseismal lines).

You often see contours where the numbers appear to be upside down. This is because the numbers are shown for the reader looking up the slope (see Fig. 12.23).

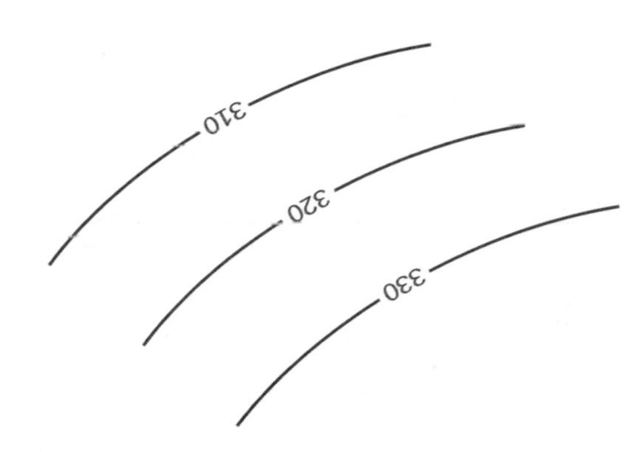

Fig. 12.23 Contour intervals (figures on contours read looking up the slope of the land)

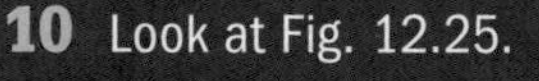

LEARNING TIP The height of the top of a hill is not the height of the last contour. On Fig. 12.24, the top of the hill is between 210 metres and 220 metres above sea level - not 210 metres.

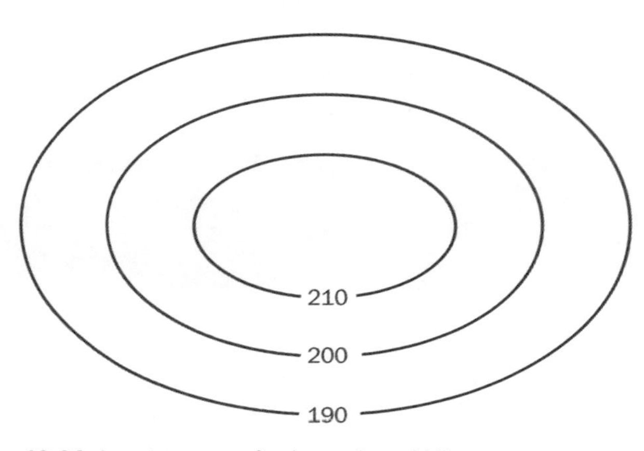

Fig. 12.24 A contour map of a dome-shaped hill

10 Look at Fig. 12.25.

a What is the contour interval on this map?

b What is the height above sea level of the water tanks (WTs)?

c What is the height above sea level of the school to the east of the map extract?

(Remember to give the units in your answers.)

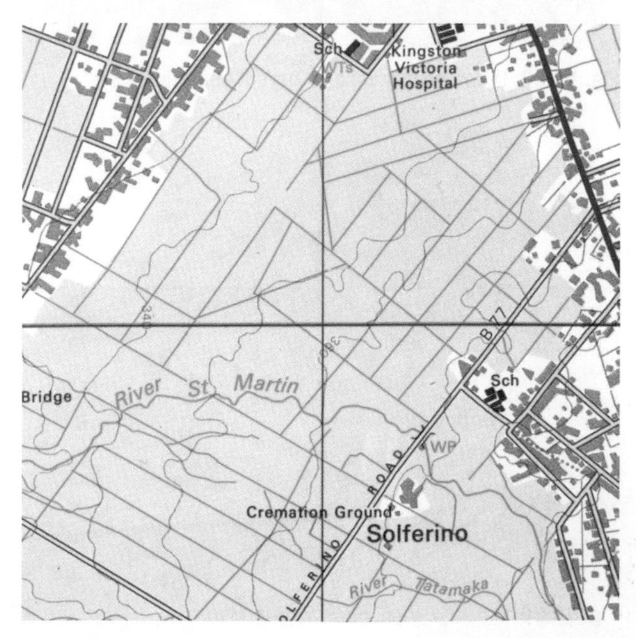

Fig. 12.25 Contours on a map of part of Mauritius

Relief features

The geographical term relief means the height, steepness and shape of the ground surface.

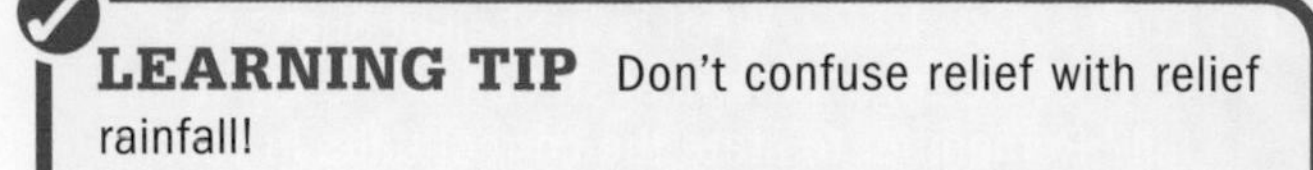

LEARNING TIP Don't confuse relief with relief rainfall!

Slopes

The closeness of the contours shows the steepness of the slope. Closely spaced contours mean a steep slope and widely spaced contours mean a gentle slope. The absence of contours may indicate flat land.

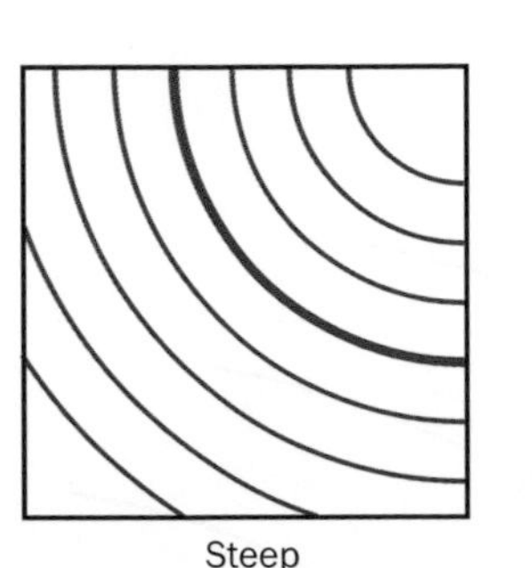

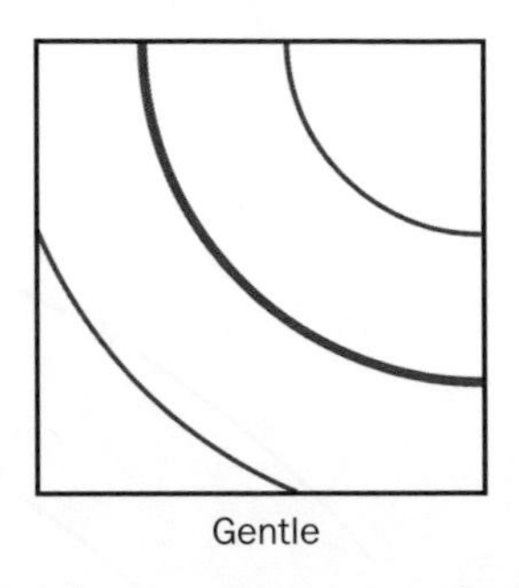

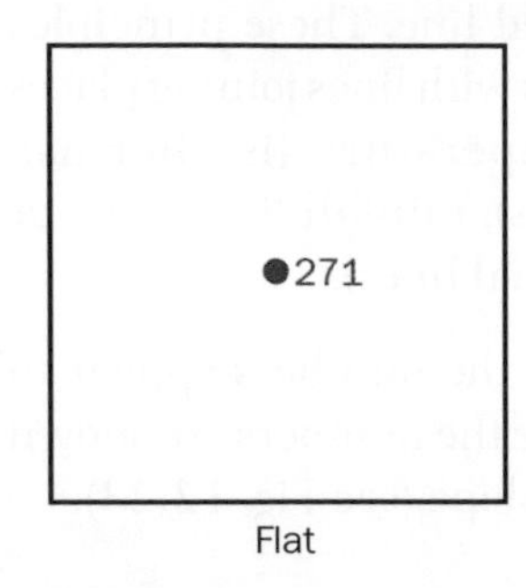

Fig. 12.26 A steep slope, a gentle slope and flat land

On some maps cliffs are shown by a separate symbol (Fig. 12.27).

Fig. 12.27 The cliff symbol

LEARNING TIP When describing relief, only use the term cliff when the cliff symbol is used on the map.

Uplands and lowlands

The contour heights and spot heights on the map show the height above sea level, and can be used to show the higher and lower areas on a map. There is no precise definition about how high or low an area has to be to be classified as highland or lowland. In some areas of the world entire countries are high above sea level.

11 Draw a sketch map of the area shown in Fig. 12.28, dividing it into areas of different relief. Use the labels *higher, lower, steep, gentle, flat.*

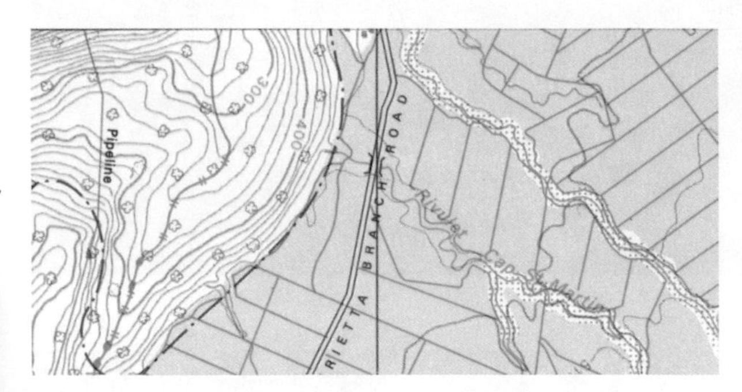

Fig. 12.28 Differences in relief in part of Mauritius

Valleys and floodplains

Small valleys without a flat floor are shown on maps by a 'V' shape in the contours. The V always points to high ground. There may or may not be a river in the centre (see Figs. 12.29 and 12.30).

Fig. 12.29 A narrow V-shaped valley in the Cederberg Mountains, South Africa

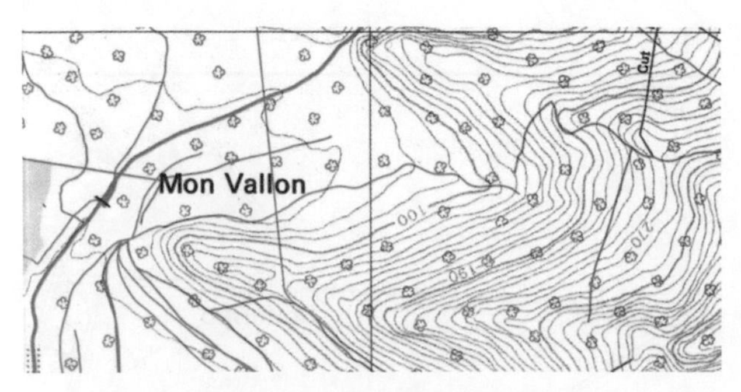

Fig. 12.30 A narrow V-shaped valley in Mauritius

Fig. 12.31 A valley with a flat floor or floodplain in the Western Cape of South Africa (the locations of the rivers are shown by the winding bands of darker vegetation)

> **12** Look at Fig. 12.32. The contour printed in bold nearest to the river is 900 m above sea level.
>
> Draw a sketch map to show the relief of this area and the river. Label the river, the valley sides and the valley floor (floodplain).

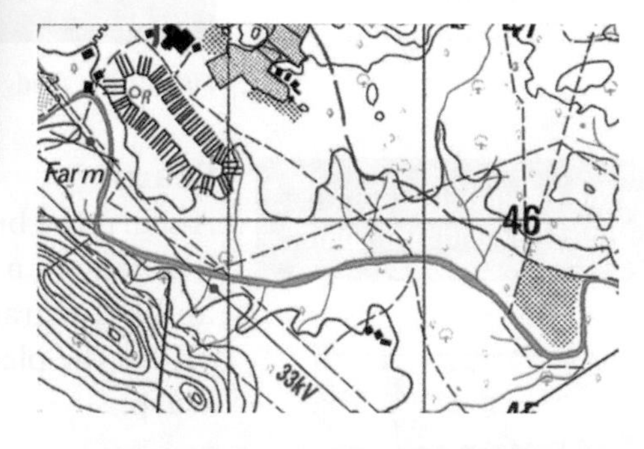

Fig. 12.32 A river valley in Zimbabwe

Plateaux

These are areas of land that are high and flat.

Fig. 12.33 A plateau in Namibia (in the background)

> **13** Draw a sketch map to show the relief of the area shown in Fig. 12.34. Label the plateau.

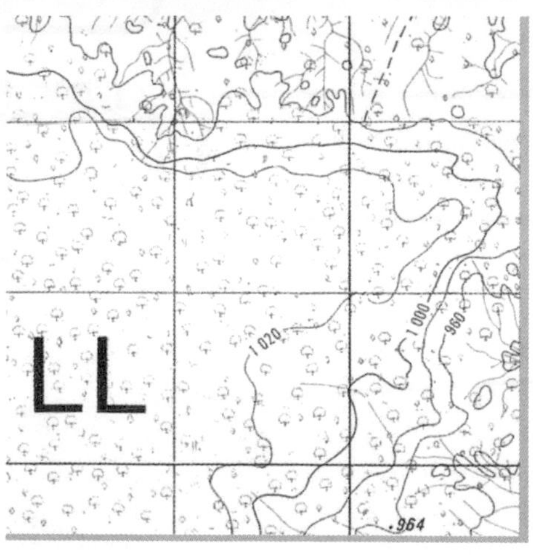

Fig. 12.34 A plateau in Zimbabwe, near the Victoria Falls

Ridges

A ridge is a long, narrow area of high ground, rather like the spine of an open, upturned book.

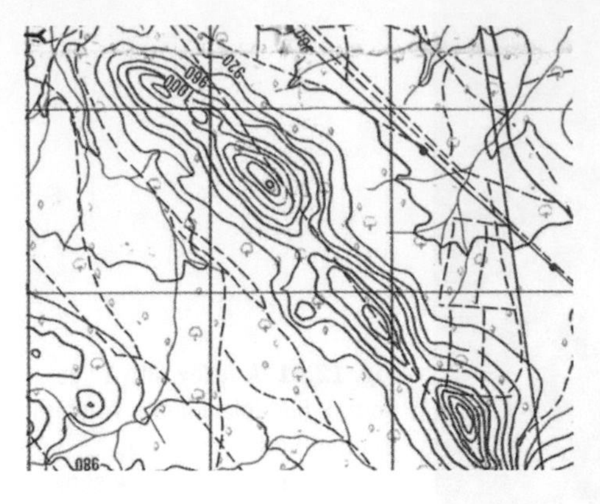

Fig. 12.35 A ridge in Zimbabwe

Fig. 12.36 A ridge near Chamonix in the Alps mountains, France

Spurs

A spur is a ridge where the spine slopes down from high ground to low ground. A spur is shown by a 'V' shape in the contours, where the V points to low ground. Look at Fig 4.12 (page 113), which shows a valley in South Africa. A prominent spur runs down to the valley from the right.

14 Draw a sketch map to show the relief of the area shown in Fig. 12.37. Label the valley and two spurs.

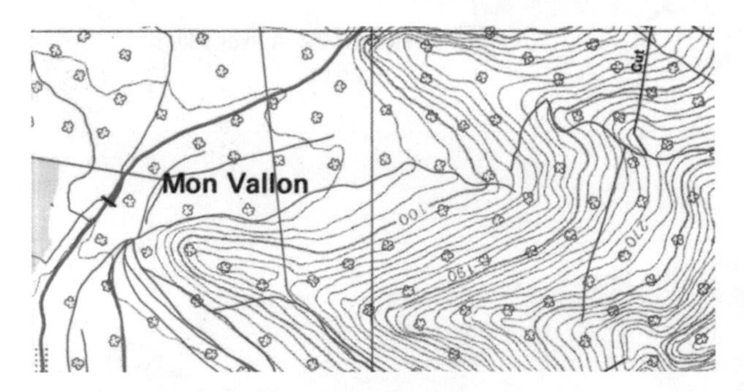

Fig. 12.37 A valley and spurs in Mauritius

LEARNING TIP Make sure you can tell the difference between the contour patterns of valleys and spurs.

Scarps

A scarp is a broad, steep slope. It could be the sides of a plateau or a ridge, as seen on the previous diagrams and photographs. The slopes may include cliffs. The side of the plateau in Fig. 12.33 is a good example of a scarp.

LEARNING TIP When describing relief on part of a map extract, describe:

- any areas of highland or lowland, including giving the height of the highest point
- any areas of cliffs, steep slopes, gentle slopes, or flat land
- any specific relief features, such as valleys, spurs, ridges, and plateaux.

Discussion point

Describe the relief in the area where you live. Which of the geographical terms mentioned in this chapter apply to your area?

Drainage

Drainage means the water features shown on the map (usually in blue). This includes rivers and streams and their features, lakes, and ponds. Marsh might be considered to be a feature of the drainage (i.e. poorly drained land), or it could also be considered a feature of the vegetation (where the plants have adapted to those conditions). Drainage may also include features produced by human activity, such as drainage channels or reservoirs (known as dams in many parts of the world).

Drainage density

Drainage density is the total length of the rivers and streams in a location (in kilometres), divided by the area (in square kilometres). Locations with high drainage density have lots of surface water, and locations with low drainage density have very little. This is often due to permeable rocks, such as limestone, which cause the water to seep underground.

Channel shape

Meandering, straight and braided channels are described in Chapter 4. They can often be identified on survey maps.

Stream (drainage) patterns

The patterns which might be identified on survey maps are described in Fig. 12.38.

Dendritic	Radial	Trellised
A tree-like pattern. This usually develops where there are no differences in the underlying rocks. Fig 12.31 shows this pattern.	Where streams flow outwards from a central high point, such as on a volcanic cone.	Where streams tend to meet at right angles – forming a rectangular pattern. This often forms when there are inclined layers of rock of different hardness.

Fig. 12.38 Types of drainage pattern

LEARNING TIP When describing drainage on part of a map extract, refer to:

- the density of drainage, e.g. 'lots of small streams'
- the direction of flow of the main rivers and streams
- specific features of the rivers and streams, such as meanders, islands, tributaries, waterfalls
- the gradient of the river (although this is often not easy to tell). Remember that a steep gradient does not necessarily mean that the flow will be rapid.
- any lakes, ponds or marshes
- channel and stream patterns.

If a question asks for *physical* features, do not refer to bridges (they are *human* features). If it asks you to describe the river, do not describe the relief features of the valley as well. Many examination candidates wrongly think that tributaries run out of rivers (and lose marks as a result). Tributaries *join* rivers.

15 Describe the drainage in the area shown on Fig. 12.39. Rivers are shown in blue – the green lines are tracks and not drainage features. The blue symbol in the west is for marsh or land liable to flooding. Are there any differences between different parts of the map?

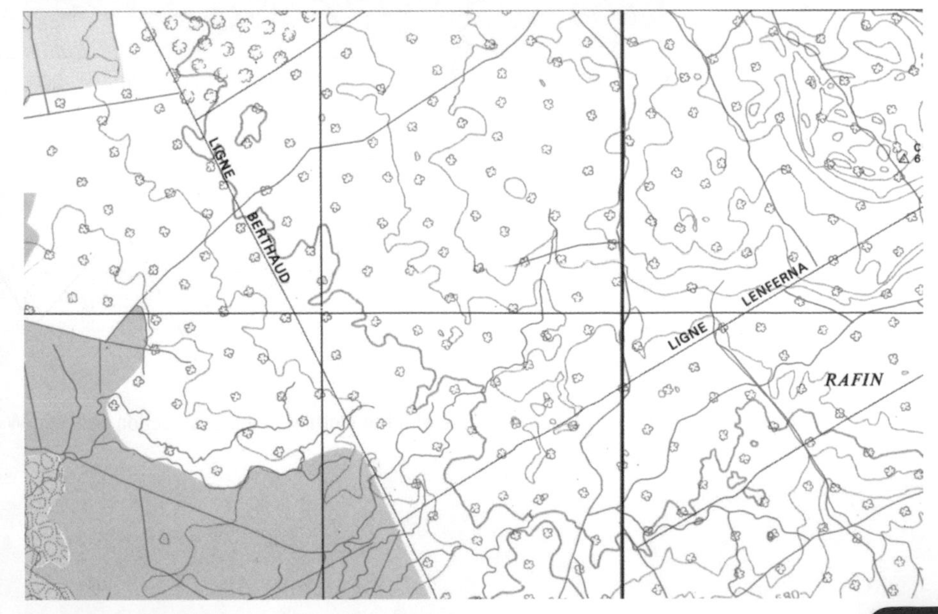

Fig. 12.39 Drainage in an area of Mauritius

Cross-sections

A cross-section is a type of diagram often used in examination questions. It is as if the landscape has been sliced vertically along a line and pulled apart. Cross-sections are drawn to scale. The horizontal scale is generally the same as the map scale, but the vertical scale is made bigger (vertical exaggeration). This is so that features like hills and valleys show up better. The position of features on the ground surface can be shown with labelled arrows.

In examination questions you may be asked to complete a partly drawn cross-section or to label map features on the cross-section.

LEARNING TIP Remember that the arrows that you use for labels must point exactly to the ground surface and not above or below it.

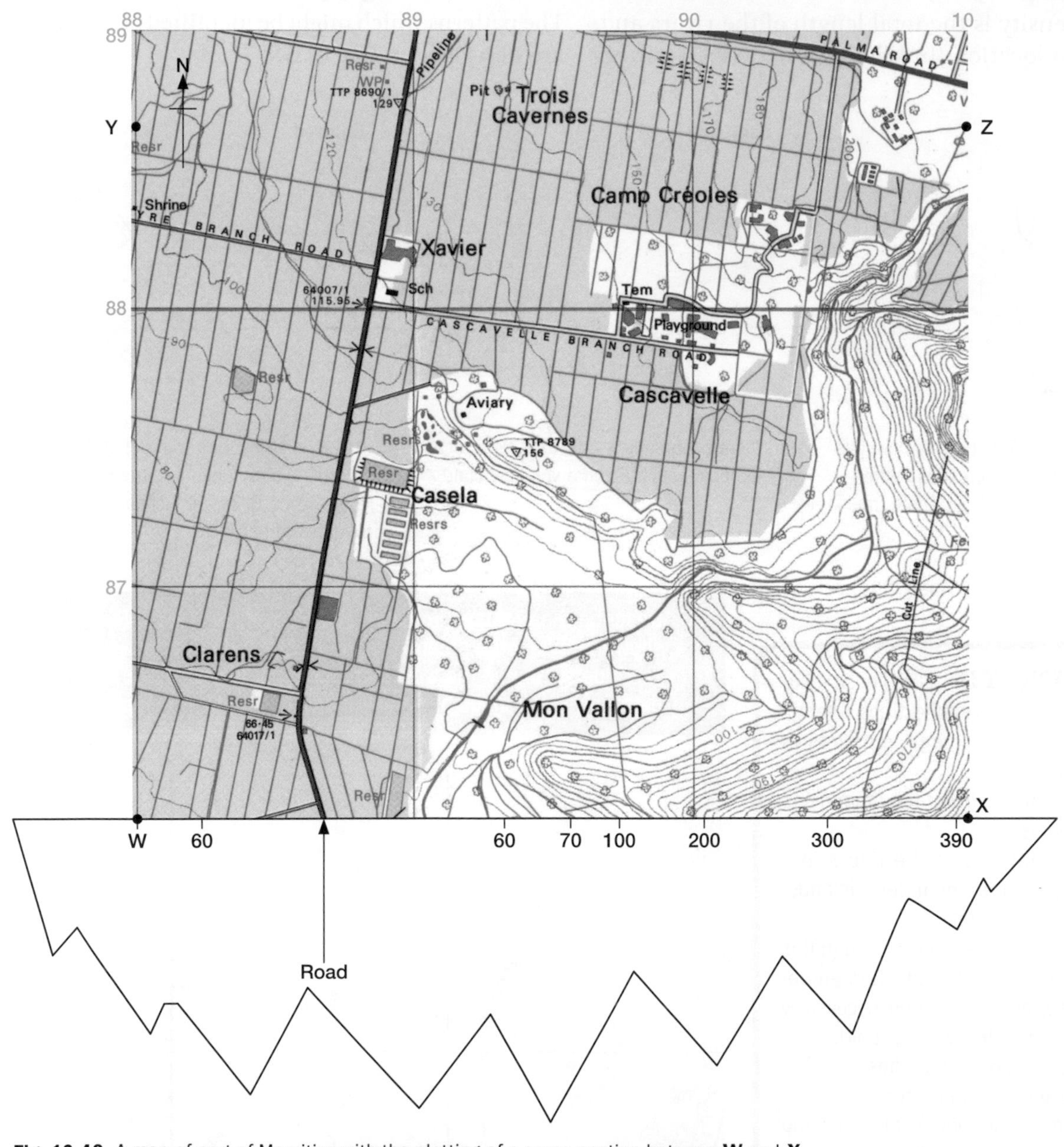

Fig. 12.40 A map of part of Mauritius with the plotting of a cross-section between **W** and **X**

How to plot or complete a cross-section

1. Place the edge of a piece of paper along the line of the cross-section on the map.
2. At the edge of the piece of paper, mark the end points of the cross-section, e.g. **W** and **X**.
3. At the edge of the piece of paper, mark and label the positions of important contours and other features. This is shown on Fig. 12.40.
4. Now take your piece of paper and place it horizontally on the frame for the cross-section, making sure that **W** and **X** are at the ends of the section. In exam questions, the frame will be provided for you.
5. Mark the positions of the contours on your piece of paper on to the frame using dots. Make sure that these are at the correct heights.
6. Join the dots to produce the profile of the ground surface. Take particular care with the tops of hills and the bottom of valleys. This is shown on Fig. 12.41. For example, if the contour interval is 10 m and the highest contour is 200 m, this means that the height of the top of the hill is more than 200 m but less than 210 m.
7. Label any key features on the cross-section with vertical arrows touching the ground surface.

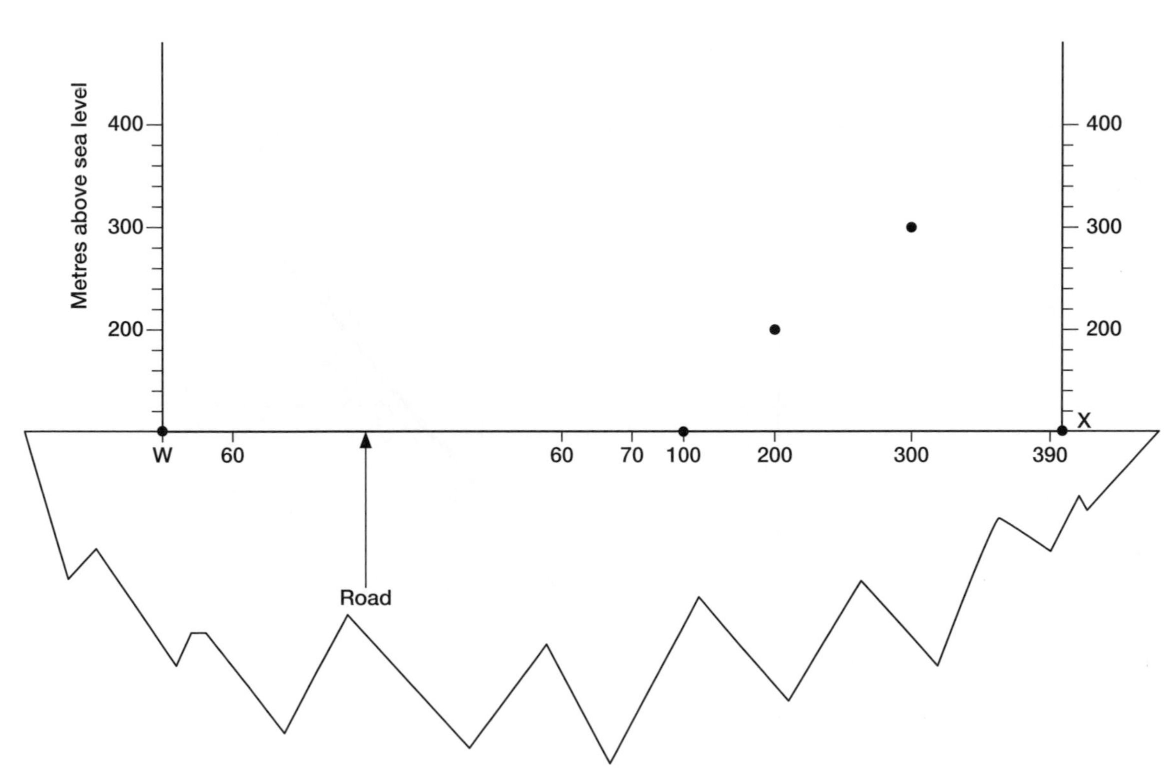

Fig. 12.41 Plotting a cross-section from **W** to **X** on Fig. 12.40. So far, the 300 m and 200 m points have been plotted and the paper is in position to plot 100 m.

16 a On Fig. 12.42 complete the cross-section between Y and Z on Fig. 12.40. Remember that the contour interval is 10 m. The first part has been drawn for you.

b On your cross-section label the positions of two roads.

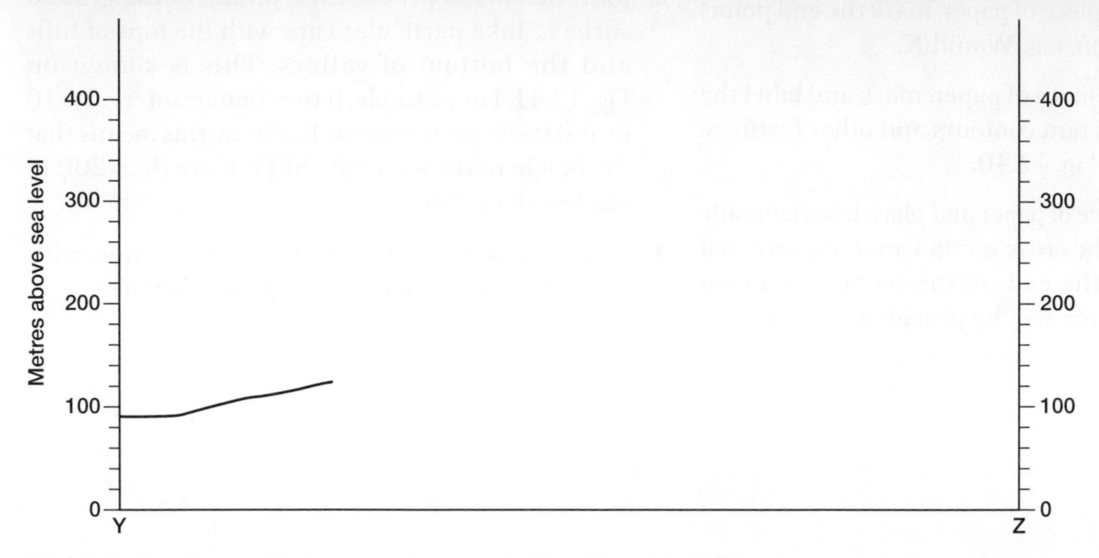

Fig. 12.42 A partly completed cross-section between **Y** and **Z** on Fig. 12.40

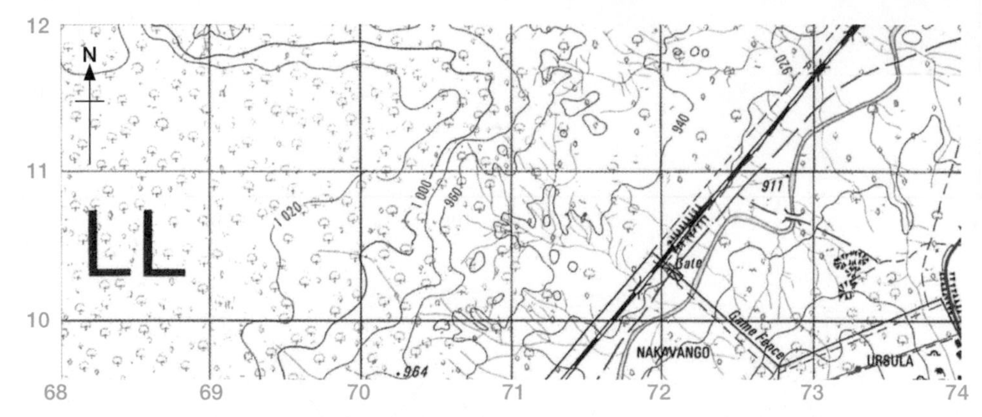

Fig. 12.43 A map of part of Zimbabwe

Fig. 12.44 A partly completed cross-section along northing 110 on Fig. 12.43

17 a Use information from Fig. 12.43 to complete the cross-section on Fig. 12.44.

b On your cross-section label the position of the railway, a steep slope, a plateau.

Physical features of coastlines

The physical features of coastlines are described in Chapter 5. You might wish to study this chapter before attempting these questions.

18 Make a copy of the coastline shown on Fig. 12.45. On your map, label the position of:

- **a** a bay
- **b** a headland
- **c** a cliff
- **d** a stack.

19 Make a copy of the coastline shown on Fig. 12.46. On your map, label the position of:

- **a** a coral reef
- **b** a coastal marsh
- **c** a spit
- **d** a beach.

20
- **a** A wave-cut platform is not shown on either of the maps. How could you distinguish this feature from a cliff on a map?
- **b** Coastal sand dunes are not shown on either of the maps. How could you distinguish this feature from a sandy beach on a map?

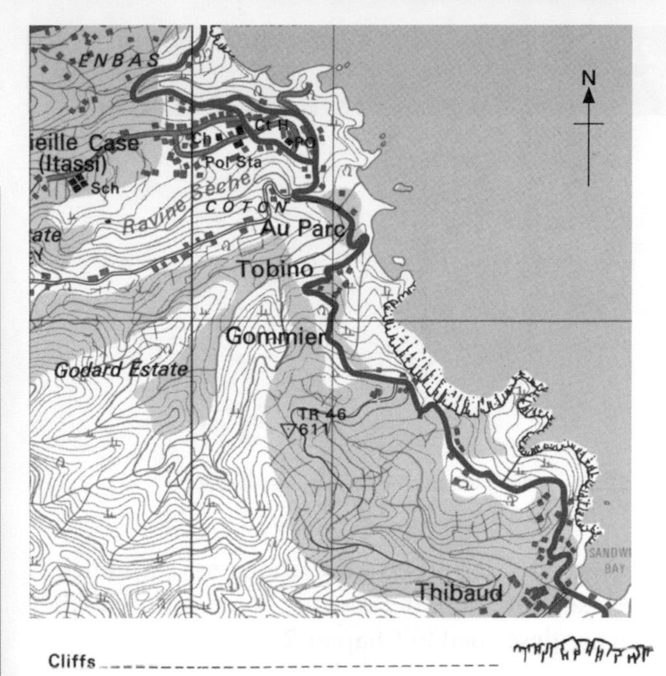

Fig. 12.45 Part of the coastline of the island of Dominica in the West Indies

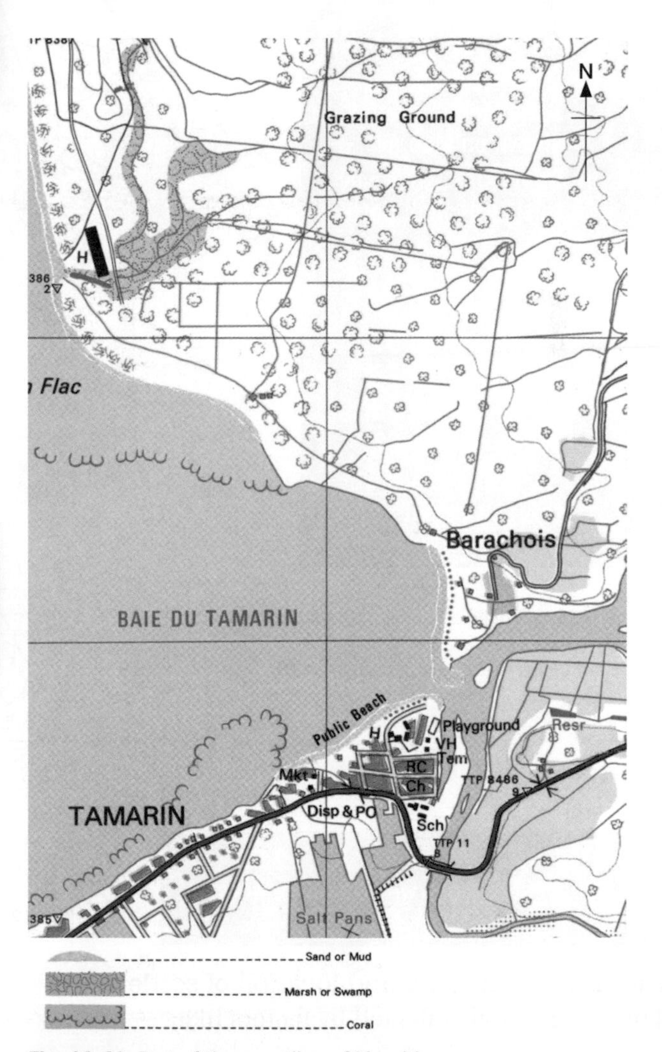

Fig. 12.46 Part of the coastline of Mauritius

Settlement

Survey map questions on settlement are based around either urban settlements (towns and cities) or rural settlements (villages and isolated dwellings). These settlements are described in Chapter 2.

Urban morphology

Urban morphology refers to the form of towns and cities, or the variations in land use within them. Different urban zones can be identified on survey maps.

21 Look at Fig. 12.47.

- **a** Which urban zone is shown on the map?
- **b** Give map evidence for your answer.

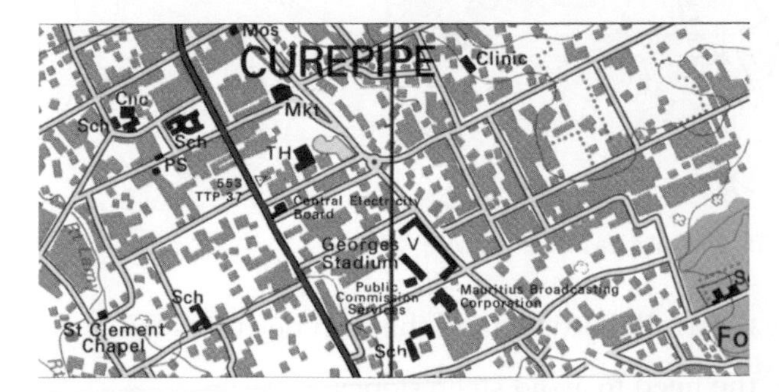

Fig. 12.47 An urban zone in Curepipe, Mauritius

22 Fig. 12.48 shows a residential zone. Give map evidence for this.

Fig. 12.48 Another urban zone in Curepipe, Mauritius

23 Look at Fig. 12.49.

a Which urban zone is shown on the map?

b Give map evidence for your answer.

Fig. 12.49 An urban zone in Zvishavane, Zimbabwe

Nucleated, dispersed and linear settlements

These settlement patterns and the reasons why they form are described in Chapter 2.

24 Fig. 12.50 shows settlement patterns in three areas of Zimbabwe.

a What type of settlement pattern is shown on Map A?

b Suggest reasons why this settlement pattern has developed.

c What type of settlement pattern is shown on Map B?

d What type of settlement pattern is shown on Map C?

e Suggest reasons why this settlement pattern has developed.

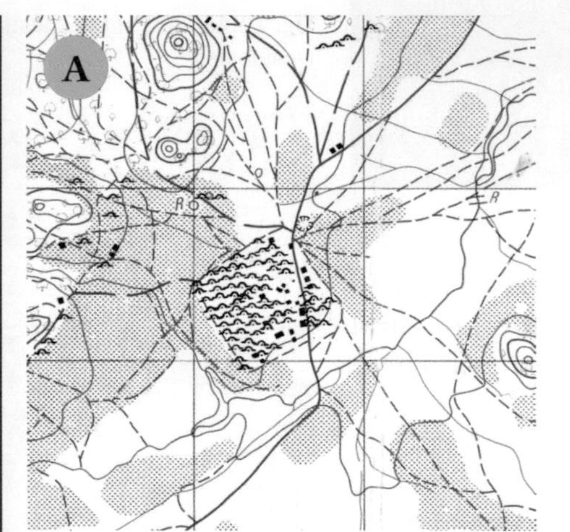

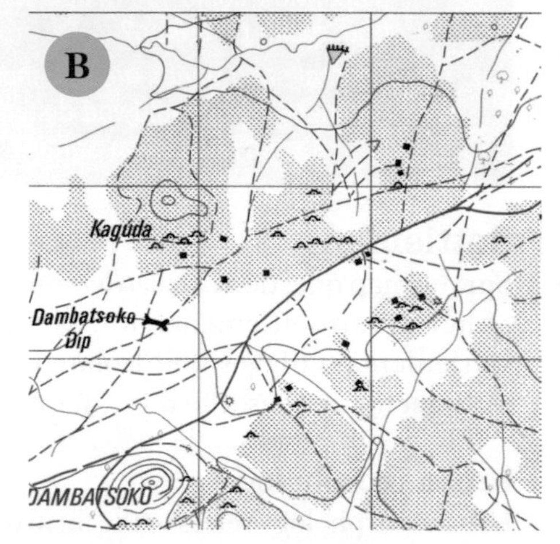

" Gravelor Earth, Bridge

Huts, Staff Quarters

Sparse Bush

Cultivation

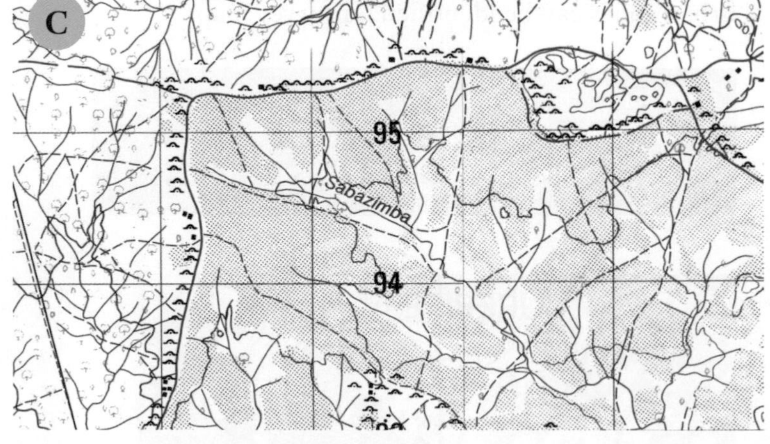

Fig. 12.50 Three rural areas in Zimbabwe

25 Look again at Map C in Fig. 12.50. Suggest reasons why there are huts in some areas but not in others.

Distribution, density and location of settlements

These are usually affected by factors like:

- the availability of communications
- the accessibility of points like road junctions (route centres) and bridge points
- the availability of land that can be cultivated
- the need to avoid steep slopes
- the need to avoid land that is liable to flooding and which may also be affected by pests and disease.

Communications

Communications on maps are generally the different types of roads, tracks and railways and, occasionally, ports and airports/airstrips. Care should always be taken to read the map key carefully to identify these features correctly.

Links with physical and human features

Roads usually try to follow gentle slopes. They try to avoid steep slopes and areas liable to flooding. For this reason, they often follow valleys – at the bottom of the valley sides, avoiding the floodplain. When steep slopes are encountered, roads may zig zag and have hairpin bends to reduce the gradient, like the road shown in Fig. 12.51.

Railways need very gentle gradients. They often follow cuttings or tunnels through hills, or cross lowland areas on embankments. Make sure that you know the difference between a cutting and an embankment and the symbols used for them.

Fig. 12.51 Hairpin bends on a road in South Africa

> **26** Fig. 12.52 shows a road and a railway close to the Victoria Falls in Africa. Describe how the routes of the road and the railway are related to the relief.

> **27** Fig. 12.53 shows a road on the island of Dominica in the West Indies. Describe how the route of the road has been affected by the relief and settlement.

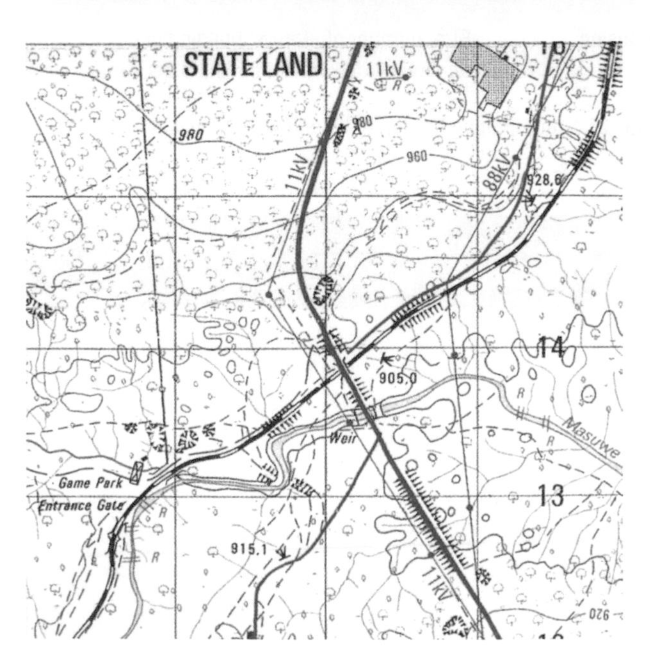

Fig. 12.52 A road and railway in Zimbabwe

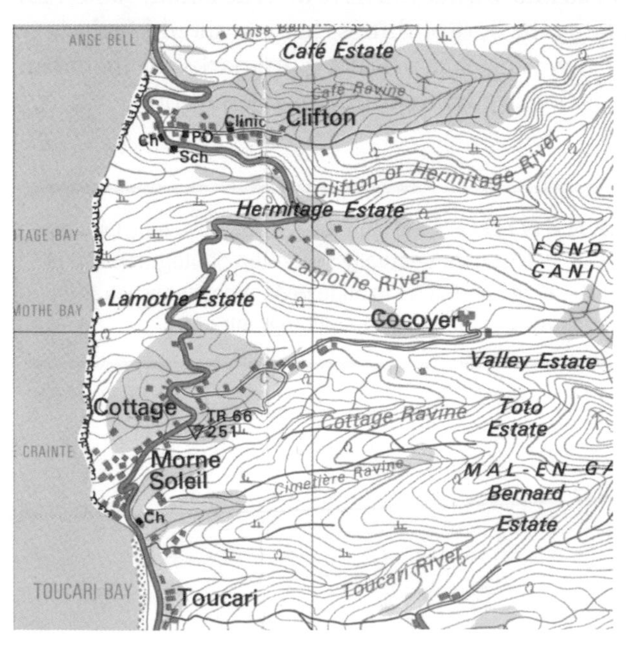

Fig. 12.53 A road in Dominica, West Indies

Land use

The land-use symbols used on maps vary greatly from country to country. Typically they show natural vegetation, types of cultivation and settlement.

28 Look back at Map C in Fig. 12.50.
- **a** Draw a sketch map to show the distribution of land-use types in the area.
- **b** Suggest how the distribution of land use has been affected by relief and other factors.

29 Look at Fig. 12.54.
- **a** Draw a sketch map to show the distribution of land-use types in the area.
- **b** Suggest how the distribution of land use has been affected by relief and other factors.

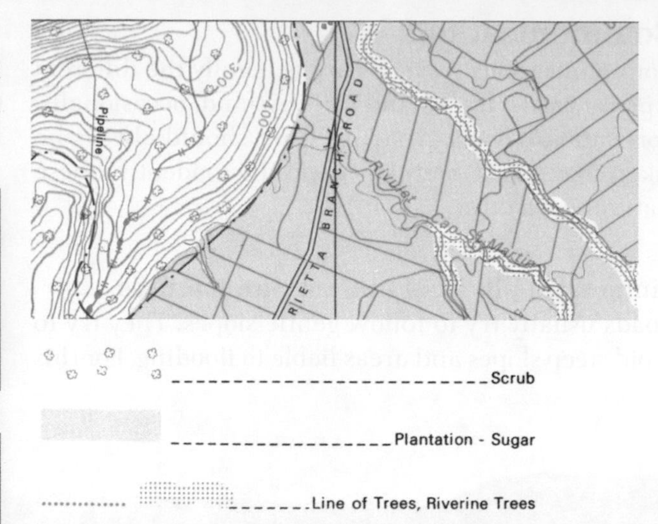

Fig. 12.54 Land use in an area of Mauritius

Term	What to describe in your answer
Physical features	Relief, drainage, vegetation
Human features	Features of buildings and settlement, agriculture, industry, transport
Relief	Features of the height and shape of the ground surface, including the names of any features you can identify
Drainage	Rivers, streams, lakes, and their features
Agriculture	Animals, the plots of land, fences, what is in the plots, e.g. grass, ploughed land, bare land, crops, any farm buildings and machinery that you can see
Settlement	Features of the buildings themselves (as listed for housing below), the types of buildings, the use of the buildings, and the spacing of the buildings and whether they are nucleated, linear, or dispersed
Housing	Size, number of storeys, building materials, quality, windows, the building plots

Table 12.1 Interpreting photographs and what to include

Small-scale maps

The 1:50 000 and 1:25 000 survey maps (topographic maps) are examples of large-scale maps. They show relatively small areas in great detail. Small-scale maps show much larger areas - for example, whole countries or continents - but in less detail. There are examples of the use of small-scale maps in Chapters 1, 2, 9, and 10. For Cambridge examination questions, you should be able to describe patterns from small-scale maps.

Photographs

Photographs tend to be used in all IGCSE® and O Level examination questions. In Paper 1 they might be used to introduce a topic, or to give you ideas. In Paper 2 they test your skills in interpreting the photograph. It is important to read the question carefully to be sure that you know what you have to write about. Table 12.1 explains the sort of things that you should include in your answers. You will not gain marks for including things that are not required by the question.

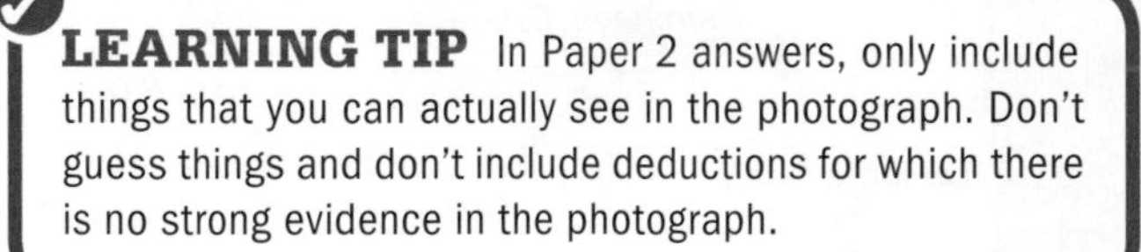

LEARNING TIP In Paper 2 answers, only include things that you can actually see in the photograph. Don't guess things and don't include deductions for which there is no strong evidence in the photograph.

30 Fig. 12.55 shows two residential areas. Describe the buildings in each photograph.

Fig. 12.55 Housing in South Africa (A and B)

31 Look at Fig. 12.56. Describe the buildings seen in different parts of the photograph. Which urban zones are shown?

32 Describe the agricultural features shown in Fig. 12.57.

Fig. 12.56 Part of Cape Town, South Africa

Fig. 12.57 An agricultural area in South Africa

33 Describe the relief of both photographs in Fig. 12.58, and also the relief of Fig 12.51. If the question had said describe the physical features, what else could you include in your answer?

Fig. 12.58 An area in Namibia (C) and an area in France (D)

Data tables

Data tables are often used in examination questions. You should be able to look at the data and identify any patterns or trends.

LEARNING TIP When describing trends or patterns from data tables and graphs, do not simply repeat the figures item by item. Instead, look for overall patterns – such as increases or decreases. You could use figures to illustrate your answer, depending on the precise wording of the question.

34 Describe the pattern of births and deaths shown in Table 12.2.

Year	1951	1961	1971	1981	1991	2001	2011	2021	2031	2041	2051
Births	790	940	900	740	790	670	780	780	770	830	840
Deaths	610	630	650	660	640	600	560	560	630	720	760

Table 12.2 The number of births and deaths (thousands) in the United Kingdom (after 2011 the figures are projected rather than actual)

Diagrams and graphs

Question 34 is not easy to answer. Trends and patterns do not stand out clearly in data tables. Diagrams and graphs help to show trends more clearly, and different types of diagram are used for different types of geographical data.

Line graphs

Line graphs are a simple form of graph. They are used when there is a continuous change in data. For example, they show the way in which something changes over time (e.g. population or crop yield). Like all graphs, it is normal to put the independent variable on the *x*-axis (horizontal) and the dependent variable on the *y*-axis (vertical). In Fig. 12.59, the years therefore go on the *x*-axis and the population on the *y*-axis.

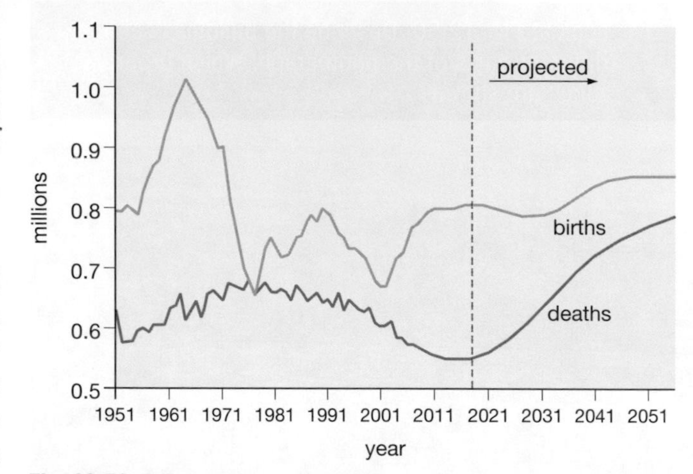

Fig. 12.59 A line graph

Bar graphs

Bar graphs can usually be used in similar circumstances to line graphs. The use of a bar graph would be preferred when there are a number of separate items that cannot really be joined, because they do not appear in a definite order. In Fig. 12.60, it would not be sensible to join the numbers for the different continents. The two axes could be reversed in this case.

In climatic graphs, it is conventional to use a line graph for temperature and a bar graph for rainfall. A bar graph where the length of the bars is shown by a number of symbols is called a **pictogram**.

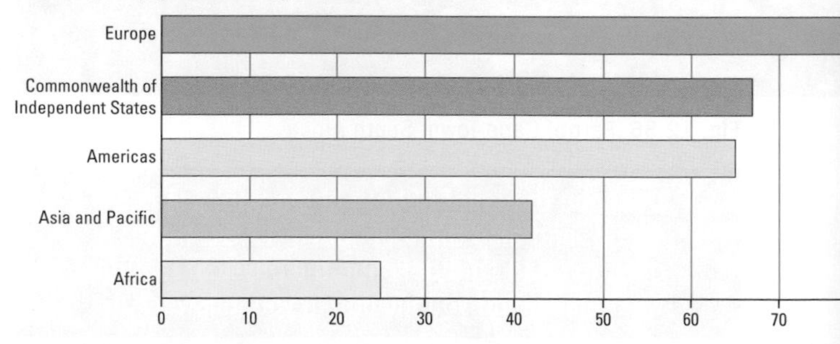

Fig. 12.60 A bar graph

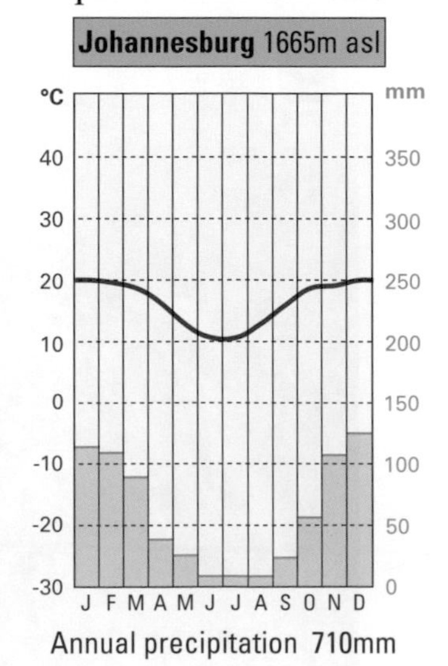

Fig. 12.61 Temperature (line graph) and rainfall (bar graph) for Johannesburg, South Africa

Histograms

Histograms are more complex. They look similar to bar graphs, but they are used for a different purpose. They show what are technically called **distributions** or **frequencies**. On the x-axis is a series of ranges of numerical values. The values do not overlap, e.g. 0-0.99, 1.00-1.99, 2.00-2.99, etc. On the y-axis is the number or percentage of items of each value. Fig. 12.62 shows the sizes of a sample of pebbles from a beach. The modal size range (the one with the most pebbles) is 8.0-11.9 cm.

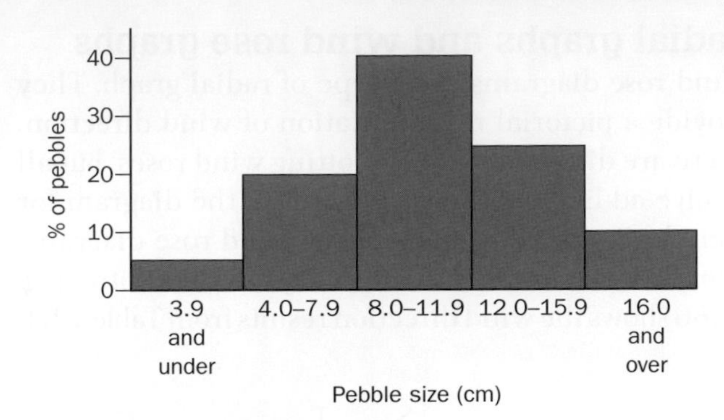

Fig. 12.62 A histogram showing the sizes of pebbles from a beach

Divided bar graphs

Divided bar graphs are used to show how a set of data can be broken down into fractions or percentages. A good example would be the percentages of different ethnic groups within a population. These graphs have the advantage of being very simple to plot and use.

Population by continent, 2006 millions of people

Europe 728 | Asia 3958 | Africa 916 | Oceania 33 | North America 518 | South America 376

Fig. 12.63 A divided bar graph showing the population of continents

Pie graphs

Although they look very different, pie graphs can be used in exactly the same way as divided bar graphs. They require care in plotting by hand. If the values are percentages, they can be converted into degrees by multiplying each percentage by 3.6, to give a total of 360^0 (as shown in Fig. 12.64).

Where two or more pie charts are used together, the size of the circle can be used to show the total population. This technique is not often used today, but care must be taken when comparing circles of different sizes.

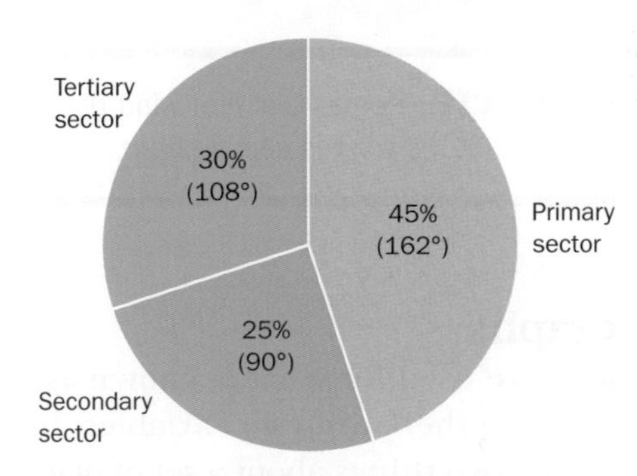

Fig. 12.64 A pie graph showing the employment structure of Morocco

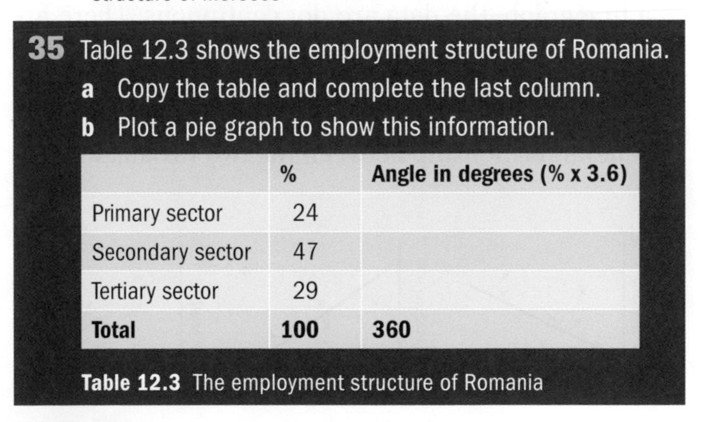

35 Table 12.3 shows the employment structure of Romania.

a Copy the table and complete the last column.

b Plot a pie graph to show this information.

	%	Angle in degrees (% x 3.6)
Primary sector	24	
Secondary sector	47	
Tertiary sector	29	
Total	**100**	**360**

Table 12.3 The employment structure of Romania

Flow diagrams

Flow diagrams are used to provide an illustration of things like traffic or pedestrian flows, population migrations, or world trade (see Fig. 12.65). They are usually based on a map, but the map might be in diagrammatic form. The flow arrows might be diagrammatic (as in Fig. 12.65). However, the thickness or width of the arrows usually indicates the size of the flow, and a scale is sometimes indicated for this.

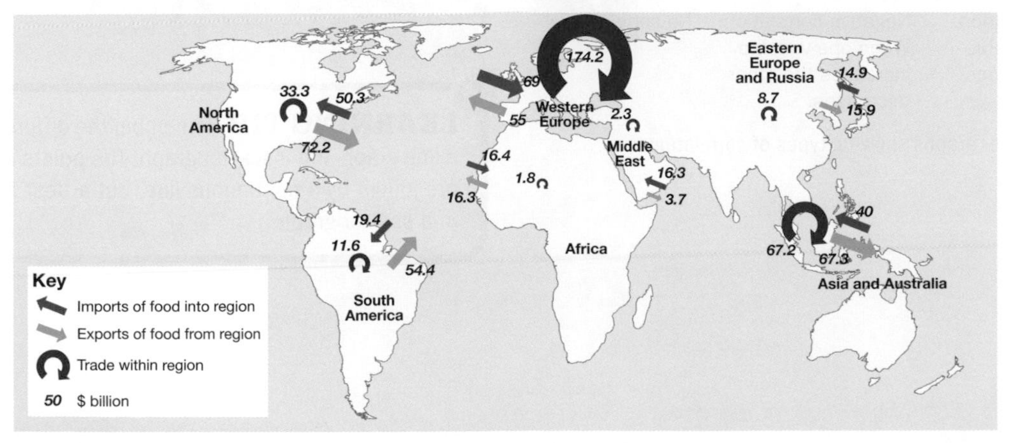

Fig. 12.65 A flow diagram showing world food imports and exports

Radial graphs and wind rose graphs

Wind rose diagrams are a type of radial graph. They provide a pictorial representation of wind direction. There are different ways of plotting wind roses, but all involve adding one measurement to the diagram for each day's wind direction. Some wind rose diagrams also show the number of calm days in the centre. Fig. 12.66 shows the wind direction results from Table 12.4.

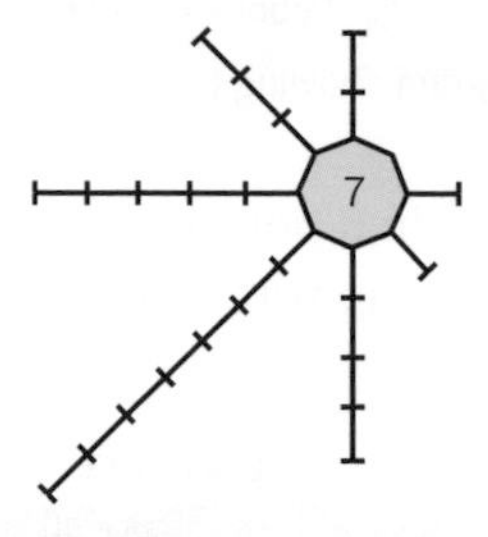

Fig. 12.66 A wind rose diagram for the month of September

Direction	Number of days
N	2
NE	0
E	1
SE	1
S	4
SW	7
W	5
NW	3
Calm	7
Total	**30**

Table 12.4 Wind pattern results for September

Direction	Number of days
N	6
NE	11
E	5
SE	0
S	0
SW	0
W	3
NW	3
Calm	3
Total	**31**

Table 12.5 Wind pattern results for October

36 Draw a wind rose to represent the data in Table 12.5.

LEARNING TIP Remember that wind directions are the direction that the wind is coming *from*.

Scatter graphs

Scatter graphs are used for what is known as paired data. This is when there are two variables. In other words, you know two things about a set of places. The scatter graph shows how the variables are related. Unlike a line graph, the data are not continuous. There are three possible situations for scatter graphs, as shown in Fig 12.67. Often a **best-fit line** is drawn between the points. This does not join the points, but shows the general relationship between the two variables.

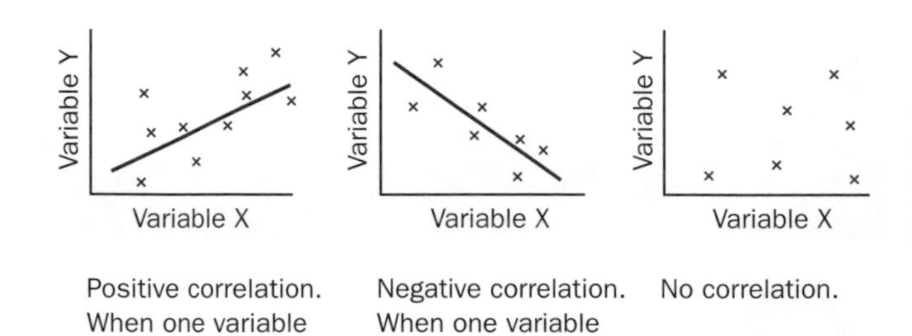

Positive correlation. When one variable increases so does the other.

Negative correlation. When one variable increases the other decreases.

No correlation.

Fig. 12.67 Scatter graphs showing types of correlation

37 Plot a scatter graph for the data in Table 12.6. Your graph will show the altitude and rainfall for each place. Which variable should be on which axis? Add a best-fit line. What relationship is there between the two variables?

Location	Altitude above sea level (m)	Annual rainfall (mm)
1	20	508
2	27	613
3	197	935
4	289	1128
5	347	1359
6	313	1246
7	5	453
8	153	892
9	101	663
10	237	1063

Table 12.6 Rainfall at ten locations at varying altitudes above sea level

LEARNING TIP Remember the difference between a line graph and a scatter graph. The points in a line graph are joined by a continuous line, but a best-fit line is used in a scatter graph.

Triangular graphs

Triangular graphs are used when you have a set of data for three variables which add up to 100%. Pie charts and divided bar graphs can show this information for one place, but a triangular graph can show it for a number of places at once. A typical use would be to show the employment structures of a group of countries (see Fig. 12.68). This graph reads in a clockwise direction, but other examples read anticlockwise.

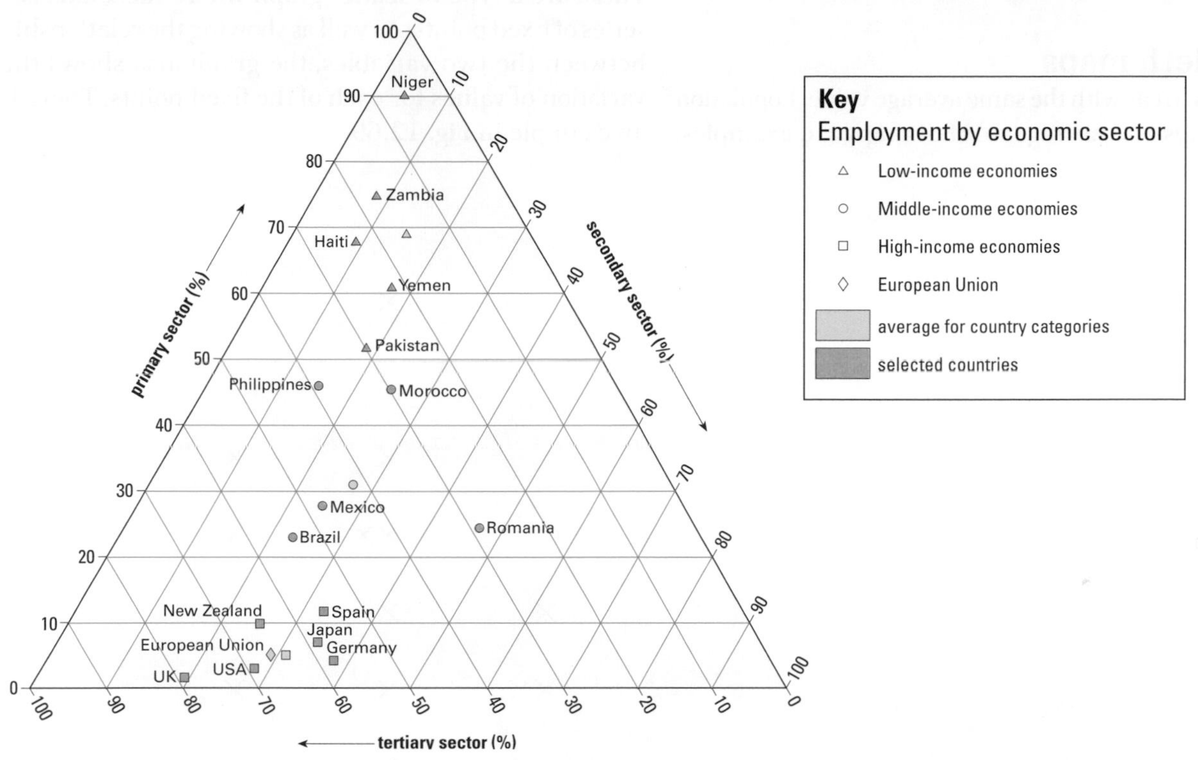

Fig. 12.68 A triangular graph showing employment by sector for a group of countries

> **LEARNING TIP** Make sure that you know how to read the axes of a triangular graph. The angle of the numbers on the axes is a guide to reading the correct lines. Remember that the values must add up to 100%.

Population pyramids

Examination questions may include population pyramids. These are described in detail in Chapter 1.

38 Six types of graph used in geography are:

- a radial graph (rose diagram)
- a pie graph (pie chart)
- a scatter graph
- a line graph
- a bar graph
- a triangular graph.

For each of the examples **a–h**, name the most appropriate type of graph from the above list.

a A graph to show the percentages of tourists arriving in a country by air, road, rail, and sea.

b A graph to show changes in the production of a crop over a number of years.

c A graph to show the relationship between crop yields and the amount of fertiliser used in a group of countries.

d A graph to show the variation in wind direction in one month at a weather station.

e A graph to show rainfall totals for months of the year.

f A graph to show changes in birth rate over a number of years.

g A graph to show the relationship between birth and death rates in a number of countries.

h A graph to show the percentages of primary, secondary, and tertiary employment in a group of countries.

Isoline maps

Isolines are drawn on a map to show places with equal values. Contours showing height above sea level are an example. Examples in this book include Figs. 2.9, 2.13, and 6.43.

Choropleth maps

These show areas with the same average value. Population density maps and average rainfall maps are examples. There are many examples in this book, including Figs. 1.7, 1.20, and 1.22.

Dispersion graphs

These are a type of scatter graph where the *x*-axis is a series of fixed points. As well as showing the relationship between the two variables, the graph also shows the variation of values for each of the fixed points. There is an example in Fig. 12.69.

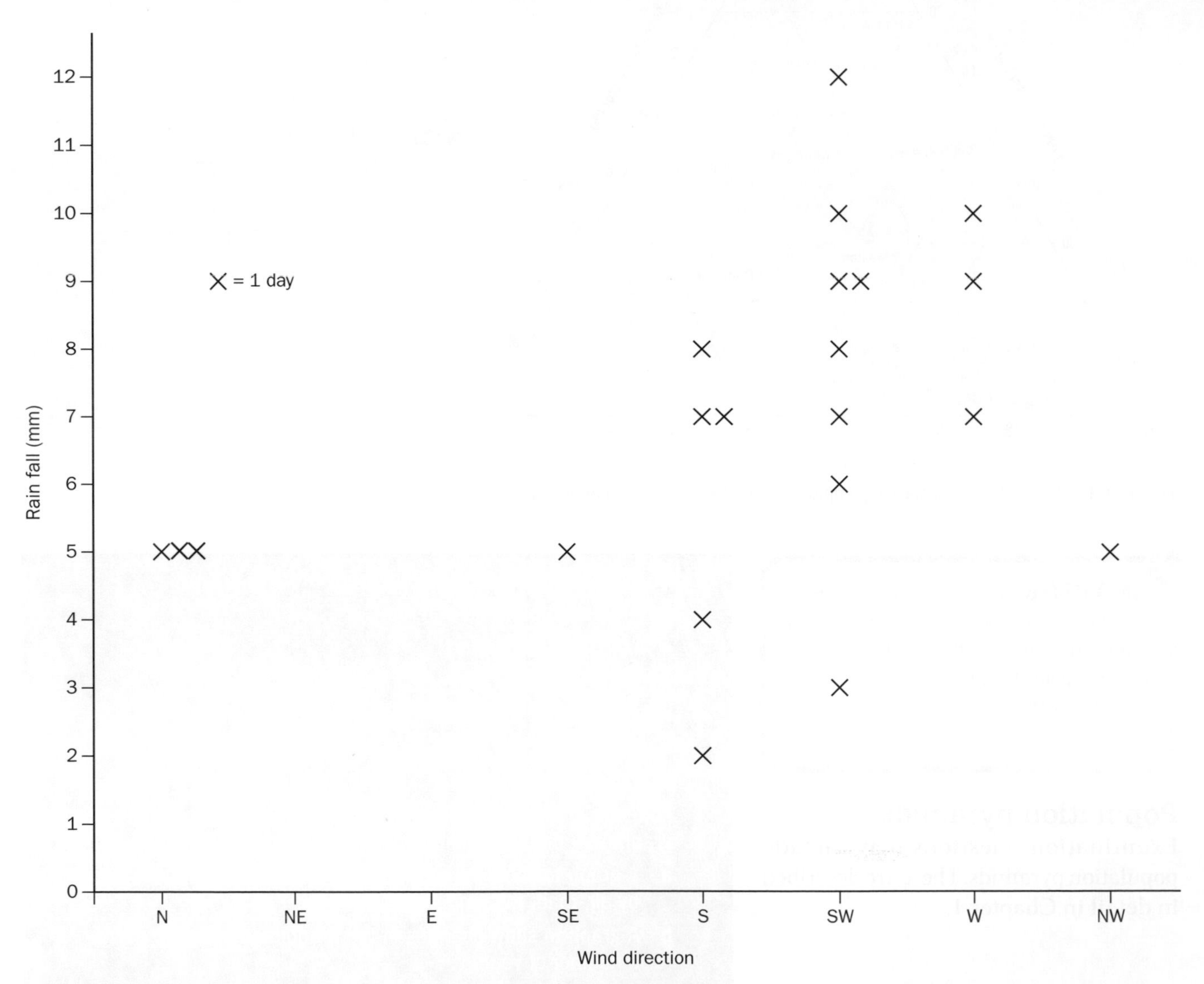

Fig. 12.69 A dispersion graph showing the amounts of rainfall received at a weather station during one month when the wind was blowing from different directions

39 What does the dispersion graph on Fig. 12.69 tell us about wind direction and rainfall at the weather station?

There is another exercise on dispersion graphs in Chapter 6.

Kite diagrams

These show the density and distribution of features. They are commonly used to show how the density of different plant species varies with distance.

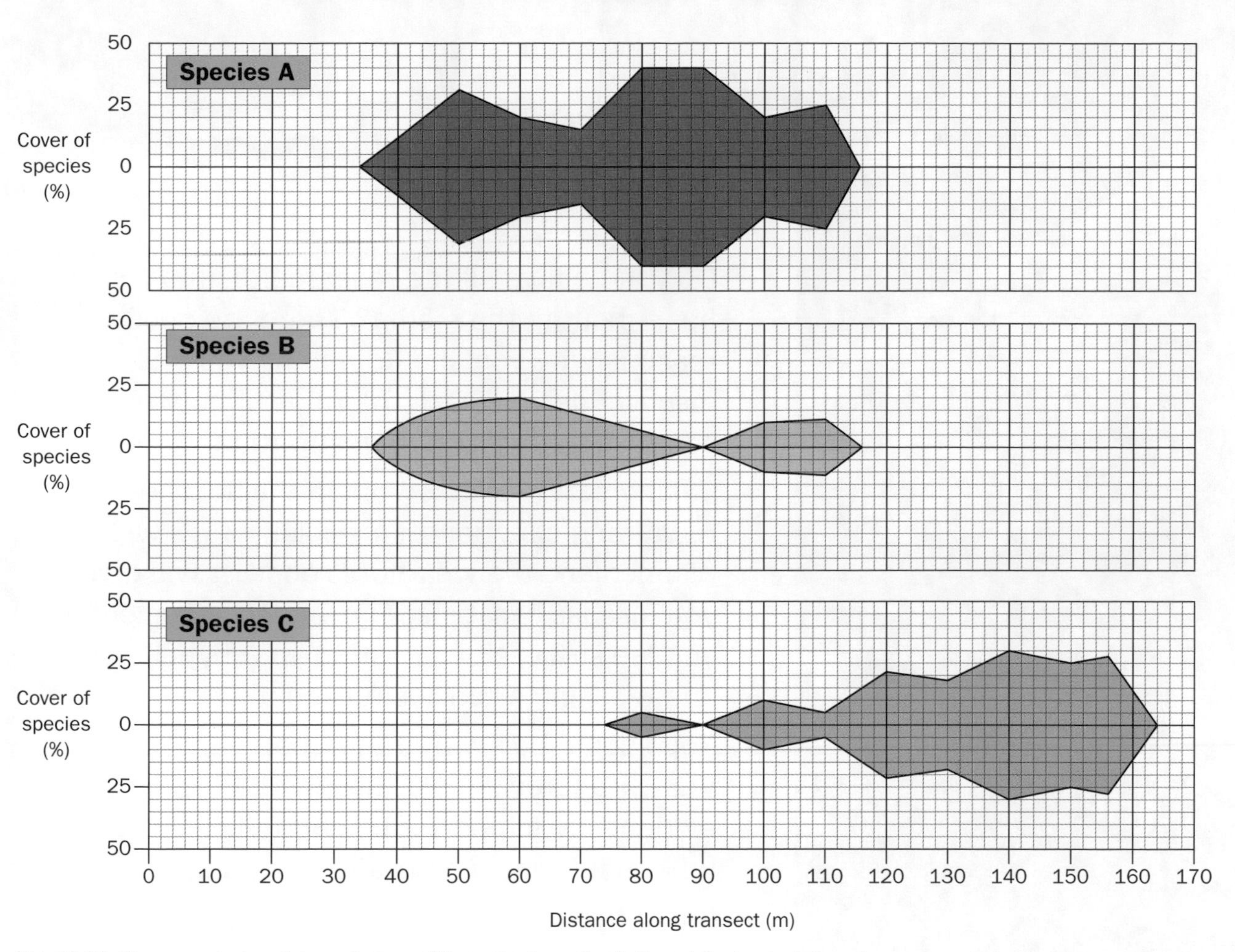

Fig. 12.70 The percentages of ground cover of three plant species A, B, and C, along a transect

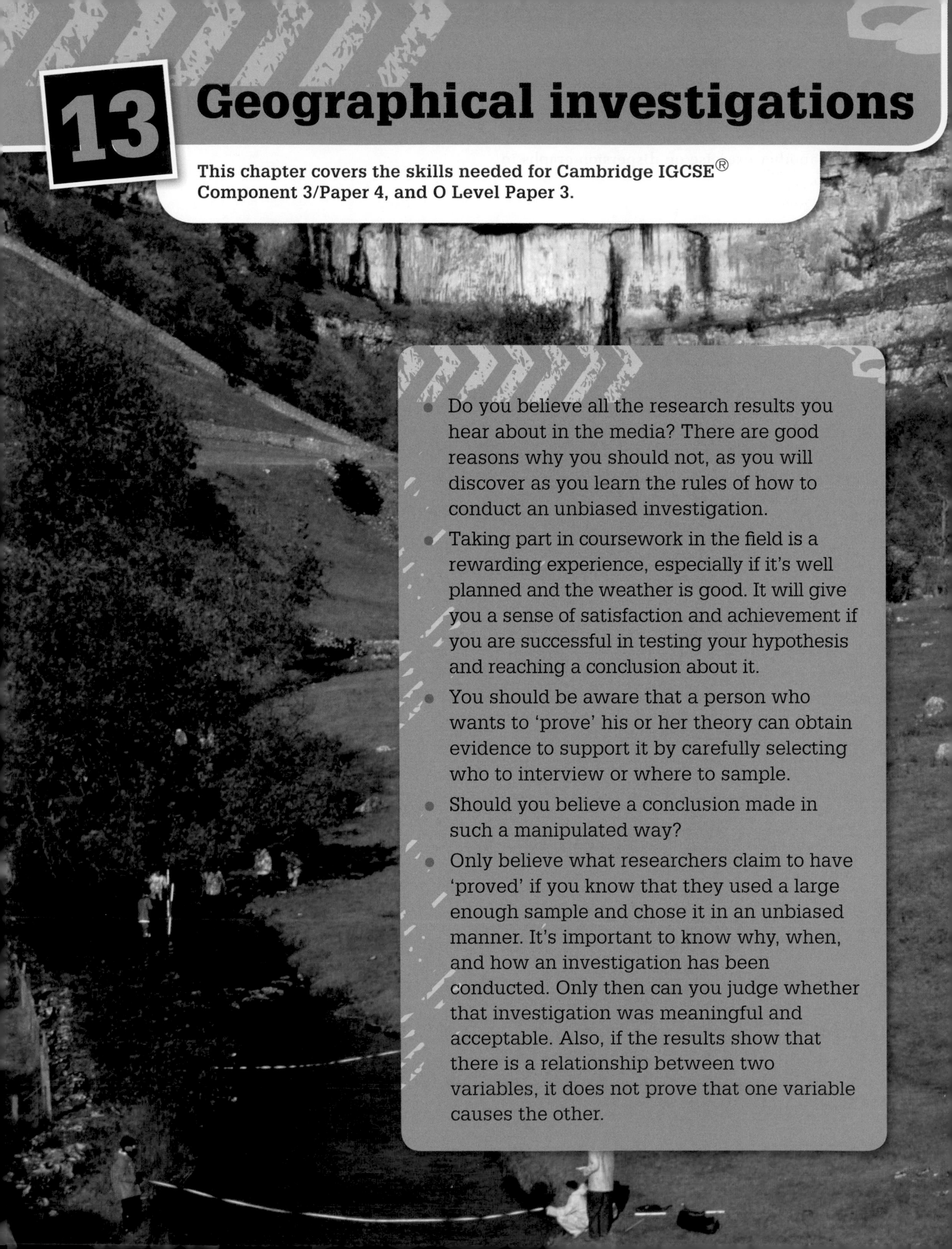

13 Geographical investigations

This chapter covers the skills needed for Cambridge IGCSE® Component 3/Paper 4, and O Level Paper 3.

- Do you believe all the research results you hear about in the media? There are good reasons why you should not, as you will discover as you learn the rules of how to conduct an unbiased investigation.
- Taking part in coursework in the field is a rewarding experience, especially if it's well planned and the weather is good. It will give you a sense of satisfaction and achievement if you are successful in testing your hypothesis and reaching a conclusion about it.
- You should be aware that a person who wants to 'prove' his or her theory can obtain evidence to support it by carefully selecting who to interview or where to sample.
- Should you believe a conclusion made in such a manipulated way?
- Only believe what researchers claim to have 'proved' if you know that they used a large enough sample and chose it in an unbiased manner. It's important to know why, when, and how an investigation has been conducted. Only then can you judge whether that investigation was meaningful and acceptable. Also, if the results show that there is a relationship between two variables, it does not prove that one variable causes the other.

In this chapter you will learn about:

- → conducting fieldwork investigations
- → answering examination questions about geographical investigations.

Fig. 13.1 summarises what you might need to consider, depending on your chosen topic. As it is difficult to include every possible study within the confines of one chapter, the information, exercises and past examination questions have been selected to give the widest possible coverage. Information about more specialised methods and techniques can be researched on the Internet.

1 Formulate a hypothesis or devise a question for the title of your investigation.

2 Plan the investigation.
- What sampling method should be used?
- What health and safety considerations are there?
- Should permission to visit be requested?
- What equipment will be needed?
- What is the minimum number of students needed?
- Would a pilot study be useful?

3 Collect the data.
Data may be primary or secondary, subjective or objective.
- Designing a recording sheet
- Counting methods
- Sampling
- Bipolar surveys
- Measuring accurately
- Designing a questionnaire
- Surveying a slope profile
- Recording observations in the field

4 Present the data.
- Select appropriate diagrams and graphs.
- Write an account of your investigation.

5 Analyse and interpret the data.
- Use simple statistical analysis (rank, range, mean, mode, median).
- Look for trends.
- Look for patterns, relationships, and anomalies.
- Suggest explanations for your findings.

6 Draw conclusions based on your results.
- Can the hypothesis be accepted?
- To what extent was it accurate?

7 Evaluate your investigation.
- To what extent was it successful?
- How could it be improved?
- How could it be extended?

Fig. 13.1 The steps involved in carrying out a geographical investigation

Choose a **topic for investigation** that will provide opportunities to measure a variety of **data**, and that can be completed in the **time** available by the number of students participating. The **location** for the study must also be accessible and safe. The topic should be one that is likely to succeed. For example, it would not be advisable to plan a study of velocity on a stream which sometimes dries up.

Formulating a hypothesis

Investigations are most likely to be successful if they test a hypothesis or answer a question. The subject of the investigation should also be specific. A vague title like 'A study of the shops in Delhi' is too broad and the area chosen too large. An appropriate title in the form of a question would be 'Do the types of shops in the CBD of town X change as distance from the centre of the CBD increases?'

The questions asked in IGCSE® Paper 4 or O Level Paper 3 are usually based on **hypotheses**. A hypothesis is a statement about your topic of study that can be proved or disproved by testing. The chosen topic must be one where data needs to be collected (preferably using a range of techniques and allowing a variety of different forms of presentation). It should, of course, be based on the geographical knowledge you have already gained during your studies.

'The size of shops decreases as distance from the centre of the CBD increases' is a hypothesis that tests an expected relationship. While such a simple hypothesis might be the subject of an exam question on IGCSE® Paper 4 or O Level Paper 3, it may be too narrow a topic for an IGCSE® Component 3 coursework assignment (depending on the size of the town to be investigated). Investigating 'types of shops' would allow research into other aspects of shops, as well as size, and also allow different methods of data collection and presentation to be used.

The syllabus has a list of suitable coursework assignments which can be formulated into hypotheses or questions.

1 In groups of four to six students, discuss what needs to be considered when deciding what topic to study, so that it's practical to undertake it. Work out your answers and then discuss your ideas with the whole class.
2 Criticise the following hypotheses:
   a 'There is more rainfall when there is a full moon.'
   b 'There is more erosion when the river is in flood.'
3 Formulate a sensible simple hypothesis about rainfall.
4 Formulate a sensible simple hypothesis about land values in a town.

A Component 3 assignment will need sufficient breadth or depth. Weather could be studied instead of rainfall, and investigating how land uses vary according to land values in a town would give more breadth to a study about land values.

LEARNING TIP When researching a relationship, keep all other variables (such as the time of the survey) the same, so that they cannot influence the results.

Planning the investigation

Sampling techniques

A **sample** is a group selected from a larger '**population**', where population means the whole of whatever is being sampled.

As it's impossible to research all rivers, crops or factories in an area, or to ask all people in a town questions, coursework usually involves sampling. The aim is to investigate the smallest number that is still large enough to be truly *representative of the whole population.* To be a fair test, the sample must be determined without **bias**. This means that *every individual in the population must have an equal chance of being included in the investigation.* The investigator must not, for example, select which people should be asked to answer questions. This is because he or she might choose those who look most friendly, or are of a similar age to them.

LEARNING TIP Do not use 'accurate' when commenting on the results of sampling. Total accuracy can only be achieved if the entire population is investigated and the work of all the students involved is also accurate. A good sample will be as representative of the population as possible, and will avoid bias. It is also acceptable to describe it as a 'fair test' or 'reliable', provided that you explain why it is so.

As a general rule, if the sampling is well designed, the larger the sample, the more reliable the results will be. It is normally considered that a sample of 30 is sufficient when a relationship is being investigated. This is also an easy number to use for the purpose of mathematical analysis (such as calculating averages). It is important to use the most appropriate sampling method for each part of the investigation.

When sampling an area on a map or in the **field**, you will need to decide whether to sample points, areas, or along lines:

→ **Point sampling** would be appropriate for a pedestrian count.

→ **Area sampling** using a square **quadrat** would be suitable for sampling vegetation cover.

→ **Line sampling** could be used to investigate changes in pebble size on a beach from the low- to high-water marks by sampling at intervals along the line.

5 What are the advantages of sampling?
6 Why would line sampling be unsuitable for use when sampling types of shops in towns with a rectangular road pattern?

Fig. 13.2 A sample from a population of cows. If it was significant to the investigation, the herd could be divided into subgroups of different breeds and each one could be sampled.

Random sampling

To remove bias from an investigation, random number tables can be used.

39 26 02 11 98 55
58 07 46 60 77 04
17 83 29 32 41 36
48 65 08 93 55 69

Fig. 13.3 Part of a random number table

The numbers can be read in any direction, so long as you are consistent. For example, if you were investigating pebbles along a line on the beach - and reading the random number table horizontally - you would pick up the 39th pebble, then the one 26th after the first and so on. Ignore any numbers which are too large for the population.

Numbers in a telephone directory could be used instead of random numbers. The starting point and direction of a random line sample can be determined using a random number table. Individuals (e.g. shops or people) could each be given a number. Then random number tables could be used to select which ones to include in the survey.

Systematic sampling

This is sampling using a regular pattern or order, such as asking questions to the seventh person who passes after each interview, or sampling pebbles every five metres along a line. It is important to choose a sensible interval, e.g. if ground vegetation is being investigated, too large an interval might cause a species or a bare patch to be missed.

If a map does not have grid lines on it, draw them on yourself and number them to enable locations to be identified.

Stratified sampling

In order to make the sampling as representative of the population as possible, it is sometimes advisable to ensure that different groups or types are represented in the same proportions as they exist in the total population. For example, if 30% of a town's residents are young, 40% of working age, and 30% elderly - and age is significant to the investigation - a stratified sample would involve three young people, four of working age, and three elderly (or multiples of these numbers). It is also usually necessary to use the same number of females and males, or a number proportionate to their percentage of the population.

Stratified sampling can also be used on maps or in the field. For example, if 33% of the land is steep and 67% gentle - and soil depth is being investigated - 33% of the samples would need to be taken on the steep slopes and 67% on the gentle slopes. Stratified sampling has the important advantage that all parts of the area are included. It allows significant differences to be noticed between the subgroups, and it can be random or systematic.

7 Explain how you would use random number tables to:

- **a** determine in which direction you should investigate.
- **b** determine the different distances at which samples should be taken.
- **c** find grid references of points to investigate on a map.
- **d** find grid squares to investigate on a map.
- **e** determine the beginning and end of lines to be sampled.

8 Imagine that you are interviewing shoppers in a shopping centre. Why would the use of random number tables not be an appropriate way of selecting which people to interview?

9 How could you use a map grid for:

- **a** quick systematic point sampling?
- **b** area sampling?
- **c** line sampling?

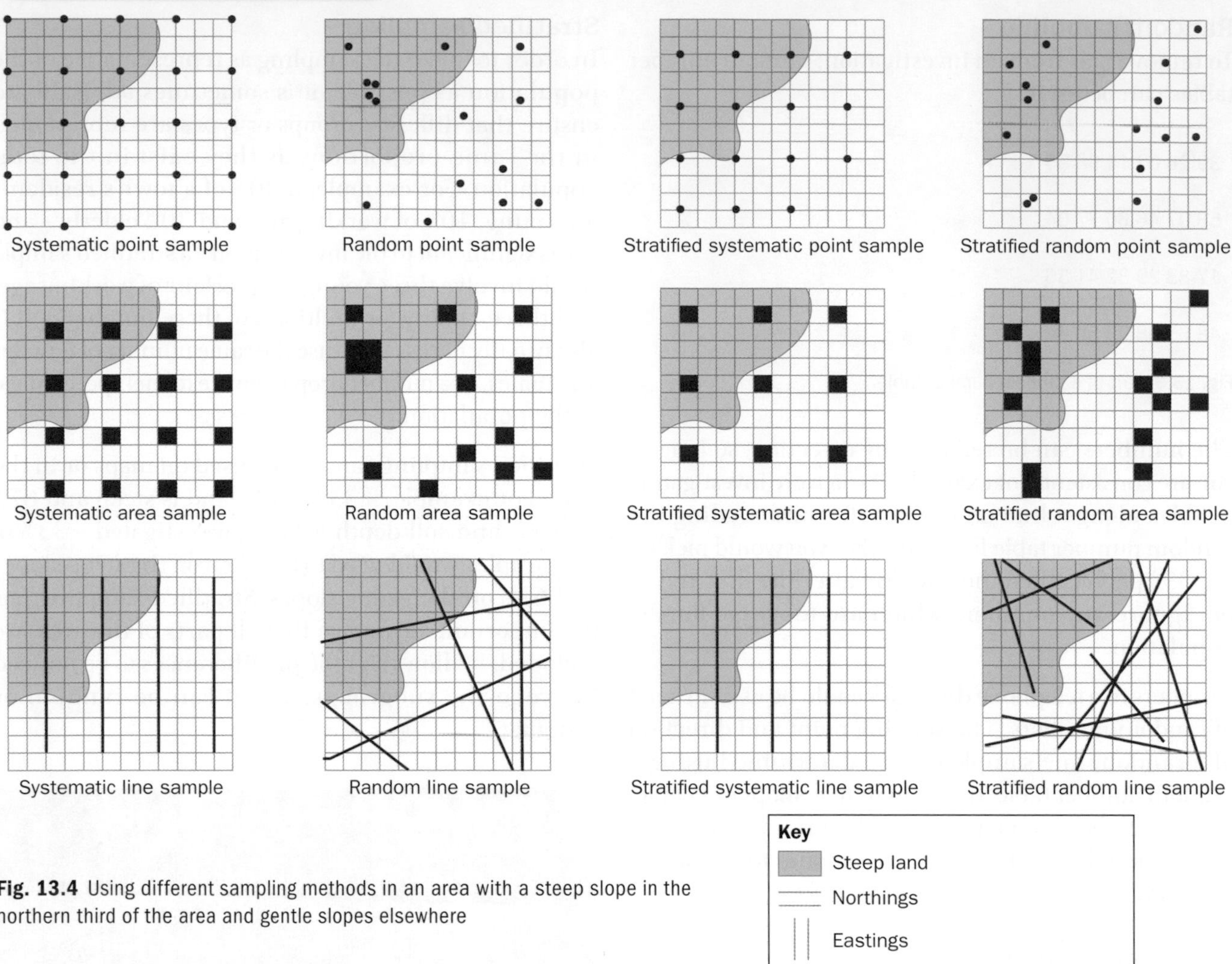

Fig. 13.4 Using different sampling methods in an area with a steep slope in the northern third of the area and gentle slopes elsewhere

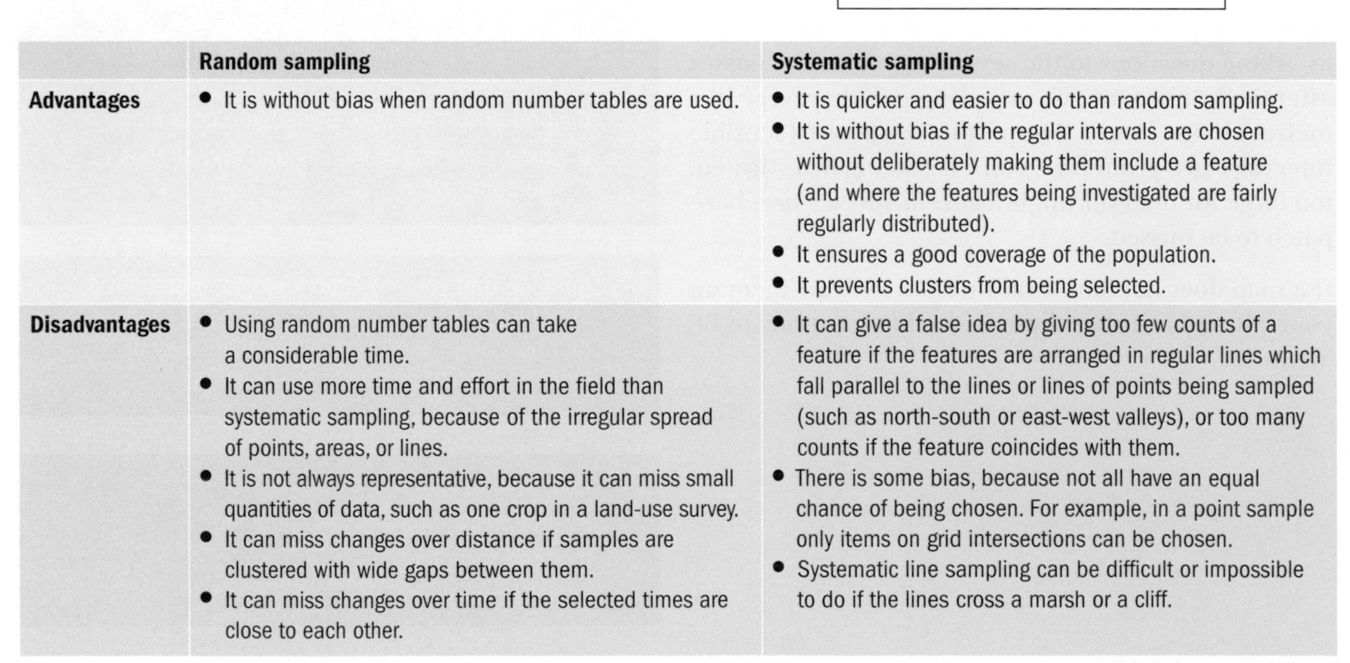

	Random sampling	Systematic sampling
Advantages	• It is without bias when random number tables are used.	• It is quicker and easier to do than random sampling. • It is without bias if the regular intervals are chosen without deliberately making them include a feature (and where the features being investigated are fairly regularly distributed). • It ensures a good coverage of the population. • It prevents clusters from being selected.
Disadvantages	• Using random number tables can take a considerable time. • It can use more time and effort in the field than systematic sampling, because of the irregular spread of points, areas, or lines. • It is not always representative, because it can miss small quantities of data, such as one crop in a land-use survey. • It can miss changes over distance if samples are clustered with wide gaps between them. • It can miss changes over time if the selected times are close to each other.	• It can give a false idea by giving too few counts of a feature if the features are arranged in regular lines which fall parallel to the lines or lines of points being sampled (such as north-south or east-west valleys), or too many counts if the feature coincides with them. • There is some bias, because not all have an equal chance of being chosen. For example, in a point sample only items on grid intersections can be chosen. • Systematic line sampling can be difficult or impossible to do if the lines cross a marsh or a cliff.

Table 13.1 A comparison of sampling methods

Discussion point

Imagine that you want to 'prove' (by unfair means) a theory that 'exposure to the sun is good for people'. What group of people would you deliberately exclude and what group would you definitely include in your research?

Health and safety considerations

It is important to check at the earliest possible stage that the area you wish to study is safe. Dangers vary according to location. For example, there are many dangers when working in rivers and on riverbanks:

- banks and rocks can be very slippery after rain
- the depth of the bed can suddenly change, as can velocity after heavy rain.

Therefore, it might be necessary to change a point chosen for sampling to ensure that it's safe. Remember that a river at a meander is deeper and faster flowing on the outside bank, so it would be safer to sample a straight stretch if that would still serve the purpose of the study. It is advisable to wear a life vest. Waterborne and water-bred diseases are also common, as well as dangerous animals in many parts of the world and polluted water which can cause serious illness.

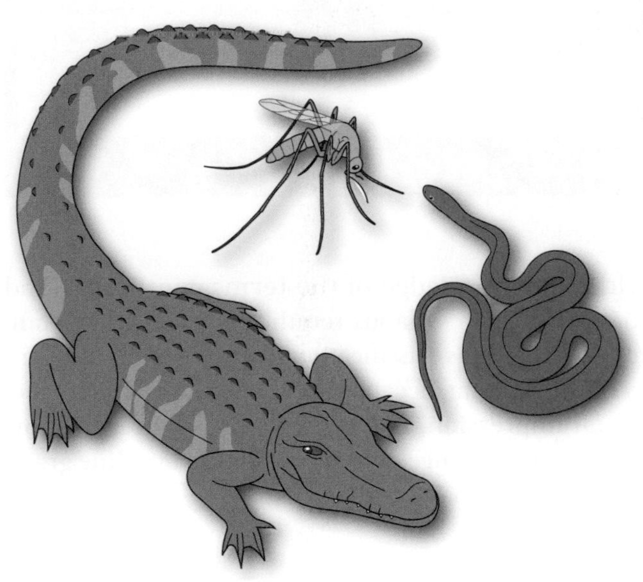

Fig. 13.5 Some fieldwork dangers

Before undertaking fieldwork on a beach, find out when high tide is due and make sure that you complete the work at low tide. Avoid areas at the foot of weak cliffs, because of the danger of rock falls or cliff collapse.

Other safety precautions include the following:

- In many areas it's wise to wear insect repellent and not to work alone.
- Conduct questionnaires in pairs and choose safe locations and times.
- Take contact phone numbers for home and school, and a mobile phone if possible, with you.
- Keep well away from any military areas.

Should you wish to enter private land, you must seek permission from the landowner by letter - stating the date and time that you would like to visit. Take the written approval with you when undertaking the fieldwork.

Always choose easily accessible sites.

Discussion point

Why is it important to behave both responsibly and cooperatively when taking part in work in the field?

10 What footwear would be most sensible for working in a rocky or uneven area?

11 If working on a hot and sunny day, what precautions should you take?

12 State two weather conditions that would make fieldwork difficult.

13 What precautions should you take if your work makes it necessary to be near the foot of a cliff?

Equipment and number of students needed

Deciding what equipment to take and how many students you need for your chosen fieldwork is important at the planning stage, but it depends on the investigation and will be dealt with later in the chapter.

LEARNING TIP If samples are to be compared, they must all be taken at the same time, so that time is not a variable that can affect the results. This will partly determine how many students are needed to carry out the research. The exception to this would be if time were a variable under investigation.

Pilot study

In a **pilot study** or **practice survey**, the methods to be used in the full investigation are tested beforehand in a small-scale survey. For example, a few samples may be taken or a questionnaire may be used with a few people in order to make sure that the methods or questions work, and to ensure that the investigation will be successful. If faults are found with the methods, instruments, or equipment at this stage, changes can be made before the full study is undertaken. It also allows you to experience working as part of a team and to practise the skills you will use during the full investigation.

Collecting the data

Types of data

Primary data is collected in the field by the person or group doing the investigation - counting, measuring, observing, sketching, or photographing.

Secondary data is data already obtained and published by a person or persons unconnected with the investigation. Investigations can, when circumstances make working in the field impossible, be entirely based on secondary sources (e.g. research on the Internet or in books), but some work in the field should always be done if possible.

For a weather study, secondary data could be obtained from the local weather station. Primary weather data, collected using the school's instruments for a month, could be compared with statistics from a secondary source, for example, with the mean figures for the same month taken over a number of years (usually 30). These could then be analysed to try to explain any differences or similarities. The sources of secondary data must always be referenced and acknowledged in the written account of the investigation.

LEARNING TIP Answering the question asked is the key to examination success. Underline the key words. Is the question asking for an *example* of secondary data or a *source* of secondary data?

Subjective (qualitative) data is obtained using the judgment or opinion of a person, whereas **objective (quantitative)** data is measured or counted. Being in numerical form, quantitative data is invaluable for analysis.

Continuous data is when all the values between two measured values are passed through before the second value is reached. An example of this is temperature data - a change from 7 °C to 10 °C involves a continuous increase through 8 °C and 9 °C. **Discrete data** is the opposite. Two values exist separately without any progression or regression of values between them. For example, if the rainfall in one month is 7 cm and three months later it is 10 cm, it is not expected that the two months in between will have totals of 8 cm and 9 cm.

LEARNING TIP When quoting data, it is important to state the units, such as metres, minutes and so on, as well as the values.

14 Copy and complete Table 13.2 to classify the type of information listed by putting two ticks in each row.

	Primary	Secondary	Objective	Subjective
Census information				
Your traffic count				
Your word description of litter				
Description from TV				
Your field sketch				

Table 13.2 Types of data

Although knowledge of the terms continuous and discrete is not a syllabus requirement, it is important to be aware of this classification when choosing diagrams to represent data, because they should give the correct visual impression. For example, it is better to show temperature change as a smooth curve on a line graph and rainfall in bar graph form.

Designing a recording sheet

Obviously, the design of the sheet on which to record information in the field will vary according to the type of investigation, but some general guidelines can be given here.

At the top it should have a title, followed by headings for date (and the day if it has extra significance), time, location, and name of the recorder, with spaces by each to be filled in at the time of the investigation. A space for other factors likely to affect the study, such as weather, should also be included where relevant.

The recording sheet needs to be carefully planned, so that there is sufficient space for all the recordings. The recording sheet should be used with a clipboard for support and kept as dry as possible. Take a large plastic bag into the field to protect it during showers.

LEARNING TIP If a question asks you to design a recording sheet, leave blank all the spaces or boxes for recording the data that you would collect in the field. Do not enter example recordings.

Measuring accurately

Obviously, measurements should be accurate but human error is always possible. There are several ways in which errors can be reduced:

→ Know how to read each instrument correctly.

→ If working alone, take an average of three or more readings or measurements.

→ If working in a group, three students should take the same reading, which can then be averaged.

→ Digital instruments should be tested for accuracy against other digital or conventional instruments before using.

Discard any reading that is obviously inaccurate or is anomalous because the site was not appropriate for the hypothesis. Other anomalies should be kept, as they may be useful in the later analysis. The accuracy of measurement can be partly determined by the method and equipment used. The best equipment may be too expensive for students or centres to purchase, but knowledge that it exists may be useful when reviewing the success of the study in the conclusion. Table 13.3 compares two ways of measuring river velocity.

LEARNING TIP One possible explanation for an anomalous result is student error in measuring or reading an instrument. One suggestion for ways in which a study could have been improved is the use of better equipment.

Discussion point

Discuss the possible reasons why student measurements might be unreliable or incomplete.

Equipment used	An orange	A flowmeter (with a propeller)
Method	It floats just below the surface, so it measures the velocity at that level (slightly slowed by friction with the air).	It is inserted into the water to the depth stated in the instructions (usually 0.6 of the depth) to obtain the average velocity of the stream at that distance from the bank (with the propeller facing upstream).
How the result is obtained	A stopwatch times how long it takes for the orange to travel between two previously measured points. This is repeated at least three times and averaged. The velocity is the distance (e.g. 10 m) divided by the average travelling time.	It is usually read from a digital display. The reader should calculate the average of several readings.
Advantages	It is cheap.	This is a quick method, which is accurate in appropriate locations (providing a precise digital reading).
Disadvantages	• There could be possible student error in starting and stopping the stopwatch at the exact moments when the orange passes the start and end points. • It takes more time. • The orange might be blown by the wind or swept by a current into vegetation at the side of the channel. • It measures the surface velocity, which is reduced slightly by friction with the air.	• It is relatively expensive. • It is not accurate where the velocity is low and the depth is shallow.

Table 13.3 Two methods of measuring stream velocity. Fig 13.24 shows a type of digital flowmeter.

CASE STUDY

Measuring infiltration rates

- Push or gently hammer an upright large bottomless tube about 10 cm into the ground. A very large tin without a bottom could be used.
- Place a ruler inside the tube as far as ground level.
- Pour a large measured amount of water into the tube.
- Time with a stopwatch how long it takes for the water to infiltrate.
- Measure the height of water in the cylinder every minute at first but less often when the rate slows. Record each reading.
- Plot the results on a line graph of time against infiltration rate in mm per hour.

Repeat on different slopes or surfaces to suit your investigation.

CASE STUDY

Choosing where to measure stream velocity

Choose a straight stretch, unless measurements round meanders are useful for the purposes of the survey. The water flow should be moving parallel to the banks and the channel should be as free of boulders and vegetation as possible.

The width of the channel can be measured and measurements of velocity made at regular intervals across it.

You will find more information about measuring stream velocity at the end of this chapter.

Counting methods

Investigating variations in pedestrian or traffic flows involves counting. The best way to record this is by making a mark on a **tally chart** or by clicking an automatic counter as each individual person or vehicle passes by. A tally chart is quick to count as groups of five are made by making the fifth mark cross the first four.

If you are doing a traffic survey at a very busy time, it is best to have one student recording the measured data while each type of transport passing by is called out by the observer. Alternatively, different students could count and record different subgroups and the total could be calculated later.

CASE STUDY

Measuring the cross-profile (section) of the bed of a stream channel and finding the area of the cross-section

The smaller the intervals between measurements, the more accurate the profile will be.

When plotting the cross-profile on graph paper, keep the vertical and horizontal scales the same to allow for easy calculation of the cross-sectional area. A less accurate method of obtaining this is by multiplying the average depth by the width.

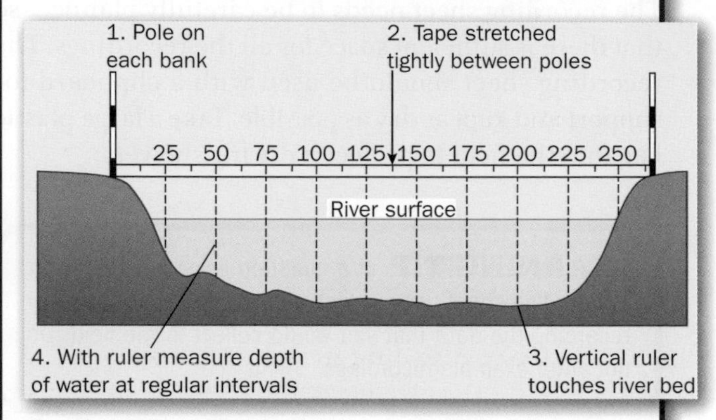

Fig. 13.6 Method of measuring the cross-profile of a stream bed

TRAFFIC RECORDING SHEET

Day: Tuesday Date: November 8th Time: 8–8.10

Street: West Street Site: after junction with Hope Avenue

Inbound/~~outbound~~ side Weather: wet and windy

Mode	lorries	vans	buses	cars	motorcycles	bicycles
Tally	卌卌 |	卌 卌	||||	卌卌卌卌 卌卌|||	卌卌卌 卌卌|||	卌卌 |||
Totals	11	10	4	33	28	13

Fig. 13.7 An example of a completed tally chart with subgroups of traffic

LEARNING TIP When undertaking traffic and pedestrian flow surveys, choose the times of the counts according to whether you want to include or exclude certain types of traffic, e.g. commuters or the school run.

CASE STUDY

Investigating traffic flows

Imagine that you are going to investigate the hypothesis 'Traffic flows in and out of the town centre vary on different main roads'.

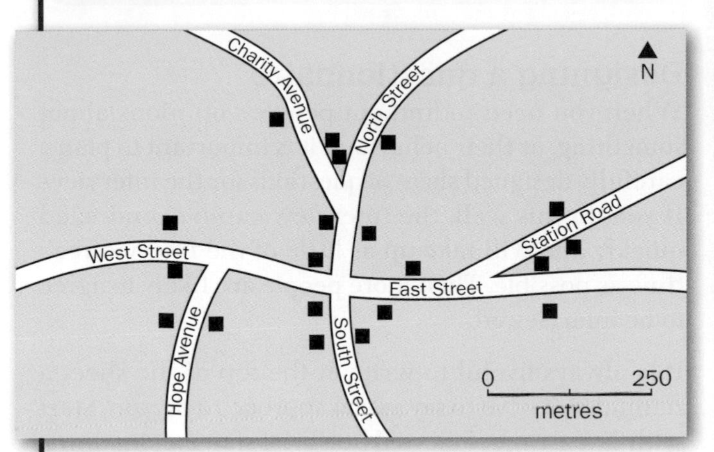

Fig. 13.8 Sites for a traffic count on main roads in a town

Preliminary planning in the classroom

- → Choose the sampling sites for this investigation subjectively, to ensure that all roads are covered at appropriate sites. The allocation of each site to a specific group of students must be done beforehand.
- → It is also important to do all counts at the same times, to ensure a fair comparison. So, before beginning the investigation, participating students need to decide what time the counts will start, how long they will last, what times of the day they will be done, and on which days. Each count should be long enough to give a representative sample for reliable data to be collected, but not so long that students lose concentration or become tired.
- → A checklist should be drawn up of the equipment needed.

If sub-groups are important to the investigation, students also need to discuss how to classify traffic into categories or subgroups. For example, it is always possible – with subjective classification – that one student might classify a vehicle as a van, while another might consider it to be a lorry. Therefore, it is important that all participants know what criteria to apply to distinguish between the categories. Subgroups allow the possibility of further analysis to be done.

Data collection during the investigation

- → At each site, at least one student must be on either side of the road – one counting the traffic moving into the town centre and the other counting the traffic moving out of it.
- → Each student needs a stopwatch, which will be started and set to stop at the pre-agreed times. Ten minutes for each count would be appropriate.
- → The counts should be done a *minimum* of three times – spaced throughout the day, to include a range of flows. The first count could be during the middle of the 'rush' to work, the second during the lunch break, and the third during the afternoon before the 'rush' home. Exact times will vary according to the location in the world.
- → More scope for analysis would be gained if more counts were added at mid-morning (to compare with the morning 'rush') and, if possible, during the evening 'rush' home (to compare with the morning 'rush'), followed by early evening (to survey the journeys for entertainment in the CBD). However, the length of the school day and the need for students to take organised transport home, might dictate when counts can be undertaken.

Follow-up in the classroom

If more than one group surveyed the same stretch of road, average the results. Enter each group's counts on a results sheet (see Fig. 13.9) and give each student a copy.

Make sure that you note any special circumstances that could have affected the results (e.g. a football match, pop concert or shops' half-day closing day). Each student will then be able to present, analyse and make reasoned conclusions using the collected data.

RESULTS SHEET

Day: *Tuesday* **Date:** *8 November*

Street: *West Street*

Site: *After junction with Hope Avenue*

Weather: *Wet and windy*

Special circumstances: *None*

Inbound vehicles

Mode	8.0-8.10	10.10-10.20	12.40-12.50	15.0-15.10	17.20-17.30
Lorries	11	5	3	4	3
Vans	10	6	2	5	4
Buses	4	2	3	2	4
Cars	33	14	16	11	16
Motorcycles	28	3	10	2	15
Bicycles	13	2	6	3	9
Totals	99	32	40	27	51

Total inbound = 249

Outbound vehicles

Mode	8.0-8.10	10.10-10.20	12.40-12.50	15.0-15.10	17.20-17.30
Lorries	4	3	4	7	9
Vans	3	5	3	7	8
Buses	4	2	4	3	5
Cars	18	7	12	19	36
Motorcycles	9	2	9	8	30
Bicycles	7	2	3	5	15
Totals	45	21	35	49	103

Total outbound = 253

Fig. 13.9 A sample results sheet for West Street after the junction with Hope Avenue

15 What are the advantages of using a tally chart?

16 Why would it be inappropriate to choose the sampling sites along the roads by:

a using random number tables to find the distances?

b systematic sampling every 100 m?

17 **a** State one advantage of choosing sites near road junctions for traffic counts.

b How should students at such locations consider pedestrians and their own safety?

18 Imagine that students sampled pedestrian flows in the CBD at 100-m intervals along the main roads leading from the centre of the CBD. Each student found his or her sampling site by pacing on the pavement the required number of paces to it. State the advantages and disadvantages of this method of (**a**) selecting the sites and (**b**) pacing the distances.

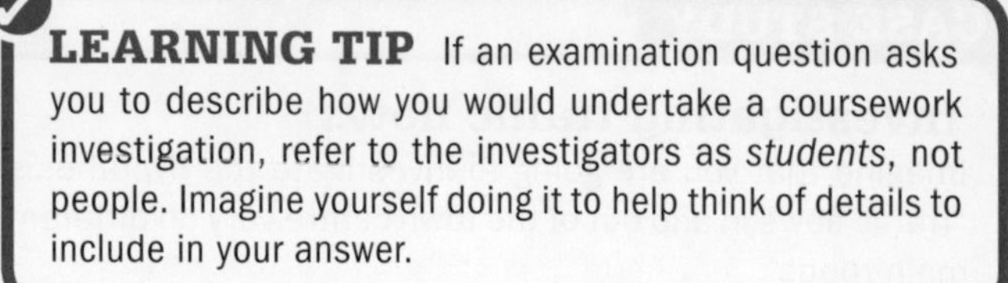
LEARNING TIP If an examination question asks you to describe how you would undertake a coursework investigation, refer to the investigators as *students*, not people. Imagine yourself doing it to help think of details to include in your answer.

Designing a questionnaire

When you need to find out people's opinions about something, or their behaviour, it is important to plan a carefully designed sheet of questions for the interview. If you do this well, the interview can be conducted quickly and will take up as little of the interviewee's time as possible. Then more people are likely to agree to be interviewed.

It is always useful to write at the top of the sheet a reminder of what to say as you approach a person. Start with 'Excuse me', followed by a brief statement naming your school or college. Then write the aims of your geographical enquiry or the aims of the questionnaire, together with a polite request to ask a few short questions to help your research. Emphasise that you will not ask for the person's identity and that the answers will be anonymous.

Use the same essential headings as for the recording sheet described earlier, but also include boxes (either before or after the questions) where you can tick your opinion about the age and gender of the person interviewed. These are sensitive issues that should not be asked directly, but they are essential for a stratified sample and might be useful during the analysis of your results.

The questions in any **questionnaire** should be essential for the enquiry - short, simply worded, unambiguous, and numbered.

- Include some closed questions - those which lead to a definite answer that will fit into one of the categories or ranges that you have put next to the question. These questions allow you to quickly tick an appropriate box.
- Begin the questionnaire with closed questions, but it might help your enquiry if you also ask some open questions at the end, so that the respondent can answer freely. Examples are: 'What are your opinions about ...?' and 'What are the reasons for your answer?'

Responses to open questions might be too long or difficult to record. They are not as easy to analyse as closed questions - where the total number of categories chosen can be counted and the percentages calculated. However, open questions can provide valuable information to use in your analysis. It is always useful to ask why a person holds a certain opinion about an issue.

When asking a closed question, it is possible to devise degrees of agreement with the question. An example would be to use 'strongly agree', 'agree', 'disagree' and 'strongly disagree'.

End the questionnaire with 'Thank you for your help and time'.

Before the research, conduct a pilot survey to test for flaws in the questions and reword them if necessary. Five respondents would be sufficient for this.

Question 1 on Fig. 13.10 is important if your research needs to ask only local people or only visitors to the area. If they are not the people you need, you can end there. Examples would be if you were finding out the benefits and disadvantages of tourism to the local area or why tourists visit it.

CASE STUDY

A questionnaire to investigate the extent and effect of aircraft noise around an airport

This questionnaire has no room for formal tick boxes, so the answers given could be ticked, circled or underlined.

Date ____________________ Location ______________________________

Time ____________________ Name of student ______________________________

Excuse me. I am a student at X School and I am investigating the effects of aircraft noise around the airport for my IGCSE Geography coursework. May I ask you a few quick questions about this? I will not ask your name and your replies will not be linked to you in any way.

1 Do you live in this village or town? Yes No

If you answered 'no', what is the nearest town or village to where you live?

____

2 Approximately how often does aircraft noise annoy you when you are inside?

every day 4–6 days a week 1–3 days a week less than once a week never

3 How often does aircraft noise disturb your sleep?

every night 4–6 days a week 1–3 nights a week less than once a week never

4 How often does aircraft noise annoy you when you are outside your house?

every day 4–6 days a week 1–3 days a week less than once a week never

5 If you have a pet, to what extent do you think it's affected by the aircraft noise?

it often shows fear it occasionally shows fear it never shows fear

6 Have you ever thought about moving house because of the aircraft noise? Yes No

7 Do you think that you will move house for this reason in the next year? Yes No

8 Do you believe that aircraft noise has affected your health or that of another person living in the house?
Yes No

9 Please give details of anyone living in the house who suffers from deafness.

Person 1 slightly deaf moderately deaf very deaf age 0–30 31–60 above 60

Person 2 slightly deaf moderately deaf very deaf age 0–30 31–60 above 60

Person 3 slightly deaf moderately deaf very deaf age 0–30 31–60 above 60

Thank you for your help and time.

Respondent's gender: M F

Respondent's age (estimated): below 20 21–45 46–65 above 65

Fig. 13.10 A questionnaire for investigating the extent and effect of aircraft noise around an airport

19 Criticise the following questions for a questionnaire:

a What is your income? ________________

b Where do you live? ________________

c How long have you lived here?
6 months 1–2 years All my life

d Do you object to the proposed factory because it will pollute the air or because it is an ugly sight?
____

20 Students suggested three different plans to carry out a questionnaire survey about the quality of life in a large squatter settlement:

Plan A Follow a transect line along roads through the centre from one side of the settlement to the other and ask people from every fifth home to complete the questionnaire.

Plan B Stand at the centre of the settlement for two hours and ask as many people as possible to complete the questionnaire.

Plan C Using a base map and random number table, select 30 homes at random. Ask an adult from each home to answer the questionnaire.

a Why might the data collected by Plan B be unreliable?

b Why is Plan C better than Plan A?

c Explain why Plan C could be used to compare different squatter settlements in the town.

d What could be possible problems in undertaking this survey?

Sampling

Questionnaires

It is often important to do a stratified sample for questionnaire enquiries. For example, the number in each age group asked should be representative of the total population. Details about the latter can be obtained from census data. For the purposes of many surveys, it would be acceptable to interview an equal number of males and females.

The time and day chosen for the survey can prevent certain groups of people from being included (giving the investigation bias): e.g. if it is conducted during working hours on a weekday, workers will be excluded. The possibility of bias should be explained in your investigation's conclusion.

Carefully consider where is the best place to stand to meet a representative number of people without causing an obstruction. Just outside and to the side of the main door to a shopping centre would be a suitable place for a shopper survey. To investigate airport noise (Fig. 13.10), it would be appropriate to sample a number of homes in each village – starting with those nearest to the airport and working away from it in at least four directions (including that of the prevailing wind and the opposite direction to it).

Land use

A land-use survey of a small CBD might be possible without the need to sample, but a land-use survey of a large town will involve sampling. This is usually done using systematic sampling along one or more transects (starting at the town centre and ending at the edge of town). Transects can be along roads radiating from the centre. Following straight lines drawn on a map would take more time, as it would involve getting from one point to another.

Depending on the hypothesis, the recording sheet may need spaces for land uses on more than one floor of a building. It is important to decide about land-use categories beforehand (Fig. 13.11).

1 Shops	
a Convenience	**d** Supermarkets and hypermarkets
b Comparison	**e** Other
c Department stores	
2 Services	
a Financial	**f** Religious
b Educational	**g** Administration
c Medical	**h** Leisure and entertainment
d Transport	**i** Other
e Food and drink	
3 Industries	
a Crafts	**c** Heavy industry
b Light industry	**d** Mining/quarrying
4 Open spaces	
a Parks	**d** Waste land
b Sports fields	**e** Other
c Demolition site	
5 Residences	
a Houses	**c** Temporary/squatter
b Flats/apartments	
6 Unoccupied	
a Residences	**c** Shops
b Factories	**d** Other

Fig. 13.11 Examples of urban land-use categories

RESEARCH There are more detailed classifications for each category so, if you want to survey just one of them, research the options in advanced textbooks or on the Internet.

LEARNING TIP Subjective observations, such as estimates by eye of the percentage of bare ground in an undivided quadrat, are never as reliable as the objective method described overleaf. To minimise error, any estimates should be made by several students and averaged.

Here are some example hypotheses about land use:

- Comparison shops will be located near to shops selling similar goods.
- The size of the frontages/floor area of buildings/number of storeys/height of buildings will increase towards the centre of the CBD. (These factors are either time-consuming to measure or difficult to judge. Height is usually estimated by averaging the number of storeys in a set number of buildings at each site. The number can be difficult to count in a very tall building on a narrow street.)
- Different land-use zones will be found at different distances from the centre of the CBD.
- Suburban shopping malls have similar types of shops and services.

CASE STUDY

Planning and conducting a land-use survey of a CBD

1. Obtain (or draw) a large-scale plan of the area on which separate buildings are shown.
2. Decide whether it is possible to record all buildings, or whether systematic sampling should be used.
3. Decide the land-use categories.
4. Decide whether to record just the ground floor, or all floors. If both, you will need to use a method to show which is which, such as X and Y.
5. Mark the uses on the base map. If you are using the categories in Fig. 13.11, record 1a for X and 5b for Y (to indicate that the ground floor is a convenience shop with an apartment above it).
6. Devise shadings for each category and plot the land uses on another copy of the map. Remember to add a key for the shading used. A separate map could be produced for each storey.

Vegetation

Studies of changes in small vegetation species, or of variations in ground coverage, involve area sampling along a transect - using a quadrat 0.5 m square or 1 m square (placed on the ground at each sample point). The point is usually selected systematically. If the quadrat has been subdivided into 100 squares, the number of squares without plants can be counted to give the percentage of bare ground. If it has 25 subdivisions, multiply by four. Random samples of vegetation can be taken using quadrats thrown over the shoulder at sites selected randomly.

CASE STUDY

Investigating vegetation change across sand dunes using systematic sampling

As dunes become older, their soil becomes more suitable for vegetation. The dunes' age increases with the distance inland. A transect from young to old dunes will show an increase in: the density of the vegetation cover, the number of plant species (unless there is a very old or grazed dune), and the height of the tallest species. Any of these changes can be investigated along a transect line.

- Place a **ranging pole** at the high-water mark. Lay out a transect from it at right angles to the shore by extending a tape measure as far as it will go. Mark the end with another ranging pole (this helps to keep the transect at right angles to the shore).
- At regular intervals, e.g. every 10 m, place the quadrat at the side of the tape and count the number of different vegetation species and/or the percentage vegetation cover. You could also use a long ruler to measure the highest species within the quadrat at each point.
- Record the results on the recording sheet.
- Move the start of the measuring tape to the second ranging pole and repeat the process until all of the sand dune ridges have been crossed.

Your results can be compared with parallel transects taken at the same time by other groups of students at regular intervals along the beach.

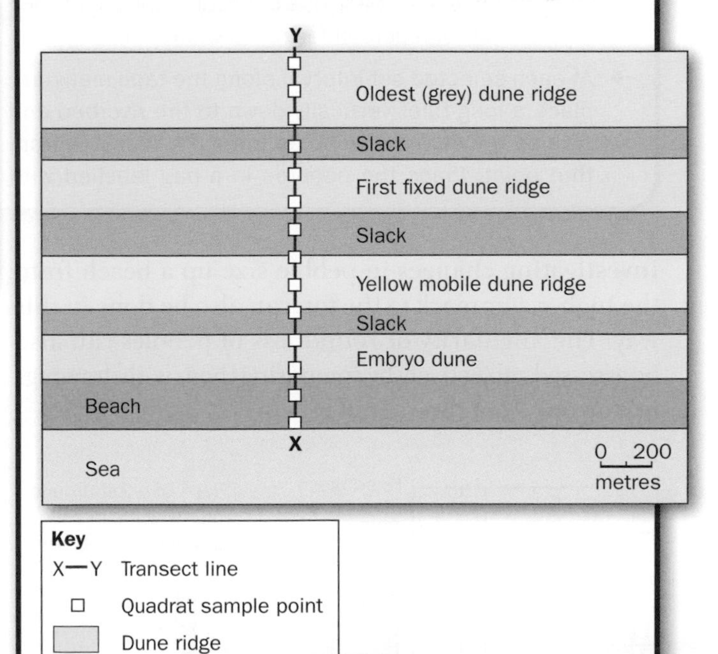

Fig. 13.12 A transect across a series of sand dunes

Recording sheet for sand dune vegetation survey

Date __________ **Location** ____________ **Name of student** __________

Transect number ____________

Distance from high water mark (metres)	Dune identity*	Seaward or inland facing slope	% cover	Number of plant species	Height of highest species
10					
20					
30					

* The dune identity includes embryo dune, yellow mobile dune, first fixed dune, oldest (grey) dune.

Fig. 13.13 Part of a recording sheet for sampling vegetation across sand dunes

Example hypotheses about vegetation include:

→ The height/density/diversity of vegetation increases from the high-water mark inland across a dune system.

→ Vegetation cover increases away from footpaths.

LEARNING TIP Use 'along', 'across', 'up' and 'down' correctly when you are describing beach surveys. Along is parallel to the shore, while across is at right angles to it. Up the beach is away from the high-water mark, while down is towards it.

CASE STUDY

Investigating pebble load size changes along a riverbed by systematic point sampling

The same technique of sampling along a transect line at regular intervals is used.

→ At regular intervals (which will depend on the length of the river), stretch a tape measure from bank to bank. Record on the recording sheet whether you are starting from the left or right bank (the left bank is the bank which is to the left of a person facing downstream).

→ At each selected set interval along the tape measure, place a long ruler vertically down to the riverbed and pick up a number of pebbles (no more than 30) from that point. Place the pebbles in a bag labelled with the identity of the sample point from which they were taken.

→ It is more accurate to measure the long axis of a pebble with callipers than to simply place it on a ruler and judge by eye. Tighten the callipers around the long axis of the pebble, remove the pebble and measure the gap of the callipers against the ruler. A device called a pebbleometer (Fig 13.14) can also be used to measure the pebble.

→ The average particle size at each sample point can be found by adding up the lengths of all the pebbles and dividing the total by the number of pebbles in the sample.

Investigating changes in pebble size up a beach from the high-water mark to the top can also be done in this way. The angularity or roundness of pebbles can also be assessed subjectively by comparing them with drawings in Powers' Roundness Index.

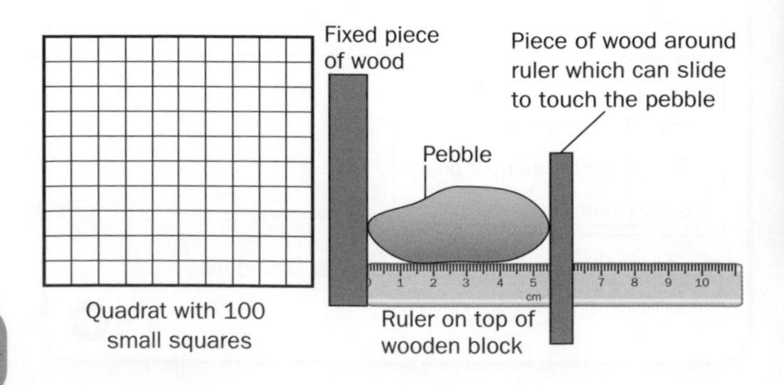

21 Why might a column for noting evidence of human impact be useful on a sheet recording vegetation change?

22 Why is a limit of 30 suggested for the number of pebbles to be measured at each sample site?

23 Why is it important to do all investigations at the same time when comparing:

- **a** shoppers at two types of shopping centre?
- **b** the downstream changes in river velocities on two rivers?
- **c** weather studies at different points in an area?
- **d** visitors to a national park?

Fig. 13.14 A quadrat and a pebbleometer

Surveying a slope profile

CASE STUDY

Surveying a slope profile – data collection and presentation

- Take the necessary equipment: two ranging poles, a clinometer, a prepared recording sheet, a pencil, and a clipboard.
- One student holds a ranging pole at the start of the slope, while a second student holds a second ranging pole at the first noticeable break of slope. (Alternatively, do this every 10 m, which is easier but less accurate.) It's important that the ranging poles are kept vertical and that they rest on the surface but do not sink into the ground as the pole length must be the same at each site.
- The first student then holds the clinometer at a known comfortable height against the first ranging pole and sights the *same height* on the second ranging pole. The second student reads and records the angle in the prepared table. Fewer students would be needed if the clinometer has a trigger. The same student who sights the height on the second pole squeezes the trigger when the instrument has adjusted to the angle of slope. The angle can then be read off.
- Measure the distance between the two poles with a tape measure stretched between them and record it in the table.
- Repeat this process until all the segments of the slope have been measured.

SLOPE PROFILE SURVEY

Date ___________ **Location** _____________

Transect number _____

Name ______________________

Segment number	1	2	3	4	5	6	7	8	9	10	11	12	13	14	15	16	17
Distance (m)	5	15	4	7.5	11	6											
Angle (°)	0	20	0	10	15	6											
Rising or falling	-	R	-	R	R	F											

Fig. 13.16 A partly completed recording sheet for a slope profile survey

This method can be used for any slope, including the long profile of a stream.

The slope can be drawn on graph paper by measuring each angle from the starting point and drawing a line at that angle for the distance recorded.

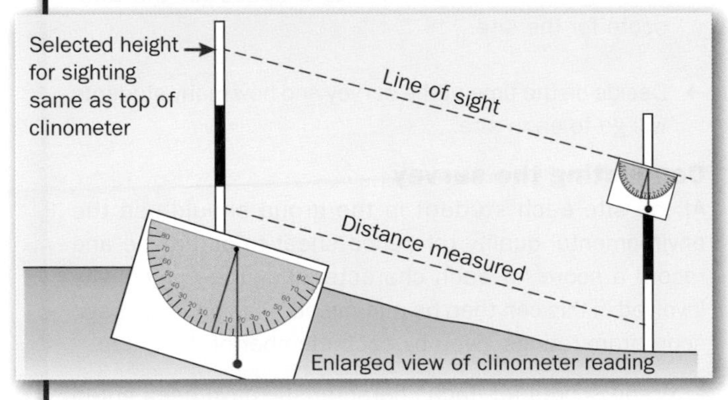

Fig. 13.15 Surveying one segment of a slope profile using a simple clinometer and regular distances

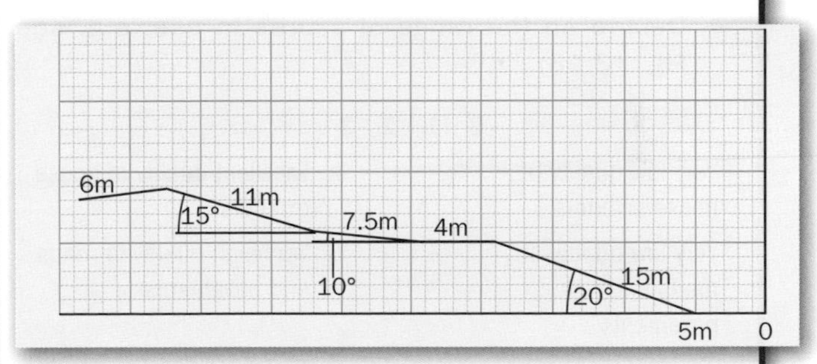

Fig. 13.17 Drawing on graph paper the slope recorded in Fig. 13.16

LEARNING TIP Examiners require detail so, if a question asks you to explain how to do a survey like this, it's important to state which equipment you would use, what you would measure with it, and where the measurements would be taken.

LEARNING TIP Many fieldwork techniques can be learned in the school grounds. For example, for a slope survey of a beach, the school boundary could represent the low-water mark and the survey conducted up a slope from it. The more experience you have of practising the techniques, the more detail you will recall for exam answers about them.

24 What is the minimum number of students needed to survey a slope?

25 Why would a teacher organise the survey so that each student in the group did each task?

Bipolar surveys

A **bipolar survey** is useful for assessing environmental quality. A range of scores is used. A simple one is from 0 to 3, where 3 represents the highest quality and 0 the lowest (see Fig. 13.18). But scales often use negative and positive figures either side of 0. The 0 represents an average situation and the negative figures show the extent of undesirable aspects and the positive figures the extent of good aspects. The scale might, for instance, range from +3 (the highest quality) to -3 (the poorest). It is difficult to make judgments based on more than seven scores, and four or five are usually adequate.

CASE STUDY

Investigating the environmental quality of a town's different residential areas – a bipolar survey

The stages involved in this are:

Planning the survey

- Choose an appropriate number of survey sites. A large number will be time-consuming and a small number will not make the investigation worth doing.
- Decide where the survey sites should be located. They might be along a transect from the town centre to the outskirts of the town. Or they might be chosen according to known differences in age and perceived differences in quality of the town's residential areas.
- Select which characteristics of the environment to measure as indicators of environmental quality. Examples include the amount of outside space individual properties have, how much litter and noise there is, the quality of pavements and roads, and the quality of house exteriors.
- Decide on the range of scores to use (e.g. see Fig. 13.18).

	0	1	2	3	
Very low quality of house exteriors					Excellent quality of house exteriors
Very low quality of roads and pavements					Very high quality of roads and pavements
A lot of litter					No litter

Fig. 13.18 A four-point scoring scale

- Prepare an environmental quality reference sheet to use as a guide to what each score for each characteristic represents. An example for the quality of roads and pavements on a five-point scale is shown in Fig. 13.19. The criteria will depend very much on the town being studied.

Characteristic	Description	Score
Quality of roads and pavements	Tarred, without potholes or breaks in the surface	+2
	Tarred, with some breaks in the surface	+1
	Tarred, with many potholes or breaks in the surface	0
	Earth road with a smooth surface	-1
	Earth road with many potholes	-2

Fig. 13.19 A five-point scoring scale

- Prepare a recording sheet for use in the field. Make sure that there is a space to write the site number. Then produce something arranged like Fig. 13.18 using your chosen type of scoring range and descriptions of each extreme at the sides for every characteristic to be judged. At the bottom add a space for the total score for the site.
- Decide on the time of the survey and how many students will go to each site.

Conducting the survey

At the site each student in the group should use the environmental quality reference sheet to determine and record a score for each characteristic. The subjectivity involved in this can then be minimised by using an average score from ratings given by each member of the group.

After the scores for each characteristic have been noted on the recording sheet, they should be totalled to give an overall environmental quality score that can be compared with those for other areas. It is then possible to rank the survey sites according to their scores. The place with the highest score is perceived to have the highest quality environment. Averages can also be calculated.

LEARNING TIP When analysing individual criteria, be careful how you interpret a score of 0 for litter or noise. It does not mean there is no litter or noise.

Assessments of noise for a bipolar survey are subjective and can vary widely. If greater accuracy is important, relatively inexpensive sound meters (decibel meters) can now be purchased and many mobile phones have them.

CASE STUDY

Investigating longshore drift over a short time

1. Paint some pebbles in a variety of bright, easy-to-see colours.
2. Choose a time when longshore drift is occurring (onshore wind and oblique wave crests).
3. One student puts one pebble in the water and stays at that point.
4. Another student watches the pebble for ten swashes and marks where it moved to.
5. Measure the distance from the start and end point with a tape.
6. Repeat. This could be combined with measuring wave frequency or wind speed, to see how they affect longshore drift. Varied pebble sizes could be used to see their effect.

Measuring the long-term effect of longshore drift

- → Prepare a map of the groynes on a beach.
- → At low-tide point measure the distances from the tops of the groynes to the surface of the beach on each side of the groyne and calculate the difference.
- → Repeat at the mid- and high-tide points.
- → On the map, plot three bars at these points along each groyne to show the difference (i.e. the higher accumulation of sediment). Plot each bar on the side of the groyne with the greatest accumulation of sediment.

CASE STUDY

Measuring wave frequency

- → Place a marker pole as a fixed point.
- → Use a stopwatch set to one minute.
- → Count how many waves pass the pole in one minute.
- → Repeat ten times and calculate the average.

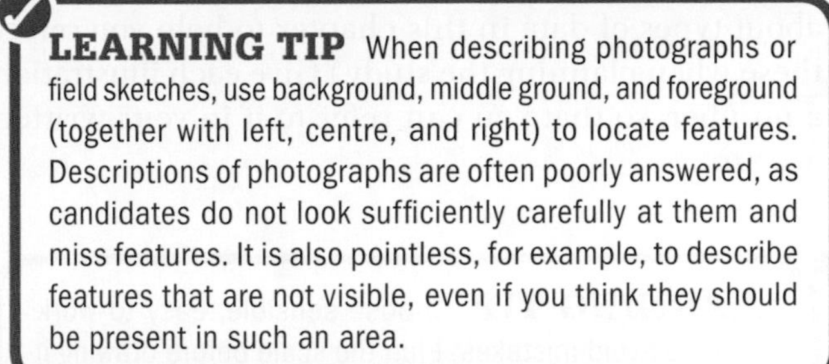

LEARNING TIP When describing photographs or field sketches, use background, middle ground, and foreground (together with left, centre, and right) to locate features. Descriptions of photographs are often poorly answered, as candidates do not look sufficiently carefully at them and miss features. It is also pointless, for example, to describe features that are not visible, even if you think they should be present in such an area.

26 Write down ten descriptive words for slopes.

27 Draw an annotated field sketch of the area in Fig. 13.20 to show the main features of the landscape.

Recording observations in the field

For some purposes it is useful to record the scene or landscape. A proper field sketch could be drawn to show the main features. The horizon and foreground are drawn in first, and then the important features are added in and labelled. A quicker way would be to take a photograph that can be annotated later from a very rough, labelled sketch. The field sketch is more subjective but is more likely to highlight the most significant features.

	Field sketch	Photograph
Advantages	• It highlights the main features.	• It is quick to take. • It can be taken in adverse weather. • It shows all the detail.
Disadvantages	• It takes longer to draw. • It is difficult to draw in very cold, very wet or extremely windy conditions. • It does not show the detail of features.	• The field of view might be too small to show all the desired features. • An undesirable foreground (e.g. a line of trees) might obscure a desired feature.

Table 13.4 Comparing field sketches and photographs

Fig. 13.20 A photograph to use for an annotated field sketch

Presenting the data

Introduce your investigation by stating its aim and the hypothesis or hypotheses. A brief description of the study area, and a map of it, helps to set the scene.

Present numerical data collected in the field in **data tables**. Draw the diagrams and graphs which will give the best visual impression of the data you have collected. (You will have used Chapter 12 and the information about types of data in this chapter to help you select these when planning the study.) Give each illustration a number, so that you can refer to it in your written account.

LEARNING TIP Choose sensible, easy-to-work scales to avoid mistakes. Plan the scale before drawing it to check that the range of data will fit on the paper. Be accurate when plotting graphs.

28 a Why should bar graphs be used to show traffic or pedestrian flows at different time periods throughout the day?

b A bar can be drawn on a map for each site where the data was collected. Why would this be useful?

c Draw two roads, A and B, meeting at a junction. The junction leads to the centre of the CBD. Along the relevant sides of each road, draw flow diagrams to show the information in Table 13.5. Show your scale and add a key.

d What type of map could be used to show degrees of noise disturbance away from an airport?

	Road A 8–8.10 a.m.	Road B 8–8.10 a.m.
Vehicles moving in	98	65
Vehicles moving out	37	26

Table 13.5 Survey results for Roads A and B

Analysing and interpreting the data

Once the data has been represented in graph or diagram form, it is easier to describe, analyse and interpret it. Also, simple mathematical analysis can often be useful for these purposes.

Simple statistical analysis

- Data can be placed in **rank order**, with the largest usually ranked one, the second largest ranked two, and so on, to give the relative importance of each feature in comparison with the other features of that type.
- The spread of the data is indicated by its **range**.
- The middle of a data set can be calculated by finding its **mean**, **median**, or **mode**. The mean is a good indicator to use when a data set has a normal distribution (without extreme values on one side that would distort the mean). Otherwise the median should be used, because it is not affected by extreme values. The mode has limited use, but the modal class can be a valuable indicator in a histogram.

Data sets can be compared using their ranges, means, medians, or modes. Reasons for the similarities and/or differences can then be suggested.

LEARNING TIP When making comparisons, either link a statement about each subject with 'whereas' or make a relative comment about them, using words ending in 'er' or 'est', such as 'larger', 'smallest', or using 'more' or 'less'. For example 'the examiner is older than you' is not as informative as 'the examiner is 45, whereas you are 16'.

Weekly total sunshine hours at a place over a nine-week period

hours (arranged in rank order)

64
60
53
44
35 → median
33
24
24
21
total 358

mode = 24
range = 21 to 64 = 43
mean = 358 divided by 9 = 39.8

Fig. 13.21 A simple statistical analysis of a set of data

Discussion point

Each member of a class writes down a number from 0 to 20. Put them together and imagine that they represent the scores made by a cricketer in 20 innings. Calculate the mean, median, and mode of the set of figures. Discuss which of these statistical measures of the middle number of the set most represents the cricketer's true form. Work out why you believe one is more suitable than the others.

Look for trends

It is usually easy to see trends, such as increases or decreases, on graphs. Wherever possible, use adjectives such as large and small to describe the trends, or quantify the changes (for example using 'halved' or 'tripled').

LEARNING TIP When describing trends, use detail, such as *slight* decline, *large* increase, etc. Support your description by quoting the amount of increase/decrease and the period of time over which it occurred. Always refer to both axes of a graph. Never simply list figures, e.g. the values for each year in turn. Figures must be interpreted and used, not simply copied or listed.

Look for patterns, relationships, and anomalies

Sometimes the examiner will ask for a description of **patterns** in the data. It may be possible to recognise patterns from data expressed in both tabular and graphical form, and to use the patterns to deduce relationships. Examples of patterns in data are that no rain falls when the cloud cover is 4 oktas or less and most rain falls when the cloud cover is 8 oktas. The positive relationship between the two variables (that the greater the amount of cloud cover the higher the rainfall amount) can then be deduced.

Patterns can also be recognised on maps. You will already be familiar with linear, nucleated, and dispersed settlement patterns, but many other patterns also exist. For example, on a map of the central area of a city, it will usually be possible to note that specialist shops are mainly located in side streets, and that offices are sited further from the peak land value intersection than shops are. You will notice from these examples that a description of pattern involves using words such as 'mainly', 'least', and 'more' and their opposites.

LEARNING TIP When describing patterns, do not describe individuals but describe common themes shown by the data, e.g. most shoppers at an out-of-town shopping centre shopped less frequently, travelled more by car and bought more high-order goods than those shopping in a suburban shopping centre.

Scatter graphs are valuable for recognising whether or not a relationship exists between two variables and whether it is positive or negative (inverse). It is very important to remember that the existence of a relationship does not mean that one variable causes the other. There may be a different factor influencing both. The closer the points are to the best-fit line, which may be curved or straight or go up and down, the stronger the relationship is. Anomalies are usually clearly visible on a scatter graph and suggest that some other factor was responsible for the anomalous value.

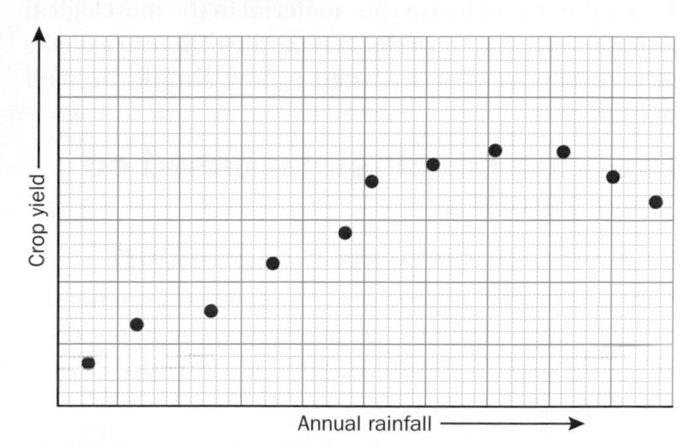

Fig. 13.22 A scatter graph showing the relationship between rainfall and crop growth. It is positive at first but becomes negative when the soil becomes too wet.

Patterns on maps reveal distributions which can then be described and explained. On Fig. 13.23, the distribution of tourist facilities can be described in detail as 'grouped in two areas of the town, the north and south west'. Areas without such facilities can also be described and explained.

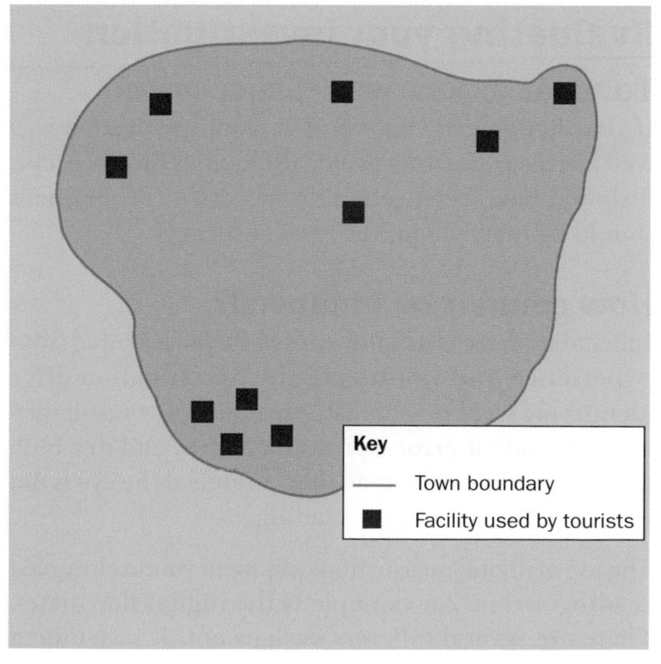

Fig. 13.23 Locations of tourist facilities in a town

Suggest explanations for your findings

You will need to use your knowledge and understanding of geographical concepts to explain the distributions, trends, patterns, relationships, or anomalies you have found, or to suggest why none were found if that was the case.

Throughout, aim to use your material in the most logical order and use paragraphs to separate different aspects of your report. The finished report should be well reasoned.

Drawing conclusions based on your results

Can your hypothesis be accepted?

You need to state in your conclusion whether or not the evidence allows you to accept or reject the hypothesis. In doing so, you should quote data to support your decision.

To what extent was it accurate?

Some investigations may reveal several anomalies which suggest that the hypothesis is partly (but not wholly) true. Again, support this conclusion by referring to one of the anomalies or a number of them.

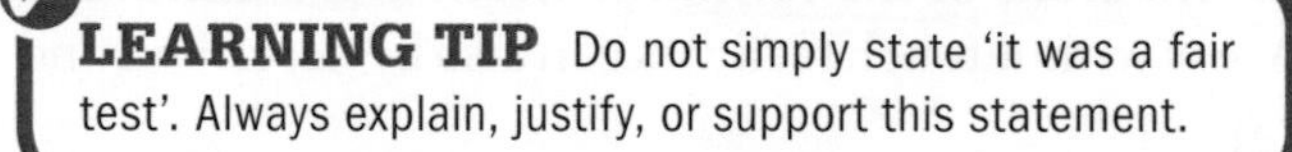

LEARNING TIP Do not simply state 'it was a fair test'. Always explain, justify, or support this statement.

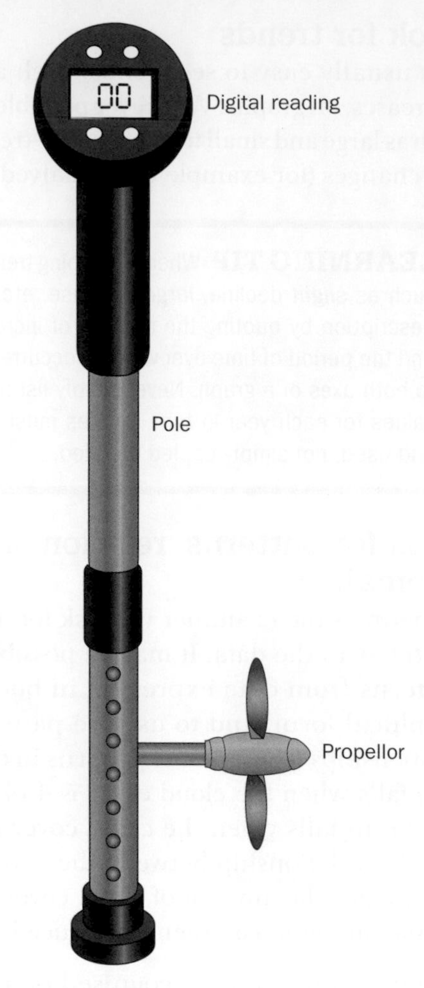

Fig. 13.24 A digital propeller flowmeter

Evaluating your investigation

To what extent was it successful?

If your hypothesis was sensible, the investigation was well planned, and unexpected difficulties did not occur, it should have been generally successful. A comment should be made about the level of success.

How could it be improved?

Inevitably, there is usually a possibility of learning from experience and improving the investigation after identifying weaknesses in it. An anomalous result may suggest student error. For example, wet and dry bulb thermometers can be read inaccurately if the eye is not exactly at the level of the reading.

The use of digital measuring equipment would eliminate reading errors. An example is the digital flowmeter. There are several different designs of this instrument and one example is shown in Fig. 13.24.

Was any part of your investigation affected by bias? Were your investigating techniques the most appropriate? Was the sample size less than 30? Doing more measurements should improve the reliability of the results, especially if an average is used.

Were the results affected by an unexpected factor, such as a road diversion during a traffic survey? More thorough research before doing the investigation should discover such problems.

Reliability can usually be improved by doing more surveys, either using more closely spaced survey points or doing the survey more frequently. Taking an average of the results of several groups doing the same counting or measuring task will minimise errors. If you do more surveys, ensure that the circumstances are the same as before. For example, an investigation done in different weather would only be useful to show how weather affected the results.

LEARNING TIP Suggestions for improvements must be practical, e.g. for a litter survey it is not practical to check the waste bins in hotel bedrooms. Neither is it sensible to lie in wait, watching to see if tourists drop litter. Do not suggest more 'observations' if more *measurements* or *counts* would be possible. Read the question carefully. Are you asked to assess the *methods* used in the fieldwork, or other aspects of it – or both?

How could the investigation be extended?

Again, suggestions must be practical, which is not the case with a suggestion to repeat the study in a named place far away. Neither is it usually possible to compare the study with the situation next year. Comparison with a past study would be valid. You could, for instance, compare your results with those for the same fieldwork if it was undertaken at your school in a previous year.

Suggestions for extending the study depend on the topic. However, it is usually possible to suggest other hypotheses about the subject, which could be used to widen the research.

Discussion point

Imagine that you cannot afford to buy a flowmeter and have to measure velocity in the way used by many students:

1. Measure a length of a stream, say 100 m long, and mark the two ends of the distance with a pole.
2. Hold an orange (which will float with most of it below the surface) on the water.
3. Release the orange when your timekeeper shouts the order to do so and starts the stopwatch.
4. When the orange passes the 100 m mark, a student there shouts 'stop' to let the timekeeper know to stop the stopwatch.
5. Repeat three times and average the results.
6. Calculate the average velocity in metres per second.

Then, discuss the following:

- The minimum number of students needed for this method compared with using a flowmeter
- The equipment needed for this method
- How you would calculate the average velocity
- What could go wrong when undertaking this method, and afterwards in the classroom
- What precautions would need to be taken when using either method
- How you would extend your research.

LEARNING TIP Don't be vague. The suggestion to 'do more studies' will not gain credit unless the type and/or location of the extra study is stated.

LEARNING TIP Try to practise all the data collection techniques at least once, so that you have the experience to answer the IGCSE® Paper 4 or O Level Paper 3 well. When answering questions, imagine being in that situation so that you can think of the difficulties, problems and good points.

29 Use the information in Table 13.6 to answer the questions. The investigation is based on secondary data.

- **a** Describe the trend shown by the data.
- **b** Describe the relationship shown by the data.
- **c** Is the hypothesis 'Mean daily temperatures are influenced by distance from the sea' correct? Use the data to support your answer.
- **d** How could you improve and extend the investigation?

Distance from the sea (km)	0	20	40	60	80	100	120
Mean daily temperature (°C)	18	21	25	23	24	31	33

Table 13.6 Data from a desert climate

Discussion point

Research into global warming in the future involves using existing temperature records and inputting possible influences on temperature into a computer model.

How important do you think the following measures would be during research to suggest the rate at which global warming is likely to occur in the next 50 years?

- The sites of the weather stations used in the study are known, correctly sited and free from outside influence.
- The intervals of time being used in the study are regular.
- The people conducting or interpreting the research are not employed in any capacity by an energy company.
- The temperature records from the past are reliable.
- The factors put into the model in order to project the future temperature are known.

Index

C

M

N

O

P

Q

R

S

U

V

W